The GSM System for Mobile Communications

A comprehensive overview of the European Digital Cellular Systems

The GSM System for Mobile Communications

Michel MOULY
Marie-Bernadette PAUTET

This book is published by the authors. Correspondence, in particular for orders, but also for comments, should be mailed to:

M. MOULY et Marie-B. PAUTET
4, rue Elisée Reclus
F-91120 PALAISEAU
FRANCE

Telephone : +33 1 69 31 03 18
Facsimile : +33 1 69 31 03 38

International Standard Book Number: 2-9507190-0-7

This book would not exist if a group of European people had not taken a common aim and worked hard together to reach it. This book is dedicated to all these mothers and fathers of GSM.

FOREWORD

by Thomas Haug,
Former Chairman, ETSI TC/GSM

At its meeting in Vienna in June, 1982, the CEPT Telecommunications Commission decided to set up a group to work out specifications for a pan-European cellular communication system for the 900 MHz band which had recently been allocated to land mobile use. The idea behind this decision was to create for the first time a system that would end the traditional European fragmentation and incompatibility in the mobile field. It was clear to the CEPT that unless the opportunity was taken there and then, the 900 MHz band would rapidly be allocated for different and incompatible systems in different countries, and that in view of the difficulty of finding another commonly available band, the opportunity would then be gone for decades to create a system with pan-European roaming.

From day one the work was directed towards a second-generation cellular system, since many countries already had first-generation systems (TACS, NMT, System C, etc.) in use or in the implementation stage. There was no decision or directive that the new system should be digital, since there was uncertainty as to what transmission mode would best meet the requirements. However, there was agreement that the new system must take into account recent developments in the telecommunication field, such as CCITT Signalling System No. 7, ISDN, OSI and other powerful innovations. Gradually, it became clear that the most likely solution would be a fully digital one, provided the radio propagation problems could be overcome, since the digital mode would be far more adaptable to the needs of a modern system than the analogue one. The final proof of the feasibility of the digital system was given after a very thorough series of measurements in various countries had been performed, above all at CNET in Paris, and the decision for a digital solution could be taken. With that as a basis, the way forward to a totally new system, fit for the 1990s, was open.

The work on the system was made possible by a united European effort, both in PTTs, research institutes and manufacturers, who participated from 1987 on. This kind of common effort towards a unified system is different from the earlier tradition in European telecommunications, where national concerns have frequently led to a protectionist attitude, with resulting incompatibility and poor economy. Also the European Community was strongly in favour of the common system and made a significant contribution in securing the legal basis for setting aside the necessary frequencies in its member states.

The result of the work by the GSM (first a CEPT working group, later on an ETSI group, and recently renamed SMG) is available through ETSI in the form of some 5200 pages of technical specifications for the first phase of the system. The work has attracted a great deal of interest even outside Europe, notably in Australia and south-east Asia. In fact, several countries have already taken firm decisions to build GSM systems.

For coming phases, more papers will follow in the years to come. Clearly, this huge amount of specifications is totally unreadable to anyone but a tiny group of specialists, just like a collection of law texts is incomprehensible to non-lawyers. This is inevitable, since an absolute requirement on a technical specification which is to be used by many different manufacturers, operators and regulatory bodies in many countries, often as a basis for legally binding contracts, is that it must be complete, consistent and unambiguous (unfortunately, it sometimes happens that this requirement is not fully met, despite all the efforts). Thus, the stringency requirement does not improve the legibility, and one may safely assume that very few people will ever master the whole set of specifications describing the system. At the same time, very many people will be involved in the implementation, procurement, operation and maintenance of the system, which means that they are in dire need of another type of description of GSM, which provides them with the necessary overview and a lot of details without resorting to the very formal description given in the specifications. This fact has been evident for a number of years, since the next lower level of description below the specification only consisted of various conference proceedings, useful in themselves, but in most cases very summary and usually without much co-ordination.

To prepare a description that goes into details on virtually every point of the system in such a way that the interested person feels tempted to read more, not less, is a very difficult task, requiring experts with a good ability to make complicated things easy to understand in plain language and above all to explain how the various parts of the system, described in seemingly unrelated technical specifications, are interconnected. It is therefore extremely fortunate that the authors of this book, both of whom are experts in the field and have made very significant contributions to the GSM work, have on their own initiative produced a description of GSM which in my view very well meets the needs described above. I wish to congratulate the authors on their achievement and to express my thanks for the very great effort they have spent. I am sure the book will be of very great help to a large community, both in and outside Europe, in the years to come.

Thomas Haug

CONTENTS

PREFACE

This book has been written with a set of goals and a few categories of reader in mind. This is not a work aimed primarily at the layman, but rather at professionals, or professionals-to-be in the sphere of telecommunications. However, it should not be a book accessible only to a handful of specialists. We have tried to waymark a route leading a reader of scientific background and interest to the understanding of a full system, from its needs to its technical choices.

The specification of the "Global System for Mobile Communications" (**G S H**)[1] and of its sibling, DCS1800, was a long process, full of sometimes lengthy and often hot debates. The major visible product of these years of work is a pack of some 5000 pages, the *GSM and DCS1800 Technical Specifications*. In drafting these specifications, the experts have tried to minimise the risk of ambiguous interpretation, rather than to ease understanding for the outside reader. The rationale behind most of the choices, as well as the alternative ideas which have not been retained in the final solution, are nowhere to be found in the specifications. Moreover, the purpose of some features and the way to implement or use them efficiently are also to a large extent out of the scope of the specifications. One of the aims of this book is to (partially) fill this gap. It offers a synthesis of the GSM and DCS1800 specifications, which will enable anyone which desires to understand what is in the *Specifications* to get a good grasp of how the whole system is designed. It will in no way replace the official specifications for developers or operators, but should increase the efficiency with which they will acquire knowledge of the system. As often as possible, topics have been presented from a different angle to the official specifications, describing the topics by a top-down approach, often for that purpose fitting together pieces found in different specifications.

GSM and DCS1800 are undoubtedly a major achievement in modern cellular telephony, and we hope this book will help to show it. But the purpose is not apologetics, neither is it advertising. A critical view (nothing human is absolutely perfect) and personal ideas (indicated

---

1   This is the commercial logotype used by GSM operators.

as such) have not been left aside. The reason for this choice is our hope that such constructive criticism will be helpful for future work, as GSM is certainly not the ultimate in its domain. GSM includes a number of firsts, but it will be followed by yet more elaborate systems. May the designers of new generations of systems find something of use in the subjective views they will find in this book.

The system is also by itself an object of interest. The standardising bodies chose not to limit the scope of the *Technical Specifications*, as is often the case for radiocommunications, to the radio interface. GSM and DCS1800 are standardised as a complete system, and much of the internal behaviour is documented, as well as the outline architecture of both the infrastructure and the mobile station. They encompass many different areas of the telecommunications field, and allow the individual presentation of each of those fields as a part of a whole. As such, GSM forms an interesting case study for students in the telecommunication field, so the book is also aimed at this readership[2]. As a consequence, a compromise has been sought between—on the one hand—using a layman's vocabulary and—on the other hand—drawing heavily from the technical jargon of the GSM committees. Good telecommunication basics are considered as the only prerequisite for reading this book.

As a consequence of this multiplicity of goals, the technical difficulty of the book varies greatly. We have spared neither basic explanations (certainly unnecessary for the specialists of each domain), nor some of the technical details which are the bread and butter of the life of us specialists, but maybe irrelevant for others. We have taken some effort to write the book so that it can be read by readers interested in only some of the topics, or who have neither time nor inclination to read the book from cover to cover. For these readers, we have compiled, at the end of the book, an index of the main terms and topics so that a reading of the entire book is not necessary to understand one subject. We have also tried to organise the text so that the difficulty of the material increases as the reader progresses. The two first chapters are of general nature, whereas the others are targeted at some specific technical areas. Each technical chapter begins with a presentation of the needs and principles, in plain terms, and becomes increasingly more detailed. The first portion of each of the technical chapters does not depend on the last portion of previous chapters, enabling two levels of reading for each of the technical domains. Where a topic has been elaborated in some finer

---

[2]   An important part of the book is indeed based on courses taught by the authors in various engineering schools and universities as well as in the context of post-graduate education.

detail within the main text, we have also made use of a smaller font, in order not to unduly interrupt the main flow for those who want the broader understanding.

Let us now have a general view of the contents. For starters, the first chapter is a general introduction. It sets the scene, including a brief history of the design of the system and how the work is conducted within the standardisation arenas. The functions which the system was designed to fulfil are described in this general presentation. The rest of the book describes the system, following more or less a top-down approach. The system is first described as a single object, as seen by the external world. Then its architecture is progressively revealed, with an increasing level of detail. Our main courses are the transmission chapters (chapters 3 to 5), which aim at describing how information is transported, and the signalling chapters (chapters 6 to 8), which describe how the constituents of the system cooperate to fulfil the functional requirements. The last chapter (chapter 9) deals with the system from the point of view of the operator: how are networks built and managed? A number of annexes are included for reference, and as a digestif. One is a detailed list of the GSM and DCS1800 Technical Specifications, aimed at newcomers who may wonder at which end to start. Another annex is an index of terms and acronyms, giving references to the book. At the end of each chapter of the main text, references to the relevant parts of the Technical Specifications are given, in order to facilitate the search for the official information.

Several menus can be suggested. Those who are interested in a general view of GSM and DCS1800 should read the two first chapters (though not easy, the second one is necessary to acquire the basic vocabulary), and browse through the beginning of each of the other chapters. Those more interested in the specificities of cellular telephony as opposed to general telecommunications will be mostly interested in Chapter 4 (the radio interface), Chapter 6 (dealing with radio resource management) and the beginning of Chapter 7 (the management of mobility). The entirety of Chapters 5 to 8 will appeal mostly to those interested in telecommunication systems. Finally, the whole work, from cover to cover, is strongly recommended to those who intend to read part (or all?) of the *Technical Specifications*.

Bon appétit!

Disclaimer

The substance of this book reflects only the understanding of the *GSM and DCS1800 Technical Specifications* by the authors. If it is true that both authors have taken part to a large extent in the design of the system, and that this book by and large tries to reflect the orthodox understanding of the *Specifications*, it can in no way be taken as the unique view of the committees who wrote the *Specifications*. The *GSM and DCS1800 Technical Specifications* are the sole reference, and the authors disclaim any responsibility for any usage of the interpretations proposed in this book.

Acknowledgements

We wish to acknowledge all the friends and colleagues, in particular from FRANCE TELECOM and MATRA COMMUNICATION, who helped us, with their expertise, to understand the variegated specialised domains that are to be travelled through when exploring the nooks and crannies of GSM.

Among all those who provided assistance, special thanks should be given to the manuscript readers, whose constructive criticisms are the source of the possible quality of this piece of work, including Christian Casenave, Nicolas Demassieux, Jean-Louis Dornstetter, Philippe Duplessis, Philippe Dupuis, Michel Lambourg, and Alain Maloberti.

As a particular mention, we wish to convey our deepest thanks to Paul Simmons and Tony Wiener, who accepted, in addition to a technical review of the book, the difficult task of correcting the weird language resulting from French people attempting to write in "English".

Remaining errors, technical or grammatical, are only to be blamed on the authors.

Lastly, we would like to thank our respective spouses for their cooperation, as well as Anne-Laure and Guillaume for their forbearance.

                                        Michel Mouly
                                        Marie-Bernadette Pautet

English Language Editors' Note

This book was written in that international stream of the English language which is being increasingly employed in Europe and other parts of the world as a medium of communication by people for whom, generally, English is not their mother tongue. Indeed, the GSM Specifications themselves were developed using a form of this language.

In editing and correcting the text we have endeavoured to retain all the flavour, inventiveness and humour expressed by the authors whilst trying to remove elements which might obstruct the understanding by a native English speaker or lead to a misunderstanding of technical content.

English is a rich, living language and we have allowed it to live a little!

Paul Simmons
Tony Wiener

1

SETTING THE SCENE

1

SETTING THE SCENE

Mass-market mobile telecommunications is certainly one of the major breakthroughs of this end of millennium. The possibility to make and receive calls through a small wireless handset, wherever you are, has an obvious appeal. The business opportunities are tremendous, since one can imagine that every person (and not only every home), including children, could be equipped provided the service is cheap enough. Many people agree that the sociological consequences will be important, much more than for videocommunications. Wireline telephony allows us to reach a place, if someone is there to answer. Mobile telephony allows us to reach a particular person, wherever (almost) he or she is. This will greatly increase the accessibility to people, and increase the feeling of security. On the other hand, this increased accessibility can in many cases be a nuisance, and widespread social acceptability of mobile telephony requires that users have a high degree of control on the calls they receive (identification of the calling party, forwarding of calls to a third party, message banks, etc.).

Mobile telecommunications is not a very recent technology, but it is a rapidly evolving one. Expensive vehicle mounted sets have been available for 30 or 40 years. A major step was made at the beginning of the 1980's, when analog cellular technology was introduced. We are now, at the beginning of the 1990's, witnessing the emergence of the next step, which will enable this technology to reach a mass-market, maybe approaching that of today's wireline telephony. The GSM system, and its offshoot DCS1800, are the European contribution to this evolution. This system was designed during the 1980's, and entered operation in various European countries during 1992.

GSM was designed internationally, in standardisation committees, by the major European telecommunications operators and manufacturers. The understanding of the gain to be obtained by combining resources, and of the business opportunity offered by mass-market radiotelephony, resulted in a substantial man-power and financial effort from the participants, thus making GSM a very dynamic project.

This design-by-committee has resulted in a public set of detailed specifications encompassing all system areas. This book is based largely on this committee output and on the authors' experience of the discussions which were held during the elaboration of the system specifications.

This first chapter will be quite general, and aims at setting the stage, the main lines of the plot and some of the main actors. The rest of the casting will be presented in the next chapter. The present chapter is divided into three parts. The first part will set the historical background of GSM. Some text is devoted to the description of how the standardisation committees worked; this will give some insight of what happened behind the stage. The second part aims to provide the basic technical foundations of public cellular radio telecommunications. Most of the addressed points are developed from different angles throughout the other chapters, recurring as leitmotivs. The last part describes the services that GSM offers to the customers as well as to the system operators. This describes the objectives of the system, and the rest of the book explains how they are accomplished.

1.1. A LITTLE BIT OF HISTORY

1.1.1. B.G. (BEFORE GSM)

Public radiocommunication requires sophisticated techniques, and therefore its evolution has always followed very closely the progress of electronics. The idea of instant communication regardless of distance is part of man's oldest dreams, and this dream became reality as soon as technique would allow it. The first implementations of radio waves for communication were realised in the late nineteenth century for radiotelegraphy. Since then, it has been a widely used technique for military communications. The first public applications concerned broadcasting (sound, then images): this is much simpler than radiotelephony as the mobile terminal is only a receiver. The real boom in two-way public mobile radiocommunication systems took place right after the second world war, when the use of frequency modulation and of

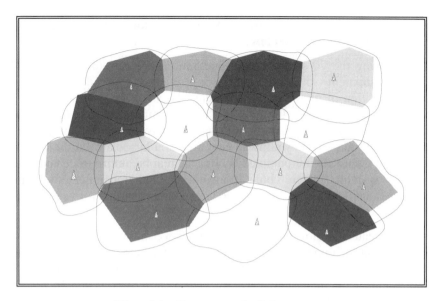

Figure 1.1 – The concept of cellular coverage

Numerous omnidirectional or sectorised antenna sites allow wide coverage,
by splitting the geographical area into overlapping coverage areas.
The lines show the limits of the overlapping coverage of the cells,
whereas the polygons represent the more usual non-overlapping representation.

electronic techniques such as the vacuum valve enabled the implementation of a real-scale telephony service for cars. The first true mobile telephone service was officially born in St Louis (Missouri, USA) in 1946. Europe, recovering from war, followed a few years later.

The first mobile telephone networks were manually operated (that is to say that the intervention of an operator was required to connect the call to the wireline network) and the terminals were heavy, bulky and expensive. The service area was restricted to the coverage of a single emission and reception site (single-cell systems). Little radio spectrum was available for this kind of services, since it was allocated with priority to military systems (this has not changed much!) and broadcasting (in particular television). As a consequence, the capacity of the early systems was small and saturation came quickly despite the high cost of terminals which deterred many a potential customer. Quality of service decreased rapidly with congestion, and the throughput sometimes fell drastically due to near-deadlock situations.

Between 1950 and 1980, mobile radio systems evolved to become automatic and the costs decreased due to the introduction of semiconductor technology. Capacity increased a little but remained too small compared to potential demand: public radiotelephony remained a luxury product for a chosen few.

During the 70's, large-scale integration of electronic devices and the development of microprocessors opened the door to the implementation of more complex systems. Because the coverage area of one antenna is mainly limited by the transmitting power of mobile stations, systems were devised with several receiving stations for a single transmitting station. They allowed coverage of a larger area at the cost of additional infrastructure complexity. But the real breakthrough came with cellular systems, where both transmitting and receiving sites are numerous and whose individual coverage areas partially overlap (see figure 1.1).

Instead of trying to increase transmission power, cellular systems are based on the concept of frequency re-use: the same frequency is used by several sites which are far enough one from the other, resulting in a tremendous gain in system capacity. The counterpart is the increased complexity, both for the network and for mobile stations which must be able to select a station among several possibilities, and the infrastructure cost due to the number of different sites.

The cellular concept was introduced by the Bell Labs, and was studied in various places in the world during the 70s. In the US, the first cellular system, the **AMPS** (Advanced Mobile Phone Service) became a reality in 1979 when the first pre-operational network was opened in Chicago, Illinois. In the North-European countries, the telecommunication administrations together with some manufacturers devised the **NMT** (Nordic Mobile Telephone) system which aimed at a Scandinavian coverage. The system started operation in Sweden in September 1981, and shortly afterwards in Norway, Denmark and Finland.

Networks based on these two sets of specifications account for the great majority of mobile networks throughout the world in the early 90's. For example, the TACS, derived from AMPS, was put in service in the UK in 1985. Most European countries have one or more cellular networks today. Table 1.1 shows the major cellular networks in operation in Europe in early 1992.

All these cellular systems are based on analogue speech transmission with frequency modulation. They all use frequency bands either around 450 MHz or around 900 MHz. Their coverage is usually nation-wide and their capacity reaches several hundreds of thousands subscribers. The largest national system in Europe (composed of two

Country	Systems	Freq. band	Date of launch	Subscribers (thousands)
United Kingdom	TACS	900	1985	1200
Scandinavia (Sweden, Norway, Finland, Denmark)	NMT	450 900	1981 1986	1300
France	Radiocom 2000 NMT	450, 900 450	1985 1989	300 90
Italy	RTMS TACS	450 900	1985 1990	60 560
Germany	C-450	450	1985	600
Switzerland	NMT	900	1987	180
The Netherlands	NMT	450 900	1985 1989	130
Austria	NMT TACS	450 900	1984 1990	60 60
Spain	NMT TACS	450 900	1982 1990	60 60

Table 1.1 – Major cellular systems in Europe in 1991

Variants of NMT and TACS are by far the most widespread
analogue cellular systems in Europe,
with Scandinavia leading the way for NMT and UK for TACS.

countrywide coverage networks) is the British TACS with more than one million subscribers by 1990. The highest population penetration is held by the Scandinavian NMT, with more than 6% of the Swedish and Norwegian population having a mobile equipment. However, these figures are much higher than the mean European values. For instance, the penetration factor in 1991 in France was about one tenth of the Swedish one.

Mobile equipment evolved very rapidly during the late 80's. At the onset, only vehicle-mounted equipment could be built. In the mid-80's portable equipment appeared, with a weight of a few kilograms and an autonomy of a few hours. Handheld equipment first appeared around 1988; not yet small enough to fit in your pocket, but fitting nicely in your attaché-case. In 1990, the smallest terminals on the market were weighing less than 400 g. and fitted in a coat pocket.

In parallel with this reduction in the size of mobile equipment, prices have decreased tremendously. Expensive car phones, accessible to a happy few, have cleared the way for pocket terminals affordable by you and me.

1.1.2. THE GENESIS OF A STANDARD

From the start of the 80's, after NMT started operating successfully, it became obvious to several European countries that existing analogue systems had limitations. First, the potential demand for mobile services, even though systematically under-estimated in the early 80's, was larger than the expected capacity of the existing analogue networks. Second, the different systems in operation offer no compatibility for mobile users: a TACS terminal cannot access an NMT network, and neither can an NMT terminal access a TACS network. What's more, the design of a new cellular system requires such a large investment that no European country on an individual basis can afford this investment if the only return expected is on its own national market. All these circumstances pointed toward the design of a new system, done in common between several countries.

The major prerequisite for a common radio system is a common radio bandwidth. This condition had already been met a few years before, in 1978, when it was decided to reserve a frequency band of twice 25 MHz at around 900 MHz for mobile communication in Europe.

The need was clear and the major obstacle removed. It remained to organise the work. The world of telecommunication in Europe always was dominated by standardisation. The CEPT (Conférence Européenne des Postes et Télécommunications) is a standardisation arena which—in the early 80's—included the European Administrations of Posts and Telecommunications of more than 20 countries. All these factors, both circumstantial and market-driven, led to the creation in 1982 of a new standardisation body within CEPT, whose task was to specify a unique radiocommunication system for Europe, at 900 MHz. The new-born "Groupe Spécial Mobile" (GSM) held its first meeting in December 1982 in Stockholm, under the chairmanship of Thomas Haug, from the Swedish Administration. 31 persons from 11 countries were present at this first meeting.

Date	Achievement
1982	"Groupe Spécial Mobile" is created within CEPT
1986	A Permanent Nucleus is set up
1987	Main radio transmission techniques are chosen, based on prototype evaluation (1986)
1989	GSM becomes an ETSI technical committee
1990	The phase 1 GSM900[1] specifications (drafted 1987-1990) are frozen
	DCS1800 adaptation starts
1991	First systems are running (Telecom 91 exhibition, ...)
	DCS1800 specifications are frozen
1992	All major European GSM900 operators begin commercial operations

Table 1.2 – GSM project milestones

The span of GSM work from the very start to commercial service extended over some 10 years, but the actual specification work did not start until 1987.

In 1990, by request of the United Kingdom, the specification of a version of GSM adapted to the 1800 MHz frequency band was added to the scope of the standardisation group, with a frequency allocation of twice 75 MHz. This variant, referred to as DCS1800 (Digital Cellular System 1800) is aimed at reaching higher capacities in urban areas for example for the type of mass-market approach known as PCN (Personal Communications Network).

The elaboration of the GSM standard took almost a decade. Major milestones are shown in table 1.2, and the corresponding stages are described in more detail in the following.

1.1.3. ORGANISATION OF THE WORK

The first two years of the GSM were dedicated to discussions of the fundamental principles. The frequency of meetings and the number of participants increased steadily. At the beginning of 1984, three "working

---

[1]   The term "GSM900" will be used to refer to the GSM standard at 900 MHz, when it is necessary to differentiate it from the DCS1800 standard. When such a distinction is not necessary, the term "GSM" will be used, encompassing both GSM900 and DCS1800.

parties" were created: the meetings were split to allow a more in-depth technical work. The amount of contributions and problems to solve increased steadily, and at the end of 1985 it became obvious that the number of meetings was insufficient. It was then decided (for the first time in CEPT) that the working parties would meet independently, reporting to the GSM plenary meetings which still endorsed the decisions. The role of the different working parties had already been clearly established for some time:

- WP1 (Working party 1), for the definition of services;

- WP2 (Working party 2), for the specification of radio transmission; this subject stayed dominant until 1987;

- WP3 (Working party 3), for all other issues, i.e., mainly network architecture, specification of the signalling protocols and of the open interfaces between network entities.

Later on, a fourth working party (WP4) was created to deal specifically with the implementation of data services.

During 1985, a detailed list of Recommendations to be output by the group, following the model of the technical Recommendations of the CCITT (Comité Consultatif International Télégraphique et Téléphonique), was discussed and settled. From 1986 onwards, work became centred around a major objective: drafting these recommendations. The list includes more than 100 recommendations, sorted in 12 series. A detailed action plan was generated to follow the progress of this huge task, and was updated at each meeting. In 1991, the list includes 130 recommendations, with a total of more than 5000 pages. These recommendations cover the full specification of the radio interface (i.e., the interface between the mobile stations and the infrastructure) as well as fairly detailed specifications of infrastructure architecture and of some of the interfaces and signalling protocols between network entities.

In order to co-ordinate the work between the working parties and to manage the edition and the updating of the recommendations, a "permanent nucleus" (PN) was created at the beginning of 1986. It consisted of a small team of full time members and was located in Paris. In 1988, the European Telecommunications Standard Institute (ETSI) was created and most of the CEPT technical standardisation activities were transferred to this new body, including GSM. Contrary to CEPT, ETSI is not restricted to Administrations, but includes members from industry and user groups as well as operators. GSM had already anticipated this need by officially allowing industry representatives to participate directly to the working parties on an ad hoc basis since 1987.

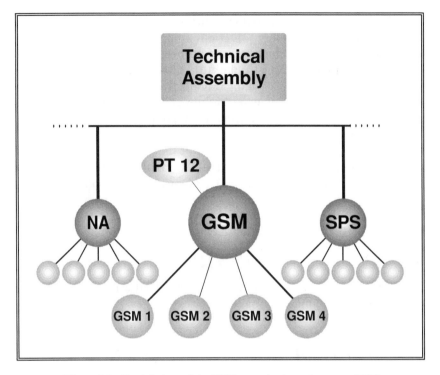

Figure 1.2 – Partial view of the ETSI organisation relevant to GSM

ETSI Technical committees such as GSM, NA (Network Aspects) and
SPS (Signalling, Protocols and Switching) report to the Technical Assembly (TA).
ETSI Project Team 12 supports the work of Technical Committee GSM.

With the transfer to ETSI, most GSM Recommendations were due
to become I-ETSs (Interim European Telecommunications Standards),
then ETSs. Becoming an ETS requires several stages of approval,
including public enquiries and voting, and this process takes several
months. In the meantime, GSM Recommendations are called "*GSM
Technical Specifications*" (GSM TS). In this book we will refer to the set
of the GSM Technical Specifications as the *Specifications*. When
integrated in the ETSI structure, the Working Parties became Technical
Sub-Committees (STCs, or "Sub Technical Committees" as the official
ETSI terminology stands; GSM itself is a Technical Committee, TC, of
ETSI) and are named GSM1, GSM2, GSM3 and GSM4. The Permanent
Nucleus became the Project Team n°12 (PT 12) of ETSI. Figure 1.2
shows the ETSI entities involved in the GSM project in 1990-1991, the
date at which the first GSM Technical Specifications were published.

At the end of 1991, activities concerning the post-GSM generation of mobile communications were added to the scope of the GSM Technical Committee, which was renamed SMG ("Special Mobile Group"), with Technical Sub-Committees SMG1 to 4 being the same as the previous GSM1 to 4, and SMG5 dedicated to the post-GSM generation, the UMTS ("Universal Mobile Telecommunication System"). A sub-group first established under the responsibility of the Permanent Nucleus to draft specifications in the area of Operation and Maintenance became SMG6 in early 1992. The name of the group was changed from GSM to SMG to distinguish it from the 900 MHz system: the term GSM is still kept to refer to the standard and to the corresponding system. Moreover, the term GSM has been chosen as the commercial trademark for the European 900 MHz system, meaning in that context "Global System for Mobile communications", with the corresponding logo:

$$\mathsf{GSM}^{\cdot}$$

The meaning of the dots in this logo, if any, is unknown to the authors and is left to the reader's imagination.

1.1.4. THE GSM MoU

In parallel with the drafting of technical specifications within the GSM committee, European public telecommunication operators (most of them GSM-operators-to-be) recognised the importance of co-operation for commercial and operational aspects, and signed a Memorandum of Understanding in Copenhagen, on September 7th, 1987. This memorandum, generally referred to as the GSM MoU, covers areas such as time-scales for the procurement and the deployment of the system, compatibility of numbering and routing plans, concerted service introduction, harmonisation of tariff principles and definition of accounting procedures. The memorandum was later signed by more operators, and amended in mid-1991 to accept members from non-CEPT countries and extend its scope to cover co-operation agreements with non-signatory bodies. The European operators which have signed the MoU in early 1992 are shown in figure 1.3. In addition to operators, regulatory bodies also signed the GSM MoU, such as the DTI (Department of Trade and Industry) in UK. DCS1800 operators have their own association, which holds close relations with the GSM MoU.

From 1990 onwards, GSM started to spread outside Europe: a number of other countries, such as The United Arab Emirates, Hong

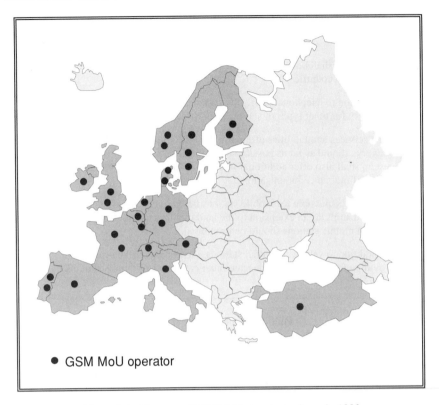

● GSM MoU operator

Figure 1.3 – European GSM MoU signatories in early 1992

In most European countries, two or more operators have been licensed for GSM
and have signed the Memorandum of Understanding.
Competition is becoming the rule and more countries will deregulate the area.

Kong, New-Zealand and Australia, envisaged adopting the GSM
standards. In 1992, Australian operators officially became the first non-
European signatories of the GSM MoU.

1.1.5. TECHNICAL CHOICES

Some of the aims of the system were clear from the start: one of
these aims was that the system should allow free roaming of the
subscribers within Europe. Practically speaking, this means that a
subscriber of a given national network may access the service when
travelling abroad. The same GSM mobile station must enable its user to
call or be called anywhere within the international coverage area.

Services

- The system shall be designed such that mobile stations can be used in all participating countries.

- In addition to telephone traffic, the system must allow maximum flexibility for other types of services, e.g. ISDN related services.

- The services and facilities offered in the PSTN/ISDN and other public networks should as far as possible be available in the mobile system. The system shall also offer additional facilities, taking into account the special nature of mobile communications.

- It should be possible for mobile stations belonging to the system to be used on board ships, as an extension to the land mobile service. Aeronautical use of GSM mobile stations should be prohibited.

- In addition to vehicle-mounted stations, the system shall be capable of providing for handheld stations and other categories of mobile stations.

Quality of service and security

- From the subscriber's point of view, the quality for voice telephony in the GSM system shall be at least as good as that achieved by the first generation 900 MHz analogue systems over the range of practical operating conditions.

- The system shall be capable of offering encryption of user information but any such facility should not have a significant influence on the costs of those parts of the system used by mobile subscribers who do not require such facility.

Radio frequency utilisation

- The system concept to be chosen shall permit a high level of spectrum efficiency and state-of-the-art subscriber facilities at a reasonable cost, taking into account both urban and rural areas and also development of new services.

- The system shall allow for operation in the entire frequency band 890-915 MHz and 935-960 MHz.

- The 900 MHz CEPT mobile communications system must co-exist with earlier systems in the same frequency band.

Network aspects

– The identification plan shall be based on the relevant CCITT Recommendation.

– The numbering plan shall be based on the relevant CCITT Recommendation.

– The system design must permit different charging structures and rates to be used in different networks.

– For the interconnection of the mobile switching centres and location registers, an internationally standardised signalling system shall be used.

– No significant modification of the fixed public networks must be required.

– The GSM system shall enable implementation of common coverage PLMNs.

– Protection of signalling information and network control information must be provided for in the system.

Cost aspects

– The system parameters shall be chosen with a view to limit the cost of the complete system, in particular the mobile units.

Table 1.3 – Basic requirements set out by GSM
(original text as written by the committee in 1985)

These system objectives were agreed and distributed
to the telecommunications industry, with the objective of providing
a clear framework for progressing the technical design.

It was clear as well that the capacity offered by the system should be better than with existing analogue network, in order to take over without reaching saturation too quickly.

As early as 1982, the basic requirements for GSM were stated. They were slightly revised in 1985, as reproduced verbatim in table 1.3, and are still rather up-to-date. Since then, portable and handheld mobile stations became the main objective, over vehicle-mounted stations.

1.1.5.1. ISDN as Godfather

At about this time, the specification of the Integrated Service Digital Network (ISDN) was close to completion and the telecommunication trends were such that GSM was necessarily designed as a system offering access to a large variety of telecommunication services: speech, data, voice and alphanumeric messaging, videotex, and so on.

From its origin in 1984, and under the strong impulsion of its chairman, the work in GSM3 was inspired by the principles of ISDN and its access protocols. Their influence is obvious when reading the *Specifications*, both in the area of services definition and in the description of signalling protocols. The use of a layered model such as the OSI (Open System Interconnection) model for the definition of protocols is but an example of such an influence, and with regards to this aspect GSM differs from the foregoing analogue radiocommunication systems. The kinship between GSM and ISDN allows some optimism when considering their integration, GSM networks being just another way to access the general telecommunication networks.

1.1.5.2. Born Digital

General trends in the telecommunication arena made it unofficially clear from the start that the system would be based on digital transmission, and that speech would be represented by a digital stream at a rate of the order of 16 kbit/s. The official decision, however, was not made until 1987.

From 1984 to 1986, GSM focused on the means to compare different technical possibilities for transmission (digital or analogue), in particular with regards to their respective spectrum efficiency. It was decided to compare several technical proposals on the basis of prototypes allowing actual radio transmission. In 1985, the French and West German P&T Administrations joined their efforts to order four studies leading to such prototypes. Comparative testing of eight prototypes, including these four plus four more Scandinavian prototypes was conducted in December 1986 at the CNET laboratories (Centre National d'Etudes des Télécommunications) near Paris, under the control of the Permanent Nucleus. All these prototypes made use of digital transmission, and most were proposed by industrial companies.

1.1.5.3. The "Broad Avenue"

The results of the comparison were reported at the beginning of 1987. The discussions were difficult, because of the prestige and the advance that would be conferred to a proponent were its solution to be chosen. To circumvent this problem, none of the proposed solutions was selected. Only the key features of the transmission method were decided: they were summarised in what was called the "broad avenue". The narrowing would have to take place afterwards, within GSM2. The key features were the following:

- medium-sized band (200 kHz carrier separation), to be compared to narrow-band systems (12.5 kHz or 25 kHz as in existing analogue systems) or wide-band systems (one of the candidates proposed a 6 MHz carrier spacing);
- digital speech transmission at a rate not exceeding 16 kbit/s;
- time multiplexing of order 8, with future evolution towards multiplexing of order 16 once a second-generation speech coder has been defined at a smaller rate;
- slow frequency hopping capability.

1.1.6. THE GSM TECHNICAL SPECIFICATIONS

Originally, the work was planned in three phases: specification writing, validation, then field tests with operational equipment. This phasing was, however, modified little by little. At the beginning, the specification of all features of the GSM were scheduled to be ready at the same moment, in time for validation and tests before an opening of service in 1991. There was one big exception, the half-rate speech coding. Because it was clear that speech encoding techniques evolve quickly, the plan was to have a first speech encoding algorithm at around 16 kbit/s (it uses 13 kbit/s), and a few years later, a more efficient scheme using half as many bits. About twice as many subscribers could then be accommodated within the same spectrum allocation, with a radio channel for carrying 13 kbit/s speech (a full rate channel) being replaced by two radio channels to carry speech encoded with the future algorithm. The whole transmission system was designed from the start to support the future half-rate speech coding as much as possible. The date of introduction of this speech coding algorithm was not precisely planned.

The idea of separating the validation phase from the specification writing phase quickly disappeared. As time went, it was more and more obvious that it would not be adequate to write all the specifications first, i.e., in a very short period of time. The idea of a phase 2 step as a functional enhancement of phase 1 gradually crystallised. Around 1988, the idea matured, and it was agreed that the launch in 1991 would not be with the full palette of services. At this date, services were split between two phases: the most important items left for phase 2 were many supplementary services, phase 1 being limited to the most common ones (such as call forwarding and call barring), and alternate services (for instance speech alternating with data within the same communication). Later on, the capability of offering data services on half rate channels was transferred to phase 2, though all the required specifications existed. This last point was decided at the end of 1991, and was a side effect of the ongoing work on half rate speech coding: it was considered better to open the possibility to change the specification of the half rate radio channel, in order to allow more flexibility in the transmission improvements. Such a change would be very difficult to implement if half rate channels were already used for data services.

During the work concerning DCS1800, the need for new functionalities (such as national roaming) appeared. Some of these functionalities are included in DCS1800 phase 1, which is thus slightly richer than GSM900 phase 1. The added functionalities will be incorporated in the phase 2 (a single set of the phase 2 GSM Technical Specifications will cover both DCS1800 and GSM900).

In the middle of 1988, first drafts of the GSM Technical Specifications were available. From this date up to the middle of 1991, the *Specifications* evolved under a control procedure of "change requests": each change had to be justified in terms of objectives, complete (including impacts on all other applicable GSM Technical Specifications), and precise. Such a mechanism was necessary to trace all changes, in particular at a period when manufacturers were already deeply involved in the development process. The rules for accepted changes became more and more strict, so that at the end the only accepted changes were those necessary to correct malfunctions or dangerous ambiguities. Ideas for enhancement, not acceptable at a late date simply because not absolutely necessary, were not discarded, but generally forwarded to the phase 2 stage.

Nowadays (at the date of writing, middle of 1992), a new set of GSM Technical Specifications is scheduled for approval soon. These phase 2 Technical Specifications will include not only the specifications of the new services (supplementary services and facsimile), but also numerous minor and major improvements. One of the big issues that came to the fore in 1991 was upward compatibility. Mobile stations designed according to the phase 1 *Specifications* will exist for some time, and they will therefore coexist in the network with phase 2 mobile stations during several years. Similarly, infrastructure equipment throughout the whole system will not be updated to phase 2 specifications during one night. Upward compatibility is basically the problem of how to specify phase 2 mobile and infrastructure equipment so that this coexistence is possible. The notion of an evolutive standard emerged quite late, and a number of small points in the phase 1 *Specifications*, not considered from this vantage at the time of decision, rendered upward compatibility difficult or even impossible in some areas: these parts were then frozen out of future evolution. This problem limited quite severely which changes could be accepted for phase 2, and often made the technical solutions more complex. However, awareness of the cross-phase compatibility issue grew with time and a few general mechanisms have been introduced to reduce constraints in future phases.

Phase 2 is not however the end of the story. More enhancements are scheduled. Half rate speech is not ascribed to any phase, and will then introduce another step. Its specifications are foreseen to be completed by end 1993.

To sum up, the original thinking was of a one shot specification work. Nowadays, GSM is an evolving standard. In this domain also, GSM marks a boundary. The reasons for the change are the level of complexity of the *Specifications* and the ever accelerating rate of technical evolution. Future VLSI and microprocessor performances will allow more and more complex systems. The time between successive technical generations is now shorter than the life time of a system, the latter being constrained by financial considerations: the infrastructure cannot be replaced before having been paid off. This trend has no reasons to slacken off, and the GSM story tells us how important it is to manage a standard as an evolutive object from the start.

This book will deal mainly with the phase 1 *Specifications* of GSM, since these are the only officially-approved standards at the date of writing. Now and then, some future schemes, or some evolutive trends, will be hinted at (in particular in relation with half rate channels, because full specifications exist for the management of these channels and for their application to data services, even though these specifications may still be changed), but not described in details.

1.2. CELLULAR SYSTEMS

1.2.1. GENERAL ASPECTS

Even though the term of network is often used when speaking about the GSM architecture ("PLMN", Public Land Mobile Network), it would be more proper to consider the system, as designed, as an access to existing telecommunication networks, aimed at users with GSM mobile stations. A GSM PLMN, as presented by the *Specifications*, is indeed not able to establish autonomously calls other than local ones between mobile subscribers. For all other call configurations, the PLMN relies on fixed networks to route the calls. Most of the time the service provided to the user is the combination of the access service provided by a GSM network, and of the service provided by some fixed network. This approach has an important impact on the *Specifications*: for example, no switch hierarchy is defined, and transport of information between GSM machines is based on the usual 64 kbit/s units multiplexed onto 2 Mbit/s links, or higher rate multiplexes. However, nothing in the technical choices really limits GSM to be an access network. One can imagine in the future, in some countries without extensive wireline-access infrastructure, that GSM, complemented with a suitable switch hierarchy, would be used as the basic telecommunication network.

Considering GSM as another "local-loop" leads us to compare its characteristics with wireline telecommunication access, highlighting two major differences:

- first, the wide-area mobility of subscribers leads them to change their point of access to the network; this poses a serious problem for the routing of calls toward subscribers; this is the realm of **mobility management**;

- second, the link between the subscriber terminal and the fixed infrastructure is not permanent and is subject to fluctuating transmission requirements; this is the realm of **radio resource management**.

The consequences of both aspects will now be described, after some preliminary considerations on the cellular coverage.

1.2.2. CELLULAR COVERAGE

The major problems with radio distribution arise from electromagnetic wave **propagation**. With a decreasing weight and an increasing autonomy, the mobile terminals have a limited transmission range. Every telecommunication engineer will remember that the power of radio waves decreases with the inverse of the squared distance (d^{-2}); however, it must be remembered that this applies only in empty space. Between two stations close to the ground, interference-creating reflections from the ground cannot be neglected, and it is very likely that obstacles intervene on the direct path between them. As a consequence, propagation at ground level in an urban environment is more difficult, and the received power varies typically with d^{-4}!

Figure 1.4 – Cellular coverage representation

Hexagons nicely pave the plane without overlapping and are commonly used
for calculating theoretical frequency reuse in cellular systems.

A second problem is **spectrum scarcity**; the number of simultaneous radio communications supported by a single fixed station (a base station) is therefore limited.

Cellular coverage allows a high traffic density in a wide area despite both problems, at the expense of infrastructure cost and of complexity. Because of the limited transmission range of the terminals, cellular systems are based on a large number of reception and transmission devices on the infrastructure side (the base stations), scattered over the area to cover, and each one covering a fairly small geographical zone called a cell. The underlying image is the one of a membrane composed of epithelial cells. A minimum density of fixed stations allows low-power mobile stations to access the system anywhere within a wide area: they are never very far from a base station. Cells are often represented by hexagons, in order to model the system by paving the plane with a single geometrical figure (see figure 1.4). However, a typical coverage looks more like figure 1.1.

Spectrum scarcity is circumvented by the reuse of radio resources. Frequencies used in a given cell are reused a few cells away, at a distance sufficient so that the unavoidable interference created by the close use of the same spectrum has fallen to an acceptable level, which depends in particular on the transmission method. This concept of frequency reuse is the key to high capacity. As an example, if the same frequency may be reused in every ninth cell, a spectrum allocation of N frequencies allows N/9 carriers to be used simultaneously in any given cell. The total system throughput, often expressed in number of simultaneous calls per km² per MHz, can therefore be increased by reducing the cell size, notwithstanding the limited spectrum available. In practice cell reduction has some other effects, and the design of a cellular system rarely copes for all cell sizes from 0 to infinity. In the case of GSM, the design was aimed at the beginning at medium-sized cells, of a diameter expressed in kilometres or tens of kilometres. Yet, the lower boundary is difficult to determine: cells of more than one kilometre radius should be no problem, whereas the system may not be fully suitable to cells with a radius below, say, 300 meters. One source of limitation is more economical than due to physical laws. The efficiency of the system decreases when cell size is reduced, and then the ratio between the expenditure and the traffic increases, and eventually reaches a point where economical considerations call for a halt. Another important point is the capacity of the system to move a communication from one cell to another rapidly, and GSM requires too long a time to prepare such a transfer to cope with fast moving users in very small cells. The cell size upper bound is more obvious: a first, non-absolute, limitation in GSM is a range of 35 kilometres. Cells of bigger size are possible, but require specially designed cell-site equipment and incur some loss in terms of maximum capacity.

The number of sites to cover a given area with a given high traffic density, and hence the cost of the infrastructure, is determined directly by the reuse factor and the number of traffic channels that can be extracted from the available spectrum. These two factors are compounded in what is called the spectral efficiency of the system. Not all systems allow the same performance in this domain: they depend in particular on the robustness of the radio transmission scheme against interference, but also on the use of a number of technical tricks, such as reducing transmission during the silences of a speech communication. The spectral efficiency, together with the constraints on the cell size, determines also the possible compromises between the capacity and the cost of the infrastructure. All this explains the importance given to spectral efficiency.

Many technical tricks to improve spectral efficiency were conceived during the system design and have been introduced in GSM. They increase the complexity, but this is balanced by the economical advantages of a better efficiency. The major points are the following:

- the **control of the transmitted power** on the radio path aims at minimising the average power broadcast by mobile stations as well as by base stations, whilst keeping transmission quality above a given threshold. This reduces the level of interference caused to other communications;

- **frequency hopping** improves transmission quality at slow speeds through frequency diversity, and improves spectral efficiency through interferer diversity;

- **discontinuous transmission**, whereby transmission is suppressed when possible, allows a reduction in the interference level of other communications. Depending on the type of user information transmitted, it is possible to derive the need for effective transmission. In the case of speech, the mechanism called VAD (Voice Activity Detection) allows transmission requirements to be reduced by an important factor (typically, reduced by half);

- the **mobile assisted handover**, whereby the mobile station provides measurements concerning neighbouring cells, enables efficient handover decision algorithms aimed at minimising the interference generated by the call (whilst keeping the transmission quality above some threshold).

We will come back to this subject after the technical aspects of the standard have been described, in Chapter 9 concerned with the engineering of the system.

1.2.3. RADIO INTERFACE MANAGEMENT

Since the number of available radio channels is much smaller than the total number of potential users (a safe assumption!), channels enabling bi-directional communications are only assigned at need. This is a major difference with standard telephony, where each terminal is continuously linked to a switch, be there a call in progress or not. Such perenniality of the link allows rather simple call set-up procedures: some device in the switch continuously monitors the line for changes between the "on-hook" and "off-hook" status in order to detect outgoing calls, whereas the user terminal is always ready to detect a ringing tone carried by the subscriber line when an incoming call is received.

In a mobile network, radio channels must be allocated and released dynamically, on a call basis. This function is additional to the usual fixed network call handling procedures.

Moreover, reaching the subscriber is not an easy problem. In GSM as well as in most other cellular systems, the user, when not engaged in a call, learns about an incoming call by listening to a specific channel. This channel carries messages called "paging messages": their role is to indicate that a given mobile subscriber is being called. Such a channel is broadcast in every cell, and the problem of the network is to determine in which cells a given subscriber should be called when needed.

The setting up of any call, whether mobile originating or terminating, requires specific means, by which the mobile station may access the system in order to get a channel. In GSM, this access procedure is performed on a specific mobile to base channel. This channel, together with the base to mobile broadcast channels transporting in particular the paging messages, is known as a common channel in GSM, since it carries information to and from many mobile stations at the same time. Conversely, channels allocated to a single mobile station for some period of time are called dedicated channels. Based on this distinction, two "macro-states" of the mobile station may be defined:

- in **idle mode**, a mobile station listens to broadcast channels; it has no channel of its own.

- in **dedicated mode**, a bi-directional channel is allocated to the mobile station for its communication needs, allowing it to exchange point-to-point information with the infrastructure in both directions.

The access procedure is the particular function which allows the mobile station to reach the dedicated mode from the idle mode.

1.2.4. CONSEQUENCES OF MOBILITY

1.2.4.1. Location Management

The mobility of users in a cellular system is the source of major differences with fixed telephony, in particular for incoming calls. A network can route a call towards a fixed user by simply knowing the network address (e.g., the telephone number) of this user, since the local switch to which the subscriber line is directly connected does not change. However, in a cellular system, the cell in which contact may be established with the user changes when the user moves. In order to receive incoming calls, a mobile user must first be located, i.e., the system must determine in which cell he currently is.

In practice, three different methods may be used to gain this knowledge. In the first method, the mobile station indicates each change of cell to the network: this is systematic **location updating** at cell level. When a call arrives, a paging message needs to be broadcast only in the cell where the mobile user is known to be. A second method would be to send a paging message in all cells of the network when a call arrives, removing the need for the mobile station to advise the network of its current location: this is ubiquitous paging. The third method is a compromise between the two first extreme methods, introducing the concept of **location area**. A location area is a group of cells, each cell belonging to a single location area. The identity of the location area a cell belongs to is sent in the cell on a broadcast channel, thus enabling mobile stations to be informed of the location area they are in. When a mobile station changes of cell, two cases may arise:

- both cells are in the same location area: the mobile station does not send any information to the network;
- the cells belong to two different location areas: the mobile station informs the network of its change of location area (location updating).

When an incoming call arrives, a paging message needs only be sent in those cells belonging to the location area where the mobile station has last performed location updating. This third method allows to balance the amount of paging messages (which increases when the location area includes more cells) with the amount of location updatings (which increases when the location area includes less cells). GSM supports this method.

1.2.4.2. Handover

The preceding section (location management) deals with the consequences of mobility in idle mode. In dedicated mode, and in particular when a call is in progress, the mobility of the user may also induce the need to change the serving cell, in particular when transmission quality drops below a given threshold. With a system based on large cells, the probability of such an event is small and the loss of a call in such conditions may be acceptable. However, the achievement of high capacities requires the reduction of cell size, and maintaining the calls despite user mobility becomes an essential requirement to avoid a high degree of customer dissatisfaction.

The process of automatically transferring a transaction in progress (a call in particular) from one cell to another to avoid the adverse effects of user movements is called **handover**. This process requires first some means of detecting the need to change cell while in dedicated mode (handover preparation), and second the means to switch a communication from a channel in a given cell to another channel in another cell, ideally in a way not noticeable by the users, and at least keeping user disturbance to a minimum.

1.2.5. ROAMING

In telecommunication systems accessed through a fixed link, the choice of which network provides the service is done (when choice is possible!) at subscription, once and for all. When mobility is introduced, new horizons emerge. Because a mobile terminal is not on a leash, different networks can provide service to a given customer, depending on where he is. When different network operators co-operate, they can use this possibility to offer to their subscribers a coverage area much wider than any of them could do on its own. This is what is called roaming, and it is one of the major features of the pan-European GSM, whose subscribers can enjoy European-wide coverage, whatever their national network of subscription.

Roaming can be provided only if some administrative and technical constraints are met, and these points will be addressed in various parts of this book. From the administrative point of view, issues like charging, subscription agreements, etc. must be solved between the different operators. The free circulation of mobile stations also requires regulatory bodies to agree on the mutual recognition of type approvals. From the technical point of view, some topics are a consequence of the administrative matters, such as the transfer of call charges or the transfer of subscription information between networks. Others are needed simply for roaming to be possible at all, such as the transfer of location data between networks, or the existence of a common access interface.

This last point is probably the most important one. It requires a subscriber to have a single piece of equipment enabling him to access the different networks. To this avail, a common air interface has been specified, so that the user can access all the networks with the same mobile station.

Other GSM-based (or DCS1800-based) systems will be created. Roaming between these systems and the European GSM may not always be possible with the same mobile station. A possible limitation is bandwidth. European countries have agreed to use a common part of the spectrum, at 900 MHz. Another band is already possible, at 1800 MHz. Roaming with the same mobile equipment is not possible between GSM900 and DCS1800, except with dual band mobile station, of as yet unproved commercial interest. Similarly, the European 900 MHz band may not be available in other parts of the world, and using a different bandwidth would preclude using the same mobile equipment for roaming. But in all such cases, it is possible to envisage another kind of roaming, based on using the subscriber-specific part of the mobile station only, in connection with a radio access part specific to each network. Such a combination will enable users to have a single subscription and be reached through the same directory number, whatever network they be roaming into. A way to achieve this aim is already included in GSM and will be expanded upon later on in this chapter, when the SIM (Subscriber Identity Module) is introduced.

1.3. GSM FUNCTIONALITIES

In this section we will describe the functions of GSM as seen by the users, that is to say the services that are provided to users, abstracting the details of how it is done.

1.3.1. GSM: A MULTISERVICE SYSTEM FOR THE USER

GSM is a multiservice system, allowing communications of various types, depending on the nature of the transmitted information as perceived by the end users. By tradition, one distinguishes speech services from data services: in speech services, the information is voice, whereas the term "data services" groups everything else, such as text, images, facsimile, computer files, messages, and so on. GSM provides a large palette of the services offered to fixed telecommunication users, as will be described further on in this section. GSM provides in addition a non-traditional set of services, the "short message services", closer to the paging services (one-way radio messaging services) than to any service provided in fixed networks. These services will be described separately from the other data services.

More generally, the definition of a telecommunication service includes more than just the nature of transported information. Other characteristics of the communication are also relevant, such as the transmission configuration (point-to-point or point-to-multipoint, half or full-duplex) and the potential partners, but also the possible "supplementary services", which refer to the possible control of various aspects of the communications by the user, and more administrative notions such as for example the charging aspects.

Service provision depends on three independent factors:

- the contents of the subscription held by the subscriber, in terms of services as well as in terms of geographical areas. The subscription packages offered by each operator will vary, as well as the corresponding subscription rates, and each subscriber will perform a choice between these packages;

- the capabilities of the network from which the user is getting service. All networks will not offer exactly the same range of services at a given date, and therefore the user might find some restrictions on the available services depending on the network in which he is currently roaming;

- the capabilities of the equipment held by the user. For instance, it is obvious that faxes cannot be sent or received on a speech-only mobile station if it is not connected to a proper fax machine.

The basic services provided by a telecommunication network, which consist in transmission media and in the means to set up calls, are distinguished from a number of supplementary features enabling users to better control the provision of these basic services, or to simplifying the daily use of telecommunications. These features will represent an important contribution to the user comfort, and include the possibility for the user to forward calls, to visualise the calling number, etc. Some of these features are performed locally by the terminal, and are not properly speaking provided by the network (though the limit is difficult to tell for the user). Other features are provided by the network, and are called the "supplementary services". Several reasons warrant the distinction between basic and supplementary services, including the fact that supplementary services apply generally to several basic services (and hence a separate presentation is better than repetition); and also because in many existing networks these supplementary services must be asked and paid for in addition to the basic services. In the future it is quite likely that some of these supplementary features will be automatically part of a service package. This statement holds in particular for those services which improve network efficiency (such as the forwarding of calls which cannot be completed).

The description of the services provided by GSM follows. We will first address the speech services, the data services and the short message services. Then the "supplementary services" will be considered. Afterwards, a small section will deal with the "local services", i.e., the facilities offered by the GSM terminals on their own. Then an important particularity of the GSM mobile terminal will be described: the Subscriber Identity Module (SIM). Finally, the security features of GSM, another complement of service for the users, will be addressed.

1.3.1.1. Speech Services

The most important service provided by GSM is telephony. This service enables bi-directional speech calls to be placed between GSM users and any telephone subscriber reachable through the general telephony network. Fixed telephone subscribers world-wide can be reached as well as mobile network subscribers or subscribers of specific networks connected to a public telephone network. With the rise of ISDN, telephony has been somewhat eclipsed by data services, which are often presented as the future of telecommunications. However, speech remains and will remain the most important service for mobile systems. The absence of a fixed wire and the arrival of handheld terminals (less than 400 g, and fit to be carried in a pocket) make cellular telephony the foremost technique for inter-personal communications, the true tele-communication method extending the most natural form of communications that human beings have used for maybe several hundred millennia.

Following the GSM official terminology, emergency calling is a distinct service, derived from telephony. It allows the user of a mobile station to reach a nearby emergency service (such as police or the fire brigade) through a simple and unified procedure, by dialling 112 (the number which has been agreed as the standard emergency number throughout Europe).

Another service derived from telephony is voice messaging. The *Specifications* do not identify this service as a separate one, but many an operator will offer it as a basic feature. It enables a voice message to be stored for later retrieval by the mobile recipient, either because he was not reachable at the time of the call, or even because the calling party chose to access directly the voice mailbox of the GSM subscriber.

1.3.1.2. Data Services

As an heir of ISDN, GSM has been designed from the start to offer many data services. Basically, most services which are provided to fixed telephony users and to ISDN users (thus covering access to other more specialised networks such as Packet Switched Public Data Networks) have been included, as far as the limitations due to radio transmission allow. This corresponds to a very large number of different cases (the *Specifications* list some 35 services), to cope with all the variants stemming from the history of telecommunications.

Data services are distinguished mainly by the potential correspondents (users of the telephone network, of ISDN, or of specialised networks), by the nature of the end-to-end information flow (raw data, facsimile, videotex, teletex, ...), by the mode of transmission (packet or circuit, end-to-end digital or making use of an audio modem, synchronous or not, ...), by the nature of the terminal, and so on.

For practical reasons, data services shall not be presented here with the full descriptive methodology used in the *Specifications* (i.e., as tables of attribute values), but shall be grouped according primarily to the type of potential correspondents, that is to say according to the type of network to which the other correspondent is a subscriber to. A summary of all data services as specified in the *Specifications* will then be given.

Connections with PSTN Users

The Public Switched Telephone Network (PSTN) is already used widely today for data transmission, with audio modems, not by vocation but simply because of its wide availability. It was then mandatory for GSM to provide its users with the same possibilities of data communication as are available between PSTN users, in particular the most popular data services: group 3 fax and videotex. Unfortunately, this cannot be done as simply in GSM as in the PSTN (which in fact provides only "speech" paths, strictly speaking 3.1 kHz audio transmission paths), because of the characteristics of the radio transmission. Voice transmission on the GSM radio interface is based on coding algorithms optimised for speech, making its representation unsuited to standard modem signals.

In GSM, the user does not need to provide a modem for such communications. The transmission between the mobile terminal and the network is fully digital, and an audio modem is provided inside the network, at the point of interworking with the PSTN. For the PSTN user wanting to communicate with a GSM user, this approach effectively limits the choice of modulations to the ones available in the GSM network. However, this is not a strong restriction, since if so desired by the operator, GSM can cope with the most widely used standards up to 9600 bit/s full-duplex, such as V.21, V.22, V.22bis, and so on, including V.32 and the modems specific to group 3 facsimile and videotex.

On the GSM user side, specific data terminals, adapted to mobility (power supply, size), will certainly sooner or later be proposed by manufacturers. Besides, the specifications cater for the connection of off-the-shelf terminal equipment (fax, personal computer, videotex terminal) to the mobile stations. This possibility enables the use of standard equipment designed for PSTN usage, that is to say designed to be connected to a telephone line (as most fax machines), or to a modem itself connected to the telephone line (as personal computers).

Thanks to its multiservice nature, GSM will provide in these cases a service a bit better than the PSTN itself, beside the obvious advantages of mobility. The palette of supplementary services will become extensive. A point worth mentioning is the possibility for a subscriber to have several directory numbers, all corresponding to the same subscription, but to different services (for instance one number for speech, and another for facsimile).

Connections with ISDN Users

Before going further, it should be noted that the distinction between the ISDN and the PSTN can be quite fuzzy. In many countries, digital ISDN-like transmission and advanced signalling methods are already used on a wide scale within the telephony network, and the transition to ISDN is very smooth. Here we will use (improperly) the term PSTN to refer to the "Plain Old Telephone System", where analogue transmission or less modern signalling methods are used.

If connections with PSTN users are a must because of the present weight of this network, connections with ISDN users are a must for a digital system because of the future expected importance of ISDN. An important point during the GSM design phase was to ease as far as possible the interconnection with ISDN, so as to enable the integration of GSM in an ISDN system, with for instance local exchanges serving both GSM and ISDN users.

Still, the 9600 bit/s upper bound of GSM data transmission puts a serious limitation on the interworking with ISDN, which provides basically 64 kbit/s transmission (and higher rates in the future with broadband ISDN). The weight of history being about the same for ISDN as for GSM, the former must also cope with the provision of data services with PSTN users. This opened a possibility of interworking between ISDN and GSM, each network considering the other as it does PSTN. Interworking between GSM and ISDN can then be done using the standardised ISDN digital formats developed for PSTN adaptation and enable it to carry low rate digital streams over the 64 kbit/s links. This leads to a somewhat wider range of services than that possible with audio modems through the PSTN, and includes, e.g., asynchronous data transmission at rates over 2400 bit/s, for which no audio modems are specified internationally.

The extreme form of this approach consists in using the ISDN as the PSTN (using the 3.1 kHz audio transmission mode and hence audio modems), and with this method the same services are available for communications between GSM users and ISDN users as with PSTN users. This possibility is opened in the *Specifications*, as an interim method.

Connections between GSM Users

GSM has not been designed as an independent network which provides primarily services between its users and, only as a side dish, with users of other networks. On the contrary it has been designed as an access network for the fixed telecommunication networks, and the services provided between GSM users have nothing specific. Indeed, in most cases an external network intervenes in a communication between GSM users. The goal is that the ISDN be this interworking network. At the beginning, it will often be the PSTN instead, and the provided services will be those that can be supported by PSTN transmission, that is to say the same as are provided between GSM users and PSTN users.

When it is certain that the communication is digital end-to-end, using a 64 kbit/s link, the service range is wider, the same as in the case of GSM to ISDN. This happens when the transit network is a modern PSTN, or ISDN, or when there is no transit network (for instance for communications local to a GSM exchange). In the last case the *Specifications* open the possibility for an original service, not provided otherwise than between GSM users since no external network supports it, the transmission of raw data at 12 kbit/s (this possibility is deemed to be marginal, since it will take some time before corresponding terminals are provided, and because data communications between GSM users will be a very marginal traffic).

Connections with Packet Switched Public Data Network Users

A Packet Switched Public Data Network (PSPDN), such as the French TRANSPAC network, is a general purpose data network using the packet transmission techniques, as opposed to circuit techniques as used for instance in the PSTN. PSPDNs are used primarily for communications with or between computers. They are also often supporting services like message handling systems, or remote data base interrogation.

A PSPDN can typically be accessed in three different ways:

- using a direct connection for the subscribers (X.25 access) of such a network;
- through the PSTN or ISDN, via a PAD (Packet Assembler Disassembler, for access via a modem using an asynchronous modulation –X.28 access)
- through the PSTN or ISDN, via synchronous access (X.32 access, that is to say X.25 plus some complements in particular for the identification of the user);
- and through the ISDN, using the capabilities of ISDN to transmit packet data.

As a result, several different methods are proposed to a GSM user which desires to communicate with a PSPDN subscriber. These different methods of interconnection are shown in figure 1.5. The differences are mainly technical, but impact the user in the type of terminal he can use, the available data rate, in the way he has to enter the called number, in the need or not to subscribe to the PSPDN and consequently in how the calls are billed.

The less specific method consists in acting as a PSTN subscriber, and to access a PAD through the PSTN, via asynchronous audio modems supported by the PSTN. This allows use of a simple data terminal on the GSM user side (case (a) in figure 1.5). For the GSM PLMN, this type of communication is perceived just as a plain data communication with some PSTN "subscriber", and charged as such. This requires the GSM

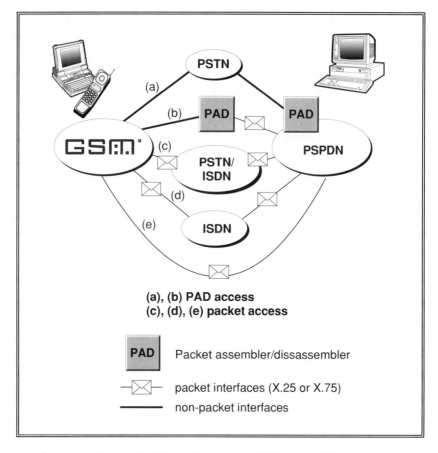

Figure 1.5 – Ways of accessing a PSPDN from GSM

The services offered to a GSM user for communication with a PSPDN user
differ according to the capabilities of the GSM user's terminal,
to the possible subscription he holds with the PSPDN,
as well as to possibilities offered by network implementations.

user to provide the PAD PSTN number himself, and also to be registered
on the PSPDN, so that he is charged for the PSPDN connection (except if
reverse charging is used by the called PSPDN subscriber). The directory
number of the desired correspondent is provided by the user afterward,
once in contact with the PAD (double numbering). The communication
can be set up only from a GSM user toward the PAD, not in the other
direction. The approach is general, and enables users to access PSPDN
from other countries (including the home country when roaming). The
service is called "basic PAD access" in the GSM terminology.

Next is what is called the "dedicated PAD access" (case (b) in figure 1.5). This corresponds to a GSM service distinct from basic data communication through the PSTN. The subscriber must have a GSM subscription for the service, but is not required to have a registration with the PSPDN. The PSPDN part is billed to the GSM operator, and the full communication is charged by the GSM operator to the subscriber. The GSM user need not indicate any PAD number: he just indicates the directory number of his correspondent and that he wants a "dedicated PAD access" communication, and the PLMN takes charge of all the details (single numbering). As in the previous case, only mobile user originating calls are possible. Usually a given PLMN will provide this type of access only for a national PSPDN, and hence the service will not be available when roaming. The access rate can be up to 9600 bit/s.

Then we find the possibility to access a PSPDN in packet mode (X.32) via the PSTN or the ISDN, used as raw data carriers (case (c) in figure 1.5, and the variant (d)). As with the "basic PAD access", the GSM user must also be registered with the PSPDN, and must provide a PSTN or ISDN number corresponding to the PSPDN access unit he wants to communicate with (double numbering). There also the PLMN (and the PSTN or ISDN) does not see a specific communication, just a data communication toward some "subscriber", which is charged as such. The PSPDN part of the communication is directly billed to the user, in his quality as PSPDN subscriber. The transmission mode is synchronous, and depends on the access unit. GSM provides synchronous transmission at 2400, 4800 and 9600 bit/s. A specific terminal has to be provided by the user to support the X.32 protocol (manufacturers may provide in the future integrated mobile stations including the support of X.32). Calls can be set up from the mobile station as well as toward the mobile station, depending on PSPDN possibilities. As in the first case, an advantage of this approach is the possibility to easily access PSPDNs in foreign countries (including the home country when roaming).

The final case is the direct interworking between the PLMN and the PSPDN (case (e) in figure 1.5). This is close to the previous case, but here the PLMN fulfils the functions of interworking with the PSPDN. The advantage is that the user does not need to be registered with the PSPDN. The PSPDN part of the connection is charged to the PLMN, which in turn puts it on the bill of the subscriber.

The access to a PSPDN, according to one or several of the means described above, is a service which can be offered to the users as such, but is also the support of some end-to-end services, using specific terminals, such as teletex and message handling. These services may be offered by PLMN operators separately from the simple PSPDN access.

Connection with Circuit-Switched Public Data Networks Users

Circuit-Switched Public Data Networks (CSPDN) are general purpose data networks, such as PSPDNs, and by and large aiming at the same kind of services. They differ by using a circuit transmission approach, and as such are more fitted to the cases of intensive end-to-end traffic. CSPDNs are likely to disappear, replaced by ISDN whose basic data services are of the circuit type.

A CSPDN is typically accessed directly from GSM, with a user-network interface according to X.21. Another possibility consists in transiting the ISDN, in which case the user-network interface is according to the ISDN standards.

GSM can provide access to a CSPDN in the same way as ISDN can. However, in the GSM case, the transmission modes are limited to synchronous transmission at 2.4, 4.8 or 9.6 kbit/s.

1.3.1.3. Short Message Services

The different data services listed in the previous section are not really adapted to the mobile environment. They are simply extensions to GSM subscribers of the services available to fixed subscribers. One of the problems is that these data services normally require rather bulky terminals, compared to the size of a handheld. They are then suited to semi-fixed usage, such as temporary installations, or for use with vehicle-mounted terminals (e.g., fax in the car). But none of the data services is totally fit for an integrated implementation in the handheld case (they require a complete computer keyboard and a comfortable-sized screen).

Still some data services are of interest for persons on the move and not desiring to be encumbered with a bulky terminal. An example of such services is the paging service, where simple messages, a few tens of octets long, can be received on a very small terminal. Since its designers recognised this need, GSM was designed to support such a service, in order to spare GSM subscribers the trouble of carrying two terminals, one for speech, the other for paging.

GSM enables the transmission of point-to-point short messages, and distinguishes between the "Mobile Terminating Short Message Service, Point to Point" (SMS-MT/PP), for the reception of short messages, and the "Mobile Originating Short Message Service, Point to Point" (SMS-MO/PP), enabling for instance a GSM user to send such a message to another GSM user. Another short message service is the "Cell Broadcast Short Message Service" (SMS-CB), enabling short messages of a general nature to be broadcast at regular intervals to all subscribers in a given geographical area. Let us examine in more details what is provided, by taking the example of a GSM user called Kevin.

Point-to-Point Short Messages

These services enable alphanumerical messages of several tens of characters to be sent from or to Kevin. In the mobile terminating case, the message is typically displayed on Kevin's GSM terminal upon reception. The service provided by GSM is akin to paging, but with many enhancements, which make use of the other capabilities of GSM, in particular the possibility to have a bi-directional dialogue between the mobile station and the network (paging services are based on unidirectional links). For instance, the network is informed whether the message has been received or not by the mobile station; and messages can therefore be kept in the network in cases of delivery failure, and repeated once Kevin is known to be reachable. Another improvement is that the originator of the message can be advised of the outcome, which is not the case in standard paging systems. On the mobile station side, the last received messages can be stored in a non-volatile memory.

The way the message is sent from the original sender to the PLMN is operator dependent. This service is not an extension of some fixed network service, and as a result does not correspond to any standardisation prior to GSM. Generating methods as widespread as keying on a dual-tone multi-frequency phone allow many PSTN subscribers to access the service; however, the message contents are in that case restricted to numerical characters. Other access possibilities include simple videotex terminals such as the French Minitel. Of course, any computer accessible through a packet data network would do, but access to the service would not be so popular! Delivery through a human operator could also be envisaged, but is not very cost effective for the PLMN operator. In the same field, voice recognition machines may be a futuristic solution to obtain a widespread access through the PSTN.

GSM subscribers are luckier than paging subscribers, since the Mobile Originating Point-to-Point short message service enables them to send short alphanumerical messages to other parties. For the time being, these other parties are not precisely defined. They include certainly the other GSM users (which will receive them according to the SMS-MT/PP service), but these other parties could also be subscribers to a paging system, an electronic mailbox or alternatively one could foresee gateway devices transforming a short message into, e.g., a facsimile or any kind of format suitable for the equipment of the recipient. Interworking between GSM PLMNs will certainly be possible.

Broadcast Messages

The service called "Cell Broadcast" in GSM is another specific feature of the system. It consists in broadcasting digital information messages cyclically towards mobile stations in a given geographical area. A mobile station designed for this service can monitor continuously the broadcast messages and present them to the user, except when engaged in a bi-directional communication with the network. A classic example for cell broadcast usage is road traffic information.

In the current specification state, these messages are neither addressed nor ciphered: any mobile station equipped for the service may receive and decode them. As a consequence, this service will usually not require a specific subscription.

The *Specifications* leave open the way in which these messages are provided to the network; it is up to the network operators to define their own rules for the generation of such messages, which could be reserved for instance for public authorities or allow wider usage such as advertising.

Bearer service	Partner networks	Offered rates
3.1 kHz ex-PLMN	ISDN	
Data circuit duplex asynchronous	PSTN, ISDN	300, 1200; 1200/75, 2400, 4800, 9600 bit/s
PAD access circuit asynchronous (basic or dedicated access)	PSPDN	"
Data circuit duplex synchronous	PSTN, ISDN CSPDN	1200, 2400, 4800, 9600 bit/s
Alternate speech/unrestricted digital	PSTN, ISDN	one of the above, corresponding to the data part of the service
Speech followed by data	PSTN, ISDN	"
Data packet duplex synchronous	PSPDN	2400, 4800, 9600 bit/s
12 kbit/s unrestricted digital	–	12 kbit/s

Table 1.4 – Bearer services in the *Specifications*

Bearer services offered in GSM are similar to those offered in ISDN, with a restriction on the available rates (9600 bit/s max. with external networks).

Data Services: a Summary

The *Specifications* define services in the same way as ISDN does. They distinguish "bearer services" which correspond to the transportation of user data between two "terminal-modem" interfaces, from "teleservices" which are complete end-to-end services, including terminal capabilities. The list of bearer services appearing in the *Specifications* in given in table 1.4.

Though identified as a separate bearer service, the basic "PAD access circuit asynchronous" service requires in fact nothing specific in the GSM networks compared to the "data circuit duplex asynchronous" service. The same sort of remark also applies to several teleservices, as can be derived from table 1.5. For instance, the specificity of a teleservice such as videotex, compared to the relevant bearer service, does not concern the GSM domain. Though videotex is identified as a separate teleservice in the *Specifications* (at least for phase 1), the GSM networks need not implement anything specific to offer their customers access to videotex, on top of the basic support of the "data circuit duplex asynchronous, 1200/75 bit/s" bearer service. Other examples are X.400 message handling systems and teletex. Conversely, some teleservices are supported by GSM in the same way, though not mentioned by the *Specifications*. Examples are group 4 facsimile or access to voice messaging centres.

Teleservice	Corresponding bearer service
Telephony	-
Emergency calls	-
Short message service MT/PP	-
Short message service MO/PP	-
Short message cell broadcast	-
Advanced MHS access (X.400)	Data circuit duplex synchronous
Videotex access (profiles 1, 2 or 3)	Data circuit duplex asynch. 1200/75 bit/s
Teletex	Data circuit duplex synchronous
Alternate speech/facsimile group 3	-
Automatic facsimile group 3	-

Table 1.5 – Teleservices offered in GSM

True GSM teleservices include speech, fax and short messages.

1.3.1.4. Supplementary Services

The "supplementary services" modify and enrich the basic services, mainly by allowing the user to choose how calls toward or from himself are treated by the network, or by providing him with information enabling an intelligent usage of the services. These functions are not specific to GSM nor to cellular radio: most of them are directly inherited from the fixed network, with minor modifications when needed to adapt to mobility. The aim of GSM is to offer eventually the widest palette of supplementary features. However, in phase 1 only a few of them are fully specified, and this limited set does not give a fair image of what supplementary services are. Many more are incorporated in phase 2, and their specifications are sufficiently stable at the date of writing so that we will include their description in this book (but not the details of their implementation).

In this section we will list and describe these features as they are perceived by the user. Though this approach may be considered superficial, many difficulties lie in this domain (including for the user!), in particular when several features interfere.

Two aspects of supplementary services must be distinguished. The first one is how they modify or complement the calls. The second one is how the users can set and ask for the various alternatives. As for this second point, there is a kind of "profile" for each subscriber, which determines the network behaviour as far as his calls are concerned. The user has access to this profile and can modify it (at least if he has the subscription entitlement for that). This is done through commands issued from the terminals. A non-sophisticated approach consists in entering abstruse key sequences (this is what is usually available in the PSTN or with many PABXs). The availability of a display and of a more comprehensive keyboard than on standard phone sets will enable mobile manufacturers to provide more friendly user interfaces, such as a menu-driven command structure.

The supplementary services are often grouped for presentation according to their commonality of implementation, as seen by manufacturers. For a change, we will present them according to the moment they impact a call, as seen by the user. We will then look at a call, and see how the treatment can be adjusted by the users (the originator of the call as well as the other party). This will leave aside the features impacting established communications, which will be dealt with afterwards.

Figure 1.6 – The puzzle of supplementary services

Commercial departments will need to make simple packages for subscribers
out of all the supplementary services combinations and abbreviations in GSM,
if they want to avoid situations such as the one shown here!

Still within the preliminaries, it should be noted that each feature
owns an official esoteric name, as well as an official abbreviation. Both
will be introduced, though, as suggested in figure 1.6, these abbreviations
will hopefully be banned by the commercial teams trying to sell the
service.

So let us proceed with a call between GSM subscribers, and see all
the steps where some supplementary feature may intervene.

a) First Check

To start a call, the GSM subscriber (say Björn) must be entitled to.
He can ask the network to block all or some of the call attempts made
under his subscription. This seemingly curious request is better
understood when one thinks about lending one's terminal to somebody
else (a teenage daughter for instance). It could be useful in such cases to
restrict the usage to the reception of calls, or to low-cost calls. The
activation and deactivation of such a blocking can be protected by a
password, allowing the control to be kept by Björn himself. There are
several varieties of this call barring facility.

The prohibition from setting up any call is the **"Barring of All outgoing Calls" (BAOC)**. Another classic service in this area is the barring of international calls (which have certainly the highest potential for heavy bills), but in a roaming environment, this can be misleading: whether a call is international or not depends on the user location. So in GSM there are two possibilities. The **"Barring of Outgoing International Calls" (BOIC)** allows one to forbid call attempts toward a country other than the one where the user is currently located; and the **"Barring of Outgoing International Calls except those directed toward the home PLMN country" (BOIC-exHC)** will only allow calls toward a party in the country of subscription.

Another obstacle against starting a call is the fact that another is already in progress. Usually, the only solution is to end this call before starting a new one. But in some cases the user may wish to keep the existing communication, just briefly interrupting it. This is possible and is called putting the call "on hold". This corresponds to the **"Call hold" (HOLD)** facility. The other party may receive an announcement indicating that the call has been put on hold, in order that he does not hang up; but there is no such announcement when the communication resumes.

b) Treatment of a Call Toward a GSM User, the Second Stumbling Block

Let us now consider how the call proceeds, when it happens to be directed toward a GSM subscriber (let us call her Nina) in order to cover the called party aspects. As a GSM subscriber, Nina can ask the network beforehand to react to received calls according to three possibilities:

1. The network tries to establish the call toward her. This is the normal case when no special facility is acting;

2. The network simply rejects the call: this is what happens if **"Barring of All Incoming Calls" (BAIC)** has been asked for by Nina; or if she is known to be reachable through a PLMN outside her country of subscription, and has asked for **"Barring of Incoming Calls when roaming outside the home PLMN country" (BIC-roam)**;

3. The network redirects the call to a third party, say Hans, chosen in advance by Nina (and the call is then back to case 1, with a new direction). This is what happens if **"Call Forwarding Unconditional" (CFU)** has been asked for by Nina.

To simplify the picture, the directed-to numbers can be different for different services (e.g., one number for telephony and another for fax).

c) Third Obstacle

Assuming that the call still exists after the earlier obstacles, the network will try to set up Björn's call intended for Nina to some subscriber, which will normally be Nina, but may be Hans if CFU has been applied. Let us consider the case when the call is still routed towards Nina. The network has then to try to contact Nina's mobile station. But this may happen not to be possible. The network can in some cases be immediately aware that Nina cannot be reached. In GSM, the mobile station can (and should if so decided by the operator) indicate to the network that it will be switched off, or that it enters an area where service cannot be provided to the subscriber (this is the concept of "IMSI detach", which is explored in Chapter 7). In other cases, an effective attempt to contact the mobile station is done, and fails. When the network knows of this situation, the call is either:

1. Terminated, with a "sorry" message or a tone toward Björn (basic case);

2. Forwarded to a third party, let us say Hans again, chosen in advance by Nina (the call is back to square 1). This is what happens if **"Call Forwarding on mobile subscriber Not Reachable" (CFNRc)** has been asked for by Nina. This feature has no counterpart in a fixed PSTN.

d) Fourth Impediment

The next possible problem arises when the destination is found to be already busy, e.g., Nina is already engaged in a communication. There again, Nina can have programmed the network in advance for one of several possibilities:

1. The call is terminated, with a lovely voice saying how sorry the network operating company is not to have been able to complete the call, or, more often, with a busy tone. This is the basic case;

2. The call is forwarded to a third party, chosen in advance by Nina. This happens when **"Call Forwarding on mobile subscriber Busy" (CFB)** has been asked for;

3. Nina is warned of the call, and can then choose what to do. This happens if **"Call Waiting" (CW)** has been asked for. Nina can then either "reject" the call, which is then treated according to 1– or 2–, as chosen in advance; or she stops (totally or momentarily) her ongoing call to accept the new one. The user can be helped in this choice, as explained in the following item.

e) Fifth, and Penultimate, Obstruction

Supposing the call has cleared all the previous hurdles, the final destination (say Hans) is alerted. Then there are still two possibilities: Hans answers, and the call is completed, or he does not, either because he is absent, or because he does not want to. In the good old telephony network, there is usually no reason not to answer selectively, for lack of information. Moreover, how to reject the call is the source of a dilemma: either Hans waits until the caller runs out of patience (but the ringing can be quite nerve-racking), or he lifts the receiver and hangs up at once (but this is not deemed very polite). GSM provides the means for better manners.

First, Hans can have asked for the **"Calling Line Identification Presentation" (CLIP)**, in which case he (or the machine) recognises who is calling, and can then take his decision as a consequence. Now, this is only true if the calling party (Björn still) has not asked for **"Calling Line Identification Restriction" (CLIR)**, in which case he may have required that his number is *not* presented. In some cases, involving authorities, this restriction can be lifted.

Second, Hans has means to actively reject the call (mechanisms exist in the GSM protocols to do so, and it is a matter for the terminal to enable this feature).

Whether on time out (if the called party does not react), or by active rejection, a "rejected" call can be forwarded to another party, chosen in advance by Hans, if he has asked for the **"Call Forwarding on No Reply" (CFNRy)**. It is worth mentioning that the number of forwardings in a row is limited to prevent infinite looping of a call. For instance, if Hans's country limits this number to 1, and if Hans has asked for CFNRy, the original call from Björn will be cleared since it has already been forwarded from Nina to Hans.

f) Last Problem

Call forwarding is fine, but may not be what Björn was really asking for. He may wish at this stage to abort the call in case of its being forwarded. This would be more polite (and less expensive) if done before the final destination (Hans) hangs up. But, to achieve this, the knowledge of the forwarding is necessary. Such information can be provided to Björn by telling him the directory number he finally gets, i.e., Hans's number. This happens if Björn has asked for **"Connected Line Identification Presentation"** (CoLP). And, of course, if Hans did not forbid this presentation by asking for **"Connected Line Identification Restriction"** (CoLR).

This steeplechase through a call allowed us to present most of the features, but some escaped. They will now be described.

Charging

Both the connected-to and the calling party can ask to be informed in real time of the progress of the call cost. This is the **"Advice of Charge"** (AoC) facility. This feature in fact covers different issues. It may be a simple indication, not guaranteed to the last decimal to be what will appear on the bill if the call was stopped at this moment; or it can be used for a real time charging (payphone application, the payment being by coins, credit or debit card), in which case the actual charge levied on the user is exactly what is indicated.

Accuracy is a real problem in an international multi-operator environment. The roaming possibility means that the PLMN in charge of the call (and issuing the toll ticket) is not necessarily the one which bills the subscriber. The call charge and the billed charge can be different, for instance in different currencies. The relevant information on tariffs, exchange rates, ... must be exchanged between operators to ensure that the advice of charge reflects accurately what the subscriber will later discover on his bill.

At the time of writing, discussions are still ongoing on the provision of the advice of charge facility when a user is roaming out of the network which holds his subscription. It is likely that this facility will not be offered initially to roaming subscribers.

Multiparty

"**Multiparty**" (**MPTy**) is a facility enabling a user to merge several communications, so that everybody hears what everybody else says. This applies only to speech communications. The way to proceed is to start with an established communication, to put it on hold, to establish a second call, and to ask for the conference. The process can be repeated so that a given user can merge up to 5 communications in which he takes part. Because the others can do the same thing, there is potentially as many correspondents as wanted together. Each call keeps its identity, and can be separated temporarily, or terminated independently from the others.

Closed User Groups

The notion of "**Closed User Group**" (**CUG**) does not cover a simple atomic facility as the others do. It refers to a complex set of facilities centred around the concept of a group of users who want to restrict their usage of the network to communications inside the group. The typical application is for companies, providing terminals and subscriptions to their employees for professional usage. It is then possible to limit for a given subscriber the outgoing calls, or the incoming calls, or both, to calls inside the group. A password mechanism allows to check the belonging to the group. Things are even more complex because a given subscriber can be part of several groups. A detailed description of all the possibilities would take too long, and the reader is referred to the *Specifications* or to the CCITT Recommendations.

Phasing of the Supplementary Services

Out of all the features presented above, only the call barring and the call forwarding features are provided for in phase 1 (namely BAOC, BOIC, BAIC, CFU, CFB, CFNRy, and CFNRc, plus as a network option BOIC-exHC and BIC-Roam). The other features cited here will be available in phase 2.

1.3.1.5. Local Features

A GSM mobile station is quite a complex piece of machinery, and includes the capacity of a small computer. As an intelligent terminal, it can offer a number of functions locally, without the help of a network. Examples include the dialling of abbreviated numbers, the storage of received short messages, the edition of short messages, the automatic repeat of failed calls, the automatic answering of calls, and so on. In some cases, like in the latter example, the same function can be fulfilled locally, if the terminal implements it, or by the infrastructure, if a voice messaging facility is provided.

The standard places few constraints on the local features. Mobile station manufacturers may or may not include them in their products. Some are specified by the standard simply because they are provided in fact by the SIM (see next section), and not by the part built by the mobile station manufacturer. In fact the only real imposed constraints pertain to the automatic repetition of call attempt, for which a number of restrictions are put, to diminish the risk of overloading the networks with useless attempts.

Another point worth noting is the existence of the '+' key, which is specified as a harmonised shortcut replacing the international prefix, whatever the convention of the network the user happens to get service from. For instance, when in Sweden, a GSM user can call somebody in Italy by dialling +39 followed by the national number, instead of dialling 00939... Another important advantage in so doing is that the stored "+39..." number will be recognised correctly by all GSM PLMNs (including in Italy), and therefore remains valid irrespective of roaming.

1.3.1.6. The Subscriber Identity Module

A mobile station in any cellular network must be personalised, i.e., associated with a given subscription. This is needed since the identity of the subscriber is not in a one-to-one correspondence with the physical medium used for access as in a wireline network. The usual approach is to store in a permanent memory of the machine the required information, such as a subscription identifier. This is what is done in most analogue cellular networks (an exception is the German C network). The approach in GSM is different.

A GSM mobile station is split in two parts, one of which contains the hardware and software specific to the radio interface, and another which contains the subscriber-specific data: the Subscriber Identity Module, or SIM. The SIM can be either a smart card, having the well-known size of credit cards, or alternatively it can be "cut" to a much

Figure 1.7 – The two types of SIMs

The plug-in SIM has been designed to enable smaller handhelds to be built,
and will not be removed as often as the card-sized SIM.

smaller format, called "plug-in SIM" (see figure 1.7). This smaller format
was introduced to put less constraints on the design of handhelds. The
SIM is a kind of key. Once removed from the terminal, the latter cannot
be used except for emergency calls (if the network permits), that is to say
it cannot be used for any service which will impact the subscriber's bill.

This view must be somewhat qualified, because the insertion and removal of the
SIM is not necessarily easy with all mobile stations. Since its small size does not make it
easy to manipulate, it is not foreseen that plug-in SIMs will be easy to remove, and in
some cases mobile manufacturers have even secured them in the handheld station by a
screwed lid. But the possibility still remains for the user to change it.

The possibility to remove the SIM presents many advantages for
the user beside its role as a key. For instance, if his mobile station fails
and must be taken to repairs, another one can be used for the interim
period. It suffices to remove the SIM from one equipment and to put it in
the other. Another example is the case of urban users, which have only a
handheld, for reasons of economy. When needed, they can borrow a more
powerful station to be used in the countryside, or rent a car equipped with
a vehicle-mounted station. In all cases, they can use their own SIM, in

order for calls toward their personal number to be routed to the rented terminal, and for the call charges to be put on the same bill as for the calls made through their handheld.

The SIM is also the custodian of much information involved in the local provision of services to the user. The SIM can be protected by a password, a PIN code (Personal Identity Number), similar to the (typically 4-digit) PINs of credit cards. Unlike many credit card PINs, the GSM PIN may be chosen by the subscriber. The SIM may also contain a list of abbreviated dialling numbers, with the corresponding alphanumerical index (for the name of the correspondent for instance) and the type of call (speech, fax, ...). The SIM can also be used to store short messages, in particular those received when the user is not present. A more technical application is the storage of a list of preference for the choice of a network when several are possible. Since the user will have to choose which network he will get service from, for instance when he crosses an international boundary, the SIM stores information to make this choice automatically, taking into account the user's preferences. When real-time advice of charge becomes available on networks, the SIM will also be able to memorise this charging information, to keep the subscriber informed of his expenses.

An interesting development for the user is the potentiality to read and modify part of the personal information stored in the SIM. This can of course be done using the keyboard of a mobile station, but a more comfortable approach could also be offered, using a card reader connected to a personal computer, and relevant software to enter abbreviated dialling numbers, to archive short messages on the computer, etc. Of course, this only holds for part of the data stored in the SIM, since most of the information is protected against alterations and in some cases even against reading. The scope of the SIM can even be extended beyond GSM, and the concept of a multi-application card is emerging. The compatibility of the SIM specification with internationally-recognised ISO standards in this domain makes the GSM application a good candidate for inclusion into a multi-application card. The concept of the SIM is yet in its infancy, and will undoubtedly become a basis for a better interworking between a user and a terminal.

Another portentous aspect of the SIM is related to roaming. We have already seen how roaming can be achieved by using the same mobile equipment to get service from two different networks with a single subscription. We will call this kind of roaming "MS-roaming" (MS standing for mobile station), since there is another possibility. The interface between the SIM and the rest of the mobile station is standardised in the *Specifications*, and this standard could provide a basis for roaming between PLMNs having different Air Interfaces, which will

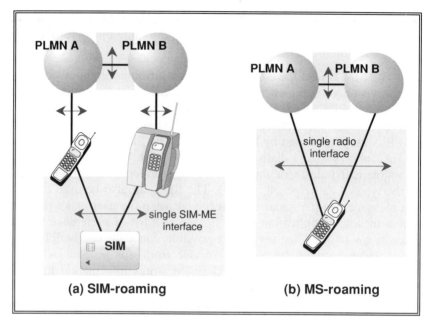

Figure 1.8 – PLMN interfaces

Inter-PLMN interfaces must be fully standardised to allow any kind of roaming;
in addition, MS-roaming requires a standardised air interface,
whereas SIM-roaming requires a standardised SIM-ME interface.
GSM offers the most flexible form of roaming, with full MS– and SIM-roaming.
(the greyed rectangles indicate the areas where standardisation is required).

be referred to as **SIM-roaming**. SIM-roaming does not offer the fully automatic network selection as MS-roaming does (except with dual- or multiple-mode mobile stations), but it allows inter-operability at a much larger scale between systems based on different radio techniques. Instead of carrying his mobile station, a user would only take his subscriber card with him and use a different mobile equipment to adapt it to the networks he wants to access (see figure 1.8).

Moreover, SIM-roaming does not present any technical obstacle to the extension to any kind of telecommunication network, wire-accessed or radio-accessed, since the network aspects of the roaming issue do not depend on the access scheme used in each network. The SIM appears then as the technical vector for personal numbering, that is to say a means to provide each user with a single telecommunication number whatever the network the user happens to be connected to. This is an important topic at the date of writing, with the studies concerning UPT (Universal Personal Telecommunications). Undoubtedly, the *Specifications* in this domain prefigure a future world-wide telecommunication system, in which the user simply equipped with his SIM will be able to access any telecommunication system.

1.3.1.7. Security Functions

A radio accessed network is inherently less secure than a fixed network. This comes from the possibility to listen to and to emit radio waves from anywhere, without tampering with operator's equipment. To correct a little this state of affair, several types of security functions have been introduced in GSM in order to protect the network against fraudulent access and to ensure subscriber privacy. These functions include:

- authentication of the subscriber, to prevent access of unregistered users;
- radio path ciphering, in particular ciphering of all subscriber information to prevent third-party tapping;
- subscriber identity protection, to prevent subscriber location disclosure.

These facilities are not subscribed to, and are not under control of the user. The reason for this is that it would be much more costly and complicated to manage a per-subscriber or per-call protection as far as ciphering and subscriber identity protection are concerned. As far as authentication is concerned, the issue is not even relevant since authentication is of benefit to everyone apart from the subscriber being authenticated.

All security functions involve the SIM, which is in fact the real subject of authentication (a correct approach is to say that what is authenticated is the SIM, and not the subscriber or the mobile station). A first consequence is that none of these functions are provided when the SIM is not inserted. Another consequence is that the physical presence of the SIM is absolutely necessary to get most services, the only exception being the emergency call. It is not possible for a mobile station manufacturer to provide a mobile equipment which reads the SIM once, and then is able to provide services under the guise of the subscriber when the SIM is removed. Moreover, SIMs are so designed that it is very difficult to duplicate them (except by their issuer, a network operator). Thus the combination of the security functions and of the SIM provides a high protection of the users and of the networks against fraudulent access.

1.3.2. GSM: A SYSTEM FOR THE OPERATOR

The user's view of a telecommunications network may be restricted to the provision of means of transmission. This is however not quite the whole picture. In the same way as a bus company consists of more than just motors and passenger seats, a telecommunication network is more than just switching offices and transmission links. Taking this image a bit further, a bus company can be described as including driving controls (steering wheel, gear lever, ...), drivers, technicians, breakdown tests and alarms, spare parts, a ticket sale system, traffic observation systems to adapt the routes or the schedule to passenger needs, and so on. All these components have their equivalent in a telecommunications system. They correspond to the operator companies, possibly also service provider companies, and a set of machines and software. Their collective purpose is referred to as "operation and maintenance" (O&M).

Operation and maintenance appears often as a hotchpotch. It can be sorted into three main areas:

- **subscriber management** includes what is needed to manage subscriptions. The most obvious tasks is to log subscribers in and out of the system. A GSM subscription can be rather complex, with the multiplicity of services and of supplementary features. All the corresponding parameters have to be accessible to the operator (or the service provider).

  A related aspect of operation (maybe the foremost for the operator!) is call charging. Call tickets have to be built, collected at one point for each subscriber and bills have to be sent out.

- **network operation** consists in "driving" the network. It includes features enabling the operator to observe the behaviour of the network such as system load, blocking rates, number of handovers between two given cells, and so on. This knowledge permits the operator to check the overall quality of service perceived by the users, and to find out where bottlenecks are. Operation includes also the means to modify the network configuration, in order to palliate problems detected by observation functions for instance, to prepare for a foreseen increase of traffic, or to extend coverage. Network modifications can be "soft" changes, done by signalling, e.g., the modification of handover parameters to change the relative boundary between two cells, but also "hard" modifications, necessitating in-the-field interventions, e.g., to add transmission capacity or install a new site. In a modern telecommunication system, monitoring and parameter adjustment are computer assisted tasks which are typically centralised at a single site.

- **maintenance** aims at detecting, locating and correcting faults and breakdowns. It bears some relationship with operation. The difference can be presented by analogy. On a car dash-board, some displays are aimed to assist the driving, such as the speed dial, or the head-lights indicator. The driver may react on these displays by accelerating or decelerating, or by switching off the head-lights. This is operation. Now, other displays aim at indicating failures, or impending failures, such as water temperature or oil-level. The driver reacts to these displays by immediate elementary maintenance actions, such as stopping the car for some time or adding oil, or alternatively by taking the car to a mechanic. This is maintenance.

The machines in a modern telecommunication system are able to detect some of their own failures, or even to forecast impending failures through self-testing, and to react by themselves. In a number of cases, redundancy has been built into the machines so that a failed part can be deactivated and its role taken by another part. This transition can be automatic. Other palliative actions can be commanded remotely by the operator, after possibly a remote localisation of the failed part. Maintenance includes also local actions to replace failed machines.

The issue of operation and maintenance obviously goes far beyond the scope of GSM or of cellular networks. Most of the work done in this field within the telecommunication community takes a broader view, and aims at all telecommunication systems. The focus of the field is the concept of TMN, the Telecommunication Management Network, which features ways to interconnect all infrastructure machines to a management network and ultimately to centralised sites where work stations enable human beings to control the whole system. GSM is designed to fit with the TMN concepts, but its specifications do not include the whole management network. However, a full series of the *Specifications* (the 12 series) is devoted to the application of these concepts to specific issues such as the GSM architecture and its specificity as a cellular network. On the other hand, some operation and maintenance features are intermingled with the service provision to the user, and are tackled by the technical specifications of the GSM interfaces.

In the field of subscription management, a specific problem is the transfer of the toll tickets (i.e., records describing a call for charging purposes) between networks when users are roaming. GSM provides technical means for this together with the means to enable in the other direction the transfer of subscriber information so that visited networks can deal with roamers.

In the field of operation and maintenance proper, some effort has to be put into specifying the control interfaces for the machines close to the radio. The operation of switches and of terrestrial links is of general application, and does not require specific study. On the other hand, radio cellular coverage introduces a number of new aspects to manage. Each cell has its own set of attributes, such as the frequencies it uses, the channels it allocates, the list of neighbour cells and other parameters needed to make optimal decisions for handovers, and so on. These attributes must be managed, not only locally, but taking into account interactions on a large scale. For instance, the frequency planning, i.e., the choice of the frequencies for each cell, must be decided globally so as to minimise inter-cell interferences. The automation of these functions calls for sophisticated centralised operation equipment. If the methods to perform these tasks are out of the scope of the *Specifications*, means to carry control information to and from the radio related machines are dealt with, enabling operators and manufacturers to design ultimately an efficiently controlled radio coverage.

In the structure of this book, we have also linked up operation and maintenance with some side topics which are not usually so. Operation as described above concerns exclusively the infrastructure equipment. But the mobile stations are an integral part of the system, and are substantial contributors to the quality of service as seen by the user. The operators' problem is that when something goes wrong the subscriber is usually not able to distinguish whether the fault lies with the mobile station or with the network, and generally the network is assumed to be the culprit. While the operators have full control of the SIM, as part of subscription management, they have only an indirect control on the main part of the mobile stations. Indirect methods provided by GSM include observations concerning the mobile stations which can be performed through the infrastructure, such as activity or call tracing, so as to detect and localise mobile stations creating problems. They also include the possibility to introduce databases so as to centralise information relative to machines, as opposed to subscribers. But the most important indirect control means is preventive and consists in testing the mobile stations to be put on the market before allowing their use. This is called "type approval", and is still at the date of writing a controversial issue in the standardisation work in SMG, in particular because it is considered one of the braking forces on the take-up of GSM on a large scale.

The problems linked with type approval are of two different natures. One is technical, and comes from the complexities of GSM as compared with previous analogue cellular systems and even more with wireline devices such as telephone sets or modems. Testing GSM mobile stations is a rather complex task. The test specifications are included in the *Specifications*, but many people in the specification committees did not realise until late in the process the importance of the task. The test specifications were drafted by a separate committee and underwent modifications quite a long time after the main specifications were frozen. Similarly, the development of the test devices was no easy task.

Other problems of a non-technical nature concern administrative issues, and are related to roaming. One of the aim of GSM designers, as well as operators and international and national regulation bodies, was that type approval can be done only once, and not once for each operator or country. One of the reasons behind this will is simply of a practical nature, but another is to prevent the introduction of bias within the industrial competition. The definition of a suitable type approval method then must follow a very narrow line between different objectives. For instance, one issue at the heart of the discussion concerns the degree of testing. On one side, some regulation bodies wish as free a competition between mobile station manufacturers as possible, testing only that mobile stations do not jeopardise the functioning of other machines. On the other side, some operators wish to get some additional guarantee that mobile stations will provide a correct quality of service, testing their full conformance to the standard as well as some performance aspects.

The *Specifications* do not resolve these issues, and it is not their goal. Still, they provide, first through the interface specifications which clearly define the properties of the mobile stations relevant to the quality of service, and second through explicit mobile station test specifications, the technical basis on which the type approval specification can be built.

SPECIFICATIONS REFERENCE

The 02 series of the *Specifications* is devoted to the presentation of the services which may be offered to the user by a GSM PLMN. In particular, telecommunication services are defined in: **TS GSM 02.02** for bearer services; **TS GSM 02.03** for teleservices; and **TS GSM 02.04** for supplementary services in general, together with more details in **TS GSM 02.82** for call forwarding services and in **TS GSM 02.88** for call barring services. Other TSs in the GSM 02.8x series will complete the set of supplementary services for phase 2.

Security aspects are described in **TS GSM 02.09**.

A set of mobile station local features is given in **TS GSM 02.07**, together with the status of each feature (mandatory or optional). This set is by no means exhaustive. Similarly, a man-machine interface for the mobile station is described in **TS GSM 02.30**. There again, requirements are kept to a minimum to leave manufacturers some freedom of design, leaving the market to decide what is best for the user.

2

ARCHITECTURE

2

ARCHITECTURE

GSM as a modern telecommunications system is a complex object. To the multi-service aspect it shares with ISDN, it adds all the difficulties coming from cellular networks. As such, its specification, its implementation and its operation are no simple tasks. Neither is its description. In the course of the specification of GSM, much effort was expended to sort out this complexity. In this book, we have built upon this foundation, so as to follow a structured approach for the presentation of the system.

The purpose of the chapter is first to present this structure, both of the system and of the book. GSM will be analysed in terms of subsystems, and the main machines and functional domains will be identified. This is the first step of the system description, of which the subsequent chapters can be considered as refinements.

Another aim of this chapter is to present the structuring method, which is of interest in itself. The question of how to deal with complex systems has been encountered in many different domains such as electronics, biology, ecology, economics, and of course telecommunications, where the Open System Interconnection (OSI) model for data networks represents one step towards a structured approach of complex meshed networks. While the authors do not claim mastery of this field, they think that the study of the structure of a concrete system like GSM and the general study of systems can be mutually helpful. Hence this chapter will introduce concepts such as functional planes, interfaces and protocols.

A last, but important, aspect of this chapter is to introduce many terms of the GSM jargon. These terms and abbreviations are used heavily in the *Specifications*, in the GSM literature in general, and in the rest of this book. Though we try to limit the use of the technical jargon, it is unavoidable; when a concept or an object has to be referred to every third line, it is best to give it a name.

2.0.1. THE THREE DESCRIPTION AXES

From one point of view, a telecommunications system is a collection of electronic boards transferring analogue electrical signals through wires or electromagnetic waves. But there is more to a system than this "reductionist" approach, equating the system with electromagnetic field values or transistor states. The opposite viewpoint would consist in looking at the system as a black box, seen only through its interfaces with the external world. Though not satisfactory either, this approach is helpful in raising two fundamental questions:

- what is part of the system (here, GSM) and what is not?
- what does GSM interface with?

These two questions will be answered in turn. But it would be somewhat frustrating to stop there, only looking at GSM from the outside. A middle way has been thought in this book between this black box approach and sheer reductionism. On this route, the system will be divided into pieces; each of these pieces will be described as a black box in itself, and the overall operation of the system will be presented as the interactions between the pieces. This approach results in a description of GSM as a set of interconnected co-operating sub-systems.

But, as already expressed, GSM is more than a concatenation of sub-systems: some areas involve many pieces of equipment and cannot be described satisfactorily by looking at each sub-system independently. Therefore, we must additionally look at how GSM operates from two different viewpoints: a static one and a dynamic one.

The **static** viewpoint enables us to identify and describe several *functions* which are fulfilled through the co-operation of several *machines*. The term machine is used here to refer to an assembly of interconnected system components, physically close to each other, working together to perform identifiable tasks. The term function is often used in the technical literature (and in the *Specifications*) to refer to some abstract machine. The use here is closer to the basic meaning of the word. A function is something to fulfil, an activity. In a sound architecture, a machine corresponds to some function. However, other structures should

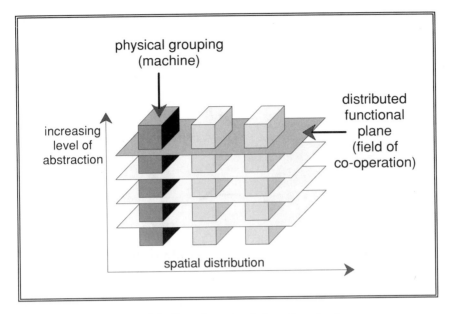

physical grouping
(machine)

distributed
functional
plane
(field of
co-operation)

increasing
level of
abstraction

spatial distribution

Figure 2.1 – Two-dimensional view of a network

Physical groupings (machines or entities) are represented by vertical blocks,
whereas co-operating functions are grouped in horizontal layers,
each one corresponding to a different functional domain.

also be identified, which group similar functions in a system. These functional groupings will put together elementary functions, possibly from different machines, which fulfil by co-operation the same goal. The usual representation, such as used in the layering method of the Open System Interconnection model, consists in showing functional groupings over several machines as "horizontal" layers, a machine being a "vertical" structure in this representation, as shown in figure 2.1.

The **dynamic** view consists in looking at the events affecting the system. Events happen at many scales, from microseconds for transmission aspects, to years when one views the deployment of a network. The description of these events, their organisation and of the way they trigger other events in turn is very important in understanding how the system performs its functions.

The GSM architecture, which is the subject of this chapter, will first be described in terms of machines, then through a functional layer view. The subsequent chapters (i.e., the bulk of the book) will be devoted

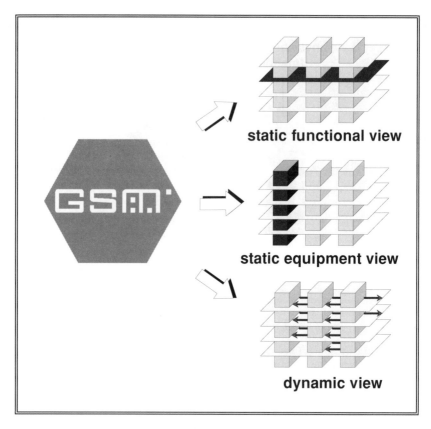

Figure 2.2 – The three axes of the description

GSM functions can be described along several axes,
each one from a different and complementary viewpoint.

to the study of functional planes in detail, going deeper into the role of each machine within each plane, with a substantial part of these chapters containing a description of event sequences. Having thus covered the three complementary axes (as shown in figure 2.2), the description should enable the reader to get a full—and consistent—picture of GSM.

Now, although this three-axes concept helps in structuring the description of the system, it is not sufficient to tackle the overall complexity. The number of possible physical groupings (that is to say machines or abstract portions of machines), or of planes, or of event sequences, is still rather big. The second trick we will use is a recursive description, or top-down approach. Taking the horizontal axis as a first example, this approach consists of describing the system in a few (less than 5 when possible) sub-systems, analysing the interactions at this level whilst taking each subsystem as a black box, and then applying this method for each subsystem. Similarly, functional planes will be composed of sub-planes, themselves composed of sub-sub-planes, and so on. Likewise for the temporal sequences, which can be analysed from the large scale down to the small scale. This allows us to identify first general, compound events, of substantial duration (e.g., system deployment, a site installation, a communication), resolved into finer events (e.g., the call set-up, its release), then stepping up the resolution (e.g., procedures, like channel allocation, start of ciphering), and so on down to elementary message exchanges, and finally to transmission and bit modulation. This is the "reductionist" part of the methodology, which must be balanced at each step by the analysis of the relationships between the sub-parts, in order not to lose the large-scale view. This will be helped considerably by the three-axes approach, each axis embodying relationships between entities along the other axes.

In this book, this recursive splitting is used as a descriptive approach. In the case of telecommunications systems like GSM, this approach is also used, more or less systematically, at the design stage. Another application, which we will see in the last Chapter, is the information model used for the design of the network management software. When applied to description, but even more so when applied to design, the method requires a lot of care for the location of each split. The basic rule to reach a successful splitting is to group things in such a way that there is more interworking within one group than between different groups. When correctly applied, this method leads to a sound architectural model, which eases greatly specification and implementation. Some effort has indeed been invested in the GSM standardisation bodies to obtain a clear-cut architectural model, which will be used in this book.

The architecture of GSM described in the *Specifications* will be referred to as the "canonical" architecture. This architecture does not describe the actual machines which will be found in a GSM implementation, first because some freedom has been left to manufacturers in terms of physical grouping choices, second because the *Specifications* cover only a small part of the specifications of an actual machine. The canonical architecture can be seen as the description of a network model, serving as a template for an effective implementation. However, the interfaces, that is to say what happens between two canonical machines, are described and specified without abstraction. The latitude left to manufacturers consists most often in the possibility to group canonical entities in a single machine (in which case canonical interfaces disappear), or to split an entity into several distinct, possibly distant, machines (thus creating "proprietary"—manufacturer specified— interfaces). Throughout this book, we will describe the system on the basis of the canonical architecture; additional information will be given on the implementation choices left to manufacturers.

Physical groupings and their borders are two sides of the same coin. In fact, borders between machines are extremely important in a system such as GSM, since the *Specifications* specify in fact mainly the behaviour of the system as seen on interfaces between machines and not the internal working of these machines (though this can be derived to a large extent from the external behaviour). It is then an important function of the architectural model to define the system interfaces, and the last section of this chapter will be devoted to this aspect. In this book, an interface represents the frontier between two machines which are in contact via a transmission *medium*. It should be noted that the *Specifications* use in some places a wider meaning for this term, as for instance when referring to the MSC to HLR interface (where the "interface" may well include a full signalling support network). In the following, both the machines and the interfaces involved at each stage will be indicated, with a particular emphasis on the interfaces for which the exchange of information is specified in the *Specifications*.

Interactions between functional layers can also be described in terms of interfaces. Because these interactions happen inside machines, they are not to be specified. However, a detailed and formal description of such interactions can often be found in the *Specifications* (as the notion of primitives between layers). In some cases this description takes up a substantial portion of the *Specifications*, and is justified by the quest for as little ambiguous specification as possible. It should be understood that in the *Specifications* the description of the interactions between layers is a model, which does not constrain implementation in any way, though it can be a useful guideline. In this book we will not refer to this formalism any more.

It is now time to apply these fine principles, and we will start by the first splitting steps, first along the vertical axis (machines), then across the horizontal axis (functional domains). This analysis is then refined in the rest of the book, where the dynamic view is the main subject of the signalling chapters.

2.0.2. FRONTIERS OF THE SYSTEM: WHERE ARE THE BORDERS OF GSM?

The first task is to identify the extent of GSM itself as a whole, thereby also identifying its external interfaces, i.e., its interfaces with the rest of the world.

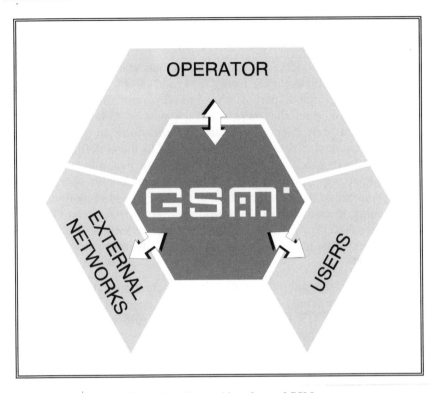

Figure 2.3 – External interfaces of GSM

When looked at as a black box, GSM shares borders with three major realms:
users of the system obtaining service through their mobile station,
other telecommunication networks through which calls transit, and
the operating company controlling the GSM domain.

When looked at from a distance, GSM is part of the Global Telecommunications Network, itself a part of the human organisation. As such, GSM is in direct contact with users (human beings or machines which are being provided with telecommunications services through GSM), with other telecommunication networks (e.g., the global telephony network) and with the personnel of the operating companies. These are indeed the three main external interfaces of GSM, as shown in figure 2.3.

Other interfaces with the external world exist, such as the contact of machines with air, ground and power supplies (which we may term environmental aspects) as well as with other systems using the radio spectrum (electromagnetic compatibility—EMC—aspects). These pragmatic aspects, which are far from negligible for manufacturers and operators, if not for users, are not directly related to the provision of telecommunication services and will not be dealt with here. Let us now look at the three main border lines, with respectively users, other telecommunication networks and operators.

On the user side, the limit lies somewhere between the user himself, who can be excluded from GSM, and the radio interface which represents the principal part of the *Specifications*. But is everything between the user and the radio interface part of GSM? The mobile stations are only partly specified by the *Specifications*: an example is that terminal equipment functional entities similar to those defined in ISDN are not defined specifically for GSM; another example is the man-machine interface of mobile stations, which is in no way specified in a binding manner in the *Specifications*, and could include functions which have nothing to do with GSM. The point of contact between GSM and the user lies therefore somewhere inside the mobile station. We will come back to this point when addressing the internal mobile station architecture.

Now for the interface with other telecommunications networks. GSM is specified mainly as an access network, enabling the setting up of calls between GSM subscribers and subscribers to other telecommunication networks. For practical reasons, machines belonging to GSM are most often kept separate from machines belonging to other networks: this comes from the division, which is now the general rule, between GSM operators and the fixed PSTN/ISDN operator(s), even in companies of marked administrative origin such as FRANCE TELECOM or DEUTSCHE BUNDES TELEKOM. Other choices exist: one could imagine telecommunication switches performing GSM functions as well as managing PSTN/ISDN subscribers. This is not excluded by the *Specifications*, but the canonical GSM architecture does not consider this possibility, and the interface between GSM and other telecommunication networks is clearly defined.

This interface includes three aspects. The first and major one concerns the point where communications transit between GSM and another network: the GSM machine (a switching exchange) at the corresponding contact point is referred to as a "Mobile services Switching Centre" (MSC). The second aspect of interfacing with fixed networks concerns the provision of basic telecommunications transport between GSM machines. In many instances, regulations do not allow GSM operators to operate the terrestrial links between their machines, often leaving them no other choice but leasing lines from a fixed network operator. This second point will be ignored, since in practice it has no effect on the functioning of GSM. A third aspect of interconnection with fixed networks, which is also within the realm of using external networks as support for GSM functions, concerns the routing of non-call related signalling between different GSM networks. For instance, the management of location data for GSM users roaming abroad, requires an exchange of data with no direct relation with given calls to be transmitted through an international signalling network. There again, GSM operators of many countries are forbidden to access the international signalling network directly, and must therefore interface with a fixed network operator in their country to transmit this signalling information between GSM entities (MSCs, location registers) of different networks.

Although the point of contact (at least for the transit of calls) is well defined, the interfaces between GSM and external networks are not specified in the *Specifications*. CCITT recommendations, as well as ETSI standards based on them, include such specifications, but usually national variants exist. Fixed network operators have their own peculiarities, and the machines in contact with their networks must be customised to meet the exact interface requirements specific to each network. In this book, as in the *Specifications*, the basis for description will be the CCITT Recommendations related to ISDN or telephony.

Now comes the border between GSM and operator personnel. As in the case of the user interface, the border is basically between machines and the human employees themselves. The set of machines intervening between this boundary and the machines handling the telecommunications traffic (as well as some parts of these machines) are globally referred to as the Operation and Maintenance Sub-system (OSS). It includes various entities such as workstations (or terminals) handling the man-machine interface with operator personnel, and dedicated computers managing a number of tasks required for operation and maintenance of the system, as well as parts of the software of traffic handling machines themselves.

Most of the OSS aspects are not specified by the *Specifications*. Only a small part of the interfaces between traffic handling entities defined in the GSM architecture is related to OSS functions, and interfaces between these entities and OSS machines are only partially specified. However, the whole OSS area must be considered part of GSM—and this would indeed be the opinion of many an operator—since what would a complex system such as GSM be without means to drive and maintain it?

2.0.3. Internal GSM Organisation

This quick study of the GSM borders already hints at a first split of the internal GSM domain into sub-systems. The mobile station (MS) and the OSS have already been identified as manifest sub-systems. The remaining part consists of infrastructure machines whose roles are to

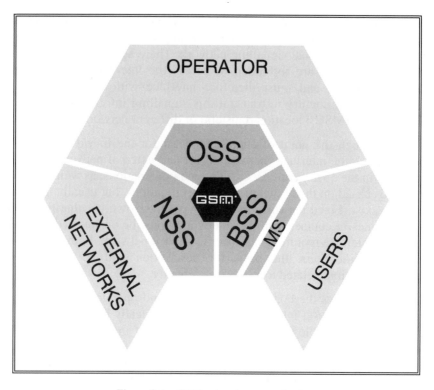

Figure 2.4 – GSM subsystem organisation

Following logically the three borders of the GSM domain, GSM can be defined
as composed of subsystems which interact between themselves
and with the outside world along the white border lines shown.

provide and to manage transmission paths between mobile stations and
either fixed networks or other mobile stations, and to provide the means
for the users to set-up communications along these transmission paths.

The canonical GSM architecture distinguishes two parts: the BSS
(Base Station Sub-system) and the NSS (Network and Switching Sub-
system[1]). The BSS is in charge of providing and managing transmission
paths between the mobile stations and NSS machines (namely the MSCs),
including in particular the management of the radio interface between
mobile stations and the rest of GSM. The NSS must manage the
communications and connect mobile stations to the relevant networks or
to other mobile stations. The NSS is not in direct contact with the mobile
stations and neither is the BSS in direct contact with external networks.

---

[1]   The term NSS is used by many operators and manufacturers, though not in the
        Specifications.

The interface between the BSS and the mobile station is the already introduced radio interface, whereas the interface between the BSS and the NSS has been named the *A interface* in the *Specifications*. The MS, BSS and NSS form the operational part of the system, whereas the OSS provides means for the operator to control them. This model is shown in figure 2.4.

On this scale, the interactions between the subsystems can be grouped in two main categories. The bottom part of the figure corresponds to a chain:

external networks ⇔ NSS ⇔ BSS ⇔ MS ⇔ users

whose business as a whole is to provide transmission paths and means to establish them. This is the operational part of the system, handling the telecommunications traffic. Above it we find the control part, composed of the OSS and the operator, which interacts with the traffic handling part by observing and modifying it so as to maintain or improve its functioning.

2.1. SUB-SYSTEMS

This section will deal with each subsystem in turn. The main purpose is to introduce a number of terms, not to exhaust the subject. The level of detail here is just sufficient to get a general idea of the functional splits. More complete architectural considerations will be found in each of the other chapters of the book, where a more thorough description of the required functions will be found.

2.1.1. THE MOBILE STATION (MS)

 The mobile station usually represents the only equipment the user ever sees from the whole system. Examples taken from the first types of GSM mobile stations to be on the market are shown in figures 2.5 and 2.6. Mobile station types include not only vehicle-mounted and portable equipment, but also handheld stations, which will probably make up most of the market.

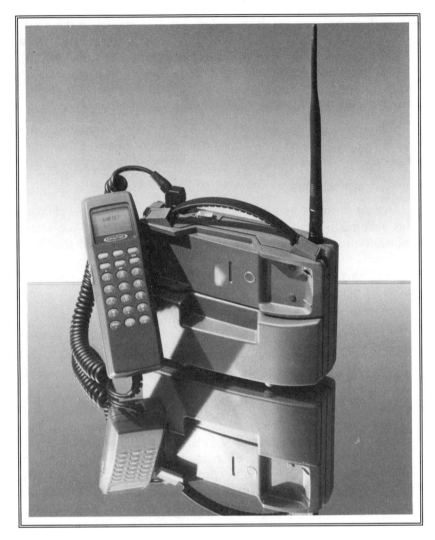

Figure 2.5 – A GSM portable mobile station (by courtesy of Orbitel)

The first GSM mobile stations on the market were portable mobile stations, weighing around 2 kilograms, and also capable of being installed in a vehicle.

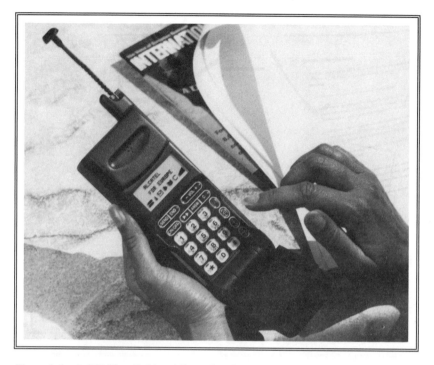

Figure 2.6 – A GSM handheld mobile station (by courtesy of Alcatel Radiotéléphone)

With their ever-decreasing weight and volume,
handheld mobile stations represent a very attractive product for the user.

But what does a mobile station involve? Beside generic radio and processing functions to access the network through the radio interface, a mobile station must offer either an interface to the human user (such as a microphone, loudspeaker, display and keyboard for the management of speech calls), or an interface to some other terminal equipment (such as an interface towards a personal computer or a facsimile machine), or both of them. An effort has been made to allow off-the-shelf terminal equipment to be connected to mobile stations (for instance group 3 facsimile machines designed for connection to the telephone network), and specific terminal adaptation functions have been specified for this purpose. However, all implementation choices are possible and left open to manufacturers, enabling fully integrated compact mobile stations to coexist with mobile stations featuring standard interfaces.

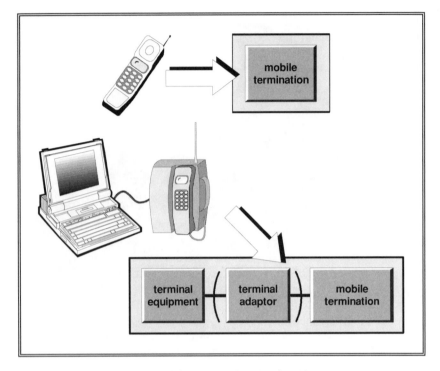

Figure 2.7 – Mobile station functional architecture

The mobile station may be a standalone equipment for certain services
or support the connection of external terminals,
either directly or through relevant adaptation functions.

This leads to the identification of three main functions, as shown in figure 2.7:

- the terminal equipment, carrying out functions specific to the service, without any GSM-specific functions: e.g., a fax machine;

- the mobile termination, carrying out, among others, all functions related to transmission on the radio interface;

- possibly a terminal adaptor, which acts as a gateway between the terminal and the mobile termination. A terminal adaptor is introduced when the external interface of the mobile termination follows the ISDN standard for a terminal installation, and the terminal equipment has a terminal-to-modem interface.

The functional split between mobile termination, terminal adaptor and terminal equipment is very much related to the transmission needs of each service, and will be detailed in Chapter 3.

Another, more significant, architectural aspect of the mobile station relates to the concept of subscriber module, or SIM (Subscriber Identity Module, a slightly restrictive name, as more than identity is involved). As described in Chapter 1, the SIM is basically a smart card (or a cut-out thereof), following ISO standards, containing all the subscriber-related information stored on the user's side of the radio interface. Its functionalities, besides this information storage capability, relate also to the confidentiality area. The rest of the mobile station contains all the generic transmission and signalling means to access the network. The interface between the SIM and the rest of the equipment is fully specified in the *Specifications*, and is simply referred to as the "SIM-ME" interface (ME stands for "mobile equipment"). The functions of the SIM will be studied in detail in Chapter 7.

In this book as well as in the *Specifications*, the term Mobile Station (MS) shall generally include a Mobile Equipment and a SIM, although a rare case exists where a mobile station could be operated without a SIM (i.e., reduced to the mobile equipment) for the handling of anonymous emergency calls when so permitted by the network.

The concept of a removable storage device for subscriber data has far-reaching consequences. In previous cellular systems, except for the German C-network which introduced the smart card concept at the time when it was making its way in GSM committees, the personalisation of the mobile station required a non-trivial intervention, only possible for technical specialists and not for the operator's administrative clerks. This situation lead to several drawbacks. A mobile station could only be sold by specialist dealers, able not only to install the equipment in a vehicle, but also to act as an intermediary between the user and the service provider to personalise the equipment. Should the mobile station fail (unfortunately not such a rare event), it was difficult to provide the user with a replacement during the repair period, and almost impossible to allow the user to keep the same directory number during this time.

The removable SIM simplifies these issues, and also brings other benefits. A potential user may of course buy a mobile equipment, but he may also lease or borrow it for any period of time, and change it as he wishes without a lot of administration. All he needs is his own SIM, obtained through an operator or a service provider, independently of any equipment choice. The last steps of the SIM personalisation can be done easily through a small computer and a simple adapter. Mobile equipment will be for sale on a much larger-scale than ever before, since their acquisition will not require the intervention of an operator or a service provider. Car phones will still require installation in the vehicle, but portables or handhelds will encourage users to buy their mobile equipment from any store.

More advantages can be envisaged. For instance, rented cars could be equipped with a mobile equipment usable with any SIM, whether user-owned or also rented. The reverse situation may also bring benefit to subscribers, if not to operators: a subscriber may change his serving operator without replacing his ME. But most of all, as explained in Chapter 1, this personal chip secured in its plastic case and called SIM is the first brick in the building of a personal communication system enabling wide-ranging mobility between different telecommunications networks.

2.1.2. THE BASE STATION SUB-SYSTEM (BSS)

Largely speaking, the Base Station Subsystem groups the infrastructure machines which are specific to the radio cellular aspects of GSM. The BSS is in direct contact with mobile stations through the radio interface. As such, it includes the machines in charge of transmission and reception on the radio path, and the management thereof. On the other side, the BSS is in contact with the switches of the NSS. The role of the BSS can be summarised as to connect the mobile station and the NSS, and hence the mobile station's user with other telecommunications users. The BSS has to be controlled and is thus also in contact with the OSS. The external interfaces of the BSS are summarised in figure 2.8.

According to the canonical GSM architecture, the BSS includes two types of machines: the BTS (Base Transceiver Station), in contact with the mobile stations through the radio interface, and the BSC (Base Station Controller), the latter being in contact with the switches of the NSS. The functional split is basically between a transmission equipment, the BTS, and a managing equipment, the BSC.

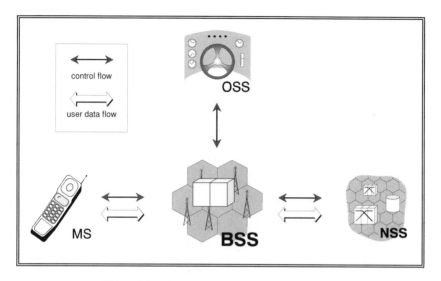

Figure 2.8 – The external environment of the BSS

The BSS bridges the space between the mobile stations on one side
(through the radio interface), and the switching functions on the other.
It is controlled by the operator through the OSS.

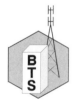 A BTS comprises radio transmission and reception
devices, up to and including the antennas, and also all
the signal processing specific to the radio interface.
BTSs can be considered as complex radio modems, and
have little other function. A typical first-generation
BTS consists of a few racks (2m high and 80 cm wide)
containing all electronic devices necessary for the
transmission functions, as shown in figure 2.9 for a GSM900 BTS and
figure 2.10 for a DCS1800 BTS. The antennas are usually a few tens of
meters away, on a mast, and the racks are connected to it through a feeder
cable. A one-rack first-generation BTS is typically able to handle three to
five radio carriers, carrying between 20 and 40 simultaneous
communications. Reducing the BTS volume is important to keep down
the cost of the cell sites, and progress can be expected in this area.

An important component of the BSS, which is considered in the
canonical GSM architecture as a part of the BTS, is the TRAU, or
Transcoder/Rate Adapter Unit. The TRAU is the equipment in which the
GSM-specific speech encoding and decoding is carried out, as well as the
rate adaptation in case of data. Although the *Specifications* consider the

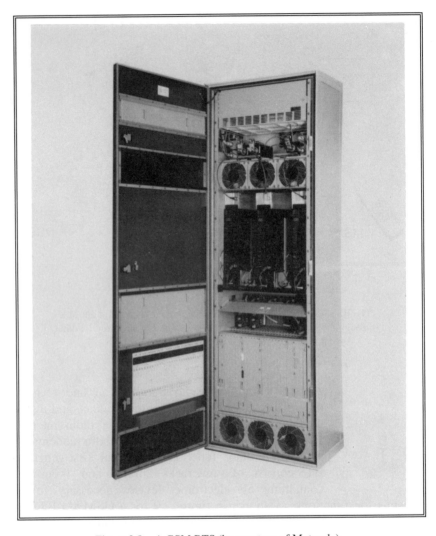

Figure 2.9 – A GSM BTS (by courtesy of Motorola)

A one-rack BTS, such as the one shown, is typically able to handle up to 5 carriers. The picture shows the rack equipped for 3 carriers.

TRAU as a sub-part of the BTS, it can be sited away from the BTS, and even more so since in many cases it is actually between the BSC and the MSC. Its remote position allows the advantage of more compressed transmission between BTS and TRAU, and its impact will be discussed in detail in Chapter 3.

Figure 2.10 - A DCS1800 BTS (by courtesy of Nokia)

The small rack shown is designed to be used outside,
typically below the antenna mast.

The internal structure of the BSS is represented in figure 2.11. On top of the BTS, it shows the second "canonical" component of the BSS, the BSC. The BSC is in charge of all the radio interface management through the remote command of the BTS and the MS, mainly the allocation and release of radio channels and the handover management. The BSC is connected, on one side, to several BTSs and on the other side, to the NSS (more exactly to an MSC).

A BSC is in fact a small switch with a substantial computational capability. Its main roles are the management of the channels on the radio interface, and of the handovers. A typical BSC consists of one or two racks, as shown in figure 2.12 and can manage up to some tens of BTSs, depending on their traffic capacity.

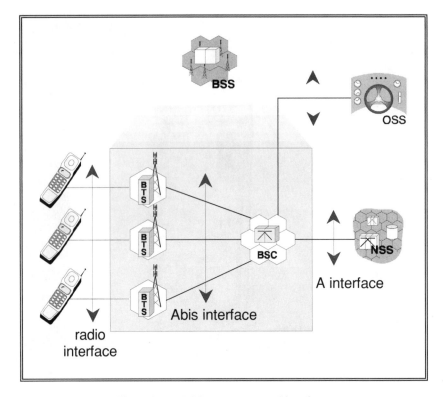

Figure 2.11 – BSS components and interfaces

The Base Station Sub-system consists of BTSs, situated on the antenna sites,
and of BSCs, each one in control of several BTSs.

The concept of the interface between BSC and MSC, called the A
interface, was introduced fairly early in the GSM standard elaboration
process. Only later was it decided to also standardise the interface
between BTS and BSC, and this interface therefore bears the (not any
more meaningful than A!) name of "Abis" interface.

In the GSM vocabulary, *a* BSS means the set of one BSC and all
the BTSs under its control, not to be confused with *the* BSS as the sub-
system including all the BSCs and BTSs.

Figure 2.12 – A GSM BSC (by courtesy of Matra Communication)

The BSC shown here consists of two cabinets:
a control cabinet holding the duplicated central control and switching functions,
and a cabinet handling the interfaces.

2.1.3. THE NETWORK AND SWITCHING SUB-SYSTEM (NSS)

The Network and Switching Sub-system, or NSS, includes the main switching functions of GSM, as well as the data bases needed for subscriber data and mobility management. It is also sometimes called the switching sub-system, which is indeed more appropriate since a GSM network includes the BSS as well as the NSS. The main role of the NSS is to manage the communications between the GSM users and other telecommunications network users.

Within the NSS, the basic switching function is performed by the MSC (Mobile services Switching Centre), whose main function is to co-ordinate the setting-up of calls to and from GSM users. The MSC has interfaces with the BSS on one side (through which it is in contact with GSM users), and with the external networks on the other. The interface with external networks for communication with users outside GSM may require a

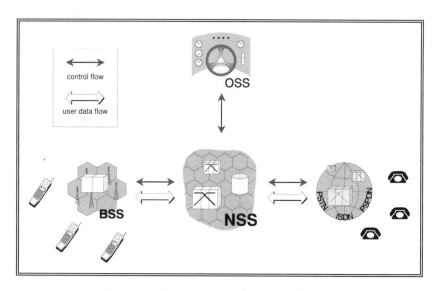

Figure 2.13 – The external environment of the NSS

The NSS, through its MSC, is in contact with the BSS and external networks.
Like the BSS, it is controlled by the operator through the OSS.
NSS entities are also in contact with NSS entities of other GSM networks
for the exchange of data through SS7 signalling networks.

Figure 2.14 – A GSM MSC (by courtesy of Matra Ericsson Telecommunications)

An MSC includes several rows of cabinets, one of which is shown here.

gateway for adaptation (Interworking Functions, or IWF), the role of which may be more or less substantial depending on the type of user data and the network it interfaces with. The NSS also needs to interface external networks to make use of their capability to transport user data or signalling between GSM entities. In particular, the NSS makes use of a signalling support network, at least partly external to GSM, following the CCITT Signalling System n°7 protocols (and therefore usually referred to as the SS7 network); this signalling network enables co-operative interworking between NSS machines within one or several GSM networks. The external interfaces of the NSS are represented schematically in figure 2.13.

As a piece of equipment, an MSC controls a few BSCs and is usually a rather big switching machine. With a medium population penetration percentage, a typical MSC at the date of writing is suitable for covering a regional capital and its surroundings, totalling say 1 million inhabitants. Such an MSC includes about half a dozen racks. Figure 2.14 shows a GSM MSC.

The interconnection of the MSC with certain networks requires the adaptation of the GSM transmission peculiarities to those of the partner network. These adaptations are the Interworking Functions (IWF). This term refers by extension to the functional entity in charge of them. It basically consists of a transmission and protocol adaptation equipment. It enables interconnection with networks such as PSPDNs (Packet-Switched Public Data Networks) or CSPDNs (Circuit-Switched Public Data Networks), but it also exists when the partner network is simply the PSTN or the ISDN. Interworking functions may be implemented together with the MSC function, or they may be performed by a separate equipment. In the second case, the interface between MSC and IWF is left open by the *Specifications*.

Besides MSCs, the NSS includes data bases. Subscriber information relevant to the provision of telecommunications services is held on the infrastructure side in the HLR (Home Location Register), independently of the actual location of the subscriber. The HLR also includes some information related to the current location of the subscriber. As a physical machine, an HLR is typically a standalone computer, without switching capabilities, and able to handle hundreds of thousands of subscribers. A functional subdivision of the HLR identifies the Authentication Centre, or AuC, the role of which is limited to the management of security data for the authentication of subscribers.

The second database function identified in GSM is the VLR (Visitors Location Register), linked to one or more MSCs, and in charge of temporarily storing subscription data for those subscribers currently situated in the service area of the corresponding MSC(s), as well as holding data on their location at a more precise level than the HLR. In current practice, as will be explained in more detail in Chapter 7, a VLR function is always integrated with each MSC.

GMSC

But the NSS contains more than MSCs, VLRs and HLRs. In order to set up a call towards a GSM user, this call is first routed to a gateway switch, referred to as GMSC, without any knowledge of the whereabouts of the subscriber. The gateway switches are in charge of fetching the location information and of routing the call towards the MSC through which the subscriber can obtain service at this instant (the Visited MSC). To do this, they must first find the right HLR, knowing only the directory number of the GSM subscriber, and interrogate it. The gateway switch has an interface with external networks for which it provides gatewaying as well as with the SS7 signalling network to interwork with other NSS entities. The term GMSC is somewhat misleading, because the GMSC function is not by technical necessity linked to an MSC. It could be thought of as an independent equipment, or as a function integrated in a digital telephony switch. However, charging considerations explained in detail in Chapter 8 are such that gateway functions will not for some time be set outside GSM networks, and economic considerations make it undesirable to have standalone machines for this function. This state of things results in the widespread implementation of the GMSC function in the same machines as the MSC function itself.

SS7

Having seen the pieces, let us look at the glue. Depending upon national regulations, a GSM operator may or may not be allowed to operate the full SS7 network between NSS machines. If the GSM operator has the full control of this signalling network, then Signalling Transfer Points (STPs) will probably be part of the NSS functions, and could be implemented either as stand-alone nodes or in the same machines as the MSCs, in order to optimise the cost of the signalling transport between NSS entities (MSC/VLRs, GMSCs, HLRs, ...).

Similarly, depending upon the terms of its license, a GSM operator may have the right to implement its own network for routing calls between GMSC and MSC, or even for routing outgoing calls as near as possible to the destination point before using the fixed network. In this case, Transit Exchanges (TE, not to be confused with "Terminal

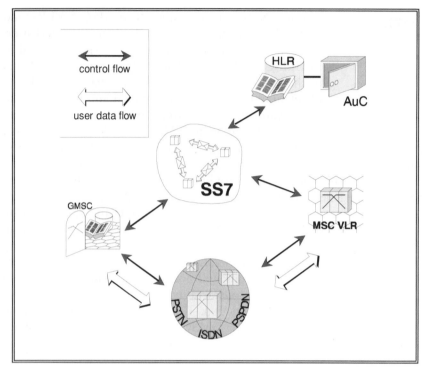

Figure 2.15 – Internal structure of the NSS

Beside switches such as the MSCs and GMSCs, linked between themselves
through a fixed network structure which may or may not be part of the GSM network,
the NSS includes databases (such as the HLR/AuC)
interconnected through an SS7 signalling support network.
This picture shows the case when the VLR database is integrated with the MSC.

Equipment" as used for the mobile station architecture) may be part of
the GSM network as well, and there again may be implemented as standalone nodes or together with some MSCs.

As a summary, figure 2.15 shows the main components of a GSM
NSS and the interconnections between them.

2.1.4. THE OPERATION SUB-SYSTEM (OSS)

OSS

As we described it in Chapter 1, the OSS has various tasks to fulfil. All these tasks require interactions between some or all the infrastructure machines such as in the BSS or in the NSS and members of the operator company or of the service providing companies. Network operation and equipment maintenance concern all machines (including by the way the OSS machines), whereas subscription management has impact on at least the HLR.

The canonical architecture described in the *Specifications* is not as specific on operation and maintenance aspects of the GSM as it is on the rest. A wide latitude is left to operators and manufacturers. One sound reason for this is that the issue is not specific to GSM. Operation and maintenance functions are necessarily implemented in existing networks, and a lot of standardisation work on the issue is going on for general application to telecommunications networks. Let us look at some of the ways to conceive the OSS.

Up to some point in the past, operation and maintenance actions were performed locally, by intervention on the site of each machine. To this avail, each equipment was provided with some man machine interface, such as through a local terminal. The co-ordination of the actions on different machines was managed by human beings. Such an approach could still be adopted for GSM, at least at the very beginning, for experimental networks. For instance, subscription management could be done by manually entering the subscription data on an on-site terminal connected to the HLR. With the local-only approach, the OSS functions are simply spread in the BSS and NSS machines, and the only "OSS" machines are the man-machine devices such as terminals.

Now, with the evolution of the technology and of the complexity of telecommunications systems, the range of possible actions on the system as well as the quantity of information to be processed have increased tremendously. The local-only approach is then inefficient as soon as the number of machines to operate becomes significant. Some centralisation is required, and this calls for separate pieces of equipment intervening between several of the traffic handling machines and the man-machine devices. Moreover, these pieces of equipment can perform some of the co-ordination functions instead of human operators, thus offering a better guarantee of consistency between the configuration of

. the different machines. The ultimate centralised approach is the concept of TMN (Telecommunication Management Network), where all operation and maintenance machines compose a network which as a whole is linked to all the traffic handling machines.

Once some centralisation is applied, interfaces between traffic handling machines and OSS machines appear, requiring some specifications. It is at this level that most of the substance related to operation and maintenance can be found in the *Specifications*. This part of the standard has been designed with the ideas of TMN in mind, in order to enable a smooth integration of GSM networks with advanced operation and maintenance machines. We will now look in very general terms at the OSS architecture, and then focus, as the *Specifications* do, on the portion of OSS close to the traffic handling machines, especially to the BSS.

In Chapter 1, we have already noted that operation and maintenance covers functions in several rather independent domains. The architecture of the OSS reflects these domains. For the needs of this chapter, operation and maintenance functions are best split in three domains, corresponding to three distinct sets of OSS machines.

Network operation and maintenance proper calls for mediation between the operator personnel and all the machines. If TMN principles are followed, the operation network is linked on one side to the telecommunications machines (MSCs, BSCs, HLRs, and others, but not the BTSs which are accessed through the parent BSC). On the other side, the operation network is linked to workstations acting as man-machine interfaces. The *Specifications* identify the OSS machines directly in contact with BSS or NSS machines. These dedicated machines are called the OMCs, or Operation and Maintenance Centres, and are linked to a few telecommunications machines of the same category (and usually of the same manufacturer), e.g., an "OMC-R" (or OMC-Radio) would be in charge of a few BSCs and, through them, the corresponding BTSs. In a simple management architecture, the OMC is autonomous, and includes the man-machine interface for the control of the traffic handling machines it is linked to. Such an OMC is typically a standalone workstation. The OMCs can also become simply the gateways to an overall management network, acting as mediation devices, following the terminology of TMN.

Subscription management corresponds to tasks which are independent from the other operation and maintenance functions, and which may then be supported by machines separate from those involved in network operation. Subscription management has two facets, subscriber data management and call charging. The *Specifications* do not address at all the first aspect, and only lightly the second. Different architectural approaches will be adopted by the different operators.

Subscriber data management involves only the HLR (including the AuC for the security related data) and dedicated OSS machines, for instance a network linking the HLR and the man-machine devices in the commercial agencies where subscribers are dealt with. This network, if it exists, can be autonomous, in contact with the rest of the system only at the HLR. The SIM is also affected at this stage. It has to be initialised consistently with the information held in the infrastructure. The SIM has then two phases in its life. In a first "administrative" phase, it is dealt with by the subscription management system, but is not active in the traffic handling part of the system. Then, once initialised, it is operational and can be inserted into a mobile equipment.

The second facet of subscription management is call charging. In a mobile environment, the call tickets relative to a given GSM subscriber can be issued by many different machines, including all the MSCs the user may visit. GMSCs are also, as will be seen in Chapter 8, a source of call tickets. Then machines are needed to centralise the billing data for each subscriber. Once again, the *Specifications* do not address these machines. However, the HLR is an obvious candidate to be the centralised equipment for charging information, if only to act as the gateway with the subscription management system to which it is linked. Some steps have been taken to enable the SS7 network to be used to carry charging information between the MSCs and GMSCs and the HLR. This solution could be a convenient one for the exchanges between networks of different countries, but others can be envisaged.

Finally, the last domain of operation and maintenance is the management of mobile stations. A part of it is done within network operation, through the infrastructure machines. There is however one machine identified in the canonical GSM architecture and specific to mobile station management, the EIR, or Equipment Identity Register. It is a database which stores data related to Mobile Equipment. As explained earlier, a mobile station consists of a SIM and a Mobile Equipment (ME). Subscriber data management concerns the SIM and is handled by the HLR and VLR. Because SIMs can be moved between MEs, overseeing subscribers does not imply overseeing mobile stations. Though the management of the mobile equipment is not absolutely necessary for the operation of telecommunications services, it holds some interest, e.g., when searching for stolen mobile equipment or when monitoring mobile station misbehaviour. For this purpose, the EIR is in charge of storing the relevant ME-related data. It interfaces with other NSS entities and with

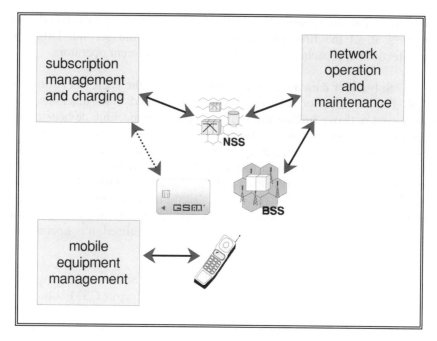

Figure 2.16 – OSS organisation

Three main areas are part of OSS:
– network operation and maintenance of telecommunications machines;
– subscription management, charging and billing; and
– mobile equipment management.

the network operation system. The interface with NSS machines is, again, through the SS7 signalling support network. This is the reason why the EIR is often considered as part of the NSS, though its functional role sets it within the OSS.

Figure 2.16 summarises the general organisation of the OSS. The whole subject will be addressed in more detail in Chapter 9.

2.2. FUNCTIONAL PLANES

Up to this point, our description of the GSM architecture has focused on the physical grouping of functions, i.e., the underlying question was *where* are the functions implemented? The present section will take another angle, looking at the functions to fulfil, grouping them according to their closeness in purpose. The question is *what* functions for what purpose? Of course, the division of the system into three subsystems already bears a strong functional flavour. This reflects the fact that the choice leading to grouping some functions in the same machines depends on the closeness of those functions. However, a number of functions are by essence distributed, and can be fulfilled only by the co-operation of distant machines. This is obvious in any telecommunications system, since the basic function, the management of communications, is distributed. As a consequence, most functional entities described so far perform tasks in several spread functional areas. For instance, machines such as the MSC or the BSS have necessarily some operation and maintenance functions; another example is the involvement of the MSC in BSS-related functions such as handover.

2.2.1. LAYER MODELLING

In the telecommunications domain, a powerful method to obtain a functional grouping is to use the Open System Interconnection model. Functions are grouped in functional planes, represented as stacked one upon the other. The lowest plane, devoted to the physical transmission of information between distant entities, relies on physical hardware media, whereas the highest one represents the view of external users. Each plane (or layer) provides services to the next layer up, these services being themselves enhancements of the services provided by the next layer below. Machines or entities are represented vertically, the intersection between entity E and layer L corresponding to the functions fulfilled by E to contribute to the objectives of L.

Besides this hierarchical organisation, based on the notion of service provided by one layer to another, there is an underlying temporal organisation. In general, lower layers correspond to functions having a short time scale, whereas the upper layers will group long time scale functions.

Within each layer, the entities co-operate to provide the required service, through information exchanges. The rules of these exchanges are specified at reference points where the information flow crosses an interface between different entities. These rules are called the signalling protocols.

The distinction between an interface and a protocol is important. An interface represents the point of contact between two adjacent entities, and as such it can bear information flows pertaining to several different pairs of entities, i.e., several protocols. For instance, the radio interface in GSM is the transit point for messages pertaining to several protocols:

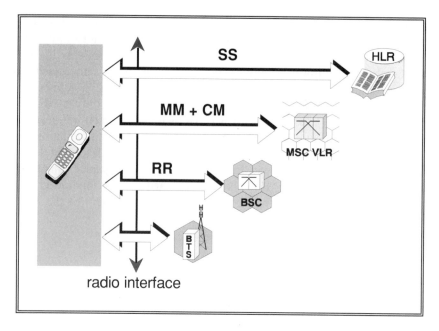

Figure 2.17 – Protocols versus Interfaces

The example of the radio interface shows the distinction between an interface
(here, the point of contact between MS and BTS)
and protocols between peer entities, some of which may be far away.
Each of the GSM interfaces typically transports several protocol flows.
RR refers to the protocol used for the management of the transmission over the radio
interface, MM to the management of users' mobility, CM to the management of the
communications, and SS to the setting of Supplementary Services parameters.

between MS and BTS (for transmission), but also between MS and BSC
(for the management of the transmission over the radio interface), MS
and MSC (for the management of users' mobility, and of the
communications), or even MS and HLR (for setting Supplementary
Services parameters), as shown in figure 2.17.

This is an example of the analysis of an interface as a protocol
stack, each element of the stack being related to the intersection of a
functional plane and of the interface. Still, signalling messages pertaining
to a given protocol may be visible on several interfaces along their path,
if the corresponding peer entities are not adjacent. The protocol then
appears on several interfaces. Thus, the specification of a protocol is
somewhat distinct from the specification of an interface. The
specification of an interface can be reduced to the description of its
protocol stack. This conceptual distinction is ill resolved in the
Specifications, where often what is called an "interface" specification is
in fact a "protocol" specification.

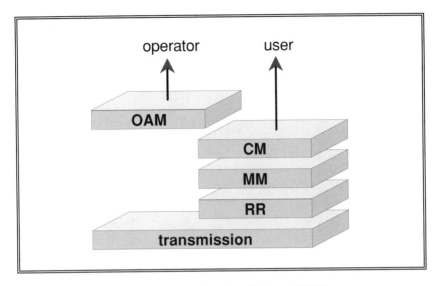

Figure 2.18 – The functional planes of GSM

The GSM functions are modelled here in 5 functional planes (Transmission,
Radio Resource management, Mobility Management, Communication Management,
and Operation, Administration and Maintenance),
which offer the user and the operator a large palette of services.

For the purpose of protocol specification, the slicing of functions in planes must lead to fairly thin "slices", in order for each one to be consistent and also to escape from too big a complexity for the corresponding protocols. Let us now look at these planes, identifying on the way their intersections with machines and the relevant protocols.

Five planes will be distinguished in this book, as shown in figure 2.18. At the bottom lies the basis of any telecommunications system, i.e., the transmission plane. It provides transmission means for the communication needs of the users, as well as for information transfer between the co-operating machines. Transmission is a domain for very short time scale events, from microseconds (e.g., bit modulation) to seconds (for message transmission).

The next plane up is concerned with the management of transmission resources. In telecommunications networks, these functions are usually grouped with the communication management functions, because fixed circuit management represents a small portion thereof. However, in the case of a cellular system such as GSM, the management of transmission resources on the radio path is a complex issue and it warrants its own functional plane. This is called the radio resource management layer, or RR layer. The RR layer provides stable links

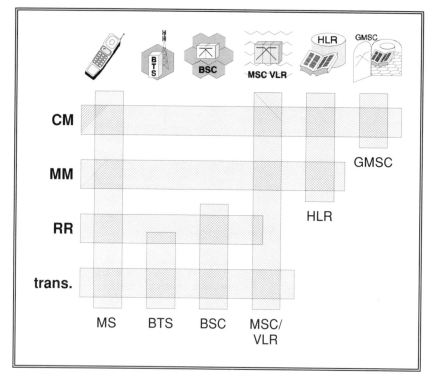

Figure 2.19 – General protocol architecture of GSM

This schematic view, which will be refined in the course of this book,
shows how the GSM machines co-operate
in each of the traffic handling functional domains.

between the mobile stations and the MSCs, coping in particular with the
movements of the users during the calls (handovers). The BSS performs
most of the RR functions. From a temporal point of view, this plane and
the two next ones deal with events on the scale of the call, that is to say
from seconds to minutes.

Next comes a small functional plane, which has not been grouped
with communication management because of its strong GSM specificity.
This Mobility Management layer, or MM layer, is in charge of managing
subscriber data bases, and in particular the subscriber location data. An
additional task of the MM layer is the management of confidentiality
aspects, such as authentication. The SIM, HLR and AuC are examples of
machines mostly involved in MM activities. The MM layer adds to the
transmission functions provided by the lower layers the means to track
mobile users when not engaged in communication, and the security
related functions.

The next plane is much less specific to GSM. It makes use of the stable basis provided by the RR and MM layers to provide telecommunications services to the users. We will call it the Communication Management layer, or CM layer. It consists of several independent components, depending on the type of service. The NSS, mainly the MSC, obviously has a strong involvement in the CM layer.

The RR, MM and CM layers provide the users with a high grade of service. But, to make the picture complete, another plane should be added, which makes use of the transmission facility. This is the Operation, Administration and Maintenance plane (OAM plane), which provides the means for operator actions. This layer is not strictly above the other ones in terms of services, since it does not enhance directly the service they provide for the user. It is characterised by a larger time scale compared to the other layers, ranging typically from hours or days to years. Its kinship with the OSS is obvious. As will be explained later, many interfaces in other sub-systems carry some OAM functionality.

Let us now describe further the respective relationships of each plane with the GSM traffic handling machines, the main principles of which are summarised in figure 2.19 (with the exception of the OAM plane).

2.2.2. TRANSMISSION

Some of the GSM machines are concerned with transmission only. An obvious example is the transcoder and rate adaptor unit (TRAU), which is only concerned in adapting speech or data representations. Another example is provided by a transit exchange, whose role is limited to the routing of signalling exchanges between distant NSS entities. But most other GSM machines also play a more or less complex role in transmission. The mobile station obviously does so, and so does the BSC, the MSC and the interworking function (IWF), which may all be along the transmission path between two users. Conversely, some of the machines have no relation to transmission, except for the minimum needs concerning signalling with the other machines. These include the data bases (HLR, VLR, EIR) and the OSS in general.

As already mentioned, the transmission plane includes two more or less independent functions. The first one is to provide the means to carry user information (whether speech or data) on all segments along the path followed by a communication. The second one is to provide the means to carry signalling messages between entities. The transport of signalling is needed between adjacent machines (i.e., MS to BTS, BTS to BSC, BSC to MSC), but also through networks such as the SS7 network used between NSS machines.

Included in this plane are aspects indeed traditionally called transmission, such as modulation, coding, multiplexing, but also other aspects such as low level protocols to format data, to ensure proper sequencing, to correct errors through repetitions and to route information throughout networks. Three chapters in this book are dedicated to the transmission plane (Chapters 3, 4 and 5). Chapters 3 and 4 deal with the "traditional" transmission aspects, also referred to as "layer 1" or "physical layer" in the *Specifications*, in accordance with the Open System Interconnection model. Chapter 4 is entirely devoted to the radio interface, because of its prime importance in GSM. Chapter 5 deals specifically with the additional functions provided by "link layers" or "network layers" (according to ISO terminology) in order to transport user data and signalling messages between communicating entities.

2.2.3. RADIO RESOURCE MANAGEMENT (RR)

The role of the radio resource management layer is to establish and release stable connections between mobile stations and an MSC for the duration of a call for instance, and to maintain them despite user movements. It must cope with a limited radio resource (and the corresponding terrestrial resources) and share it dynamically between all needs. The functions of the RR layer are mainly performed by the MS and the BSC. In addition, since the responsibility for the handover process lies entirely within the RR layer, part of the functions implemented in the MSC are within the RR domain, in particular the ones related to inter-MSC handovers.

The detailed study of the RR layer functions and protocols is to be found in Chapter 6, which is devoted to this subject.

2.2.4. MOBILITY MANAGEMENT (MM)

The machines concerned with mobility management are mainly the mobile station (and more precisely the SIM inside the mobile station), the HLR and the MSC/VLR. The management of the security functions are done by the same machines, and more particularly by the AuC inside the HLR. The BSS is not concerned with the MM plane.

The detailed description of the MM protocols, including mobility management issues and security management issues, is given in Chapter 7.

2.2.5. COMMUNICATION MANAGEMENT (CM)

The functions of the communication management layer, or CM layer, consist in setting up calls between users at their request, as well as of course maintaining these calls and releasing them. It includes the means for the user to have some control over the management of the calls he originates or receives, through the "Supplementary Services". The variety of the Communication Management functions makes it easier to describe as three sub-domains.

2.2.5.1. Call Control

The MSC/VLRs, GMSCs, IWFs and HLRs, through basic call management functions, are able to manage most of the circuit oriented services provided to GSM users, including speech and circuit data. This functional core represents a sub-part of the CM layer, and is called CC (Call Control) in the *Specifications*.

An important aspect of communication management, beside establishing, maintaining and releasing calls, is the routing function, i.e., the choice of transmission segments linking distant users and their concatenation through switching entities. GSM mostly relies on external networks to perform this task, interfacing these networks through MSCs and GMSCs. The IWF may also have a switching function for communications to and from the networks it interfaces with. Call management requires access to the subscription data, in order to check the profile of the subscriber, and therefore the HLR also intervenes in the CM layer.

2.2.5.2. Supplementary Services Management

A second aspect of the CM layer concerns the management of the so-called supplementary services. As explained through examples in Chapter 1, users in GSM have some control on the way their calls are handled by the network. This potentiality is described as "supplementary services", each one of them corresponding to some specific variation of the way the basic service is rendered to the user. The impact of supplementary services on calls is mainly a CC function. However, the management of the supplementary service status itself, i.e., the modification or checking of their actual configuration, can be done through GSM independently of the calls. It is the object of a separate sub-part of the CM plane: the SS management part.

The entities involved in SS management are very few: the mobile station and HLR are the only entities involved.

2.2.5.3. Short Message Services

The last aspect of the CM layer is related to the point-to-point short message services (SMS-PP). For the purpose of these services, GSM is in contact with a Short Message Service Centre (SM-SC). A service centre may be connected to several GSM networks. In each of these, one or several functional entities are in charge of interfacing the SM-SC. They are basically gateway functions, with the general definition we gave previously. However, the *Specifications* have special terms for the gateway functions when applied to Short Messages. They define two types of such entities: the SMS-GMSC for Mobile Terminating Short Messages (SMS-MT/PP) and the SMS-IWMSC for Mobile Originating Short Messages (SMS-MO/PP). The role of the SMS-GMSC is identical to the role of the GMSC for incoming speech or data calls. The role of the SMS-IWMSC is much less obvious and adds little value to the service except providing a fixed GSM point of interconnection for an SM-SC, rather than enforcing its connection to the SS7 network which would enable information transfer with any MSC.

All CM layer functions will be addressed in detail in Chapter 8, the last of three chapters dedicated to signalling issues.

2.2.6. OPERATION, ADMINISTRATION AND MAINTENANCE (OAM)

The OAM plane includes the functions which enable the operator to monitor and control the system. In one direction, it mediates the observation flow from machines to the operator. In the other direction, it enables the operator to modify the configuration of machines and functions. As a functional plane, it hovers over all the others, whilst not using the services provided by the other planes except the basic transmission functions for the exchanges between the concerned machines.

The kinship between the OAM plane and the OSS is obvious. The OSS is an integral part of the OAM plane, but all the machines in the

BSS and the NSS also contribute to the Operation and Maintenance functions. There are a variety of small tasks incumbent on these machines; they are often those of the smallest time scale and scope. For instance the raw information which forms the basis for the observation of the system's behaviour is clearly issued inside the traffic handling machines themselves. The data are then transferred to OSS machines. Another example is the first level of automatic maintenance, such as self testing or immediate actions such as activation of redundant parts in case of failure, and this is also done by the machines. Some other examples can be found, such as call tracing.

A more interesting point in the canonical GSM architecture is the operation of the BTSs. As defined in the *Specifications*, the BTSs are not in direct contact with OSS equipment, but are operated through their parent BSC. The BSC appears then as an OSS device, in charge of centralising the operation of several other machines.

In terms of interfaces and protocols, the organisation of the OAM plane has many different aspects, and most of them are outside the scope of the *Specifications*. Completely excluded are the independent network, if any, dedicated to the support of subscription management, and most of the network operation and maintenance network. TMN-related work may be the basis for the interfaces and protocols in these domains.

In many cases, the interfaces and even the protocols identified for traffic handling are used to support OAM related functions. We have already seen that the EIR fulfils its functions by communicating with NSS machines through the SS7 network. The corresponding protocol functions are in fact specified within the MAP. In another area, many OAM functions are intermingled with traffic related functions. Examples are call tracing or the management of terrestrial links. This explains why protocols such as the BSC-MSC protocol includes OAM related procedures.

Finally, some protocols dedicated to OAM are included within the *Specifications*. The first one appears on the Abis interface, and supports the functions needed for the operation of BTSs through the BSC. The others appear on the interface between the BSC and the OSS proper.

All OAM related functions will be addressed in details in Chapter 9.

2.3. INTERFACES AND PROTOCOLS – AN OVERVIEW

The details of the architecture of the protocols inside the different functional layers will be found in each of the specialised chapters. However, because we use in this book a model and a terminology slightly different from the one used in the *Specifications*, some general explanations may be useful for the readers already used to the GSM literature. Unfortunately, this is difficult to present without hinting at or using notions or terms not yet presented, and can be skipped in a first reading. Basically, this section lists the main GSM protocols, with our terminology.

The *Specifications* do not always give an abbreviation or a name to the protocols. This is the main reason for our introduction of new terminology, in an area already riddled with acronyms and other jargon terms. However, there is a need for a short term used consistently to refer to each protocol. In particular this name will be used systematically before a message name in the signalling chapters, so as to enable an easy distinction between messages. The reader unaware of the problem will understand better the need for the mention of the protocol when he reads that CIPHER MODE COMMAND and ENCRYPTION COMMAND are two messages fulfilling basically the same purpose, but belonging to two different protocols.

Figure 2.20 shows an overview of the signalling architecture in the machines of a GSM transmission chain, as far as the telecommunications functions are concerned (i.e., without showing explicitly the OAM functions). The horizontal axis corresponds to spatial distribution, starting with the mobile station on the leftmost part of the diagram and going through the various infrastructure machines on the way. The vertical axis corresponds to the functional planes, starting from the bottom with transmission and going up through the different layers described in the previous section.

The functions of these protocols will be described in the relevant chapters. Firstly, let us just briefly consider the stack of protocols on the radio interface, which identifies many of the important GSM protocols. At the very bottom, all transmission functions use protocols between MS and BTS. Then the RIL3-RR protocol enables MS and BSC to co-operate for the management of radio resources (RIL3 stands for Radio Interface Layer 3, following the name of TS GSM 04.08 where the RIL3-RR is specified together with two others). This protocol also appears on the Abis interface. Upper layer protocols (RIL3-MM and RIL3-CC) define

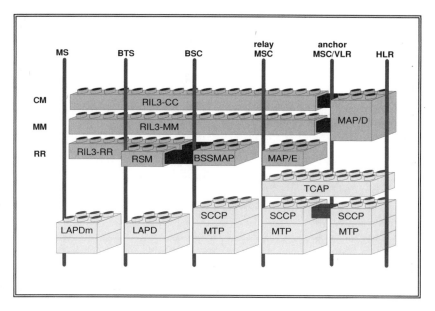

Figure 2.20 – GSM signalling architecture

GSM machines (shown as vertical lines)
and functional layers (shown as horizontal layers of bricks)
demarcate protocols, a stack of which can be defined on each of the interfaces.

the rules for signalling exchanges between the MS and NSS entities. These protocols also appear when looking at the Abis and A interfaces. Taking this last case as an example, the BTS and BSC are "transparent" to these signalling exchanges, i.e., they just act as carriers for information whose semantics is irrelevant for them.

Inside the NSS, each machine has in fact a single interface with the SS7 signalling support network. The corresponding stacks of protocols share the same lower layers: the protocols used for signalling transport on an SS7 network and referred to as MTP (Message Transfer Part). Above MTP, the protocols may vary depending on the corresponding peer entities concerned. Call-related signalling between MSCs and external networks makes use of TUP, ISUP or national variants of these protocols. Non call-related signalling corresponds to many different protocols, which are grouped together in the MAP (Mobile Application Part). In this book we will distinguish the different protocols by names such as MAP/B, MAP/C, and so on from B to I. For instance MAP/C is the protocol between a GMSC and an HLR. All these MAP/x protocols use the services provided by the SS7 protocol TCAP (Transaction Capabilities Application Part), itself using the service offered by the SS7

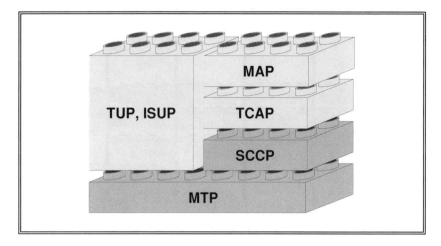

Figure 2.21 – Stack of protocols on the SS7 interfaces

All SS7 interfaces make use of MTP, but the MAP/x protocols require SCCP
and TCAP on top, whereas call-related signalling for standard call management
use TUP or ISUP related protocols.

protocol SCCP (Signalling Connection Control Part). The different
MAP/x protocols are not stacked up, but used independently in parallel.
The corresponding diagram is given in figure 2.21.

In the *Specifications*, the presentation is a bit different. The MAP/x
protocols are specified as a single "protocol", named the MAP, and what
we call here "protocols" are termed "interfaces". These "interfaces" are
referenced with letters (B to G, in no order of prominence). To keep a
strong relationship with the *Specifications*, we have used the letter
referring to an interface to denote the corresponding protocol. No H and I
interfaces appear in the MAP specification. Still, we have defined
MAP/H as the protocol for transfer of short messages although the
corresponding MAP "interface" in the *Specifications* is interface "E"
between MSCs, which is also used for the management of handover. In
the same way, MAP/I was introduced here as the MS to HLR protocol for
supplementary services management, which only appears in the
Specifications on the "D" interface between MSC and HLR, in order to
be closer to the conceptual approach of this book. All the MAP/x
protocols referred to in this book are summarised in figure 2.22.

As can be seen from this figure, MAP includes a great variety of
functions. The corresponding procedures will be described in the relevant
functional chapters of this book, each individual MAP/x protocol being
considered within the scope of the relevant CM, MM or RR planes.

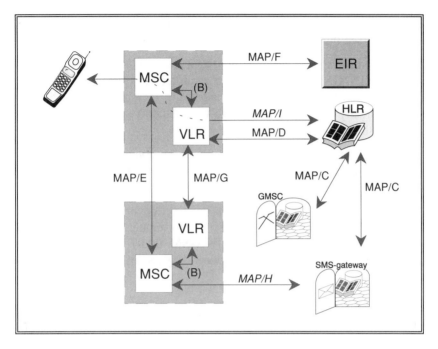

Figure 2.22 – MAP/C to MAP/I protocols

The MAP interfaces defined in the *Specifications* bear letters which have been used
as a basis to describe separate MAP/x protocols
for the sake of compliance with the conceptual guidelines of this book.

This summary of interfaces and protocols does not go into any detail as far as the transmission layer is concerned. In particular, both user data and signalling make use of specific transmission techniques and "link layer" protocols which have not been described here. This will now be the purpose of the next chapters of this book, dedicated to transmission.

SPECIFICATIONS REFERENCE

In fact there are very few *Specifications* devoted to general explanations of the general structure and architecture of GSM. Pieces can be found here and there, but the main source of information (beside this book!) is various articles in specialised conference proceedings and magazines (see the bibliography).

The only specification which appears as clearly devoted to the subject is **TS GSM 03.02** (Network architecture). Its scope is almost entirely limited to the infrastructure aspects, principally of the NSS. This TS is basically descriptive, with few explanations. One of its useful aspects is the reference to other TSs.

The internal structure of the mobile station in terms of terminal equipment, terminal adaptor and mobile termination is presented in **TS GSM 04.02**.

There is no TS introducing the distinction between the RR, MM and CM planes. Part of the subject, unfortunately written in a very technical jargon, can be found in **TS GSM 04.07**, will deals with the modelling of the interactions between layers of the corresponding protocols.

3

TRANSMISSION

3

TRANSMISSION

The first goal of a telecommunication network is to provide a means of transmission between end users. What exactly is to be provided, and how, varies, according to the different kinds of information to be transported, and to the specific constraints of the different interfaces to be crossed.

Both points are relevant to GSM. The multi-service nature of GSM requires that it interconnects with various kinds of external networks, each with their own transmission requirements. As far as internal interfaces are concerned, the radio interface is usually the focal point in a cellular network. GSM is no exception, and the specifications of its radio interface include more original features than any other public radio interface yet developed. Though transmission on fixed links is more constrained by existing standards, some new features have been introduced on the terrestrial links between GSM infrastructure machines.

Despite this variety, the whole system must provide consistent end-to-end transmission paths, taking into account different optimisation schemes on the successive segments along the way. This calls for translation functions between some of the transmission segments, and is a source of complexity.

The purpose of this and the next two chapters is to give an overview of transmission, with a focus on the specific aspects of GSM. A top-down approach will be followed. This chapter will start with a presentation of how end-to-end transmission paths are structured. The way in which user data is transformed along the way will be addressed; this includes the GSM digital speech coding, and the various rate adaptation schemes for data services. Then transmission between the

GSM infrastructure machines will be detailed. It was felt that the radio interface deserved a chapter of its own, which follows this one. Both chapters will focus on the circuit transmission mode, which covers almost all of the services offered by GSM. This leaves aside the packet mode transmission, which is used in GSM for the exchange of control information between machines, and for the Short Message services, which will be described in Chapter 5, the third and last chapter dealing with the transmission functions.

3.1. MODELLING PRINCIPLES

When looked at as a whole, the transmission functions of GSM give an impression of complexity. Because GSM is aimed at providing multiple services, there are a number of different types of information flow to be transmitted, and a variety of networks to interface with. Three broad types of information can be identified: speech; various information formats such as text, images, facsimile, computer file, messages, and so on, grouped under the vague term of data; and finally the internal signalling messages. Depending on the possible transit network, and on the network of the correspondent, the same kind of information may be transmitted in different ways. Moreover, the transmission methods also vary within the system, from one interface to another.

In fact (luckily), this seemingly complex jungle can be structured in two directions. A first "horizontal" structure stems from the need for consistent end-to-end connections. A number of them are defined, according, typically, to the type of transmitted information and to the final network or equipment. A second "vertical" structure corresponds to a layering design approach. With such an approach, a given portion of the transmission path uses machines which need not have a full knowledge of the type of data to be transported. The basic principle is that the meaning of the information transported on a given physical connection can be peeled little by little like the different layers of an onion, each layer representing a different level of knowledge.

If we represent the layering along a vertical axis, with the raw information at the bottom, and the distilled product at the top, and if the transmission path follows a horizontal axis, we obtain a representation such as the one in figure 3.1. Intermediate nodes need not be concerned with the knowledge of upper layer semantics. As a consequence, the specifications of the transfer mechanisms on the individual interfaces can

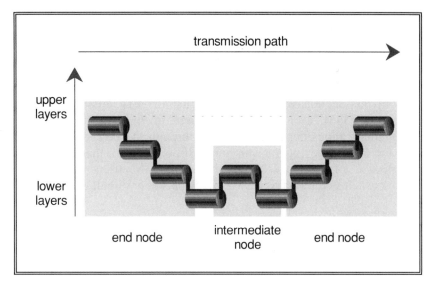

Figure 3.1 – Layered approach

Higher layers, dealing with the contents and presentation of the information,
are typically dealt with at both ends of the transmission path, whereas lower layers
may use different protocols on different segments between intermediate nodes.

be much simplified, by taking into account only the information attributes
relevant for transporting it.

If we apply this model, a first split along the vertical axis consists
in separating two domains, which will be described in the two main
sections of this chapter:

- the upper and outermost end-to-end domain is concerned with
  the information as dealt with by the final users. This domain is
  by nature rather "horizontal", i.e., each type of information
  warrants its own individual description (speech, different kinds
  of data, ...). When describing the end-to-end transmission path,
  special emphasis will be put on the interworking between GSM
  and other networks. At this stage, GSM will be considered as a
  "black box" obeying certain rules;

- the lower and innermost transmission domain is concerned with
  the way in which the information, regardless of its upper-layer
  semantics, is transported inside GSM. In the corresponding
  section, the GSM "black box" will be opened, gradually
  unfolding its internal organisation and interfaces.

3.2. AN END-TO-END VIEW OF TRANSMISSION

In this section we will address first the transmission of speech between a GSM user and another user in the GSM or in some telecommunications network accessible through the PSTN or the ISDN. Then the other types of transmission will be adressed, that is to say the transmission of non-speech data between GSM users and users in networks such as the PSTN, GSM, ISDN, packet or circuit data networks.

3.2.1. SPEECH

Telephony is by far the most popular service offered by public networks, including fixed networks and mobile cellular networks. After a general presentation of how voice is transported from mouth to ear through these networks, we will attempt to describe the environment with which GSM must interwork.

3.2.1.1. General Overview

Let us take the example of a speech call between a GSM user called Bernard and a subscriber of a fixed telephony network, called Fred. The function of the transmission plan is to transfer speech signals from Bernard's mouth to Fred's ear, as well as from Fred's mouth to Bernard's ear. The transmission path goes through GSM equipment from Bernard to some interworking point, and from there to Fred through the PSTN.

Starting with Bernard, the first item encountered is the microphone of his mobile station. Inside this mobile station, the analog voice signal is transformed into a digital information stream at a rate of 13 kbit/s, which represents the speech signal. Other processes in turn change this digital bit stream into a high frequency analog signal transmitted over the air.

After being detected by a base station antenna, this radio signal is processed to recover the digital signal representing speech, which is transported over coaxial cables towards a speech transcoder. From this incoming 13 kbit/s input, the speech transcoder derives another digital representation of the speech signal, at a rate of 64 kbit/s (the standard used in fixed network transmission). It is routed through the mobile

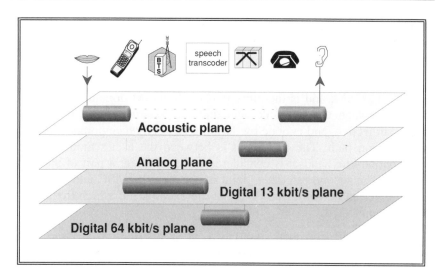

Figure 3. 2 – Speech representations

Transmission between two end users takes place on several transmission planes, each one corresponding to a different representation of the speech signal.

services switching centre (MSC) and various links and switches in the PSTN until it reaches the local switch to which Fred's telephone is connected. There (or possibly elsewhere inside the PSTN), the analog speech signal is reconstructed from the digital 64 kbit/s flow, and transported on Fred's subscriber line until it reaches his telephone, where the loudspeaker enables Fred to hear an acoustic signal which should be recognised as Bernard's voice.

This elementary description of a telephony call involves several transmission modes for speech, and therefore several transmission planes.

Starting with the GSM subscriber's end, these transmission planes unfold as follows (see figure 3.2): acoustic transmission, analog transmission, digital transmission at 13 kbit/s (this transmission being performed in two different ways over the radio path and between the base station and the speech transcoder), and finally another digital transmission mode in which speech is represented by a 64 kbit/s signal.

Closest to the end user is the acoustic transmission plane; digital transmission at 13 kbit/s and 64 kbit/s as well as analog transmission are found further down the transmission path. The transmission means vary from one interface to another, even within the same transmission plane.

3.2.1.2. Transmission Segments and Interworking Functions

In this description, the transmission mode at 13 kbit/s is clearly specific to GSM, and is therefore part of the GSM domain. The precise border between the GSM domain and the external world is however rather subjective, and is done by defining two reference points within the transmission path. The first point is between the user's mouth and the microphone: the handset is considered to be an integral part of the GSM domain. The second reference point is between the MSC and the switch of the external network (the PSTN in the example) to which the MSC is directly connected.

This split considers the GSM as a distribution network. The transmission path is then separated into:

- the local "loop" relating to Bernard which goes from him to his MSC. It is a part of GSM and is therefore specified in the *Specifications*;

- the rest includes a long-haul segment, which goes from Bernard's MSC to Fred's local switch, and Fred's local loop, which goes from him to his local switch. Both segments belong to the general telephony network, and may include national networks of several countries in the case of an international call.

The segments may be combined in different ways depending on the type of call. For example, a call between two GSM users that are both being served by the same MSC will only involve two distribution segments: these segments will be connected inside the MSC switching matrix. A call between two remote GSM users will include two GSM distribution segments and a long-distance segment between the two MSCs. In this case the two MSCs are connected as if the other was part of an external network, typically in the same way as with ISDN.

The functions to be fulfilled in order to adapt between the outer world and the inner GSM transmission are very simple on the mobile station side, since the reference point is at the boundary of the acoustic transmission segment. On the infrastructure side, the network interworking functions (or IWF, though the term is used rarely for speech) depend on the mode used for representing speech in the transit network. With the future ISDN, and with modern PSTN equipment, speech is transmitted digitally, coded in PCM (Pulse Code Modulation), at 64 kbit/s. But in the case of older PSTN equipment, speech is transmitted by analog means. The sole example described in the *Specifications* is however the 64 kbit/s case, other possibilities being considered only as country-dependent interim solutions.

The 64 kbit/s digital coding is part of every basic telecommunication curriculum. The analog signal is sampled at a rate of 8 kbit/s; this operation limits the bandwidth to 4 kHz. Each sample (which has an analog value) is given an integer value after the application of a logarithmic compression law known as A law. Each value is coded as an 8 bit symbol. The outcome is then a flow at a rate of 8 kbytes/s, i.e., 64 kbit/s. The transcoding between the analog signal and its digital A law representation includes an analog process (the pre-emphasis), sampling, a linear analog-to-digital conversion of the samples giving a result on 13 bits, and finally a coding process which transforms the 13 bit samples into an 8 bit code. All these processes are specified in more detail in the CCITT Recommendations, in particular in Recommendations CCITT G.711/G.714.

The effect of the speech transmission methods used in the PSTN and even in the ISDN cannot be neglected by the GSM transmission, since they do not provide a true reconstruction of the original acoustic signal. In both cases, the high frequencies of the signal are filtered out. Fortunately, this does not raise a problem. The lower part of the spectrum is also distorted, and this has some undesirable consequences. In the analog case, the 0-300 Hz band is completely filtered out. Even with digital transmission, this band usually disappears in the fixed correspondent's handset. The main consequence is that the two directions, from a GSM user to a PSTN user and vice versa, are not identical. In the first case, the signal is processed first by GSM machines, before any distortions introduced by the other network. The result is that transmission quality is, barring other differences, better in the mobile to fixed direction.

Another network interworking problem for speech communication is the problem of echo. The final segment in the PSTN uses a two-wire cable, and there is necessarily a 2-wire/4-wire adaptation somewhere, which is a first source of echo. The other sources are local to the terminal installation. As will be seen, the GSM transmission introduces a large delay, amounting roughly to half that which is encountered with a communication involving a satellite link. Since it is considered that echo is particularly disturbing to the user when delay is more than 25 ms, an echo control function must then be provided at the boundary between GSM and the PSTN to avoid the negative impact of a delayed echo.

3.2.2. NON SPEECH SERVICES

Non-speech services, or data services, cover the exchange of a lot of different types of information. Data transmission encompasses the exchange of text, of drawings, of computer files, of animated images, of messages, and so on. An important part of the information processing is done at the two extremities, in a machine most often outside the scope of the *Specifications*. We call such a machine "terminal equipment" or more simply "terminal", though in some cases it can be a complex installation, such as a videotex server or a message handling system.

The main functions performed by a data terminal in the realm of end-to-end information are the following:

- source coding, which transforms back and forth text, images, sound, etc. in the international currency of the world of information which are the binary digits;

- end-to-end protocol, dealing with the organisation of the communication, juggling with such concepts as pages, sessions, languages;

- and, most important, the presentation of the information to the user, by display, sound generation, printing, and so on.

In most cases it is possible to confine such processes to the ends of the transmission chain. This enables the reduction of the number of different cases which need to be taken into account by eliminating the need to study the intermediate transmission devices. The relevant characteristics distinguishing the different cases are few, and include the bit rate, the acceptable transmission delay (fixed or variable) and the maximum acceptable degradation due to transmission errors. The concept of *bearer capability* is used to describe and to refer to what is provided by the intervening equipment, i.e., the transportation of information between two user-network interfaces. A bearer capability is then characterised mainly by the attributes listed above.

Conversely, the concept of teleservice corresponds to the full chain, including end-to-end processing of specific information. The detailed functions in the data terminals are outside the scope of this book, just as they are outside the scope of the *Specifications*. This is natural, GSM being basically a distribution network. In most cases, only one of the two terminals at the ends of the transmission path is within reach of the *Specifications*. The terminal functions, and the corresponding end-to-end protocols between the terminals, are by necessity those that are used in the networks GSM interworks with. For instance, the way images are transported through GSM between two fax machines is the one used between PSTN users, because one of the two machines may be in the PSTN. A terminal equipment can be for instance a facsimile device, a personal computer, a computer terminal or a videotex terminal.

We will concentrate on the bearer capabilities used between the terminals. Looking closely at the structure of the transmission path, we immediately detect an important point: the boundary between GSM and the external network. GSM can be connected to a variety of external networks, since we are not yet in the promised land of broadband ISDN where a single international long-distance network supports all possible telecommunication services. Examples of networks include the good old PSTN (this ubiquitous telephone network which is still the principal carrier of data transmissions), Packet Switched Public Data Networks (PSPDNs), such as TRANSPAC in France, or Circuit Switched Public Data Networks (CSPDNs).

The existence of an external network divides the transmission path into two segments. The segment between the GSM user terminal and the boundary point is entirely within GSM. But the other segment, from the boundary point to the other terminal, is entirely outside the control of GSM, and follows transmission rules that are specific to the external network.

To reduce the number of cases dealt with by transmission equipment within GSM, despite the variety of interworking cases, two generic functions are inserted on each side of the GSM segment, as shown in figure 3.3. These functions enable GSM to deal with a small amount of internal transmission modes, and still accommodate the various interworking needs. The adaptation function at the boundary between GSM and the external network is called the *network interworking function*, most often reduced to the last two terms, and often further reduced to "IWF". On the GSM user side, the functional part of the mobile station which performs the adaptation between a specific terminal equipment (TE) and the generic radio transmission part is called the Terminal Adaptation Function, or TAF.

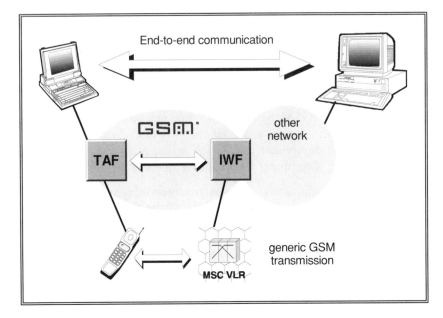

Figure 3.3 – Data transmission planes

Data transmission can be studied on three levels,
the end-to-end transmission plane between terminal equipment,
the TAF-IWF plane inside GSM,
and the generic GSM transmission plane.

The TAF on one side and the IWF on the other side act as entry points into the GSM world. Their functions depend on the type of end-to-end service. Conversely, GSM entities in-between TAF and IWF are not concerned with the end-to-end service, but solely with the bearer capabilities required to transport the corresponding data flow. Figure 3.3 shows how data transmission can be looked at on three different levels, the end-to-end level, the TAF-IWF level and the generic GSM transmission level.

As already noted, most of the end-to-end domain is out of the scope of the *Specifications*. But the adaptation functions (in TAF and IWF) are of direct concern to GSM. For most of the data services, the tasks fulfilled by the adaptation functions can be inferred from the single knowledge of the bearer capability. There is however an exception of importance, which is the facsimile teleservice. In this case, the adaptation includes additional functions dedicated to facsimile, and will be further discussed later.

So, with the sole exception of facsimile, the adaptation functions and the general configuration of the transmission paths depend mainly on the bearer capability and on the external network. It appears that the key factor, at the origin of most of the differences, is the external network. As a consequence, we chose to present the different cases not service per service, but external network per external network, starting with the PSTN.

3.2.2.1. The PSTN Case

Data (i.e., binary digit flows) can be transmitted over the speech-oriented Public Switched Telephone Network (PSTN) via the use of audio modems, which transform the bit stream in an analog signal which is constrained to the 300-3400 Hz bandwidth carried by the PSTN. Most of the data services, including fax, videotex, are transported by the PSTN. A data connection between two PSTN users has the configuration shown in figure 3.4, case a.

When considering a connection between a GSM user and a PSTN user it is clear the we must keep the half of the configuration concerning the PSTN user as if he was in connection with another PSTN user. An audio modem must then appear somewhere between the GSM user and the PSTN-GSM interworking point, as shown in figure 3.4, case b. The

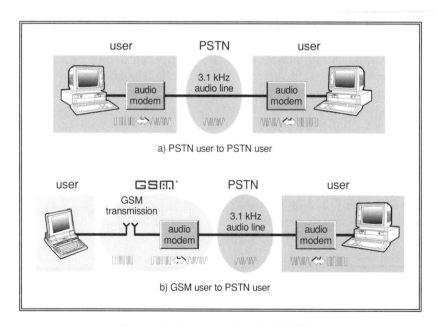

Figure 3.4 – Interconnection with the PSTN

GSM appears as an extended terminal-to-modem interface
for connection to the PSTN.

problem is then to design the GSM part of the transmission path, with the constraints imposed by the radio interface. For different reasons, chiefly because of the difficulty in designing an efficient and robust transmission scheme over radio for audio modem signals, the choice was to put the required modem on the infrastructure side, in fact in the interworking function. This means that, between the TAF and the IWF, the GSM network is only required to transport digitally provided data.

Putting modems on the infrastructure side necessarily restricts the freedom of the user to use any kind of audio modems (as in the PSTN), since they are constrained to use the types offered by the operator. However, in general, only standardised modems are used in the PSTN, and among all possible modulations, only a handful (mercifully!) have survived the standardisation process. Moreover, the available rates are in a limited series (defined by CCITT Rec. X.1), in which the typically used values are the bi-directional symmetric 300 bit/s, 600 bit/s, 1200 bit/s, 2400 bit/s, 4800 bit/s, and 9600 bit/s, and the asymmetric 1200/75 bit/s used in particular for the videotex service.

Between the two modems, the stream formatting falls into two categories, the asynchronous format, and the synchronous format. The difference of names comes from the fact that in the asynchronous format the instant of transmission of the bits are not aligned on a regular time base, whereas they are in the synchronous case. In fact, the difference is more profound, as the asynchronous transmission is a small protocol by itself, grouping bits into characters, and providing flow control for example. The unit of transmission is a character, a grouping of 7 to 9 bits. A character is preceded and followed by special signals, the start and the stop "bits". The data rate represents the rate of transmission of bits within a character, but successive characters may be separated by a period of any length, as shown in figure 3.5. In the synchronous case, bits are

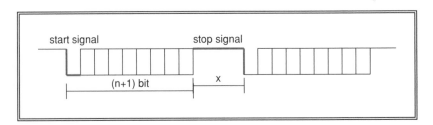

Figure 3.5 – An asynchronous data flow

The "start" and "stop" signals help the receiver to decode the flow of characters,
where bit beginnings are not synchronised with a clock edge.
The length x of the stop signal is not a multiple of the bit duration,
but must be greater than a minimum value.
Typical values of n are 7, 8 or 9.

Modem type	Rate	Mode of transmission
V.21	300 bit/s	asynchronous
V.22	1200 bit/s	asynchronous, synchronous
V.22bis	2400 bit/s	synchronous
V.23	1200/75 bit/s	asynchronous
V.26ter	2400 bit/s	synchronous
V.32	4800, 9600 bit/s	synchronous

Table 3.1 – Audio modem types supported by the *Specifications*

The most widely used audio modem types are supported by GSM,
including both asynchronous and synchronous modems
at rates up to 9600 bit/s.

transmitted regularly and continuously. In this case, the unit of transmission is the bit, and it is up to end-to-end conventions to group bits in one way or another.

These aspects are important because the start/stop format, used by the asynchronous services are still, despite their ancestral character, the most widely used form of data transmission, being used for instance by the modern videotex, on the serial port of personal computers, for the connection of most computer terminals, or in the interface with smart cards (including the SIM-ME interface in GSM).

A finite and limited list of modem types, rates and transmission modes which can be supported for the interworking between GSM and the PSTN has then been established. It covers all the rates previously listed, with in each case, the modes allowed by the CCITT Recommendations. The list is given in table 3.1.

Not only do the CCITT Recommendations specify the modulation method used between two modems over an audio line, but also the interface between the terminal and the modem (the DTE/DCE interface). All the cases belong to a wide family which, in various ways, use the signals defined by CCITT Rec. V.24. On the mobile station side, the interface between the TE and the transmission part follows the standard corresponding to the modem with which it interworks. As a result, the user to network interface is not at the same point as it is for a PSTN user. The interface is at the boundary of the Data Terminal Equipment and the rest of the mobile station, and the Terminal Equipment is connected as if to a modem. The GSM transmission segment appears in this case as being inserted between the TE and the modem. The problem is then to be as unobtrusive as possible. A few problems are encountered at this level, and some steps are necessary for the adaptation to the GSM transmission. These functions are fulfilled in the TAF and the IWF.

The classical interface between a modem and a computer includes several wires, on which several information flows are transmitted in parallel. Two of these wires carry the user data bit streams (one in each direction), and, in the case of a synchronous transmission, two which carry the bit clocks. But the data exchanged between a modem and a terminal includes not only the data streams, but also additional signals, whose general purpose is to allow the terminal to control the modem. For example, the modem may work only in half-duplex, and a signal is then needed to set it to receive or transmit mode. Because in GSM the modem is remote from the terminal, these signals must be transported over the GSM transmission system. These signals account for the difference between the user bit rate (e.g., 9600 bit/s) and the carrying capability of GSM (e.g., 12000 bit/s). Because GSM is basically a serial means of transmission, a function for multiplexing and demultiplexing the basic rate and the added signals must be fulfilled in the TAF and in the IWF.

A second problem comes from the asynchronous transmission mode. The transmission system of GSM is basically synchronous. This raises the problem of the adaptation between an asynchronous character-oriented flow and a synchronous one. Some function must be put to this avail in the TAF (on the mobile side) and the IWF (on the network side).

Conversely a problem exists in the case of a synchronous transmission, because the clock between the modem and the terminal may, in some cases, be independent from the GSM clocking system. This leads to a possible slight difference in frequency, a drift, which can add up to a non negligible value after some time. This issue is referred to as "Network Independent Clocking".

The specification of these functions was not something new, because the same problem of interconnection with the PSTN exists in ISDN, and was already tackled within the scope of ISDN standardisation when GSM started to work on the issue. The solution adopted in GSM is derived from the ISDN specifications, with many parts used without modifications, both for simplicity and for ensuring an easy transition between GSM and ISDN. Moreover, most of the problems arise when interworking with other networks. This is why the details of how they are solved will be explained later, so as to allow a discussion of the characteristics of some of these other networks and, in particular, ISDN.

Facsimile

Facsimile is considered as a key service for the marketing of GSM, but it is the source of many technical difficulties. Its specification was delayed, and even removed from phase 1 for some time before being re-introduced in 1991. Some of the difficulties come from the long transmission delay incurred by the transmission over the radio path, which appeared not to be compatible with the protocols used by group 3 facsimile machines. Another problem is the absence of a standardised terminal interface other than the plain analog two-wire connection for PSTN connection. The consequence of this is the need for a specific adaptation function, both in the mobile station and in the IWF. The GSM model distinguishes in fact a "generic" Terminal Adaptation function (the one used for synchronous bearer capabilities) and an additional "fax adaptor" which is inserted between the facsimile machine and the generic TAF.

The interface between the fax adaptor and the TAF is a terminal to modem digital interface, which supports if need be automatic dialling and answering, and the various synchronous data rates used by fax group 3. The interface between the terminal proper and the fax adaptor is a 2-wire analog telephony interface, enabling the connection of existing facsimile

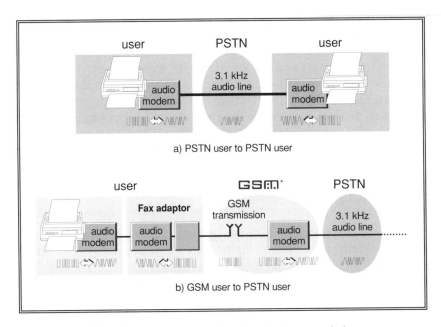

Figure 3.6 – Reference configuration for fax transmission

Beside the "generic" terminal adaptation function of the GSM transmission chain,
a fax adaptor is required to accommodate standard PSTN fax terminals.

machines. As a consequence, the fax adaptor includes the complex audio modem (V.21, V.27ter, and V.29) required to communicate with the terminal, and the overall transmission chain has the complex configuration shown in figure 3.6.

This is the functional point of view. Concrete implementations of this model will certainly include integrated products, where all functions will be grouped inside a single piece of equipment. The market will decide on the opportunity of such developments, which are certainly attractive from the technical point of view. For instance, integrating the facsimile terminal and the fax adaptor would remove two audio modems at once.

But the fax adaptor is not reduced to a modem. In addition, it manages a number of higher level functions. In a fixed facsimile communication, the end-to-end protocol specified in CCITT Recommendation T.30 enables the facsimile terminals to co-operate for the management of functions such as the choice of the modulation speed, the mutual identification, the page delimitation, the management of half-duplex, etc. This protocol is tampered with by the fax adaptor and by the IWF, in particular to avoid the problems raised by the long transmission delay.

The specific functions required for the management of the facsimile teleservice are confined to the fax adaptor (on the mobile station side) and the IWF. They can be considered as an additional layer on top of the basic synchronous transmission capabilities of GSM, with the latter not being impacted by facsimile transmission. The specific management of facsimile would require quite some development if it was to be described in detail. It is certainly an interesting item, but it is quite specialised, and we chose not to deal with it any further within this book. The interested reader is referred to the TS GSM 03.45 and 03.46 for details, or the papers referenced in the bibliography.

3.2.2.2. The ISDN Case

The interconnection with the Integrated Services Digital Network (ISDN) is a must for a modern digital telecommunication system. Speech raises no problem, but this is not the case for data services. The basic data service supported by the ISDN uses the capacity of a bi-directional 64 kbit/s channel. By deliberate choice, GSM is not able to provide this service. Because it would use at least four times as much spectrum as a speech channel (and eight times in the future with the half rate speech coding scheme), a heavy price would have been charged to the user of this service. In these conditions, it is difficult to imagine it would have

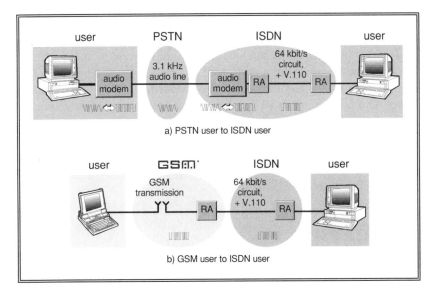

Figure 3.7 – Interconnection with ISDN

The rate adaptation scheme (RA) used by ISDN to interwork with PSTN (case a)
has been implemented in GSM, leading to the diagram of case b.

been much used. Moreover, it was clear from the start that the inclusion of a 64 kbit/s capability would have seriously impacted the design of the transmission system and increased the total system cost.

The interconnection between GSM and ISDN for data services is nevertheless possible, but in a devious way. To understand how, let us first note that compatibility with the good old PSTN is a constraint, not only for GSM, but also for ISDN. The configuration of a connection between an ISDN user and a PSTN user is shown in figure 3.7, case a. The end-to-end stream is rate-adapted ("RA" box in the figure) to be transported on 64 kbit/s circuits. This is the subject of CCITT Recommendation V.110 (or I.463), which specifies how an ISDN subscriber equipment must send data at 64 kbit/s when in communication with a PSTN subscriber, which transmits and receives data through a modem at, for instance, 4800 bit/s. The ISDN is able to carry the same data services as those which are possible between GSM and PSTN (and more). These services may then be offered for interconnection between GSM and ISDN. The result is that both the terminals on the GSM side and on the ISDN side act as if the other was in the analogue PSTN. This exemplifies the weight of history.

Even though the services are similar in both networks, the interconnection between the GSM and the ISDN, both using digital transmission, is radically different from the interconnection between either of them and the PSTN. Between GSM and ISDN there is no need for an audio modem. At the boundary point, the information is carried on a 64 kbit/s bit stream, with a portion of this rate corresponding to the bits exchanged between the end terminals, according to Recommendation V.110. The configuration of the connection, as presented in figure 3.7, case b, shows that each side is almost identical to the case where the other end is in the PSTN (the difference being the audio modems). By comparison with the PSTN to ISDN case (figure 3.7, case a), it can be seen that the GSM transmission replaces the modems and one of the rate adaptation functions.

3.2.2.3. The PSPDN Case

GSM can offer the possibility to communicate between a GSM subscriber and a Packet Switched Public Data Network (PSPDN) subscriber, or more generally between a GSM subscriber and a subscriber that can be reached through a PSPDN. This possibility can be provided by different means, depending on the terminal on the mobile station side, and on the level of intervention from the GSM infrastructure. To explain the different cases, it is simpler to start from the ways in which fixed subscribers can access a PSPDN.

There are three different means widely used to access the PSPDN, plus one of potential future utility when the user is an ISDN subscriber. The first is the direct access, where the subscriber is connected physically to the PSPDN. An example of an access interface, widely used, is the one specified by CCITT Recommendation X.25. The configuration (shown in figure 3.8, case a) usually includes a modem in the subscriber premises, but this modem is not an audio modem and it is part of the PSPDN. The terminal exchanges data with the network according to a high-level packet protocol (X.25 levels 2 and 3). The subscriber is basically identified by the access line.

The second method is a variation of the first, where the access is via the PSTN (see figure 3.8, case b). Audio modems must be used. The connection is established first through the PSTN, as a normal PSTN communication. When the PSTN subscriber initiates the call, he enters a PSTN number addressing the PSPDN entry port. Then a second number, referring to the end correspondent, will be provided from the terminal to the PSPDN (this is the notion of double numbering). The major

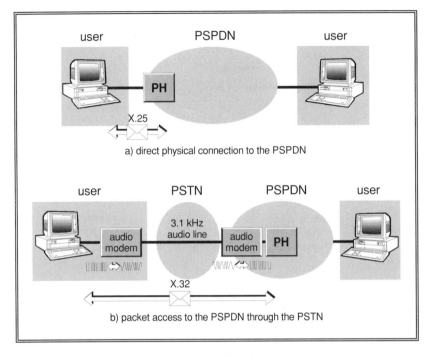

a) direct physical connection to the PSPDN

b) packet access to the PSPDN through the PSTN

Figure 3.8 – PSPDN packet access

A fixed user may access the PSPDN in a packet mode
either through a physical X.25 link to a packet handler (PH) of the PSPDN (case a),
or through the PSTN via an X.32 procedure (case b).

difference with the previous case is that it is not possible to identify the subscriber by the access line. This is why the access protocol is slightly modified in this case to convey the subscriber identity. X.25 thus modified is the protocol specified by CCITT X.32. However the terminal and the PSPDN still use a packet protocol for communication.

A third method makes use of a PAD function (a Packet Assembler Disassembler). It enables the user to have a simple terminal, which does not support a packet protocol. The major disadvantages of this solution are that the transmission data rate is limited, and that calls can be set up only at the initiative of the subscriber (incoming calls are not supported). The transmission uses an asynchronous protocol, with a character-oriented simple access protocol on top, which is used for numbering for example (an example of PAD access protocol is the one specified by CCITT Recommendations X.28). Access to a PAD can be direct, but the

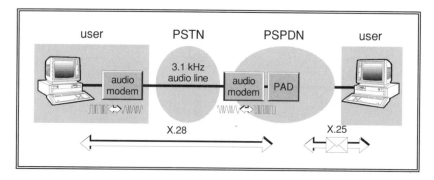

Figure 3.9 – PAD access to a PSPDN

A simple asynchronous terminal may still access a PSPDN
through a PAD, which converts the asynchronous stream of user data
into packets and manages the PSPDN packet protocol.

principle application is for the access through the PSTN. The
configuration is then as presented in figure 3.9.

The last case is the access through the ISDN. CCITT
Recommendation X.31 specifies how an ISDN subscriber can access the
PSPDN. It refers to two methods, one being in fact the X.32 access, using
the ISDN as if it was a PSTN (the configuration is then the same as in
figure 3.8, case b). With the other method (shown in figure 3.10), the
user-network interface between ISDN and the subscriber includes X.25.
The packet protocol is then used between the terminal and an ISDN
machine, which in turn interworks with the PSPDN. This ISDN machine
is in charge of the identification problem. The subscriber only has to

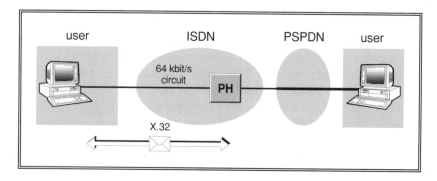

Figure 3.10 – ISDN X.25 access

ISDN offers the possibility for a subscriber to access
an X.25 service directly, through an ISDN packet handler (PH).

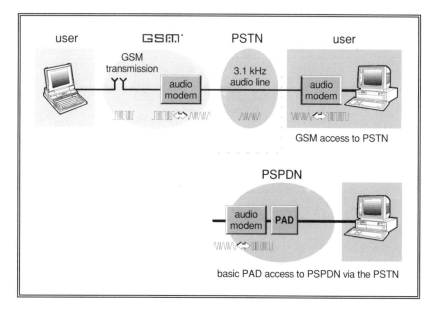

Figure 3.11 – "Basic" packet (X.32) and PAD access in GSM

Basic interconnection modes with the PSTN can be used
for X.32 or PAD access, without any specific handling by the GSM network
compared to the pure PSTN interconnection case.

enter a single number, and does not have to be registered in the PSPDN
but only with the ISDN.

The different schemes for access to a PSPDN from GSM are
adaptations of the three last schemes (direct access is not suited for users
that are by essence mobile users). The total number of combinations is in
fact more than three, first because of the possibility for GSM to intervene,
principally to suppress the need for double numbering, and second
because of the possible interposition of ISDN.

An obvious possibility is to access a PSPDN through the PSTN,
with GSM machines unaware that the call is ultimately for the PSPDN.
Both X.32 and PAD access methods are possible, as shown in figure 3.11
for the basic PAD access. The PSPDN access protocol is employed
between the terminal and some PSPDN machines, GSM being only a
carrier, as is the PSTN. This requires double numbering, and it also
requires that the GSM subscriber be a subscriber of the PSPDN for
charging reasons. This approach is of no concern to GSM, which does not
provide any additional service than it does for data transmission through
the PSTN. Yet the method is cited in the *Specifications* for the PAD case,
and is referred to as "basic PAD access".

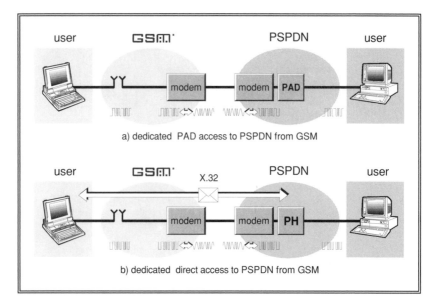

a) dedicated PAD access to PSPDN from GSM

b) dedicated direct access to PSPDN from GSM

Figure 3.12 – Dedicated GSM access to the PSPDN

GSM can act as a direct bridge between the subscriber and a PSPDN,
either via a PAD (case a) or via a packet handler (case b).
Interworking through modems is shown,
but direct digital interworking is also possible.

Another possibility is access through the ISDN, as an ISDN user, along the lines shown in figure 3.10. This possibility is referred to in phase 1 (notion of "transfer mode packet"), but will be suppressed for phase 2, and it is likely it will never be used.

The remaining possibilities are the more interesting, and mostly involve the GSM machines. They are the "dedicated PAD access", and the "dedicated packet access" (the second option does not appear explicitly in the phase 1 *Specifications*, where the packet access covers both what is described in the following paragraphs, and the access through the ISDN). In both cases, a single number is requested from the user, and no specific subscription with the PSPDN is required. The GSM interworks directly with the PSPDN, and the transmission does not necessarily make use of audio modems, depending on agreements between the GSM and the PSPDN operator. The transmission configuration is as shown in figure 3.12. The user-network interface is then between the terminal and the mobile station. The PSPDN access protocol (X.28 or X.32) is still handled between the user terminal and a machine in the PSPDN. However, the GSM machines (more precisely the IWF) are perfectly aware that it is a PSPDN access, and they do interfere with the protocol, mainly to add the required identification (which is specific to the PLMN and not to the subscriber). The transmission requirements are then identical to the basic cases, with only the network interworking functions specific to the PSPDN access. As far as the PSPDN is concerned, its subscriber is the GSM network operator, which is then in charge of dispatching (and recovering) the PSPDN charges to the relevant subscribers.

3.2.2.4. The CSPDN Case

Circuit Switched Public Data Networks are, as their name indicates, telecommunication networks devoted to data, and use circuit transmission, like the ISDN or the PSTN, as opposed to packet transmission like in PSPDNs. The standardised user to network interface access for CSPDN follows CCITT X.21 or X.21bis.

GSM provides the specifications for the support of X.21 and X.21bis terminals in the mobile station, as well as for interworking with CSPDNs. This is done in a way which is very similar to the previous cases, noting that for CSPDNs only synchronous access is provided, with rates equal to 2400, 4800 or 9600 bit/s. ISDN specifications already provide for these functions, and GSM simply inherited them. The rate adaptation specification in this case is CCITT X.30, which is very close to the synchronous case of V.110: the differences are almost entirely in the additional information signals, which are not the same in CCITT X.21 and in CCITT V.24, but the structure of the frame is exactly the same. The differences are only visible in the TAF and in the IWF, so the intervening machines are not involved.

There are two ways identified in the specifications for accessing CSPDNs. Either GSM is directly interconnected, or a CSPDN is accessed through ISDN. There is no provision for access through the PSTN. In all cases the data flow at the interface between the IWF and the BSS follows the ISDN specifications. No further adaptation is needed as the connection goes on through the ISDN. In the case of direct interconnection, the IWF performs the rate adaptation functions for the translation between the ISDN format (X.30) and the format for CSPDN,

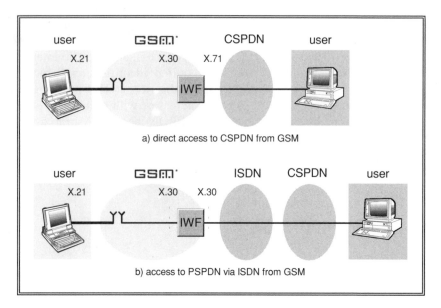

Figure 3.13 – CSPDN interconnection

Access to a CSPDN can be provided directly or through the ISDN.

which is defined by CCITT X.71. In both cases the interworking is digital, without specific modems. Figure 3.13 illustrates the two interconnection configurations.

3.2.2.5. A Summary of the Different Cases

Considering all the cases we have seen, we have two main categories of data transmission schemes from the GSM point of view, depending on the presence or not of a modem in the IWF. However, the inner portion of GSM transmission, up to and excluding the IWF, is basically the same in both cases. This will allow us to proceed with the study of the inner GSM domain in a unified approach, ignoring the nature of the external network and the presence or not of a modem inside the IWF.

3.3. TRANSMISSION INSIDE GSM

In the previous sections we have looked mainly at the aspects of transmission outside, or at the boundary, of GSM. It is time to tackle the innards of the system. The inner part of the GSM transmission system extends from a point somewhere in the mobile station (inside the TAF for data services, and where speech is an acoustic signal for the speech case) and the interworking point between GSM and external networks. Between these two points lie several machines and several interfaces. Our first task will be to present them.

3.3.1. ARCHITECTURE

Let us start by looking a little bit more in detail at the functions situated at the borders of GSM (the IWF on one side, the TAF on the other) before describing the more internal parts of GSM.

The IWF is a set of functions fulfilling the adaptations necessary between GSM and external networks. As will be seen, it can be rather limited for speech toward the PSTN, and for basic data when interfacing with ISDN. But in other cases, such as facsimile, Interworking Functions can be quite extensive. The IWF as a function lies somewhere between the MSC and the external network it interfaces with. A first implementation approach is simply to put the IWF in the MSC, and this is the usual approach for simple cases such as speech. For the complex cases, it can also be imagined to have special machines devoted only to the interworking functions, and linked to several MSCs. This centralised approach is sensible if the traffic through the IWF is but a small proportion of the overall traffic. This implementation is not precluded by the *Specifications*, but there is no standard specification of an MSC/IWF interface, and any such interfaces will be proprietary.

Let us turn to the mobile station side. The canonical GSM architecture identifies on one side the terminal (TE), in direct contact with the user, and on the other the core functionalities of a mobile station, which are common to all services. In between lie the Terminal Adaptation Functions, and in addition for facsimile the Fax Adaptor. The piece of mobile station equipment which contains the functions common to all services is called in the *Specifications* the Mobile Termination (MT).

If we turn now to concrete implementations of this functional model, we find a number of possibilities, differing by the grouping of the functions in specific machines, and on the interface specifications between these machines. The simplest case is when everything is integrated, generic functions, terminal equipment and adaptation functions if applicable. Such a machine is called MT0 (Mobile Termination type 0) in the *Specifications*. This is the approach retained for speech in all known implementations. Integrated mobile stations for other services will certainly appear sooner or later, for instance for facsimile.

The next simplest case, which will be our basis of work in the following for data applications, is when the TAF is totally integrated with the generic functions, and interfaces with the terminal through a classic modem to terminal interface. This integrated device is called MT2 (Mobile Termination type 2).

Another identified possibility is when the external interface of the Mobile Termination is the ISDN "S" interface, to which off-the-shelf ISDN terminal equipment can be directly connected. In this case, the machine is called MT1 (Mobile Termination type 1). A terminal using a modem to terminal interface can still be connected to an MT1, provided an ISDN Terminal Adaptor (TA) is inserted. In this case, the Terminal Adaptation Functions (TAF) are spread between the MT1 (where a synchronous adaptation is performed) and the TA (where for instance the synchronous/asynchronous adaptation is performed). The different mobile station configurations are illustrated in figure 3.14. In the following, we will not distinguish these different physical implementations. We will refer to the Terminal Adaptation Functions in general, whatever their implementations; similarly, we will refer to the TE-TAF interface, which can be according to the case the interface between the TE and the MT, or the interface between the TE and the TA.

Let us look now at what exists between the mobile station and the IWF. Along the transmission path, the canonical architecture of GSM distinguishes the BTS (Base Transceiver Station), the BSC (Base Station Controller) and the MSC. Between the mobile station and the BTS is a clear reference point, the radio interface, where the information crosses the space riding the 900 or 1800 MHz electromagnetic waves. The BTS/BSC/MSC split is adequate for the study of the signalling aspects. But the MSC and the BSC have little role to play in the transmission chain. Historically, the BTS and the IWF were the main actors in the transmission scene, and only basic transmission functions were found between them. Then another piece of equipment was introduced: the TRAU (Transcoder/Rate Adaptor Unit), which is definitely transmission equipment, and which was conceived to be distinct from the BSC or the MSC. The TRAU will take the starring role in this section.

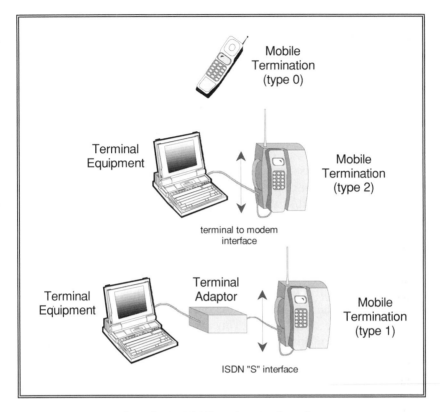

Figure 3.14 – Mobile stations configurations

Mobile stations can be either fully integrated,
or include a separate terminal equipment connected to a Mobile Termination,
through a Terminal Adaptation Function which can be either integrated
or kept as an independent piece of equipment.

The rationale behind the existence of the TRAU, distinct from the MSC and BSC, consists of several points. The implicit assumption during the elaboration of the concept of MSC was that it would be implemented more or less as a modified ISDN switch. As a consequence, the transmission at the level of the MSC is very close to that of the ISDN specifications. In particular only 64 kbit/s circuits are switched. As a consequence, the A interface must conform to the lower layers of the ISDN specifications. Indeed, the 2 Mbit/s standard multiplex structure used on the A interface (and also on the Abis interface) is not specific to GSM, but follows the CCITT G.703 standard. Their basic usage is to carry 64 kbit/s circuits compliant with the needs of ISDN. The multiplexing is based on a 125 μs cycle, each cycle transporting one octet per circuit. This structure is aimed at the transport and switching of

64 kbit/s circuits, but, in addition, enables the transport of sub-multiple rates such as 32, 16 or even 8 kbit/s. This possibility is effectively of interest for GSM, which does not require connections of more than 16 kbit/s, and where the cost of internal terrestrial links (between BTS and BSC, and between BSC and MSC), usually leased by the operator, represents a substantial part of the operational cost. A transmission method using only 16 kbit/s for user data (signalling is kept on 64 kbit/s links) was then devised, to allow this cost reduction which seems compelling despite some drawbacks. First, this introduces some extra delay for the transmission, and hence lowers the overall speech transmission quality. Second, it introduces a gateway function at the border between 16 and 64 kbit/s, which is really the purpose of the TRAU.

The late introduction of the TRAU, and the will to keep the switching capability of the MSC strictly similar to the one of an ISDN switch, is the source of its eccentric architectural location. The TRAU may be located in different places along the transmission chain, between the BTS, to which it belongs functionally, and (but not including) the MSC. One may then deduce that the only site possible when not on the BTS site is the BSC site. This is however not quite so: the implementations of many manufacturers include a remote transcoder situated on the MSC site. The BSC, as a functional unit, is then "spread" over its own site and the MSC site, and includes the link between these two sites. Conversely, the BSC-MSC interface (or A interface) is situated on the MSC site, over a very short distance. Figure 3.15 shows the positions of the TRAU relatively to the other BSS machines.

This somewhat artificial definition stems from historical reasons (i.e., reasons which were meaningful when the decision was taken, even though this meaning might have been lost through later evolution...). The definition was chosen in order to avoid the introduction of an option on the A interface, between transporting user data at 16 kbit/s or at 64 kbit/s, since operators have always been keen on limiting the number of options on the A interface to help multi-vendor inter-operability. Manufacturers also might have found an advantage in the decision of having a single rate of 64 kbit/s on the A interface, by avoiding the implementation of switching matrices at 16 kbit/s when the TRAU is put on the BTS side of the BSC matrix.

As a consequence, the *Specifications* strictly speaking do not allow the placing of the transcoder inside the MSC. Every call between two GSM users must then undergo two transformations (from 16 kbit/s to 64 kbit/s and vice versa, entailing for speech two transcoding operations between the 13 kbit/s and the 64 kbit/s representations), even if the two users are connected to the same BTS. One may imagine that such a restriction could be removed in future phases of GSM...

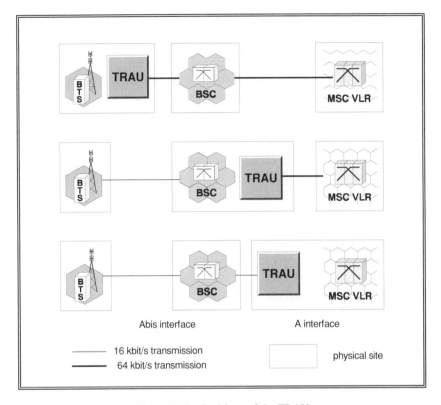

Figure 3.15 – Positions of the TRAU

The TRAU, which is functionally part of the BTS,
can be installed in a remote location (up to the MSC site),
to save link costs between the BTS and the TRAU,
thanks to the increase in transmission capacity from 16 to 64 kbit/s.

Because the TRAU is the true intermediate equipment for transmission, and because of its architectural predicament, we will not use the notion of A and Abis interfaces in this section, but instead the BTS/TRAU and the TRAU/IWF (or TRAU/MSC) interfaces. The BSC (and the MSC in the case of data) is simply ignored, as it has no special role as transmission equipment (but some as a switching equipment!). In the following paragraphs we will describe the transmission of speech, then data inside GSM, but we will exclude the details of the radio interface, since they will be studied in the next chapter. The relevant interfaces are in each case the following:

- the radio interface;

- the BTS-TRAU interface (which can be non-existent if the transcoders are situated at the BTS);

- and the TRAU-IWF interface, or more generally the interface between the transcoder and the point of interconnect with other networks.

Across those interfaces, the transmission system has the task of carrying speech or binary flows for a variety of data services. This still represents quite a few different cases, and additional treatment is performed to obtain a higher uniformity. In the rest of the chapter we will tackle these aspects and the detailed transmission schemes inside the infrastructure, first for speech and then for data. The study of speech will take us to the very specific area of digital speech encoding, whereas for data we will discover the subtleties of the numerous rate adaptation schemes.

3.3.2. SPEECH

Digital speech transmission over a radio interface in a mobile environment is quite a challenge. As already mentioned, a special digital speech coding algorithm is used in GSM, chosen for its low bit rate (13 kbit/s) and its resistance to high error rate conditions. The description of this algorithm will be the first topic of this section. This description is somewhere between the view of laymen (which the authors basically are in this field), and that of a specialist. Some emphasis will be given to some side features of voice transmission, such as vocal activity detection and discontinuous transmission, which are important for the spectral efficiency of GSM. The rest of the section is devoted to the rate adaptation which enables speech encoded with this algorithm to be carried not only over the radio interface, for which it was originally designed, but also over fixed digital links, between the BTS and the TRAU.

Taking again figure 3.2 (page 129), the GSM transmission path for speech can be divided into the following segments:

a) the mobile station;

b) from the mobile station to the base station: the radio path;

c) from the base station to the (remote) voice transcoder;

d) from the voice transcoder to the MSC.

The junction points separating the segments a) to d) described above correspond to places where a speech representation is changed to another one. These transcoding points are of major importance here, since the description of a transmission scheme is intimately related to the description of the corresponding transcoding functions. Transcoding points are the "elevators" between the different floors of figure 3.2. The following transcoding points are identified inside the GSM domain:

- Acoustic to Analogue Electric transcoding, implemented in the microphone, and the reverse Analogue Electric to Acoustic transcoding, implemented in the loudspeaker; this type of transcoding is not however specific to GSM;

- Analogue to 13 kbit/s Digital transcoding (and the reverse operation), implemented in the mobile station.

- 13 kbit/s Digital to 64 kbit/s Digital transcoding (and the reverse operation), implemented in the voice transcoder, either in the BTS or in the TRAU.

This does not mean that the signal is transported exactly in the same way on all links between two transcoding points. The signal representation is adapted to the transmission medium in intermediate processing points. The two main adaptations are:

- adaptation of the 13 kbit/s digital representation for transmission on the radio path;

- adaptation of the same 13 kbit/s digital representation to transmission on fixed links between the BTS and the voice transcoder in the TRAU.

3.3.2.1. Speech on the Radio Interface

Let us dive into some more detail and examine the digital mode used to represent the speech signal for transmission over the radio interface.

The prime concern for the design of the speech transmission means on the radio path was spectrum efficiency. The goal was to use as low a data rate as possible while providing an acceptable level of quality. Since speech is considered as the prime service, these considerations have heavily influenced the whole design of the system. On the radio interface, two types of raw carrying capabilities are defined, the "full rate" channel (which deserve this title just because it was the first to be specified), and the "half rate" channel, which indeed makes use of half as much radio resource as the previous one. The existence of these two types comes from history: at the time of definition, it was certain that it would be easy and quick to specify a digital speech coding at around 16 kbit/s with the required quality, and it was foreseen that it would be possible to do the same thing with half as many bits some years later. Because of the stress on spectrum efficiency, it was out of the question to forget this halving possibility. It was therefore decided to define the system in two steps, starting with a less efficient coding scheme, but paving the way for a twofold increase in efficiency to be introduced as soon as possible.

In the first phase of GSM, speech is then only defined for the full rate channel. The description of the coding scheme for this type of channel is presented in the following pages. As of the end of 1991, commercial use of half rate speech is foreseen for 1994 or 1995.

The Speech Coding Algorithm

The GSM speech coding scheme at 13 kbit/s is called RPE-LTP, which stands for **Regular Pulse Excitation-Long Term Prediction**. It aims at producing a speech quality similar—when no errors are added—to the fixed telephony network quality, with a much smaller rate in order to optimise the use of the radio spectrum.

General Principles

The aim of this section is to give the general principles of the GSM speech coding scheme. The bit-exact algorithms for coding and decoding are given in TS GSM 06.10; therefore a specialist shall find no better description than the one in the *Specifications*. However, the non-specialist will find here a general idea of the GSM speech representation.

The easiest way to dive into this subject is to look first at the contents of the transmitted signal, and its translation from 13 kbit/s to 64 kbit/s, which is performed on the decoder's side.

Speech is cut into 20 ms slices: in fact, rather than saying that the transmission rate is 13 kbit/s, it would be more realistic to say that speech is transmitted using groups of 260 bits every 20 ms. Synchronisation (i.e., separation of the 260 bits blocks) relies on external means and not on information transmitted within the blocks. The radio interface makes

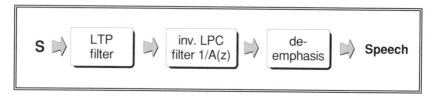

Figure 3.16 – Modelling of the 13 kbit/s speech signal

Speech can be generated from an excitation signal S by applying three filters in succession: LTP, LPC and a defined de-emphasis filter.

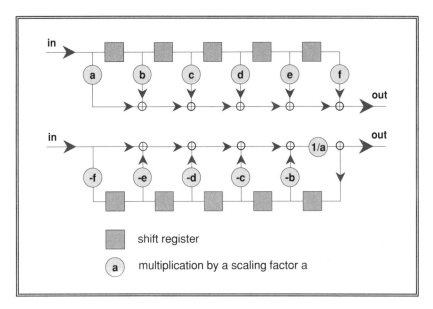

shift register

multiplication by a scaling factor a

Figure 3.17 – A linear filter and the corresponding inverse filter
A linear filter combines the input signal with delayed representations of itself and is
usually represented by a polynomial $A(z)$. The example shown would correspond to:
$$A(z) = a + bz + cz^2 + dz^3 + ez^4 + fz^5,$$
and the inverse filter (shown below the linear filter) would correspond to $1/A(z)$.

extensive use of its complex synchronisation scheme: this is why the 13 kbit/s flow, structured in frames of 20 ms, does not include any information helping the receiving entity to determine the frame boundaries.

For each block, the output signal is reconstructed by the receiver from an input signal (the excitation signal) which is filtered through a succession of filters (i.e., of linear transforms), as shown in figure 3.16.

The Long Term Prediction—or **LTP**—filter is a very simple filter, which consists in adding the signal and its delayed image multiplied by a factor b_r, the delay being N_r samples. The values of both N_r and b_r are transmitted in the speech frame, once for every 5 ms slot.

The Linear Prediction Coding—or **LPC**—filter is the inverse of an 8th order linear filter. A linear filter of nth order performs a linear combination of the signal and of itself delayed by 1, 2, .. n samples at 8 kHz. The coefficients of this filter vary from one block to another, and are transmitted in the speech frame. The structure of a linear filter and the inverse filter can be found in figure 3.17.

The **de-emphasis** filter is pre-defined, and therefore requires no parameters to be transmitted.

		Bits per 5 ms	Bits per 20 ms
LPC filter	8 parameters		36
LTP filter	N_r (delay parameter)	7	28
	b_r (gain parameter)	2	8
Excitation signal	Sub-sampling phase	2	8
	Maximum amplitude	6	24
	13 samples	39	156
Total			260

Table 3.2 – Parameters transmitted in each speech frame

In each speech frame, 188 bits relate to the excitation signal.
The other bits are the parameters of the LPC and LTP filters
applicable during the 20 ms period.

Last but not least, the excitation signal S itself is coded so that the set of all parameters—those of the above filters plus the description of the signal S—fits into 260 bits. S is sampled regularly ("RPE, Regular Pulse Excitation") at a rate of only 8/3 kHz. According to the signal processing theory, this allows to know accurately the information concerning only the lower 1.3 kHz bandwidth of S. The excitation signal at the input of the filters is reconstructed by inserting null value samples, so as to obtain a signal sampled at 8 kHz. From a spectral point of view, this results in a signal with spectral components above 1.3 kHz which are derived (second and third harmonics) from those below. The phase of the 8/3 kHz samples with regard to the 8 kHz samples can vary, and is transmitted once for every 5 ms slot.

The samples are coded using Adaptive Pulse Code Modulation (APCM). This coding is called "adaptive" because the maximum amplitude and the ratio of each sample to this maximum value are coded separately. This differs from the usual 64 kbit/s coding, where each sample is directly coded using a fixed scale.

Table 3.2 summarises all the parameters transmitted in each 260 bits frame (i.e., every 20 ms).

The Speech Decoder

The structure of the speech decoder follows very closely the modelling described above. The major processing stages are as follows:

- Production of the excitation signal by putting the 13 samples back to scale, and production of an 8 kHz sampled signal by adding 27 null samples, according to the phase indication;

- LTP filtering, involving the samples of the current 5 ms block and of the 3 previous ones;

- LPC filtering, according to the transmitted coefficients;

- de-emphasis filtering.

This is a simplified picture of the decoder; the interested reader shall find in the *Specifications* the full description of the quantizing methods, as well as the description of some additional processes which aim at smoothing the signal at the boundary between successive blocks, so as to avoid sharp transitions.

The Speech Coder

The coder is somewhat more complex than the decoder, since it must compute the different parameters so that the reconstructed signal will be as close as possible to the original one. The order of its processes is the reverse of the decoder's.

Starting from the 64 kbit/s digital signal (i.e., speech sampled at a rate of 8 kHz), the A law 8 bits samples are converted to 13 bit samples corresponding to a linear representation of the amplitude. This linear representation of the signal goes through a first filter (the **pre-emphasis** filter). 160 samples of the resulting signal, representing a period of 20 ms, are grouped together in a block; the next two stages (LPC analysis and filtering) are performed on such blocks:

- the LPC filter is chosen in such a way as to **minimise the energy** of the signal obtained as the input signal filtered by the *reverse* LPC filter. Because the speech signal is inherently very redundant, a substantial gain (in term of energy reduction) may be achieved through such a choice of LPC coefficients. When LPC filtering is applied by the decoder, the energy of the LPC filtered excitation signal will be amplified by the corresponding factor, whereas quantization errors will be amplified by a smaller factor. The tolerance for quantization errors of the excitation signal is then higher, thus allowing the use of less bits to represent it than what would be needed for the original signal itself;

- the next steps are applied to the signal obtained after filtering by the reverse LPC filter, called the "short term residual signal";

The residual signal is then processed by blocks of 40 samples representing 5 ms of speech, as follows:

- the coefficients of the long-term filter are calculated in such a way as to minimise the residual power, with the following constraint: the delay N_r must be in the range 40 to 120 samples (i.e., between 5 and 15 ms); this range has been chosen so that this delay corresponds usually to the fundamental frequency of speech, which is somewhere between 60 and 200 Hz depending on the speaker;

- the short term residual signal is then filtered by the reverse LTP filter, and sub-sampled to a third of the original 8 kHz rate, with a phase chosen as the one corresponding to the result of maximum energy; the resulting samples are coded in APCM.

Implementation

Both the speech coding and decoding algorithms described above use many calculations. As an indication, the coding and decoding of speech requires about 1.5 million elementary operations (addition or multiplication) per second. Digital Signal Processors—called DSPs—are hardware components specialising in this type of calculation and are therefore commonly used in speech coder/decoder implementations.

Speech coding raised a problem for the specification of the type approval tests of the mobile stations. Subjective testing, which is used to compare coders, is very expensive, since it involves ratings by several independent human listeners. Moreover, in real operation, the coding and the corresponding decoding are done in machines from different makes (one is in the mobile station, the other on the infrastructure side); a complete test would then require to test all possible pairs of coder and decoder! Such testing was avoided by specifying a complete implementation using integer arithmetic on 16 and 32 bits. As a consequence, the results produced by any implementation of the coder and decoder are not allowed to differ—not even by one bit—from the specified one.

Discontinuous Transmission, and Voice Activity Detection

An important side aspect of speech transmission is what is called in the GSM jargon the *DTX mode* (Discontinuous Transmission). This corresponds to what has been sometimes called in the US literature the *variable bit rate*. It aims at increasing the system efficiency through a decrease of the interference level, by inhibiting the transmission of the radio signal when not required from an information point of view. This DTX mode is an optional alternative to the normal mode. Because the DTX mode slightly deteriorates the quality of transmission, in particular when used twice along a path, that is to say in the case of a communication between GSM users, the choice between the two modes can be done by the network on a call per call basis.

In the DTX mode, the goal is to encode speech at 13 kbit/s when the user is effectively speaking, and otherwise at a bit rate around 500 bit/s. This low rate flow is sufficient to encode the background noise, which is regenerated for the listener (this is the notion of comfort noise) to avoid him thinking that the connection is broken. The low rate encoding corresponds to a decreased effective radio transmission since, to be exact, the active speech flow is of one frame of 260 bits each 20 ms, and the inactive speech flow is of one such frame each 480 ms. The choice of these values is related to other features of the radio interface.

In order to implement such a mechanism, the source must be able to indicate when transmission is required or not. In the case of speech, the coder must detect whether or not there is some vocal activity. This function is called Voice Activity Detection, or *VAD*. At the reception side, the listener's ear must not be disturbed by the sudden disappearance of noise and the decoder must therefore be able to generate some "comfort noise" when no signal is received.

One must be careful to distinguish between activity detection, which is only concerned in deciding whether some part of the information flow may avoid transmission, and discontinuous transmission. Activity detection is information dependent (speech, various kinds of data, ...), whereas DTX is bound to transmission characteristics: it raises issues such as its impact on radio measurements, synchronisation recovery, etc. This section only deals with VAD and the generation of comfort noise, which are specified respectively in TSs GSM 06.31 and 06.32. The consequences of DTX on radio resource management will be described in Chapter 6.

VAD – Voice Activity Detection

The VAD algorithm is very closely linked to the speech coding algorithm. For each output frame, the coder provides an additional bit of information indicating whether the frame must be transmitted or need not be, depending whether the algorithm decides that it contains speech or background noise.

The decision is mainly based on a comparison between a threshold and a measure of the filtered signal energy. The role of the filter is to distinguish a speech signal from background noise. Both the threshold and the filter are continuously adjusted according to the characteristics of background noise which are evaluated during non-active periods.

Generation of Comfort Noise

Experience has shown that a listener is greatly disturbed when the background noise behind the speech suddenly stops. This would happen regularly with discontinuous transmission. A mean to avoid the disturbance is to generate an artificial noise when no signal is received. The characteristics of this noise are regularly updated and transmitted to the receiving end by the speech coder situated at the other end.

These characteristics are transported by specific frames called SID frames (SIlence Descriptor frames). A SID frame is sent at the start of every inactivity period, and more are then sent regularly, at least twice per second, as long as inactivity lasts.

In order for the receiver to be able to distinguish speech frames from SID frames, the latter uses certain combinations of values which cannot be found in an error free speech frame.

3.3.2.2. Speech on the BTS-TRAU Interface

On the infrastructure side, the functional entity where 13 kbit/s encoded speech and 64 kbit/s encoded speech are translated one into the other is the TRAU. When the TRAU is physically distant from the BTS, for instance on the MSC site, the 13 kbit/s stream must be carried between them, over standard digital links. This makes use of 16 kbit/s circuits.

In fact, the 16 kbit/s bit stream contains more than the 13 kbit/s encoded speech bit stream. It also includes some auxiliary information to carry the 20 ms synchronisation (which cannot be derived from the 13 kbit/s flow) and to allow the remote control of the transcoder by the

BTS. A basic principle guiding the specification is that no additional signalling flow exists between the BTS and the remote transcoder unit (TRAU): all the necessary information is carried in-band. Before looking at the contents of this information in detail, let us study the requirements to be fulfilled.

The Requirements

Synchronisation

The 13 kbit/s speech encoded stream is structured in successive blocks of 260 bits every 20 ms. The stream itself does not contain any information allowing the receiver to determine the starting bit of a block: this must be provided separately. On the radio path, this is provided by the general synchronisation mechanism. On a 2 Mbit/s link, the longest cycle which can be guaranteed from one end to the other is the 125 μs cycle, which amounts to groups of 2 bits for a 16 kbit/s link. A part of the remaining 3 kbit/s must then be used for the 20 ms synchronisation.

Time Alignment

Both the radio path and the 16 kbit/s interface use a 20 ms structure. Moreover, in the downlink direction, transmission on the radio path can start only when a whole 20 ms block is received. There is then an optimum time relationship between the moment of the beginning of a block transmission on the radio path, and the end of the reception of a block from the 16 kbit/s link. The transmission suffers an additional delay, of up to 20 ms, if this relationship is not optimum.

For this reason, a small protocol, needing a part of the 16 kbit/s stream, allows the BTS to control the phasing of the incoming 20 ms blocks generated by the TRAU.

Speech/Data and Full/Half Rate Discrimination

In the future, when a half rate speech encoding scheme is specified, different capabilities will be required in the TRAU. Beside, not only speech, but also data (as will be seen in the next section) can be transported on 16 kbit/s links.

For all these reasons, in-band information allows the TRAU to know what kind of information is received and then what type of adaptation it must apply both for the uplink and downlink transmission.

Reception Quality

The receiver (demodulator and decoder) in the BTS is able to tell whether its correcting capabilities have been overwhelmed by the number of errors or not. The information about whether a speech frame is regenerated correctly or not is very important for the transcoder, which will ignore a frame indicated as bad. Such an indication is therefore conveyed from the BTS to the TRAU for each frame.

DTX Mode Aspects

As already mentioned, two modes of speech transmission are possible. In one mode, the speech flow is coded continuously at 13 kbit/s (one speech frame each 20 ms), irrespective of whether the user is speaking or not. In the second mode (DTX mode), the transmission alternates between speech-active phases, with a transmission of one speech frame each 20 ms, and of speech-inactive phases, during which the transmission falls to a frame each 480 ms (only the comfort noise characteristics are transmitted). The mode can be different between the uplink and downlink part of the connection.

The speech transcoder need not be told whether transmission in the uplink direction is in the DTX mode or not. A non-transmitted frame is seen as a bad frame, and the distinction between comfort noise frames and speech frames can be done on the basis of the frame contents.

On the contrary, the speech transcoder must be told by the BTS which mode to apply for the downlink stream.

In phase 1, this indication is not supported by the network signalling: this will be corrected for the phase 2, and one can assume that a number of remote transcoder implementations will anticipate this correction.

Even in the DTX mode, the transcoder provides continuously correctly formatted frames, either encoding speech or comfort noise. However, the BTS has to decide whether or not a frame must be transmitted to the mobile station. Obviously, all the frames encoding speech must be sent. In addition, from a speech transmission point of view, some of the frames encoding comfort noise must also be sent. The rule for this is that the last transmitted frame preceding a sequence of non transmission must be a comfort noise frame (see figure 3.18). The other frames (all comfort noise frames) may be transmitted or not, as the BTS

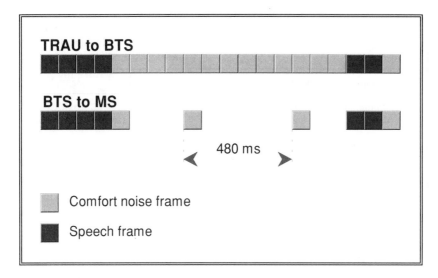

TRAU to BTS

BTS to MS

◀ 480 ms ▶

Comfort noise frame

Speech frame

Figure 3.18 – Transmission of speech frames

The TRAU continuously transmits frames, but the BTS may stop transmitting
some comfort noise frames according to some synchronisation rules,
but must take care to send a comfort noise frame before entering this mode.

wants. The distinction between speech frames and comfort noise frames
can be done by looking at the frame contents. It would have been useful
if the transcoder gave the information whether, from its point of view, the
frame must be sent or may not be sent. This is not so, and the only
auxiliary information which is provided is the redundant "SP"
information, which tells whether the frame is a speech frame or a comfort
noise frame. It is the BTS which must take care that at least one comfort
noise frame is transmitted before going to the 1 frame per 480 ms mode.

Other Information

For historical reasons, some other information, which can be
regarded as useless or redundant, is added. We have already encountered
one such case in the guise of the SP indication. Another example is the
SID indicator (SIlence Descriptor), used in the uplink direction to sort
frames into three categories according to the number of "1"s received
among the 95 bits which are all set to "0" in a comfort noise frame. Yet
another piece of redundant information is the Time Alignment Flag
indication in the uplink, used to mark the one frame each 480 ms when in
low rate. In fact, as far as can be seen, the behaviour of the transcoder is
not affected by the value of this indication.

	Number of bits in uplink frames	Number of bits in downlink frames
Frame synchronisation	35	35
Discrimination between speech and data, full rate and half rate	5	5
Time alignment	6	6
Bad frame indication	1	
DTX mode	1 (not in phase 1)	
Other information (specified, mandatory but redundant)	3 (BFI + TAF)	1 (SP)
Speech block	260	260
Spare	5 (6 in phase 1)	9

Table 3.3 – Contents of a speech block for transmission at 16 kbit/s

The block contains 316 bits, among which 260 represent speech.

The 16 kbit/s Bit Stream Structure

The 16 kbit/s bit stream is structured in successive blocks occurring on average each 20 ms. A block contains 316 bits, leaving on average 4 bits between successive blocks to cope with the variations due to time alignment. The contents of the block is summarised in table 3.3.

The details of the format, and the detailed protocol for starting and maintaining the time alignment, can be found in TS GSM 08.60.

3.3.2.3. Speech on the TRAU-IWF Interface

On a 64 kbit/s link, the standard G.711 speech transmission is used, with A law coding.

3.3.3. DATA

We have seen in the first part of this chapter the general aspects of the connection between users when circuit data services are provided. The grand result was that between the mobile user terminal and some point inside the IWF, the GSM segment was an over-stretched DTE (data terminal equipment, or terminal) to DCE (data communication equipment, usually modem) junction. This statement is true even when the IWF does not effectively include a modem, e.g., when direct digital interworking is possible. Seen like that the different cases for data transmission correspond to different cases of "modems", with various data rates and other properties. We will devote the rest of this chapter to study how the synchronous GSM radio interface has been made suitable to become part of the various types of terminal to modem connections.

The design of the GSM data transmission was led by two main issues. The first one is simply how to transport a multi-wire interface such as a standard terminal to modem interface, possibly using the start/stop transmission format, over basically a single wire and synchronous medium. A similar issue exists in ISDN, and technical answers are specified in the ISDN recommendations, such as V.110. The second problem is the radio interface itself, with its high raw error rate.

Though an important design goal for a transmission system is to reduce the number of different ways in which the data has to be transported, several connection types have been defined in GSM. They reflect two widely different approaches of how to adapt a terminal to modem interface, and warrant a detour to explain the different types of connections which have been designed between the mobile station and the IWF.

3.3.3.1. The Connection Types

In the case of GSM the will to limit the number of different transmission modes had to be pondered by transmission quality and delay considerations, and the result is a compromise. Transmission over mobile radio links is not by far as easy as over fixed lines. In particular, raw bit error rates of more than 10^{-3} are not uncommon on such links. For speech transmission, this does not cause a problem since the human ear is very tolerant to noise, but it is unacceptable for most data services. For such services, error correction protocols do in fact exist, either end-to-end or on some segments of the transmission path. Yet these protocols are designed for the error conditions that are common over fixed lines, and these conditions are far more lenient than those found on the radio transmission. Besides, such error correction schemes are not even used for some data services. It was then necessary to provide error correction schemes inside the GSM transmission path.

Information theory tells us that for a given raw bit rate and given error conditions, the characteristics of transmission when error correction is provided is a compromise between the throughput of data, the transmission delay and the remaining error rate. In the case of GSM, no single trade-off fits all the different types of services. Out of the possibilities, a short delay despite a relatively high error rate is better in some cases, whereas (for fax for instance) a long delay can be tolerated in order to achieve a better transmission quality. For these reasons, several types of connection are provided in GSM.

A fundamental division exists among the different types. In one category, the error correction is entirely done by a forward error correction mechanism provided by the radio interface transmission scheme. This case is referred to as "T" in the *Specifications*, as will be explained later. In the other category, referred to as "NT" in the *Specifications*, an additional scheme is used, where information is repeated when it has not been correctly received by the other end.

The T mode of transmission is derived from the ISDN specifications (in particular from V.110), and the path between the TAF and the IWF is seen as a synchronous circuit: the available throughput is constant, and the transmission delay is fixed. The information exchanged between the two entities include the user data, at some rate between 600 and 9600 bit/s, plus some auxiliary information, as in the case of ISDN. As we French say, *qui peut le plus peut le moins*, and it could have been imagined that a single transmission mode, able to transport the highest rate, could have been used. Simple rate adaptation schemes could have been designed for the transmission of other rates. But because better protection can be provided by forward error correcting schemes when the rate to transmit is low, the preference went to the definition of three different intermediate rates, despite the increased complexity.

The lower user rates, up to and including 2400 bit/s, are grouped into a single category, as if all were at 2400 bit/s. The two other intermediate rates correspond to user rates of respectively 4800 and 9600 bit/s. User rates up to (and including) 4800 bit/s do not necessarily require the maximum channel capacity ("full-rate" channels) on the radio interface, but the highest rate of 9600 bit/s does. The lower the intermediate rate, the better the achievable transmission performances, as shown in Table 3.4. The table also shows the different cases which will appear in the future, when "half-rate" channels are used. In the case of 9600 bit/s, or of 4800 bit/s on the half rate channel, the performance is not very satisfying, and this explains why a different approach (the NT mode) was specified in addition.

In the NT approach, the transmission on the GSM circuit connection is considered as a **packet** data flow (though the offered service, end-to-end, is most often a circuit service). Consequently, the available throughput varies with the quality of basic transmission (the

User rate	Intermediate rate	Channel type	Residual error rate
9600 bit/s	12 kbit/s	full rate	0.3 %
4800 bit/s	6 kbit/s	full rate half rate	0.01 % 0.3 %
≤ 2400 bit/s	3.6 kbit/s	full rate half rate	0.001 % 0.01 %

Table 3. 4 – Performance versus rate in the T mode

On a given radio channel, a lower data rate leaves more room for
an efficient forward error correction scheme, resulting in less residual errors.
The figures for the residual error rates are extracted from TS GSM 05.05,
considering typical urban conditions with frequency hopping.

higher the probability of error, the lower the throughput), as well as the transmission delay, but the resulting transmission quality in terms of residual errors is much better than in the T approach. The basic rate is 12 kbit/s (6 kbit/s on the half-rate channel), whatever the user rate, and with the same forward error correction scheme as for the T connection, respectively for 9600 bit/s and 4800 bit/s. Bits are considered grouped in successive frames of 240 bits. The frames include redundancy bits to enable a receiver to detect remaining errors, and this is the basis for a repetition-when-needed correction scheme, the RLP protocol (Radio Link Protocol, the expanded name as often is misguiding, so we will use only the acronym). This protocol is operated between the TAF and the IWF. In rough terms, the user data stream (including auxiliary information) is sliced in blocks of 200 bits, sprinkled with some redundancy and sent to the other end within frames of 240 bits. Each frame is numbered. When received, the frame is tested for correctness thanks to the redundancy. If found correct, the receiver acknowledges the reception (using the frame number). If not, a negative indication (either explicit, or by default) is received by the sender which tries again. RLP is basically a link protocol, such as those used for the transport of signalling messages. More details on its functions and characteristics have therefore been grouped together with the description of similar protocols, in Chapter 5.

The insertion of RLP in the transmission chain raises a major problem, which is the data rate. If the user rate of 9600 bit/s is considered as an example, the addition of the auxiliary information leads (as in the T mode) to a rate of 12000 bit/s. The question is then where to find the overhead rate which is needed for the error detection, the frame numbering, the acknowledgement indication, the repetitions, and so on. The trick was to notice that in many cases of end-to-end data an important overhead exists already. For instance, the start/stop protocol is quite inefficient: to each character of 8 bits sent by the user, the equivalent of at least 2 bits is added, in the form of the start and stop signals. This represents an overhead of at least 20%. Another case which is exploited is when a low level protocol is used, like LAPB in the case of an X.25 access or in the case of a fax transmission. The main functions of these protocols are framing, error correction by retransmission and flow control, all functions that are also fulfilled by the RLP protocol. In all cases, e.g., start/stop and LAPB, the idea is to *replace* them on the GSM segment, by some equivalent features adapted to the GSM transmission environment. Because of this replacement, a connection using RLP is referred to as "Non Transparent", or NT in short. The plain circuit case is referred to as "Transparent", or T (once again, we will use henceforth only the acronym).

An important consequence of this replacement concept is that the RLP approach can only be used in some specific service configurations (in fact only the cases cited above), for which the GSM machines know which low layer protocol is being used.

To summarise, restricting our view strictly to the TAF-IWF segment, abstracting the structure of the data flow performed by the TAF, the IWF and the external world, data transmission in GSM provides 4 different types of data connections in the first phase of the system. When half rate channels are introduced, they will be used to support the existing 2400 bit/s and the 4800 bit/s types of data connections, and one new type, the 4800 bit/s connection with RLP. In fact, the type of channel is not neutral. It influences the transmission delay and the error statistics. Because these characteristics are important for some services, it is best to take the radio channel type into account when defining the types of connection. This means that the TAF and the IWF must know, or even control, this parameter. We then obtain 7 cases, summarised in Table 3.5. Seen from the outside world, the rate 9600 bit/s is offered with two compromises of quality and delay, rate 4800 bit/s will be offered with three compromises between quality, delay and radio spectrum utilisation, and rates up to 2400 bit/s with two compromises between quality, delay and radio spectrum utilisation.

3.3.3.2. The General Issue of Rate Adaptation

Let us return to our muttons, and look at how the terminal to IWF connection is performed. Not very astonishingly, the T and NT cases are very different in this account. In the T mode, the GSM choice was to put the modem (if any) in the IWF, and to have the transmission specification

Name	Quality of service	Delay (two-way, TAF-IWF)
TCH/F9.6, T	low	330 ms
TCH/F9.6, NT	high	> 330 ms
TCH/F4.8 (T)	medium	330 ms
TCH/F2.4 (T)	medium	200 ms
TCH/H4.8, T	low	600 ms
TCH/H4.8, NT	high	> 600 ms
TCH/H2.4 (T)	medium	600 ms

Table 3.5 – The 7 GSM data connection types

Channels such as the TCH/F ("full-rate" channel, 23 kbit/s raw bit rate)
and in the future the TCH/H ("half-rate" channel) carry user data rates
of up to 9600 bit/s, with different compromises between delay and quality.
The quality estimation is only statistically indicative,
since in chosen geographical spots even the TCH/F9.6, T
may lead to very good results.

at the TRAU to IWF interface (64 kbit/s) conform exactly with ISDN specifications, in particular with V.110 or X.63 (the two being rather close, we will focus on V.110 in the following). This fulfils all the needs, whether for the remote access to a modem or for direct ISDN interworking. An advantage of this choice is that no further adaptation is needed in the IWF in the cases of interworking with the ISDN without a modem, at least on the transmission level.

For the NT cases, the transmission between the TAF and the IWF makes use of the RLP, which then has little involvement with the rate adaptation schemes of ISDN. The connection is best described in two layers. The higher layer corresponds to the conversion of the signals between the terminal and the IWF into something which can be carried by the RLP. This aspect involves the TAF, and the IWF for the reverse conversion, and the RLP transfer in between. The second (and lower) layer includes the means to transport the RLP frames between the TAF and the IWF. We will see that this lower layer uses means derived from the T mode (in particular this transport is basically a synchronous circuit).

Some knowledge of V.110 is a key to the understanding of the GSM transmission scheme. The detailed description will then start with a summary of V.110, seen as the basis for the interface specification between the TRAU and the IWF, followed by a description of how the matter is tackled in ISDN. Then we will look in turn at the T and the NT mode, before looking at individual segments along the GSM TAF-IWF path.

V.110

Let us first look at how the different issues mentioned above are solved in ISDN, that is to say

- the transport of the auxiliary information;
- the transport of an asynchronous flow over a synchronous link;
- the transport of a synchronous flow over a synchronous link using independent clocks.

CCITT Recommendation V.110 specifies how ISDN supports a data terminal equipment (DTE) equipped with a V-series interface. It contains the specification of an asynchronous/synchronous adaptation (the **RA0** function, for Rate Adaptation 0, though it is not strictly speaking a rate adaptation like RA1 or RA2, which will be described later). It contains also the definition of the auxiliary information to be carried, and the sampling rate, as well as the means to support network independent clocking. These specifications were used unmodified for the GSM application.

The Modem Control Signals

In V.110, the auxiliary information is limited to two additional signals in one direction (from the terminal), and three in the other. This additional information amounts to 8 bits every 5 or 10 milliseconds (according to the basic rate), which are multiplexed together to form a single bit stream. Table 3.6 lists the different signals, as well as their sampling rates.

The RA0 Function

As mentioned previously, an asynchronous data flow is a succession of characters, each typically 8 bits and preceded by a start "bit" and terminated by a stop "bit". On such flows, it is not required that the bit edges fall with a regular clock. Figure 3.5 (page 136) gives an example of such a flow.

In ISDN as well as GSM, data transmission is only synchronous, and the role of the Rate Adaptation 0 is to transform the asynchronous flow into a synchronous one, as shown in figure 3.19. Like a sergeant major, the RA0 must get the bits in line.

terminal to modem signals	modem to terminal signals	mean sampling rate
status of circuit 108 (data terminal ready)	status of circuit 107 (data set ready)	each 1.25 ms or 2.5 ms
status of circuit 105 (request for sending)	state of circuit 109 (line signal detector)	each 2.5 ms or 5 ms
	status of circuit 106 (ready for sending)	each 2.5 ms or 5 ms

Table 3.6 – V.110 transportation of modem control signals

The auxiliary information transported in V.110 frames corresponds to the sampling of circuits between DTE (terminal) and DCE (modem).

At first view, this is not too difficult. It seems enough to impose a short delay to the bits which are not truly in line. This works fine if the real (asynchronous) throughput is below the maximum (synchronous) capacity, so that the time lost in the delaying process does not add up. The maximum data rate may be theoretically the same before and after the adaptation, but in fact the frequency inaccuracy allowance is such that the incoming rate may be actually higher than the outgoing rate. The difference may be only a few percent or even a fraction of percent, but this is sufficient to cause trouble. The specification being there to answer all nasty problems, this case was foreseen, and it is allowed to skip now and then a stop signal to compensate. The receiving side must then detect

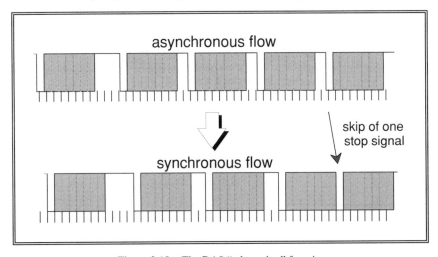

Figure 3.19 – The RA0 "adaptation" function

The role of RA0 is to convert an asynchronous flow into a synchronous one for transmission into systems such as ISDN or GSM.

the deletion, and reinsert the missing stop element. The same problem of speed adaptation exists at the receiving end. It is solved there by shortening the duration of the stop signal.

Network Independent Clocking

The exact transmission rate through digital networks such as ISDN is imposed by a network clock. Now, the exact end-to-end rate may be different when one end is in the PSTN. An audio modem used with a PSTN line may synchronise its transmission on its own clock. In this case, the frequency tolerance is 100 ppm, a very high value corresponding to one bit in excess or in default each second at 9600 bit/s.

To cope with such cases, V.110 includes mechanisms which enable a rate adaptation unit to indicate frequency corrections to the other. These mechanisms also allow an indication of when a bit has to be skipped, or on the contrary to be added and in this case the bit value.

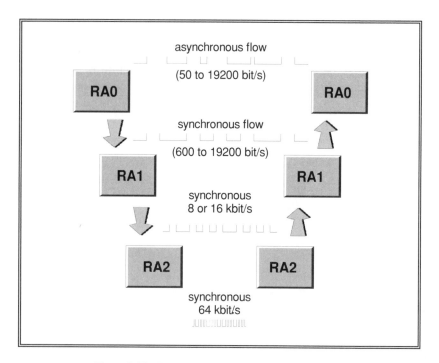

Figure 3.20 – Rate adaptation in ISDN: the three steps

Rate adaptation in ISDN proceeds in three steps:
– RA0 when necessary for the asynchronous-synchronous conversion;
– RA1 up to an intermediate rate of 8 or 16 kbit/s;
– RA2 for "padding" of this intermediate rate into 64 kbit/s channels.

These mechanisms make use of a set of 8 commands, such as "no change", "accelerate your clock toward the terminal by 20 %", "skip one bit", "add a bit of value 1", and so on.

The ISDN General Rate Adaptation Scheme

When considering the end-to-end synchronous case, or when looking at the portion of the transmission path between two RA0 functions in the case of the end-to-end asynchronous case, the data flow consists in a bi-directional synchronous flow at the nominal rate, accompanied by some auxiliary information representing a rate of a few kbit/s. Rate adaptation proceeds in two steps, called respectively RA1 and RA2. The whole process is shown in figure 3.20.

The RA1 function provides a bit flow at the **intermediate rate** of 8 kbit/s or 16 kbit/s, according to the nominal rate to transport. RA1 is not only a rate adaptation function. It also includes the multiplexing and demultiplexing between the auxiliary information (modem control plus other signals) and the main flow, as shown in figure 3.21. This is done by time multiplexing, that is to say that the multiplexed bit flow is a regular alternation of bits of the main flow and of auxiliary information bits. This requires synchronisation between the multiplexer and the demultiplexer, which is maintained by transporting additional bits. The period of recurrence for the multiplex structure is either 5 ms (for the 16 kbit/s intermediate rate) or 10 ms (for the 8 kbit/s intermediate rate), and defines successive multiplex frames. By the way, these values of 5 and

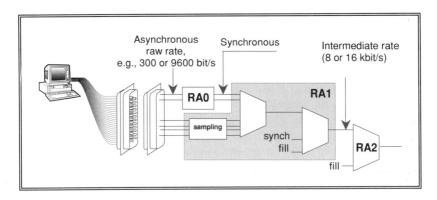

Figure 3.21 – Rate adaptation in ISDN: RA1

Pictured on the transmitter side, the RA1 function multiplexes
a synchronous user data flow with sampled auxiliary information
and synchronisation to obtain an intermediate rate
of 8 or 16 kbit/s as an input to RA2.

bit name (V.110)	information carried	comment
S1, S3, S6, S8 (or SA)	status of circuit 108 (data terminal ready) or 107 (data set ready) depending on direction	
S4, S9 (or SB)	status of circuit 105 (request to send) or 109 (line signal detector) depending on direction	
X	status of circuit 106 (ready to send)	sent twice per frame
E1, E2, E3	real data rate	This indicates the end-to-end data rate
E4, E5, E6	network independent clocking (used in synchronous cases to control the remote clock when modems are not synchronised on the transport network)	Codes a speed-up or slowing of the clock rate or bit skip or insertion
E7	40 ms synchronisation in the case of 600 bit/s only (X.30 compatibility); may also carry information relating to end-to-end flow control	

Table 3.7 – V.110 auxiliary information

8 "status" bits (S1, S3, S4, S6, S8, S9 and two X) and 7 "E" bits (E1 to E7) carry auxiliary information in each V.110 frame

10 ms are one of the reasons for the choice of 20 ms as the frame period for speech in GSM.

When seen at the transmitter side, RA1 proceeds for the lower bit rates to a first rate adaptation. This is done for nominal rates below 4800 bit/s, by repeating each bit as many times as necessary to obtain 4800 bit/s. This raises no difficulty since all the lower standardised rates are sub-multiples of 4800 bit/s. Information is added in the digital stream to indicate the effective end-to-end rate. After this step, the data rate is either 4800 bit/s, or 9600 bit/s. A frame, with a period of 5 ms for a basic rate of 9600 bit/s, and of 10 ms for a rate of 4800 bit/s, then contains 48 user data bits. The auxiliary information bits are then added through time multiplexing, as well as enough synchronisation bits to obtain respectively 16 kbit/s or 8 kbit/s. The auxiliary information accounts for

15 bits (see Table 3.7 for the list) and the synchronisation bits for 17 bits of each multiplexed frame.

The RA2 function rate-adapts the intermediate rate to 64 kbit/s, by simply adding 7 or 6 bits set to 1 in each octet. This can be seen as time-multiplexing between the intermediate-rate flow and a constant flow consisting of all "1"s. In this case, no additional synchronisation bits are necessary. This comes from the fact that the basic ISDN channel is not strictly speaking a 64 kbit/s channel, but an 8 kilo-octet per second channel. This subtle distinction means that the grouping of bits by 8, in a way unambiguous for the two end parties, is a side product of the ISDN transmission scheme. This property, usually referred to as "8 kHz integrity", allows to easily define sub-flows at rates which are an integral multiple of 8 kbit/s.

3.3.3.3. The GSM T Connections

The transmission path between the TAF on the mobile user side and the IWF is functionally totally equivalent to what appears between the terminal to "modem" interface and the 64 kbit/s circuit in the case of an ISDN connection using V.110. So the RA0, RA1 and RA2 functions will appear somewhere between the TAF (included) and the IWF (excluded). However the transmission over the radio interface must be introduced somewhere in the picture.

Data transmission on the radio interface is not done at 64 kbit/s, and V.110 obviously cannot be used in its pure form. A first idea could be to keep V.110 as it is with the exception of the RA2 function, which is very simple, and has clearly to do only with transportation over 64 kbit/s circuits. Between RA1 functions, the transmission is done at an intermediate rate, 16 kbit/s or 8 kbit/s, which could have been fitted onto the transmission over the radio interface. Yet, the problem on the radio interface is to limit as much as possible the information to be transmitted, so that the maximum part of the raw throughput can be devoted to optimised redundancy, in order to maximise the transmission quality.

When the V.110 bit stream at the intermediate rate is looked at, it becomes apparent that an important part of the exchanged bits can be removed in GSM. The first of these are the synchronisation bits. Out of the 80 bits of a V.110 frame, 17 are used for synchronisation. GSM radio transmission is based on a complex synchronisation scheme, and there is no difficulty in deriving the V.110 frame boundaries from the GSM synchronisation (thanks in particular to the choice of 20 ms as a fundamental GSM synchronisation period, which is a multiple of 5 and 10 ms). In fact, another important aspect of synchronisation comes from the forward error correcting scheme used over the radio interface. With such schemes, residual errors are grouped into bursts, corresponding to an ill-fated radio coding block. There are some advantages to map precisely the V.110 frames and the coding blocks. This is possible because the coding block recurrence has been chosen to be 20 ms. The rule is then simple: a radio coding block corresponds exactly to 2 or 4 V.110 frames.

But 63 bits still remain per V.110 frame. Out of those, three are not transmitted over the radio interface, because they can be reconstructed by the receiver. These are bits E1, E2 and E3, which indicate the true end-to-end data rate. This does not correspond to new information between the mobile station and the infrastructure, since the rate is transmitted separately by signalling means for setting-up purposes, and thus can be dispensed with. What remains consists finally of 60 bit frames, which can be seen as a simple subset of the original V.110 frame.

The resulting "intermediate rate" for GSM is then 12 kbit/s (derived from the 16 kbit/s) or 6 kbit/s (derived from the 8 kbit/s). In fact, a third and lower rate has been introduced for user bit rates below 2400 bit/s, once again to optimise the redundancy. We have seen that in ISDN, rates below 4800 bit/s are rate-adapted to 4800 bit/s by simple bit repetition. In GSM, this simple repetition is done only up to 2400 bit/s. Because the same amount of auxiliary information is kept, the intermediate rate corresponding to user data rates of 2400 bit/s or less is then 3.6 kbit/s (2.4 + 1.2). The "V.110-like" frame in this case is not 60 bits long any more, but 36 bits long. The transformation from ISDN frames at 4800 bit/s to GSM frames is done simply by taking every other user bit. The reverse transformation consists in duplicating each user bit.

Figure 3.22 shows how the radio interface is introduced in the rate adaptation chain, to be compared to the ISDN case of figure 3.21. The RA0 function is performed on the mobile side, as well as the rate adaptation inspired by the ISDN RA1, called RA1'. This includes in the synchronous cases the network independent clocking control as defined in V.110.

On the infrastructure side the RA1'/RA1 function performs the translation between the radio interface format and the ISDN format, and an RA2 function completes the ISDN adaptation, so that the data flow reaching the IWF is in a full ISDN format.

The difference between a V.110 frame and a radio rate adaptation frame is simple, and the translation between the two is easy. It is just a matter of adding (respectively removing) the synchronisation bits, synchronising the V.110 frames with coding blocks; and adding (respectively removing) bits E1, E2, E3, whose contents are known thanks to signalling inside the GSM infrastructure.

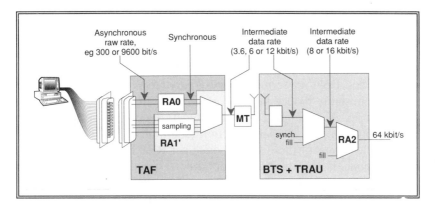

Figure 3.22 – Rate adaptation in GSM

Adaptation functions RA0 (for asynchronous data only)
and part of RA1 (called RA1')
are performed in the TAF (inside the mobile station),
whereas the complement of RA1 and RA2 are performed in the BTS/TRAU.

3.3.3.4. The GSM NT Connections

Since in any case the IWF has a lot to do for an NT connection, there is no reason why GSM NT connections need to strictly follow ISDN specifications as in the T case. The needs are basically to transport a flow of 240 bit frames between the TAF and the IWF, using a maximum total rate of 12 kbit/s. The adaptation to ISDN is done, if need be, at the ends of the connection (TAF and IWF).

However there are some advantages in using as close as possible transmission methods for different modes, and NT transmission has been designed to have a common core with the 9.6 kbit/s T connections. Indeed, the data rate effectively carried between the TAF and the IWF for 9600 bit/s T connections is also 12 kbit/s, as explained above. In addition, the distinction between user data and auxiliary information is irrelevant for the RA1' function or for the transmission over the radio interface. Let us see how these common aspects are exploited.

Let us first start with a difference requiring some explanation. A simple solution for the transmission of the auxiliary information on NT connections would have been to do the same as in the T cases. There would have been no major obstacle to this choice, and the result would have been elegant. A different choice was made. For T connections, auxiliary information adds 12 bits for each period of 5 or 10 ms. This would have resulted in respectively 48 or 24 bits per RLP frame. This was felt to be a high overhead, of which at least a third is useless (bits E4 to E7), since network independent clocking signals are not used in NT connections. Moreover the rest of the bits, which correspond to side signals to and from the terminal, rarely vary. A more complex approach has been chosen to reduce the load incurred by the auxiliary information in most cases. The key fact is that at most, three side signals from the terminal are sampled in each direction (see table 3.6). The idea was to transmit the values of these signals once per RLP frame, plus to indicate the transitions, if any, during the period corresponding to the user bits in the frame. The formatting is such that a minimum of 8 bits is consumed per frame, plus a further 8 bits to indicate a transition. So, if the signals are stable, which is the case when the connection is operational, only one octet per RLP frame is used for auxiliary information instead of 6. If the side signals are often changing, the auxiliary information may use much more than 6 octets per frame, but such cases cannot occur when effective transmission of user data takes place.

Another important point for NT connections is the need for frame delimitation. As usual, frame delimitation can be easily obtained on the radio path as a side product of the comprehensive synchronisation arrangement. Then, despite the difference of frame length (RLP frames are four times as long as V.110 frames for the 9.6 kbit/s T connection), it was possible to use exactly the same transmission scheme for the T and NT cases over the radio interface, and so it was done. (Well! This is somewhat a rewriting of history. In reality, the NT mode was contrived by starting from the T case, wondering how the latter could be used to obtain a better transmission quality!).

Unfortunately the radio path is not the only segment in the way, and the RLP frame delimitation must also be transported between the BTS and the IWF. The V.110 frame delimitation is available, but is not sufficient, the RLP frames being larger. And there like a devil out of its box we find the three E bits: they are free for any use in the case of an NT connection. They are used to convey the frame delimitation.

For T connections, the three E bits are dealt with in the RA1'/RA1 translation. NT connections use a slightly modified RA1'/RA1 translation, which manages the correspondence between the RLP frame delimitation over the radio interface and the one given by the V.110 frame delimitation and the E bits. In the NT case, these bits are used to indicate the position of a V.110 frame in a group of four constituting an RLP block. In the downlink direction, the RA1'/RA1 uses this information to put the four frames of same RLP block in a same radio interface block. The size of the RLP block has been chosen according to the radio block size, such that errors affecting one coding block affect only one RLP frame. Conversely, the E bits are set in the uplink direction according to the position of the contents of the V.110 frame in the corresponding radio block.

The next point to look at is the protocol conversion. It has already been mentioned that the RLP replaces the start/stop protocol or the packet protocol used by the terminal. The RLP between the TAF and the IWF provides the same functionality as the original protocol, but adapted to the GSM transmission. The conversion is done by relay functions in the TAF and the IWF. These functions depend of course on the terminal protocols. There are three cases: the start/stop protocol, LAPB (X.25-2) and the protocol used for fax. The fax protocol is in essence the same as X.25-2 but with additional signalling so that transmission is basically identical and the three cases are effectively only two. The *Specifications* distinguish two relay protocols, namely the L2R-COP (Layer 2 Relay Character Oriented Protocol) and the L2R-BOP (Layer 2 relay Bit Oriented Protocol).

In the asynchronous case, all the functionalities of the start/stop protocol are fulfilled by the NT mode functions. The frame synchronisation enables bits inside the frame to be constructed into octets, thus removing the need for start and stop signals. Only the 7 or 8 bits of user information are transported (1 fill bit is added in the case of 7-bit characters, so that in all cases there are 8 bits per character in the frame). At the receiver's end, the start and stop signals can be reinserted at their correct position. Note that the duration of the stop signal is lost, and as a consequence the relative timing of the characters is lost, but their order is kept. This just makes asynchronous transmission even more asynchronous, and no application is known which suffers from this. Flow control is fulfilled in the start/stop protocol either by using special characters, or by toggling modem control signals. The RLP provides its own flow control, in particular because it needs to regulate the flow when for instance too many repetitions are needed at one môment, which can happen in the case of bad luck. So the start/stop flow control protocol can be relayed by the similar functionality of RLP, once the type of flow control in use is known. A last interesting function of the start/stop protocol is the 'break' signal, which is basically a violation of the start/stop rules (a break signal is a start signal longer than a character, i.e., such that the stop signal arrives too late). The break signal is used basically as a reset mechanism at the disposal of the user, to be used when things are going strangely. A special method is provided in the relay protocol to convey an indication of the reception of a break signal, which is regenerated by the receiver.

A last point for this presentation of the start/stop relay protocol concerns the efficiency of the method. If we take the worst case, which is a 9600 bit/s asynchronous connection with 7-bit characters and 1-bit long start and stop signals, we have a maximum throughput of $9600/9\approx1067$ characters per second, that is 21 and one third characters per RLP frame period. A frame contains as a whole 240 bits, 40 of which being used for error detection, for frame numbering and for the acknowledgement and flow control protocol. A minimum of 8 is used for the modem control signals. There remains 192 bits, that is to say 24 octets (which corresponds to a data rate of 9.6 kbit/s exactly). A benefit of 2 2/3 octets is then obtained, allowing on average, one frame out of 8 to be a repeated frame. For user rates of 4800 bit/s or less, things are obviously much better.

The relay method for protocols which deal with frames, such as HDLC, is rather different. Because the RLP replaces the protocol, the digital stream coming from the external world can be stripped from the overhead introduced by the replaced protocol. In the case of HDLC, this corresponds to the link layer header (2 octets per HDLC frame), the synchronisation overhead (a minimum of 1 octet) and the error detection overhead (2 octets). The remaining data consists of chunks of variable length, which in general do not fit into the fixed length RLP frames. Frame delimitation is then needed, and this is done using a special status octet after the last octet of the content of an HDLC frame. The final gain is then 4 octets per HDLC frame. The relative gain depends on the HDLC frame length. The frame rate in RLP frames being exactly equal to the original rate, the breathing space obtained for repetition is never null, but is greater when the HDLC frames are smaller.

3.3.3.5. The BTS-TRAU Interface

We have now presented the complete connection for data. Or so it seems. A last complication has to be studied, which is the transmission at 16 kbit/s. We have seen that in order for the operator to make economies on the cost of internal links, a scheme has been contrived to transport speech on 16 kbit/s links. The same thing had to be done for data. At first view, it seems that a modified RA2 function would do. But this would not take into account the constraint that only in-band information can be used by the BTS to control the TRAU. The problem is then to distinguish speech from data with in-band information, and there is no room available in V.110 frames for this.

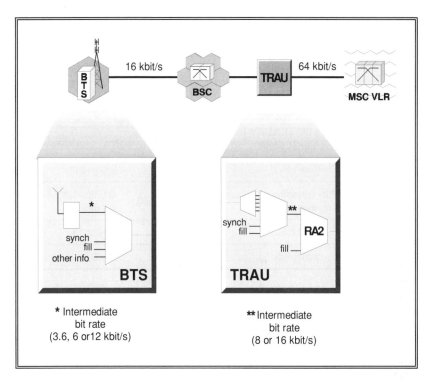

Figure 3.23 – The split between BTS and TRAU for rate adaptation

The transmission of data between BTS and TRAU takes place at a rate of 16 kbit/s, but in-band control of the TRAU imposes an additional specific multiplexing stage.

What has been chosen in order to solve this issue is to specify specific rate adaptation modes on the BTS-TRAU interface, rather different from V.110, but compatible with the BTS-TRAU interface as specified for speech. This is schematically represented in figure 3.23. The bit stream is structured in 320 bit blocks, recurring each 20 ms. The synchronisation and data/speech discrimination has already been described for the speech case (see page 163). A notable difference is that the time alignment control, introduced for speech to prevent an additional transmission delay due to lack of synchronism with the radio path, is not used for data.

To simplify the TRAU, the 63 bits resulting from the stripping of the 17 synchronisation bits from a V.110 frame are kept unmodified. Hence bits E1, E2 and E3 are sent 2 or 4 times in the blocks (whether the intermediate rate is 8 or 16 kbit/s). Thanks to this approach, the TRAU does not need to be told whether the connection type is T or NT, nor has it to bother with RLP frame delimitation.

Role of the bits	Number of bits per frame (uplink = downlink)
Frame synchronisation	35
Discrimination between speech and data, full rate and half rate	5
Intermediate rate on the TRAU/IWF interface (8 kbit/s or 16 kbit/s)	1
Data block	120 or 240
E1, E2, and E3 bits (meaning as indicated in the previous sections)	2 times 3 or 4 times 3
Fill bits	144 or 18
Spare bits	9

Table 3.8 – Contents of a data block on the BTS-TRAU interface

When the intermediate rate is 8 kbit/s, fill bits replace 120 data bits and the missing E1, E2 and E3 bits in the 320 bits block.

This results in the block contents as summarised in table 3.8. The structure is the same in both directions (uplink or downlink).

It is worth noting that, because the specification indicates that the fill bits (in the 8 kbit/s intermediate rate case) are set to "1" as in V.110, the distinction between the 8 kbit/s and the 16 kbit/s cases is totally irrelevant for the transcoder, which can perform the same operations in both cases while still respecting the specification.

SPECIFICATIONS REFERENCE

The topics addressed in this chapter relate to many different *Specifications*. We present them here in the order we think is the best to easily unfold the complexity, in particular for the data services.

The 09 series include a series of *Specifications* devoted to the interworking between GSM and other networks. They are organised by type of network.

TS GSM 09.07 concerns the interworking between GSM PLMNs and the PSTN or the ISDN. **TS GSM 09.06** and **TS GSM 09.05** deal with the PSPDN case, involving communications in an X.25 mode or via a PAD respectively. **TS GSM 09.04** deals with the CSPDN case.

The 07 series includes *Specifications* devoted to the interface between the Mobile Termination and the Terminals. **TS GSM 07.01** is a general introduction to the problem. It presents the different configurations of a mobile station, lists and describes the functions needed in the TAF in all cases, and introduces the rest of the Series. **TS GSM 07.02** and **TS GSM 07.03** deal respectively with the asynchronous and synchronous cases. They contain in particular the description of the relay protocol used in the NT case.

TS GSM 03.45 and 03.46 deal with the facsimile services, when T and NT transmissions respectively are used. This is where all the information concerning the fax adaptor can be found. **TS GSM 03.43** deals specifically with the videotex service. It does not include important specifications, but is interesting also as presenting an example of service provision from a general standpoint.

TS GSM 03.10 describes the different types of connection from TAF to IWF. The bulk of it concerns the data services.

TS GSM 03.50 presents the transmission path for speech communications. It addresses in particular the echo problem.

Then we find the *Specifications* providing the detailed description of the interfaces specific to GSM.

The whole **06 Series** in devoted to speech transmission. Its focus point is **TS GSM 06.10**, which specifies the voice coding algorithm. The other *Specifications* in the Series deal with Discontinuous Transmission, Vocal Activity Detection and comfort noise.

Rate adaptation for the data connections is specified in **TS GSM 04.21**, for the radio interface, and in **TS GSM 08.20** for the A and Abis interface when 64 kbit/s circuits are used. Transmission over 16 kbit/s circuit is described in **TS GSM 08.60**, both for speech and for data.

At a lower level the physical aspects of the transmission over the A and Abis interface are dealt with respectively in **TS GSM 08.04** and **TS GSM 08.54**. Transmission over the radio path is the subject of the next chapter.

The RLP protocol (not including the relay part specific to the start/stop or LAP protocols) is specified in **TS GSM 04.22**.

THE RADIO INTERFACE

THE RADIO INTERFACE

Among all GSM interfaces, the radio interface is probably the most important one, at least if a measure of importance is the time spent for its specification. Several reasons explain why this interface between the mobile stations and the fixed infrastructure plays a leading part in the system specification.

First of all, it needs to be completely specified to achieve a full compatibility between mobile stations of various manufacturers and networks of different operators. This is the key to one of the main goals of GSM, that is to say to obtain MS-roaming, for example allowing users to get service in different countries using a single mobile station.

Second, the spectral efficiency of a cellular system is a key economic factor, and is determined entirely by the transmission over the radio interface. The economic importance of spectral efficiency is best understood when it is defined as the number of cells needed to cover a given area and a given traffic with a given amount of radio spectrum. The better the efficiency the lower the number of cells. Spectral efficiency depends on the number of simultaneous calls which can be fit in the available spectrum, on the resistance of the transmission to interference, which determines the frequency reuse factor, and of a clever exploitation of a number of interference reduction techniques, such as emission power control, detection of silent periods within speech and optimised handover decision methods. The radio interface of GSM is an excellent example of the search for maximum spectral efficiency.

Two main aspects of the transmission over the GSM radio interface will be studied, the multiple access scheme, and the signal processing, from bits to radio waves. The multiple access scheme describes the way the radio spectrum is shared between several simultaneous communications, occurring between different mobile stations, in different cells. In GSM, after epic discussions, the choice settled on a mixed Frequency Division and Time Division multiple access (FDMA and TDMA), spiced with frequency hopping.

FDMA, which is used in particular to share the spectrum between neighbouring cells, is of a medium bandwidth type (200 kHz). The basic TDMA factor, i.e., the number of calls obtained from a 200 kHz frequency band, is 8 for data at 9600 bit/s and for speech in a first phase, and will be 16 for speech in the near future. The TDMA scheme used in GSM is an interesting feature of the system, being based on complex cycles. The final ingredient of multiple access, frequency hopping, which consists in regularly changing the frequency of transmission for a given mobile station to base station connection, has been introduced for improving quality at low speeds as well as spectrum efficiency.

Signal processing in GSM also presents interesting features. The digital modulation of the 900 MHz or 1800 MHz radio waves is quite classical, being Gaussian Minimum Shift Keying (GMSK). A less classical point is the emission and reception on a burst basis, with equalisation adapted burst per burst. GSM also makes use of various digital error correction codes with interleaving, with the intent to provide an excellent compromise between the spectrum usage of a channel and resistance to interference. This part of the presentation will also be the occasion to have a glance at ciphering.

But before dealing in details with all these technical aspects, we will first consider the needs which were to be fulfilled, and which indeed explain the various design choices. The previous chapter has already identified the different kinds of user data flow that the system is able to transport. Other transmission needs, related to the exchanges between the mobile stations and the infrastructure for control purposes, have yet to be introduced.

4.1. THE NEEDS

There are several reasons why data is transmitted to and from mobile stations. The purpose of this section is to identify them, but in doing so we will do a bit more. To make easily the link between the needs and the transmission scheme, we will introduce at this stage the concept of **channel**. Different channels are used to transmit different simultaneous transmission flows. The methodology used hereafter is to present each identified need in turn, introducing the corresponding channel types used in GSM to fulfil the need.

4.1.1. USER DATA TRANSMISSION

The very first aim of a communication system is to transport user information. The first service to be offered to GSM users is speech. The radio interface must therefore accommodate bi-directional speech transmission. In order to limit the use of the radio spectrum, speech is represented by a binary signal at a rate of 13 kbit/s, through a coding scheme specific to the system and described in the previous chapter.

Data services also require bi-directional transmission means. GSM only accommodates fairly small data rates: the raw data to be transmitted is a binary flow at a rate of 12, 6 or 3.6 kbit/s depending on the given service. These rates allow data services which are similar to those offered by a PSTN through modems using respectively 9.6, 4.8 and 2.4 kbit/s rates.

Most of the user services offered by GSM rely on one of these four transmission modes (speech and 3 data modes). One exception, however, consists in the transfer of short messages, which is implemented on the radio path by methods derived from those used to transfer signalling. Point-to-point short messages use means identical to signalling transmission, and will therefore be described together with the means which support signalling, in Chapter 5. Cell Broadcast short messages make use of a special channel, the CBCH (Cell Broadcast Channel), which will be dealt with in a next section.

For all other services than short messages, a user engaged in a communication must have at his disposal a portion of the radio interface uniquely devoted to the call for its duration. This corresponds to a special kind of channels: the **TCHs** (Traffic Channels). From the multiple access point of view, two kinds of TCHs are defined:

- the **TCH/F** (where F stands for "full rate") allows the transmission of 13 kbit/s speech or of data at 12, 6 or 3.6 kbit/s;
- the **TCH/H** (where H stands for "half rate") allows speech coded at a rate around 7 kbit/s to be transported (the specification of this coding is currently under way), or data at a rate of 6 or 3.6 kbit/s.

4.1.2. SIGNALLING

User data is not the only flow of information to be transported during a call. Signalling messages must also be conveyed; they allow the mobile station and the network to discuss the management of several issues either related to the user (e.g., call in progress indications) or concerning technical aspects of the communication (e.g., preparation and execution of handover). The establishment and release of a call also require signalling exchanges. In addition, we will see that some information exchanges between mobile stations and the infrastructure are needed even when no communication is in progress.

4.1.2.1. Signalling in Connection with a Call

In order to transport signalling data in parallel with the transmission of a user data flow, GSM offers two possibilities. Each traffic channel comes with an associated low rate channel, used for the transport of signalling: the SACCH (Slow Associated Control CHannel). This bi-directional channel may carry about 2 messages per second in each direction, with a transmission delay of about half a second. It is used for non-urgent procedures, mainly the transmission of the radio measurement data needed for the decisions concerning handover.

The other needs for associated signalling, such as the messages to indicate the call establishment progress, or to authenticate the subscriber, or to command a handover, to give only a few examples, make use of the TCH itself. This usage shall be referred to here as **fast associated signalling**, in place of the term FACCH (Fast Associated Control CHannel) used in the *Specifications*. Fast associated signalling does not indeed refer to a channel in our use of the term, but to a particular use of a TCH.

The receiver is able to distinguish both uses by reading a binary information transmitted on the TCH and called the stealing flag (see page 237). During the initialisation and release phases, no user data is transmitted and therefore signalling may use the channel without any conflict with other types of data. However, during the call, the transmission of fast associated signalling is done to the prejudice of user data: there is a loss of user data, as if transmission errors (of known location) had occurred. Hence the term "stealing".

4.1.2.2. Signalling outside a Call

In some cases, there is a need to establish a connection between a mobile station and the network solely for signalling matters, be it at the user's demand (e.g., call forwarding management, transmission of short messages) or for other management needs, such as location updating.

A TCH/F or a TCH/H may be used for that purpose. This is however wasting a lot of spectrum, since few messages are to be exchanged (typically half a dozen each way) and the channel would only be used for a very small portion of the time.

In order to increase system efficiency, an additional type of channel has been introduced. Its rate is very low and the only specified usage—so far—for this channel is signalling (and short messages transmission). This channel shall be referred to here as **TCH/8**, where "8" stands for "eighth". If a TCH/H may be considered as half a TCH/F, then this small channel is one eighth of a TCH/F. This terminology diverges from the GSM one, where this channel is called SDCCH (Stand-alone Dedicated Control CHannel, not a very enlightening term). The characteristics of such a channel are in fact very close to those of a TCH/F or H, the prevalent difference being its rate. The choice of vocabulary in the *Specifications* stems from the fact that no user data mode has been specified for the TCH/8. However, this does not appear the most pertinent distinction, given the definition of a channel (see page 193)! Moreover, nothing would fundamentally preclude an evolution of the *Specifications* to use a TCH/8 as a traffic channel, for instance for low rate data services like telemetry. A pragmatic reason for this change of terminology is to avoid in many places "TCH or SDCCH".

As the other TCHs, a TCH/8 is bound to an SACCH. The notion of fast associated signalling also applies to a TCH/8, but for the moment it is a trivial matter, since it is the only usage of a TCH/8.

4.1.3. IDLE MODE

The rarity of the radio spectrum does not allow each user of the system to have his own TCH at all times. Traffic channels are therefore allocated to the users only when the need arises. This leads us to the basic distinction of **dedicated mode** and **idle mode**, which is an essential concept in radiotelephony.

Formally, a mobile station shall be considered in dedicated mode when a TCH is at its disposal. This corresponds to the phases when full bi-directional point-to-point transmission is possible between the mobile station and the infrastructure, for instance during a call established for the user, or to perform a location updating. TCHs and SACCHs are therefore referred to as dedicated channels in the *Specifications*.

When a mobile station is active (powered on) without being in dedicated mode, it is deemed to be in idle mode. However, the mobile station is far from being idle... It must continuously stay in contact with a base station, listen to what this base transmits in order to intercept paging messages (to know if its user is being called), and monitor the radio environment in order to evaluate its quality and choose the most suitable base station. In addition, there is one telecommunications service which is provided to the mobile stations in idle mode: the Cell Broadcast Short Message service.

The transition between idle mode and dedicated mode requires some information exchanges between the mobile station and the base station (the "access" procedure). The mobile station indicates to the network that it needs a connection and the network indicates in return which dedicated channel it may use.

All of these uses require specific transmission means which are grouped under the terminology "common channels".

4.1.3.1. Access Support

In order to communicate with a base station, a mobile station must first become (and stay) synchronised with it. Two channels are broadcast by each base station to this avail: the **FCCH** (Frequency Correction CHannel) and the **SCH** (Synchronisation CHannel).

Mobile stations in idle mode require a fair amount of information to act efficiently. Most often, a mobile station can receive, and potentially be received by, several cells, possibly in different networks or even in different countries. It has then to choose one of them, and some information is required for the choice, for instance the network to which each cell belongs. This information, as well as much other, is broadcast regularly in each cell, to be listened to by all the mobile stations in idle mode. The channel for this purpose is the **BCCH** (Broadcast Control CHannel).

Let us now come to the access. We have seen that paging messages have to be broadcast to indicate to some mobile stations that a call toward its user is being set up. The access procedure itself includes a request from the mobile station and an answer from the base station, allocating a channel. Paging messages and messages indicating the allocated channel upon prime access are transmitted on the **PAGCH** (Paging and Access Grant CHannel). This terminology combines the two terms "PCH" (Paging CHannel) and "AGCH" (Access Grant CHannel) used in the *Specifications*. Referring to two channels in that case is not consistent with the fact that the partition between PCH and AGCH varies in time, hence the concatenation in notation.

All the common channels listed above (FCCH, SCH, BCCH, PAGCH) are "downlink" unidirectional channels, i.e., they convey information from the network to the mobile stations. The last type of common channel, which allows the mobile stations to transmit their access requests to the network, is an "uplink" unidirectional channel. It is called the **RACH** (Random Access CHannel). Its name indicates that mobile stations choose their emission time on this channel in a random manner. This results in potential collisions between the emission of several mobile stations, and will be studied in detail later.

4.1.3.2.   The Cell Broadcast Short Messages

Cell Broadcast short messages require the means to transmit around one 80 octet message every two seconds from the network towards the mobile stations in idle mode. This corresponds to half the capacity of a downlink TCH/8. In each cell where this service is supported, a special channel, a **CBCH** (Cell Broadcast CHannel) is used for broadcasting messages. A CBCH is derived from a TCH/8. Some special constraints exist for the design of this channel, because of the requirement that it can be listened to in parallel with the BCCH information and the paging messages.

4.1.3.3.   A Point of Terminology: What is a Channel?

As noted a few times in the previous sections, we introduced some new terms in this book, and we will avoid some other terms though they are of common use in the *Specifications* and in the GSM literature. This requires some justifications. The issue revolves around the notion of channel, which in our opinion is not used consistently in the GSM literature.

CCITT officially defines a channel as "an identified portion of an interface". Though it can be debated at length, we interpret this definition in the way described below, with which the terminology in the Specifications does not always comply.

First, a channel is defined by its characteristics in terms of transmission support, and not by its potential uses. Very often a channel is dedicated to only one usage, hence the mixture of concepts.

Second, it should always be possible, before reception, to know the channel to which a given part of a data flow belongs, without having to look at the contents of this data flow. The knowledge of the characteristics of the channel (timing, frequency) should enable the recipient of the information to know whether the data may be for him or not. This method is used, e.g., in the subscriber's loop of a fixed telephony network (the line itself identifies the subscriber). Another means to identify the destination of data would be to include explicitly the destination's address in the data. This method is the one used, e.g., for postal dispatches, where the address of the recipient is written on the letter or package, hence the term "addressee". In the latter case, we prefer not to speak of a channel for each address, but of a single channel shared between the different flows. One advantage of defining a channel independently from the information carried by the channel is when describing the allocation of channels, or a configuration of channels. In such cases, what matters is the description of the part of the interface which can be used for a given usage.

The GSM terminology mixes the concept of channel with its usage, and also mixes a channel (in the sense defined above) with a data flow identified by an internal address. In our view those approximations are detrimental to the grasping of the underlying concepts. Two major examples for the application of these principles have been encountered in the previous sections. First, the term PAGCH is used in this book for the type of channel conveying both the paging flow and the channel allocation flow, instead of the GSM terminology PCH (for the "channel" carrying all the paging flow and some of the channel allocation flow) and AGCH (for the complementary "channel" carrying only channel allocations). Second, the term FACCH used in the *Specifications* will not be used here, since it refers to a specific use of a channel (the TCH) but not to a pre-defined portion of the resource, and we will prefer to refer to fast associated signalling as a mode of use of the channel. The FACCH gives an example of the conceptual difficulties: to which channel does the stealing flag belong?

We understand that diverging in terminology from the *Specifications* (or for that matter from the established usage in some field, as our using emission where most often transmission is used) will cause some difficulties. On the other hand, we hope that the new terms help the reading of this book, and that the balance is beneficial for the overall understanding of the GSM concepts.

4.2. THE MULTIPLE ACCESS SCHEME

Many techniques have been devised to create channels out of a communication medium, in order to use them as separate links. Using the standard terminology, the radio interface of GSM uses a combination of Frequency Division Multiple Access (FDMA) and Time Division Multiple Access (TDMA), with a pinch of Frequency Hopping for flavour. But the resulting mixture is something different than just the juxtaposition of the ingredients.

A basic concept of the GSM transmission on the radio path is that the unit of transmission is a series of about a hundred modulated bits, and is called a **burst**. Bursts have a finite duration, and occupy a finite part of the radio spectrum[1]. They are sent in time and frequency windows, that we will call **slots**. Precisely, the central frequencies of the slots are positioned every 200 kHz within the system frequency band (FDMA aspect), and they recur in time every 0.577 ms—or more exactly, every 15/26 ms (TDMA aspect). All slot time limits are simultaneous in a given cell. A time interval corresponding to simultaneous slots will be referred to as a time slot, and its duration will be used as a time unit, designated **BP** (Burst Period).

---

[1]   In practice, there are specification and implementation tolerances, and some energy is sent outside the finite spectrum window defining the slot.

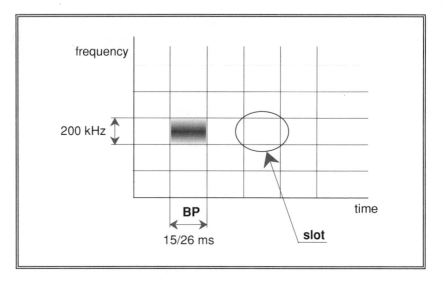

Figure 4.1 – A slot in the time and frequency domain

A transmission quantum in GSM (a burst)
fits in a time and frequency window called a slot,
which lasts about 577 μs and occupies a bandwidth of 200 kHz.

A slot may therefore be pictured in a time/frequency diagram as a small rectangle 15/26 ms long and 200 kHz wide (see figure 4.1). Similarly, we will call a frequency slot (a radio frequency channel in the *Specifications*) a 200 kHz bandwidth as specified for GSM.

The above paragraph is heavy with terms and definitions. Once again, there is some difference between the choice of terms here and the one of the *Specifications*, because the terms such as timeslot or burst are used with several different meanings in the *Specifications*. For instance, burst refers sometimes to the unit time-frequency "rectangle", and sometimes to its contents. Likewise, timeslot is used to mean either what we call here "slot", or its time value, or also the cycle using one slot every eight slots in time.

To use a given channel means to transmit bursts at specific instants in time, and at specific frequencies, that is to say in specific slots. To define a channel consists then to specify which slots can be used by, or are part of, the channel. Usually the slots of a channel are not contiguous in time. A channel has therefore a temporal definition giving, for each time slot, the number of slots which are part of the channel. For the time being—and probably for quite a while...—a channel may only consist of 0 or 1 time/frequency slot per time slot,. The temporal definition of a channel is cyclic, that is to say it repeats itself after some time. The simplest case of a cyclic definition is "1 burst every n". As will be seen on the next pages, the GSM channels are rarely as simple.

In parallel to the time definition, the frequency definition of a channel gives the frequency of every slot belonging to a channel. It consists basically of a function allocating a frequency to each time slot where a channel has a slot. There exists fixed frequency channels, for which the frequency is the same for every slot, and frequency hopping channels, whose slots may use different frequencies.

For bi-directional channels (e.g., a TCH), the two directions could have been defined in different ways. However, for simplicity reasons, the channel definitions for the two directions are always related in a very elementary manner: a fixed frequency gap (the "duplex separation") of 45 MHz (in the 900 MHz band) and of 75 MHz (in the 1.8 GHz band) and a time shift, which depends on the channel type, separate two corresponding slots of a given channel. In more academic terms, one direction is related to the other by a translation in time and in frequency.

4.2.1. THE TIME AXIS

The organisation of a channel along the time axis can be quite complex. This organisation is always cyclic, but the length of the cycle as well as the number of slots in a cycle vary according to the type of channel.

The positioning of the cycles in time is achieved through system synchronisation. Each cell provides a reference clock, defining the time slots, but also a "dating" scheme to which the cycles of all the channels are referred. In GSM, each time slot (and hence all the corresponding slots on the different frequencies) is given a number, which is known both by the base station and the mobile station, and which is part of the synchronisation information. The description of a given channel (e.g., sent by the base station to the mobile station) refers to this numbering scheme. The time slot numbering is cyclic, but of a very long cycle (3.5

hours), which has been chosen as a multiple of all cycles needed for multiple access. Within each period, any slot can be unambiguously referred to by its time slot number and its frequency slot number.

4.2.1.1. Dedicated Channels

TCH/F and its SACCH

A TCH/F is always allocated together with its associated slow-rate channel (SACCH). The resulting group of channels bears unfortunately no specific name in the *Specifications*, and is therefore often confused with the TCH/F part only. Because we found the distinction useful, and to avoid long paraphrases, we choose to introduce a non GSM term. The group TCH/F plus SACCH shall be referred to in this book as a **TACH/F**.

A TACH/F has a simple cyclic definition. It consists of one slot every 8 BP in each direction, i.e., one slot every 4.615 ms (or, better, 60/13 ms). A consequence of this definition is that it consists in the slots whose time slot number is 8 times an integer plus a value k between 0 and 7 specific to the channel. This value k is the phase modulo 8 of the numbers of the slots of the channel. It is called in the *Specifications* the *Time slot Number* (TN) of the channel. Let us point out that this meaning of *Time slot Number* is not the same as a few lines above, but TN is the GSM term and is the abbreviation of *Time slot Number*. In the following, we will use TN only in short form, to limit the confusion.

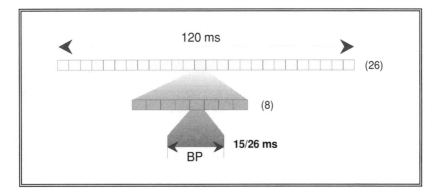

Figure 4.2 – Choice of value of a burst period

The exact value of a burst period is derived from a multiple of 20 ms by applying two multiplexing schemes (× 26, × 8).

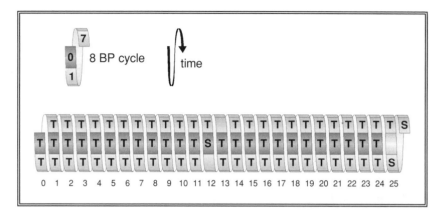

Figure 4.3 – Time organisation of a TACH/F

A full rate TACH uses a cycle of 26 × 8 slots. Within each of these cycles,
24 slots are used for the TCH and one slot for the corresponding SACCH.
In order to spread the arrival of SACCH messages at the base station,
the cycles of two TACHs using successive slots are separated by 97 BPs
(i.e., 12 × 8, plus the difference of one slot).

Eight different types of TACH/F can therefore be defined depending on their phase modulo 8 (their TN). Two TACH/F with the same phase consist of simultaneous slots.

On the network side, 8 TACH/F with different TNs may be emitted by a single transmitter: at any given moment, a single TACH/F of the group is to be emitted. This is the core of the concept of TDM (Time Division Multiplex). This is an important aspect of the system, leading to substantial savings in the base station.

A TACH/F contains a TCH/F and an SACCH. The split between these two channels is also specified in the time domain, using a cycle of 26 TACH/F slots, i.e., 26 × 8 successive slots, or a period of 120 ms. The value 120 ms is an exact one; which was chosen as a multiple of 20 ms in order to obtain some synchronism with fixed networks, ISDN in particular. We then can explain the reason of the strange value of BP: the *BP* is therefore exactly 120/26 × 8 ms, i.e., 15/26 ms (see figure 4.2).

The TACH/F 26 slot cycle includes 24 slots on which TCH/F bursts are sent, 1 slot on which an SACCH burst is sent, and one slot where no transmission takes place (see figure 4.3).

Coding follows cycles based on the grouping of 4 successive bursts, as will be explained on page 247. For the TCH/F, a cycle contains 6 times 4 bursts. However, for the SACCH, the full cycle, taking into account this grouping 4 by 4, lasts 4 × 26 × 8 = 104 × 8 BPs, i.e., 480 ms.

The position in time of the full 104×8 cycle of a TACH/F is the same for all the TACH/F of the same TN in a given cell. We will try to describe the relationship between the different TN in simple terms. The position in time of the TCH/F bursts (by opposition to the SACCH and unused slots) follows a 13×8 BPs cycle. The beginnings of this cycle are almost simultaneous for the TCH/F of different TNs, in the sense that the first slot of the 13×8 cycle of a TCH/F of TN 1 follows immediately the corresponding slot for a TCH/F of TN 0. Very exactly, they follow each other from 0 to 7. But the start of the SACCH cycle for the different TNs do not happen in the same 8 BP interval. Still, all the TACH/F have the same definition in the time domain, except for a translation in time. This translation invariance property is important for the design of a mobile station: it means that the scheduling of the treatment for all TACH/F is the same, the only impact of the TN is that the scheduling starts at different moments.

The reason for the shifting of the 104×8 BP cycles comes from load considerations in the infrastructure. If the SACCH had been almost simultaneous, the base station would have received the SACCH messages from all the mobile stations almost simultaneously every 480 ms, resulting in a very uneven load. In order to avoid that drawback, the cycle of a TACH/F using TN n+1 (for n from 1 to 7) is shifted $(12 \times 8) + 1 = 97$ BPs from the one of a TACH/F using TN n. This shift can be seen in figure 4.3; since four bursts are necessary to build an SACCH message, the base station will process the SACCH messages corresponding to 8 TACH/F of TN 0..7 at 8 different moments evenly spread in time.

Relationship between Uplink and Downlink

As seen from the base station point of view, the organisation in the uplink direction is derived from the downlink one by a delay of 3 BPs. This delay of 3 BPs is a constant throughout GSM. In fact, the convention is that the numbering of the uplink slots is derived from that of the downlink ones by a shift of 3 BPs; this choice allows the slots of one channel to bear the same TN in both directions.

But the point of view of the mobile station is affected by considerations about propagation delays which, even at the speed of light, are not negligible compared to the burst duration (the round-trip delay between an MS and a BTS 30 km apart is 200 µs). In a first step, we will ignore the problems due to propagation delays, and consider that a mobile station very close to the BTS sees the 8 BP cycle as shown in figure 4.4.

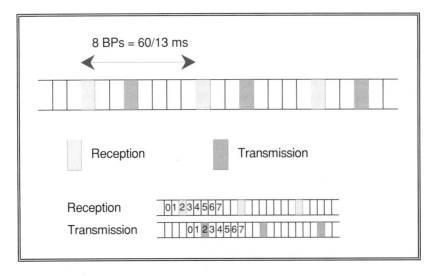

Figure 4.4 – Cycle of 8 BPs seen by a mobile station close to the base

Emission by a mobile station happens 3 BPs later than Reception.
Typically, a mobile station will receive during one time slot, then shift in frequency
by 45 MHz to emit some time later
(3 burst periods minus the correction in time for propagation),
then possibly shift again to monitor other channels, and move
to the adequate receive frequency to start the cycle all over again.

Such a choice allows mobile stations to avoid emitting and receiving simultaneously, thereby promoting easier implementations: the receiver in the mobile station need not be protected from the emitter of the same mobile station.

When the mobile station is far from the BTS, propagation delays cannot be neglected, and an exact 3 BP shift cannot be maintained both at the MS and at the BTS. But it is imperative that the bursts received at the BTS fit correctly into the time slots, and as we will see this is not very roomy. Otherwise, the bursts from mobile stations using adjacent time slots could overlap, resulting in a poor transmission quality or even in a loss of communication. The only solution is that the mobile station advances its emission relatively to its reception by a time compensating the to and fro propagation delay. This value is called the **timing advance**. The exact shift between downlink and uplink as seen by the mobile station is then 3 BP minus the timing advance. The timing advance value can be computed only by the BTS, and is then provided to the mobile station through signalling. This is why we have chosen to deal completely with this subject in Chapter 6.

TCH/H

The half-rate TCH, or TCH/H, has led an eventful life. Until the beginning of 1991, a TCH/H was fully specified for data services, but not yet for speech. The following description will refer to that specification. However, as explained in the previous chapter, "half-rate" speech coding studies (i.e., studies aiming at defining a speech coding scheme adapted to transmission on TCH/Hs), which were under way in early 1991, have shown that the time structure chosen for TCH/Hs was not necessarily optimum. It was therefore decided that the half-rate TCH was no longer part of the phase 1 *Specifications*, even for data services.

As for a TCH/F, a TCH/H is always allocated together with its SACCH, and this group of channels will be referred to as TACH/H. A TACH/H is defined in the time domain as consisting on average of one slot every 16 BPs, hence the "half-rate" compared to a TACH/F. The term "on average" is important here, because it is not true that all the TACH/Hs are defined as exactly each sixteenth slot, as the TACH/F is defined as exactly each eighth slot. It is true for half of the TACH/Hs, but not for the other, for whom the cycle is 13 × 16 BP long. The cycle is shown in figure 4.5. It should be noted however that in both cases, a TACH/H consists only of slots of the same TN. This state of affairs is unfortunate for the simplicity of the description, and for the mobile station implementation (the two categories of TACH/H differ by more than a translation). It has no clear reason, and will quite likely be modified before half rate channel specifications are completed.

As far as the time definition is concerned, there are 16 kinds of TACH/H, which are usually referenced by their TN (8 values) plus a sub channel number (sub-TN, 0 or 1). For a TACH/H of sub-TN 0, the time slots of the TCH/H are of even number modulo 16,

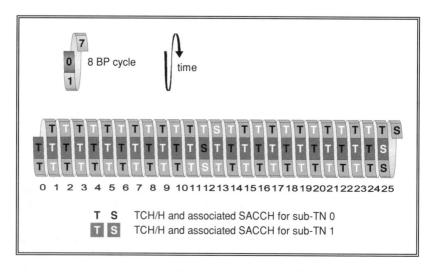

Figure 4.5 – Time organisation of TACH/Hs

A TACH/H, as defined provisionally, uses one slot every 16 in average.
On "even" TNs, they use exactly one slot every 16.
On "odd" TNs, the scheduling is not so regular,
as can be seen on TN 1 or 7 (occurrences 11, 12, 13 for the channel of sub-TN 1,
and for occurrences 24, 25, 0 for the channel of sub-TN 0).

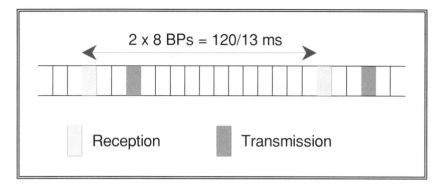

Figure 4.6 – Cycle of 16 BPs seen by a mobile station

A mobile station using a TACH/H performs the same operations as for a TACH/F
(same timing between reception and emission),
but does so only every other group of 8 bursts.

whereas they are of odd number in the case of sub-TN 1. Now TACH/H of TN 0, 2, 4, 6
are well-behaved, in the sense they use exactly each sixteenth slot, whereas the others
have the irregular cycle.

The crafting of the TACH/H was done so that the two TACH/H of the same TN
but of different sub-TN have no simultaneous slot, and then may be grouped to form the
equivalent of a TACH/F. Such TACH/H pairs and/or TACH/Fs may then be grouped in
8s for a single transmitter in a base station. A transmitter may accommodate up to 16
TACH/Hs, or more generally a combination of 8 sets, each set containing either a
TACH/F or a pair of TACH/Hs.

The split between a TCH/H and its SACCH is done for the well behaving
TACH/H along a cycle of 13 TACH/H slots, i.e., 120 ms as for the TACH/F. The cycle
contains 12 slots for the TCH/H bursts and one for the SACCH. All the slots of the
TACH/H are then used for transmission. For the other category of TACH/H, the cycle is
less regular, as shown in figure 4.5 for TNs 1 or 7.

As for the full-rate channel, a complete cycle lasts 480 ms (104 × 8 BPs) when
coding is taken into account, and the start of cycles is defined in the same way. The
reader may have noticed that the SACCHs have the same time organisation as those
associated with TCH/Fs. Half of them (those with sub-TN 0) are emitted exactly as if
they were associated with a TCH/F of the same TN, the other half (those with sub-TN 1)
being emitted during the slot which would be free in the cycle of a TCH/F of the same
TN.

The timing offsets between TACH/Hs of different TNs is similar to the one
defined for TACH/Fs: a TACH/H of TN n+1 is shifted 97 BPs compared to the TACH/H
of TN n and of different sub-TN, so that it is shifted twice 97 BPs compared to a
TACH/H using TN n and the same sub-TN. The uplink and downlink directions are also
related in the same manner as those of full-rate channels: an uplink slot follows a
downlink slot 3 BPs later (at the base station). As seen by the mobile station, the 16 BP-
cycle of a well-behaved TACH/H can be pictured as shown in figure 4.6.

TCH/8

The description of a TACH/8 (i.e., a TCH/8 and its SACCH) is somewhat more complex than the one of the full, and even of the half rate TACHs, because there exists many different kinds of TACH/8s from the point of view of the time organisation:

- some may be grouped by 8 to form the equivalent of a TACH/F: they are called SDCCH/8 in the GSM terminology;
- others may be grouped by 4 and combined with common channels to form all together the equivalent of a TACH/F: they are called SDCCH/4 in the GSM terminology.

All TACH/8 have many properties in common: they all follow a cycle of 102×8 BPs, where 8 slots are used for the TCH/8 bursts (a group of 4 slots separated by 8 BPs, then 51×8 BPs, then again a group of 4 slots separated by 8 BPs, ...) and 4 slots are used for the SACCH bursts (one group of 4 slots separated by 8 BPs). The attentive reader will have noticed that the length of the TACH/8 cycle (102) bears no simple relationship with the TACH/F cycle (26, and 4 times 26 is 104). The origin of this choice lies in the possibility to associate 4 TACH/8s with common channels; the latter follow a 51×8 BPs cycle, as will be explained in a few paragraphs. In order to keep some homogeneity between the different TACH/8s, this cycle has been used also for the TACH/8s crafted to be grouped by 8, though a cycle of 52×8 could have as easily been chosen. This difference between the length of the TACH/8 cycle and the one of the TACH/F or /H results in slightly different rates (2% difference) for the corresponding SACCHs.

The TACH/8 vary in their phase relations between the TCH slots and the SACCH ones, as well as between the uplink and downlink directions. Figure 4.7 shows the time organisation for both categories of TACH/8.

From the figures it is obvious that the TACH/8 cannot be derived one from each other by a simple translation in time. The result is that there are 12 different schedulings for mobile stations in connection on a TACH/8. In fact the figure shows 4 cases (2 in the case of grouping by 8, and 2 in the case of grouping by 4), but the notion of measurement reporting period (dealt with in Chapter 6) results in 12 classes (the same notion does not have similar impacts on the TACH/F and the TACH/H).

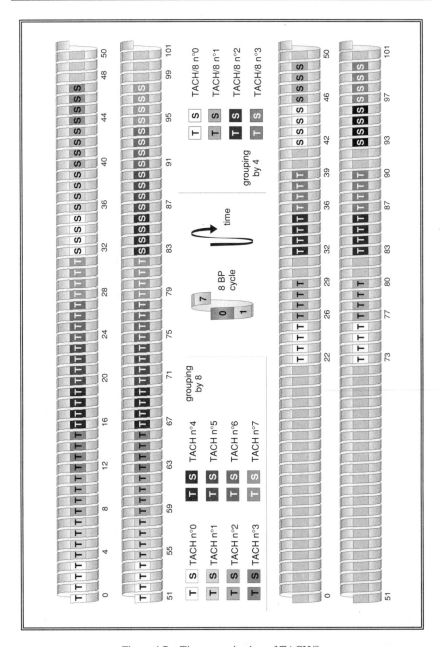

Figure 4.7 – Time organisation of TACH/8s

TACH/8s can be grouped by 4 or by 8 in a cycle of 102 × 8 BPs.
During this cycle, 2 blocks of 4 slots are used for a TCH/8
and 1 block of 4 slots for its SACCH.
The picture shows the downlink cycles.
The uplink cycles can be obtained by a simple translation of 66 × 8 BPs.

4.2.1.2. Common Channels

General Organisation

All common channels have been defined with the intention of grouping them together in few combinations. Their time definitions are therefore all based on the same cycle, i.e., 51 × 8 BPs.

This cycle and the cycle of traffic channels were deliberately chosen with different values (in fact, they were chosen not to have any common divider) in order to allow mobile stations in dedicated mode to listen to the synchronisation channel (SCH) and frequency correction channel (FCCH) of surrounding base stations, both of which carry the information needed for mobile stations to become and stay synchronised with a cell. With the numerical relationship between the cycle of the common channels and the cycle of the TACH/F or H, the bursts of the common channels file off one after the other in front of the reception windows of the mobile stations situated in surrounding cells, and in particular in front of the large window left open in a TACH/F cycle by the unused slot. These mobile stations are then able to receive—at least from time to time—a burst belonging to the FCCH or to the SCH. By so doing, they acquire the synchronisation information they need on surrounding cells, whatever the relation between time bases of neighbouring base stations (which may be anything indeed). This functionality is referred to as "pre-synchronisation" and is part of handover preparation: the reader will find more on that topic in Chapter 6.

FCCH and SCH

Both the *FCCH* and the *SCH* have the same time structure: one

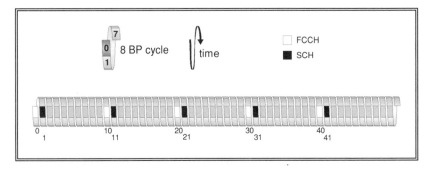

Figure 4.8 – time organisation of the FCCH and SCH

Once a mobile station has found an FCCH burst, it knows that the SCH slot can be found 8 BPs later on the same frequency.

SCH slot follows each FCCH slot 8 BPs later. Each of those two channels uses 5 slots in each 51×8 BPs cycle, as shown in figure 4.8.

A single set (FCCH + SCH) is broadcast in any given cell. In all cells, the slots of these channels have the same position within the 8 BP cycle, that is to say the same TN. This position is *by definition* called **TN 0**. Indeed no burst on the radio interface carries its TN, and the mobile station knows the TN of a slot only by reference to the FCCH and SCH. Every burst of the SCH indicates the remaining part of the time slot numbering (i.e., the remainder modulo 8), thereby enabling mobile stations to derive the numbering of all slots within the cell.

BCCH and PAGCH

Apart from the SCH and FCCH, the other downlink common channels include the BCCH and the PAGCH, introduced on pages 192 and 193. The difference between these two channels lies more in their usage than in their transmission characteristics. It is indeed possible that, in later phases of the system, their respective size may be allowed to vary, e.g., in order for the BCCH to gain some capacity at the expense of the PAGCH capacity.

Two kinds of *PAGCH* are defined, which have different capacities, hence which use a different number of slots per cycle. They do not bear a specific name in the *Specifications*; they will be called here PAGCH/F ("full") for the larger one and PAGCH/T ("third") for the smaller one, by analogy with the dedicated channels terminology.

A BCCH together with a PAGCH/F uses 40 slots per 51×8 BP cycle, all with the same TN. Figure 4.9 shows how these slots are spread. These 40 slots are built as 10 groups of 4, the four slots of one group

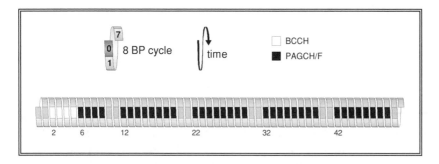

Figure 4.9 – Time organisation of a BCCH and a PAGCH/F

The BCCH uses 4 slots per 51×8 BP cycle, and 36 slots are dedicated to the Paging and Access Grant channel.

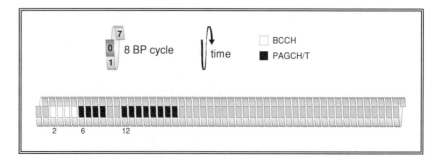

Figure 4.10 – Time organisation of a BCCH and a PAGCH/T

A PAGCH/T uses 12 slots per 51 × 8 BP cycle, i.e., one third of a PAGCH/F.

being separated by 8 BPs and containing bursts from a single coding block. The BCCH uses the first quartet and the PAGCH uses the 9 other quartets.

A BCCH together with a PAGCH/T uses 16 slots per 51 × 8 BP cycle, all with the same TN. Figure 4.10 shows how these slots are spread. These 16 slots are organised as 4 groups of 4. The BCCH uses the first quartet and the PAGCH uses the 3 other quartets (hence one third of the PAGCH/F).

RACH

As for the PAGCH, two types of *RACH* exist: the RACH/F and the RACH/H. The RACH/F uses one slot every 8 BPs, so that its time organisation is similar to the one of a TACH/F in the uplink direction.

On the other hand, a RACH/H only uses 23 slots in the 51 × 8 cycle, and its capacity is therefore slightly more than half the one of a RACH/F. The time organisation of a RACH/H is shown in figure 4.11 and allows the combination of a RACH/H with 4 TACH/8s.

Common Channels Combinations

Every cell broadcasts one single FCCH and one single SCH. As far as the BCCH, PAGCH and RACH are concerned, every cell supporting mobile access must have at least one of each (one may however imagine cells accessible only through handovers, in which case these channels would not be necessary).

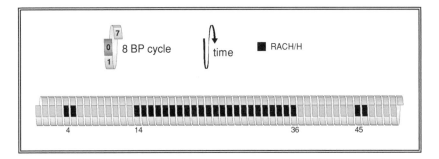

Figure 4.11 – Time organisation of a RACH/H

A RACH/H fits in the bursts left free in the uplink by 4 TACH/8s,
i.e., it uses 27 bursts during each 51 × 8 BP cycle.

In order to save spectrum, common channels are always arranged
in groups; there are 3 such possible combinations.

The basic combination includes (in the downlink direction) a
FCCH, a SCH, a BCCH and a PAGCH/F, all of the same TN, namely 0.
The uplink direction contains a RACH/F. This arrangement is shown in
figure 4.12. All these channels together use the same amount of resources
as a TACH/F, which allows a single base station transmitter to manage
such a combination plus 7 TACH/Fs (of TNs 1 to 7).

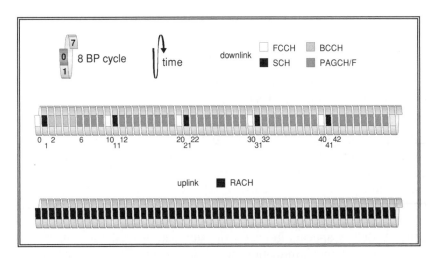

Figure 4.12 – Basic common channel pattern

A typical medium cell common channel pattern uses TN 0 of one carrier
for FCCH, SCH, BCCH, PAGCH/F and RACH/F.

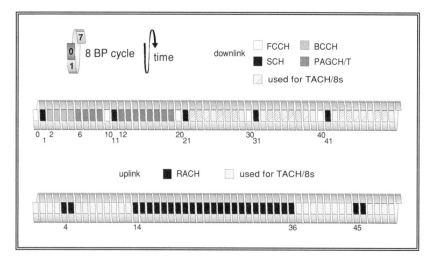

Figure 4.13 – A common channel pattern for small capacity cells

A typical common channel implementation of a small capacity cell
combines 4 dedicated channels (TACH/8s) used for signalling
with a set of common channels (FCCH, SCH, BCCH, PAGCH/T and RACH/H)

The second combination has been introduced with the small capacity cells in mind (which are usually not the smallest in geographical size...). When the capacity of a PAGCH/T and a RACH/F are not needed, the operator might be interested in combining:

- in the downlink direction, a PAGCH/T with the usual FCCH, SCH, BCCH plus 4 downlink TACH/8s;

- in the uplink direction, a RACH/H with 4 uplink TACH/8s.

Such a combination is shown in figure 4.13. As for the basic combination, this one uses TN 0.

Conversely, the third combination is used in very high capacity cells, in which a PAGCH/F and a RACH/F are not sufficient to cope with too high a traffic level. On top of a set of channels grouped as in the basic combination (i.e., with PAGCH/F and RACH/F), a cell may accommodate up to 3 extension sets, each set following the third combination. An extension set contains the same channels as the basic combination, except the FCCH and SCH (they must be unique in the cell). A BCCH appears in each extension set, at least for two reasons: first, a part of the information broadcast on the BCCH relates to the RACH of the same TN, and hence can differ from one TN to another; second, it is simpler for the mobile station to listen to bursts of a single TN only.

These extension sets are found on TN 2 (one extension), TNs 2 and 4 (two extensions), or TNs 2, 4 and 6 (three extensions). Why impose this constraint? The reasons are the following:

- first, all common channels of one cell must use the same downlink frequency (and therefore the same uplink frequency): this will be explained when exploring the frequency realm page 224;

- second, cells of very large radius may allow RACH bursts to overflow into the next slot, as explained in Chapter 6. If consecutive TNs had been allowed for extension sets, they would not be compatible with such possibilities;

- third and last, it was desirable to simplify the system complexity by minimising the number of different cases.

CBCH

A Cell Broadcast CHannel (CBCH) follows a cycle of $8 \times 51 \times 8$ BPs (lasting for about 2 seconds), where 4 times 4 slots are used. The allowed positions in the 51×8 BPs cycle, and the allowed TNs, are limited, so that there is no collision with the requirement to listen to other BCCH or PAGCH information. The CBCH can be seen as a sub-part of a TCH/8. Two cases are to be distinguished. If the common channel configuration is the small one with a PAGCH/T and a RACH/H, the CBCH can use the same TN (TN 0) and frequency as the common channels. It then uses slots that would otherwise belong to one of the four TCH/8s which can use TN 0 and the beacon frequency. A second possibility, applicable whatever the common channels configuration, is for the CBCH to use TN 0 (but not on the beacon frequency), 1, 2 or 3; the CBCH bursts must there again use a specific position in the 51×8 BP cycle, which would otherwise belong to a TCH/8.

When a CBCH is used, the first block of the PAGCH in the 51×8 cycle cannot be used for paging. All these rules ensure a minimum time between a CBCH burst and a burst belonging to a block carrying a paging message. However, in this second case, it is allowed (but not mandatory) to have a CBCH with a TN different than 0. In this case, the mobile station in idle mode has to listen regularly to bursts of different TNs, a source of complexity for the scheduling of reception. This is the sole case where such a requirement exists within one cell.

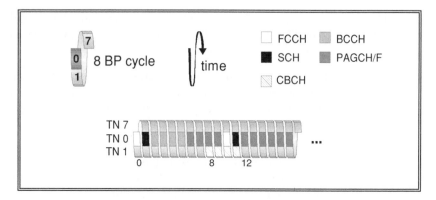

Figure 4.14 - Time organisation of a Cell Broadcast CHannel (CBCH)

A CBCH uses part of the capacity which would otherwise be allocated to a TACH/8.
The example shows the CBCH when combined with TACH/8s on TN 1.

Inside the $8 \times 51 \times 8$ BP cycle, the CBCH can be seen as a half downlink TCH/8, using four out of the eight 4-burst blocks. The example of a CBCH using TN 1 is given in figure 4.14. The four other blocks, i.e., the slots that would be used by the SACCH, and the uplink corresponding slots, are not used by the CBCH, and cannot be used for anything else. However, it is allowed to stop the transmission of the CBCH in case of congestion, and then these resources can be used for a TACH/8 during such periods.

4.2.1.3. Channel Organisation in a Cell

The above sections have mainly taken into account the mobile station point of view when describing the channels. A few ideas of how channels may be grouped in a cell have been introduced; let us now dig deeper into the management of time resources by a base station.

An elementary transceiver can emit or receive continuously, but on a single frequency slot at any given instant: it may be able to change its frequency as often as once every burst period, but it cannot emit or receive two bursts on different frequency slots during the same time slot. A base station usually contains several such elementary transceivers in order to reach the desired capacity. This concept of elementary transceiver is used in the *Specifications* (mainly in the specification of the interface between BTS and BSC) under the term *TRX* (Transmitter Receiver).

Channels	Unused slots
1 TACH/F	1 out of 26
2 TACH/H of different sub-TNs	none
8 TACH/8 of different sub-TNs	3 out of 51
1 SCH + 1 FCCH + 1 BCCH + 1 PAGCH/F + 1 RACH/F	downlink: 1 out of 51 uplink: none
1 BCCH + 1 PAGCH/F + 1 RACH/F	downlink: 11 out of 51 uplink: none
1 BCCH + 1 PAGCH/T + 1 RACH/H + 4 TACH/8	downlink: 3 out of 51 uplink: none

Table 4.1 – Possible combinations of channels of the same TN

The same capacity of one slot every 8 slots may be used for various combinations, some examples of which are given in this table.

It is therefore desirable, in order to optimise implementation costs in a base station, to choose channels so that they form groups where at most one burst is emitted at any one time, and to fill the time slots within these groups as much as possible.

In order to facilitate such groupings, the time organisation of the radio interface makes an extensive use of the 8 BP cycle. Every TRX is able to cope with 8 groups of channels, each group corresponding to a given TN. For instance, a group of channels using the same TN may consist of one of the combinations listed in table 4.1 (the CBCH is not included).

A TRX may combine eight such groups, with only the constraints listed earlier on common channels, in particular the TN(s) to use. The three combinations using only TACHs may exist on any TN. In order to complete the picture, let us give a few examples of channel combinations in a cell.

A **small capacity** cell with a single TRX will typically be organised as follows:

TN 0: FCCH, SCH, BCCH, PAGCH/T, RACH/H, 4 TACH/8;

TN 1 to 7: 1 TACH/F each.

A **medium capacity** cell with 4 TRXs may include:

one TN 0 group: FCCH, SCH, BCCH, PAGCH/F, RACH/F;
twice 8 TACH/8; and

29 TACH/F.

A **large capacity** cell with 12 TRXs may include:

one TN 0 group: FCCH, SCH, BCCH, PAGCH/F, RACH/F;
one TN 2 group, one TN 4 group and one TN 6 group: BCCH, PAGCH/F, RACH/F;
5 times 8 TACH/8; and

87 TACH/F.

4.2.1.4. Synchronisation Acquisition, or "How did it all start?"

Any reader who is not an expert in synchronisation might well at this stage ask the reasonable question: how does the mobile station manage to find the very first synchronisation with a cell, in order to read any of the channels defined in the previous paragraphs? What kind of bootstrap is there?

A few explanations might clarify the issue. As already mentioned, the FCCH and the SCH are provided for helping the mobile station to acquire the synchronisation. More precisely, the successful reception of an SCH burst will give the mobile station all the information needed for synchronisation. The problem is then to find an SCH burst. The specifications are such that an SCH burst always follows an FCCH burst 8 BPs later on the same frequency. Now an FCCH burst has a rather easily recognisable structure. A possibility is then to look for FCCH bursts, at all frequency slots and at all times.

When such a burst is encountered, the mobile station is able to get some information out of it. First (as the name FCCH evokes), it is able to correct the frequency of its internal time base in order to ease the demodulation of other channel bursts—this is however not the main point here. What is more important for synchronisation acquisition is that the mobile station is able to have a rough idea of the boundaries between slots, and of the situation in time of the slots of TN 0 (since the FCCH uses by definition slots of TN 0).

Knowing how SCH slots are positioned relatively to FCCH slots, the mobile station may then proceed to find and demodulate an SCH burst. Inside, it will find some precise information on the limits between slots, as well as sufficient information to deduce the number of the time slot in the cycle of $8 \times 26 \times 51 \times 2048$ BPs, and hence its position in all useful cycles. From this moment onwards, the mobile station only needs to maintain its knowledge of slot boundaries and to add 1 at each BP!

4.2.1.5. Frames

This small section is not aimed at adding more explanations on the time organisation of the channels. The previous sections were to give all explanations, and it is hoped they have reached their goal. Yet, the proposed description is quite different than what can be found in the *Specifications*, or in the main literature on GSM (which usually follows the same approach as the TSs). In the following paragraph we will try to bridge the gap.

In the *Specifications*, the time description of channels refer to "frames", a word which we did not use in this context. A frame is often presented as the succession of n slots. In particular, a "TDMA frame" represents a succession of 8 consecutive slots, the accent being put on the grouping of slots rather than on the 8 BP cycle. The grouping vision is quite natural when dealing with the implementation of a base station, which caters for many channels. But the cycle approach is much more natural as seen from the mobile station, which deals with few channels at the same time. This is why we preferred to put the stress on the concept of cycle (shown as an helix in the figures) rather than use the notion of "frame".

However, since the numbering of slots in GSM is very much based on frames, it is worth presenting the GSM frame hierarchy now. In the *Specifications*, a TDMA frame most often refers to a grouping of 8 time slots starting with one of TN 0. This allows to use the "TDMA frame number", which is simply the quotient modulo 8 of the time slot numbers of the time slots in the frame. But because all the channels are designed as having only slots of the same TN, the full time slot numbering is never used in the *Specifications*. Instead, one finds the TDMA frame number (FN) plus the TN.

There are other frames appearing in the *Specifications*, corresponding to the major cycles. They are shown in figure 4.15.

The "26 TDMA frame multiframe" is defined as a succession of 26 TDMA frames, and corresponds to the 26×8 BP or 120 ms cycle used in the definition of the TACH/F and the TACH/H.

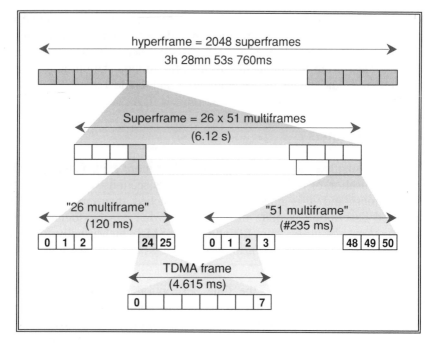

Figure 4.15 – Hierarchy of frames

The superframe is the smallest multiple of both the 51 TDMA frame cycle
for common channels and the 26 TDMA frame cycle for dedicated channels.
2048 superframes build an hyperframe, which serves as the basis for frame numbering.

Similarly, the "51 TDMA frame multiframe" is defined as a succession of 51 TDMA frames, and corresponds to the 51×8 BP cycle used in the definition of the TACH/8 and of the common channels.

The "superframe" is a succession of 51×26 TDMA frames (6.12 seconds), and corresponds to the smallest cycle for which the organisation of all channels is repeated. Note that this repetition abstracts some properties of the channel, for instance the SACCH coding period is not taken into account (the period would then be 4 superframes), neither is the internal structure of the PAGCH (which in some cases does not recur more rapidly than the hyperframe which will now be defined).

The "hyperframe" is the numbering period. It is $2048 \times 51 \times 26 \times 8$ BP long, that is to say exactly 12533.760 seconds, or 3 hours, 28 minutes, 53 seconds, and 760 milliseconds. It is obviously a multiple of all previously cited cycles, and determines in fact all the cycles in the transmission on the radio path. It is in particular the smallest cycle for frequency hopping and for ciphering.

4.2.2. THE FREQUENCY AXIS

4.2.2.1. The Available Frequencies

GSM was first devised as a cellular system in a specific 900 MHz band, called "the primary band". This primary band includes two sub-bands of 25 MHz each, 890-915 MHz and 935-960 MHz (see figure 4.16). This does not mean that the whole primary band must be used for GSM in a given country, especially at the start of the system. Moreover, a given operator is rarely given more than a portion of this band, since most countries have several operators. However, every mobile station must be able to use the full band, in order not to impose constraints upon roaming users.

In 1990, upon request of the United Kingdom, a second frequency band was specified for being used with the *Specifications*. This band includes the two domains 1710-1785 MHz and 1805-1880 MHz, i.e., twice 75 MHz: three times as much as the primary 900 MHz band. Mobile stations using this band are different from those using the primary band: using the same mobile equipment for roaming between the two variants of the system, GSM900 and DCS1800, although not ruled out, is not envisaged in the near future.

Another extension of the primary band is foreseen. It should consist of the band which is directly "below" the primary band. For instance, an 8 MHz extension would raise the 900 MHz bands to 882-915 MHz and 927-960 MHz, i.e., twice 33 MHz.

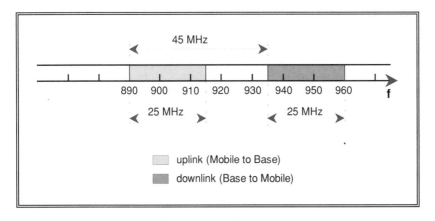

Figure 4.16 – GSM primary band

The GSM primary band includes two 25 MHz sub-bands around 900 MHz.

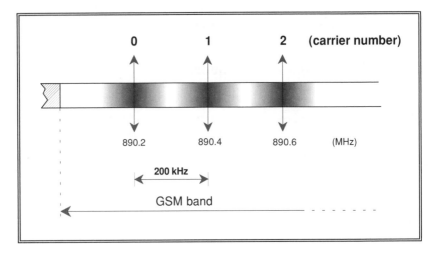

Figure 4.17 – Carriers at the border of the GSM band

Carrier spacing is equal to 200 kHz.
A guard band of 200 kHz between the edge of the band
and the first carrier is needed at the bottom of each of the sub-bands;
the carrier numbered 0 is often not used in practice.

The central frequencies of the frequency slots are spread evenly every 200 kHz within these bands, starting 200 kHz away from the band borders (see figure 4.17). 124 different frequency slots are therefore defined in 25 MHz, and 374 in 75 MHz. The modulation spectrum is somewhat wider than 200 kHz, resulting in some level of interference between bursts on simultaneous slots or on adjacent frequency slots. This is a nuisance mainly near the band borders, since this interference could disturb non-GSM applications in adjacent bands. The border frequencies are therefore usually avoided. The normal practice is not to use the frequency slots at the border (those numbered 0 and 124), except when a special agreement has been reached with the users of the adjacent band. As a consequence, the number of frequency slots which can be used in 25 MHz is usually limited to 122.

4.2.2.2. Frequency Hopping

The radio interface of GSM uses slow frequency hopping. Frequency Hopping consists in changing the frequency used by a channel at regular intervals. The origin of this technique lies in military transmission systems, where it was introduced to ensure secrecy and combat jamming. Publications in that domain distinguish *Fast Frequency*

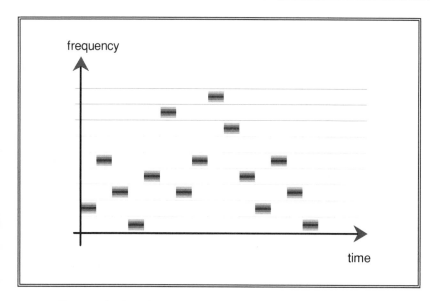

Figure 4.18 – Slow Frequency Hopping in the time-frequency domain

Frequency for a given channel may change at each burst,
and remains constant during the transmission of a burst

Hopping (FFH), where the frequency changes quicker than the modulation rate, from *Slow Frequency Hopping* (SFH). In GSM, the transmission frequency remains the same during the transmission of a whole burst; GSM belongs therefore clearly to the slow hopping case. Figure 4.18 shows an example of a time-frequency diagram for a frequency hopping channel.

Slow frequency hopping was introduced in GSM for two main reasons. The first reason is **frequency diversity**. As shall be explained later, error-correcting codes are introduced in the transmission chain. Such codes are based on redundancy: the data is made redundant in such a way that, even with a certain amount of errors, the original data may be reconstructed from what remains in the received flow. This redundancy is spread over several bursts. SFH therefore ensures that this information is sent on several frequencies, and this improves transmission performance. To explain this, a digression concerning propagation is needed.

Mobile radio transmission is subject to important short term amplitude variations when obstacles are involved; these variations are called Rayleigh fading. In most cases, the emitting and receiving antennas are not within direct sight one with the other, and the received signal is the sum of a number of copies of one signal with different phases. For instance, if the path includes a reflection on an obstacle, there

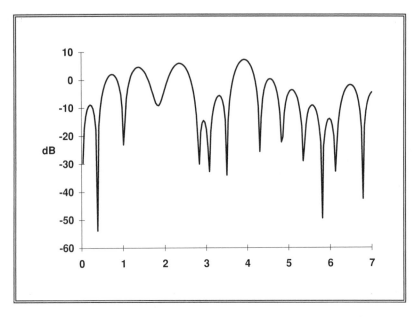

Figure 4.19 – Typical amplitude variations due to Rayleigh fading
(the time unit is the time to move through one wavelength,
e.g., 24 ms at 50 km/h for 900 MHz)

are usually many other reflected paths, in particular when the reflector is irregular and of a scale greater than the wavelength (buildings meet these criteria). The sum of a lot of phase-shifted signals with a random distribution of phases has an envelope following the Rayleigh distribution. Figure 4.19 shows an example of the variation in time of the envelope of a Rayleigh affected signal.

Now, the fading incurred by signals at different frequencies are not the same, and become more and more independent when the difference in frequency increases. With frequencies spaced sufficiently apart (say 1 MHz), they can be considered completely independent. With frequency hopping, all the bursts containing the parts of one code word are then not damaged in the same way by Rayleigh fading.

When the mobile station moves at high speed, the difference between its positions during the reception of two successive bursts of the same channel (i.e., at least 4.615 ms) is sufficient to decorrelate Rayleigh fading variations on the signal. In this case Slow Frequency Hopping does no harm, but it does not help much either. However, when the mobile station is stationary or moves at slow speeds, SFH allows the transmission to reach the level of performance of high speeds. The gain has been evaluated to be around 6.5 dB. This advantage is of prime importance for a system where a high proportion of handhelds is sought, since hand-held users are usually moving at a slow pace or not moving at all.

The second advantage of frequency hopping is **interferer diversity,** a property associated with Code Division Multiple Access (CDMA). In high traffic areas, such as large cities, the capacity of a cellular system is limited by its own interferences caused by frequency reuse. The relative interference ratio (C/I) may vary a lot between calls: C (the Carrier level) changes with the mobile station position relative to the base station, with the amount of obstacles between them, etc.; I (the Interference level) changes depending on whether the frequency is being used by another call in some nearby cell, and it also varies according to the distance with the interfering source, its level, etc. .

Since the aim of a system is usually to satisfy as many customers as possible, its maximum capacity is calculated based on a given (small!) proportion of calls subject to a noticeable decrease in quality due to interferences. Because of this concept of "worst case", the capacity of a system is better when, for a given mean C/I value, the statistical spread around this mean value is as small as possible. Let us consider a system where the interference level perceived by a call is the mean of the interference level caused by many other calls; then, the greater this number of interferers for a given total sum, the better the system. This is how interferer diversity operates.

In a system such as the current analog ones, a call potentially receives interference from a small number of other calls (typically 2 to 6, depending on the reuse pattern). At the other extreme, in a CDMA system—very fashionable these days West of the Atlantic—all calls interfere a little with all others. For the same mean interference value in the two systems, a call in the conventional system will either have a very good quality or be completely jammed, whereas a call in the CDMA system will always have some low level of interference, rarely so bad that transmission would fail. Thus the interferer diversity can be used to increase system capacity.

The major drawback of CDMA systems is that their design usually leads to calls interfering in the same cell and in adjacent cells. What is good for reducing the spread around the mean value causes the same mean value to decrease! GSM has been devised to avoid collisions inside one cell and between a certain number of adjacent cells. It allows however to spread the interference between many calls of a potential interferer cell, instead of a single one as in conventional systems.

An example should help to illustrate this principle. Let us consider a cell using 4 frequencies (cell A), and 2 potentially interfering neighbour cells (see figure 4.20). Each one of the latter two also uses 4 frequencies, two of them being part of the 4 frequencies used in the first cell. The figures listed in table 4.2 give an example of the interference levels experienced by a mobile station in cell A when traffic is maximum. A call is deemed jammed when the sum of all interferences amounts to a C/I ratio of less than 5.0 (7 dB). Without frequency hopping, the mobile station has an even chance to receive correctly the signal (if the allocated frequency is f_1 or f_2), whereas with frequency hopping, the quality is correct in all cases. This example is not a deliberate choice of a rare case; such situations arise quite often and frequency hopping really creates a significant statistical gain.

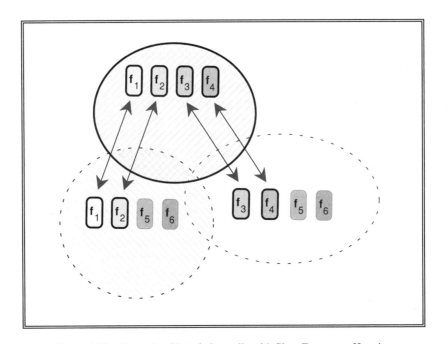

Figure 4.20 – Example of interfering cells with Slow Frequency Hopping

Cells using the same frequencies, but with decorrelated hopping sequences, lead to interferer diversity.

	f_1	f_2	f_3	f_4
Mobile to base interference level on the frequency	0.10 (C/I=10 dB)	0.14 (C/I=8.5 dB)	0.25 (C/I=6 dB)	0.28 (C/I=5.5 dB)
Mobile to base average interference level	0.19 (C/I=7.2 dB)			
Base to mobile interference level on the frequency	0.10 (C/I=10 dB)		0.28 (C/I=5.5 dB)	
Base to mobile average interference level	0.19 (C/I=7.2 dB)			

Table 4.2 – Interference levels for 4 calls in each cell of figure 4.20

(without power control; levels relative to wanted signal level)
Without frequency hopping, performance can be compared with the first and third lines;
with frequency hopping, it can be compared with the second and last lines.
Interferer diversity helps to improve capacity for a given mean quality.

4.2.2.3. Hopping Sequences

GSM allows a wide diversity—indeed almost an infinity—of different channels, when both time and frequency parameters are taken into account. A hopping sequence—i.e., the sequence of couples (TN, frequency) allocated to a channel—may use up to 64 different frequencies. Of course, the single frequency list is possible: it corresponds to a fixed frequency channel, which appears here as a "degenerate" frequency hopping channel.

Hopping sequences are described for channels using one slot every 8 BPs. A hopping sequence is then a function of the timeslot number modulo 8. For a channel using less than one slot every 8 BPs, the hopping frequencies are calculated by applying the same function. For example, if a TACH/F uses the following hopping sequence:

1   2   3   4   1   2   3   4   ...

Then the corresponding TACH/H shall use the following sequences:

sub-TN 0:   1   3   1   3   ...

sub-TN 1:   2   4   2   4   ...

Since the number of bursts in the uplink direction derives conventionally from the one in the downlink direction by a delay of 3 BPs, the hopping sequence (i.e., the function which associates a frequency to each TN modulo 8) in the uplink direction derives from the one in the downlink direction by simply adding 45 MHz.

For a set of n given frequencies, GSM allows $64 \times n$ different hopping sequences to be built. They are described by two parameters, the MAIO (Mobile Allocation Index Offset) which may take as many values as the number of frequencies in the set, and the HSN (Hopping Sequence Number) which may take 64 different values. Two channels bearing the same HSN but different MAIOs never use the same frequency on the same burst. On the opposite, two channels using the same frequency list and the same TN, but bearing different HSNs, interfere for 1/n th of the bursts, as if the sequences were chosen randomly. The sequences are indeed pseudo-random, except for the special case of HSN = 0, where the frequencies are used one after the other in order. Pseudo-random sequences have been chosen because they have statistical properties similar to random sequences.

Usually, channels in one cell bear the same HSN and different MAIOs: it is desirable to avoid interference between channels inside a cell. Adjacent cells are not interfering either, since they use disjointed frequency sets.

In distant cells using the same frequency set, different HSNs should be used in order to gain from interferer diversity. If this gain is sought, it is best to avoid HSN = 0, which leads to poor interferer diversity, even with non-identical frequency sets.

4.2.2.4. The Case of Common Channels

There exists a restriction to the use of frequency hopping: common channels (FCCH, SCH, BCCH, PAGCH and RACH) *must* use a fixed frequency. This constraint is meant to ease initial synchronisation acquisition (described on page 214): once the mobile station has found an FCCH burst, it will look for an SCH burst on the same frequency. Since this burst is too small to contain the description of a hopping sequence for the BCCH, the simplest way is to put the BCCH on the same frequency as the SCH. If the PAGCH and the RACH were hopping channels, their hopping sequences could be broadcast on the BCCH. This would however increase system complexity for little gain. The choice was that common channels on TN 0 never hop and all use the same frequency.

Similarly, extension sets of common channels are also forbidden from hopping and use the same frequency as the primary group, so that there is no need to transmit the description of their frequency organisation on the BCCH of TN 0.

Another peculiarity related to common channels is that the frequency they use must be emitted continuously, even if no information needs to be conveyed on some bursts. This is needed because mobile stations in neighbouring cells continuously perform measurements on this frequency, in order to determine the best cell they should listen to or to report measurements for handover preparation. When there is no information transfer request, a specific pattern is emitted (the fill frames). Because of these special roles of the frequency carrying the FCCH, we will refer it by a special name, the "beacon frequency" of the cell (in the *Specifications* it is in some places referred to as the BCCH frequency).

4.2.2.5. Channel Organisation in a Cell

The description of the channels has been up to now focused on the description of one channel at a time, a point of view close to that of the mobile station. The point of view of a base station is somewhat different.

We have seen in the time domain description that channels have been designed so as to use as well as possible a transmitter which is limited to one burst per burst period. This introduces the notion of TRX, which is the natural unit to measure the capacity of a base station (this unit is so natural it appears in the procedures between the BTS and the BSC). A TRX has then the capacity of 8 TACH/F, or 16 TACH/H, or 64 TACH/8 grouped by 8, or a lot of other combinations. It is perfectly possible to build equipments with smaller capacities, but it seems they have no commercial interest, and the TRX is a concept used by all manufacturers.

Another reason for this concept is the frequency allocation. A cell is usually allocated an integral number n of frequencies, and the maximum capacity of such a cell corresponds to the capacity of n TRXs. This does not mean that there is a one to one relationship between frequencies and TRXs: this would be true only if frequency hopping was not used. But in most applications, a cell is equipped with exactly as many TRXs as allocated frequencies. In fact, it may happen that a cell is equipped with less TRXs, for economy reasons, but this raises a specific problem that will be dealt with later.

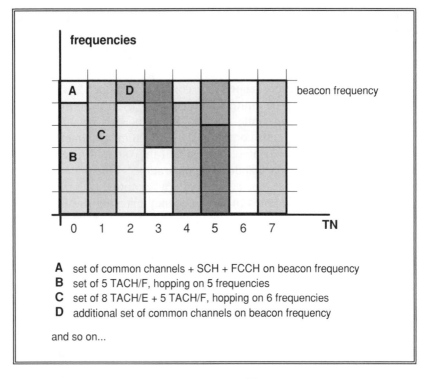

Figure 4.21 – Frequency/TN groups

The GSM hopping sequences are such that, for a given TN in a given cell,
one may define channel groups hopping on the same set of frequencies.
In the case of group C (to take an example), if all channels are not allocated,
the BTS must still manage to emit on the beacon frequency in all slots of TN 1.

Let us look at a typical cell with n frequencies and n TRXs. The
channel organisation must include one common channel group with a
FCCH and a SCH, consisting of either (FCCH + SCH + BCCH +
PAGCH/F + RACH/F), or a group with combined TACH/8 (FCCH +
SCH + BCCH + PAGCH/T + RACH/H). Other common channel groups
may be added in the first case. These choices are determined by load
considerations. The common channels are non-hopping channels, and all
use the same frequency, the beacon frequency. The rest of the resources
are distributed among TACHs, with a ratio between TACH/F and
TACH/8 which depends on load considerations.

The constraints on the hopping sequences are few. To show
various possibilities, we will use the diagram in Figure 4.21. Such a
diagram is based on the particularity that the channel configuration can
be described independently for each TN, because no channel use bursts of

different TNs. For a given TN, the channels are grouped in the frequency domain in one or several sets including the channels which have at least one frequency in common. In fact, the GSM frequency hopping sequences are defined in such a way that the only rational approach is that all the channels in a group use the same set of frequencies. Hence the diagrams, which show such frequency/TN sets.

The consequence of these possibilities is that the bursts in succession on the same frequency may belong to a lot of different channels, and that there is very little logic in the succession. This is why the notion of TDMA is somewhat misleading with frequency hopping. Channels are not sharing a frequency on a time division basis, the channels in a same frequency/TN group are sharing several frequencies.

Substantial gains in frequency and interferer diversity are obtained when at least 4 frequencies, and preferably more (say 8), are used in a hopping sequence. This causes a problem in the cases where for capacity reasons a single TRX would be sufficient in a given cell. The operator may choose for the gain of frequency hopping to allocate more frequencies to that cell than the number of installed TRXs.

A small problem then arises from the necessity to emit continuously on the beacon frequency. In cells of small capacity, the operator may choose either to let the channels of TN other than 0 hop on only as many frequencies as there are TRXs (but the gain of frequency hopping is small), or on as many frequencies as available. In the latter case, an additional transmitter dedicated to the filling of the common channels frequency is needed. Because the frequency/TN groups are not fully used by the installed TRXs, the continuous emission of the beacon frequency is not guaranteed by these TRX alone. The role of the additional transmitter is to emit on the beacon frequency when it would have been used by one of the missing channels.

4.3. FROM SOURCE DATA TO RADIO WAVES

Up to this point, we have addressed only how transmission resources are organised to be shared between users, not how they are used. In the previous chapter, we have seen that, if we restrict our view to the radio interface, all needs for user data transmission can be fulfilled with a few different transmission modes. Let us recall and name these modes, with a notation derived from the *Specifications*:

- **TCH/FS, TCH/F9.6, TCH/F4.8, TCH/F2.4** are the transmission modes over a TCH/F, respectively for full rate speech, 12 kbit/s, 6 kbit/s and 3.6 kbit/s data.

- **TCH/H4.8, TCH/H2.4** are the transmission modes which were defined over a TCH/H, respectively for 6 kbit/s and 3.6 kbit/s data (the speech mode on a TCH/H is not yet specified). There is a good chance that data modes stay as indicated here.

To these modes must be added the different needs for signalling, such as fast associated signalling, and the transmission mode on the SACCH, BCCH and PAGCH. In these cases, the source data appear as an irregular flow of blocks. Finally there is the variety of specific transmission needs coming from the RACH, the FCCH and the SCH.

We will see that, in detail, GSM transmission for the different modes presents at the same time some uniformity and some variety. The unifying concepts are the burst, which will be our first topic, and the modulation, which will be our last. On the other hand, the different modes differ in the details of the error correction and error detection coding schemes.

After the presentation of the burst, and its divers species, the presentation will follow the fate of the information from data to radio waves. Radio emission involves several successive operations in order to convert source data into the final signal. Conversely, reception implies that a reverse series of operations be performed at the receiver's side until the original data is—approximately—regenerated. The sequence of such operations for speech is shown in figure 4.22. The process is identical for other user data and signalling.

The operations described in this section are common to all transmission modes. The following operations take place on the source side:

- **channel coding** introduces redundancy into the data flow, increasing its rate by adding information calculated from the source data, in order to allow the detection or even the correction of signal errors introduced during transmission. The result of the channel coding is a flow of code words; in the case of speech, for example, these code words are 456 bits long;

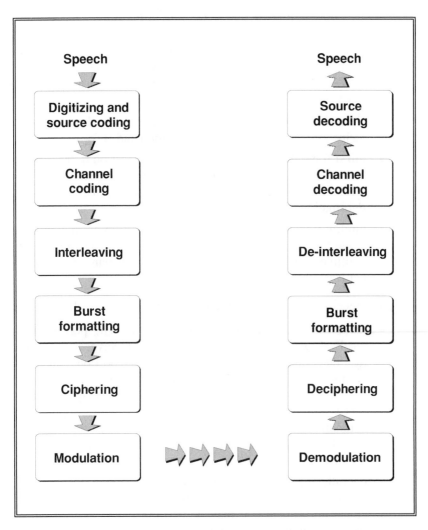

Figure 4.22 – Sequence of operations from speech to radio waves
... and back to speech

After having transformed speech into digital blocks, channel coding adds redundancy.
The blocks are interleaved and spread into pieces,
which are combined with flags and a "midamble" to build up the bursts.
Ciphering is applied to these bursts and the resulting data is used to modulate
the carriers. The reverse transformations are performed on the receiver side.

- **interleaving** consists in mixing up the bits of several code words, so that bits which are close to one another in the modulated signal are spread over several code words. Since the error probability of successive bits in the modulated stream is very much correlated, and since channel coding performance is better when errors are de-correlated, interleaving aims at de-correlating errors and their position in code words. After interleaving, the flow of information is a succession of blocks—one block for each channel burst;

- **ciphering** modifies the contents of these blocks through a secret recipe known only by the mobile station and BTS.

- **burst formatting** adds some binary information to the ciphered blocks, in order to help synchronisation and equalisation of the received signal; the output of this stage consists of binary information blocks.

- **modulation** transforms the binary signal into an analog signal at the right frequency and at the right moment, according to the multiple access rules described starting on page 215; last but not least, this signal is radiated as radio waves.

The receiver side performs the reverse operations as follows:

- radio waves are captured by the antenna. The portion of the received signal which is of interest to the receiver is determined by the multiple access rules. This portion undergoes **demodulation** with the help of the additional information introduced during burst formatting. The result may consist in a succession of binary information blocks. More sophisticated demodulators, however, are able to deliver an estimated probability of correctness for each bit: this is called "soft decision";

- **deciphering** modifies those bits by reversing the ciphering recipe; since this recipe is a bit-by-bit "exclusive or" with a ciphering sequence, it may be performed just as well when a soft decision process is applied;

- **de-interleaving** puts the bits of the different bursts back in order to rebuild the code words;

- last, **channel decoding** tries to reconstruct the source information from the output of the demodulator, using the added redundancy to detect or correct possible errors in the output from the demodulator. This operation is much more efficient when the demodulator indicates an a priori error likelihood of each bit.

4.3.1. THE BURSTS

The burst is the transmission quantum of GSM. Its transmission takes place during a time window lasting $(576 + 12/13)$ μs, i.e., $(156 + \frac{1}{4})$ bit duration. Within this time interval, the amplitude of emission rises from a starting value of 0 to its nominal value. The signal phase is then modulated to transmit a packet of bits. After that, the amplitude decreases until it reaches 0.

> This description is valid only for emission by the mobile station, and for emission by the BTS if the adjacent burst is not emitted. The *Specifications* allow (but do not enforce) the BTS to keep the amplitude constant between two adjacent emitted bursts.

Several kinds of bursts are defined with regard to their time-amplitude profile, e.g., "normal" bursts and access bursts. An example of the amplitude profile of a normal burst (used, e.g., on traffic channels) during its emission window is given in figure 4.23, together with the *time mask* defined in the *Specifications* and which gives the acceptable limits within which the amplitude profile must lie. Its constant amplitude part lasts 147 bit periods, i.e., 2.5 bits on each side of the 142 information-carrying bits.

The packet of bits used to modulate the signal phase of a burst includes in general the useful part of the information, plus a training sequence and three additional "0" bits at each end. Theoretically, the signal phase is obtained by applying the modulation method to an infinite series of bits, consisting here of the bit sequence of the burst preceded and followed by infinite series of "1".

The three "0" bits added at the beginning and at the end of each burst avoid a loss of demodulation efficiency for the extreme information bits. An interesting question is why the choice was not to set them to "1" like the infinite sequence assumed outside the burst. The specification is such that the transition from "1" to the first "0" bit of the burst, and from the last "0" bit of the burst to "1" fall exactly in the ramping portion of the burst amplitude profile. A property of the modulation is that in absence of transition, the modulated signal is shifted toward the higher frequencies, and the interference created by the ramping outside the frequency slot would then be greater than with a bit transition.

The training sequence is a sequence of bits known by the receiver. There are several such training sequences defined in GSM, as will be seen in the next paragraphs. The signal resulting from the transmission of this training sequence allows the receiver to determine very precisely the position of the useful signal inside a reception window, and to have an idea of the distortion caused by transmission. These information are of prime importance to obtain good demodulation performances.

Several burst formats are defined:

- the **access burst** is only used in the uplink direction during initial phases when the propagation delay between the mobile station and the base station is not yet known. This is the case with the first access of a mobile station on the RACH, or sometimes with the access of a mobile station to a new cell upon handover. The access burst is a short burst; it is the only kind of burst used on the RACH;

- the **F and S bursts** are used respectively on the FCCH and on the SCH. They serve solely for initial synchronisation acquisition of a mobile station in a given cell.

- the **normal burst** is a long burst used in all other cases.

4.3.1.1. The Normal Burst

A normal burst contains two packets of 58 bits surrounding a training sequence of 26 bits (see table 4.3). Three "tail" bits (set to 0) are added on each side.

The Specifications also include the guard time in the burst. The actual guard time is determined by the signal envelope (see figure 4.23). If we only consider the guard time as the period during which the signal is below −70 dB, its duration is about 30 μs. In the uplink direction, this guard time is barely enough to compensate for equipment inaccuracies and for multipath echoes if they are spread on the maximum range allowed by demodulation. The propagation delay itself is compensated by the timing advance mechanism (see Chapter 6 for this topic). In the downlink direction, the guard time could have been chosen shorter, but it has been kept at the same value for symmetry reasons.

Tail	Information	Training Sequence	Information	Tail
3	58	26	58	3

Table 4.3 – Contents of a normal burst

116 information bits are spread on both sides of a midamble,
or training sequence, of 26 bits.

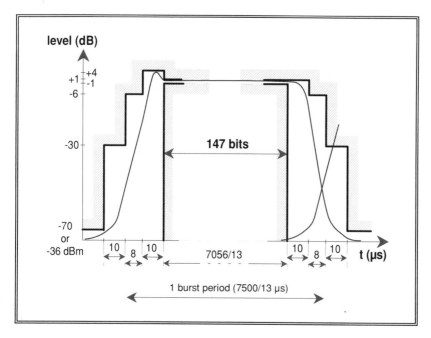

Figure 4.23 – Amplitude profile of a "normal" burst

The time mask of a normal burst is specified with a constant amplitude
during the "useful part" of the burst and power ramping at both ends.
The power level during the guard time should not exceed 10^{-7} (-70 dB)
of the "useful part" level, or $10^{-6.6}$ W (-36 dBm), whichever the highest.

The training sequence has been inserted in the middle of the burst
in order to minimise its maximum distance with a useful bit, and is
therefore sometimes called "midamble" (it has the same role as a
preamble, but is in the middle of the burst). The only drawback of that
position is the need for the receiver to memorise the first portion of the
burst before being able to demodulate it, but this is a very mild constraint
compared to the benefit gained from it.

Eight different training sequences have been specified. Why not a
single one? Let us consider the case of two similar interfering signals
arriving at the receiver at almost the same time. If their training
sequences are the same, there is no way to distinguish the contribution of
each of them to the received signal. The situation is much clearer when
the two training sequences differ, and are as little correlated as possible.
Distinct training sequences will therefore be allocated to channels using
the same frequencies in cells which are close enough to interfere with one
another.

The 8 training sequences have been chosen for their low correlation between one another as well as for the special shape of their autocorrelation function, which is meant to ease some demodulation techniques. Figure 4.24 shows the autocorrelation function of one of these 8 training sequences, calculated between the central 16 bits and the whole 26 bit sequence. All 8 sequences share the central correlation peak surrounded by 5 "0" on each side.

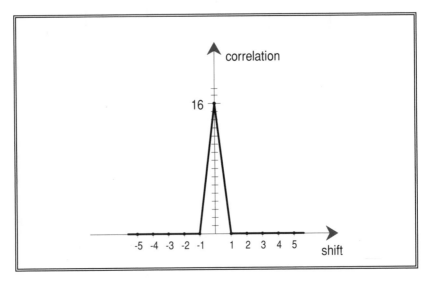

Figure 4.24 – Autocorrelation function of a GSM training sequence as a NRZ signal

Every GSM training sequence shares this autocorrelation shape,
calculated between the 16 central bits and the whole 26-bit sequence.

4.3.1.2. The Access Burst

As already mentioned, the access burst is the only short burst defined in GSM. Its envelope during the time window is constrained by the time mask shown in figure 4.25.

An access burst contains a 41-bit training sequence, 36 information bits and respectively 7 and 3 tail bits at the beginning and at the end (see table 4.4). The training sequence and the initial tail are longer than the ones of normal bursts, in order to increase the demodulation success probability: the task is quite hard indeed, since the receiver starts from

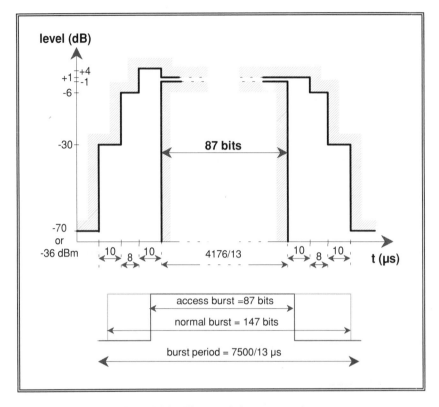

Figure 4.25 – Time mask for an access burst

An access burst has the same ramping specifications as a normal burst,
but its useful duration is much shorter in order to account for propagation time
with distant mobile stations.

scratch: it knows neither the reception level, neither the frequency error,
nor the exact reception time!

A single training sequence is specified for access bursts: the
interference probability is indeed too small to justify the added
complexity of multiple training sequences.

Tail	Training sequence	Information	Tail
7	41	36	3

Table 4.4 – Contents of an access burst

The training sequence is longer than in a normal burst,
and the useful information is very short.

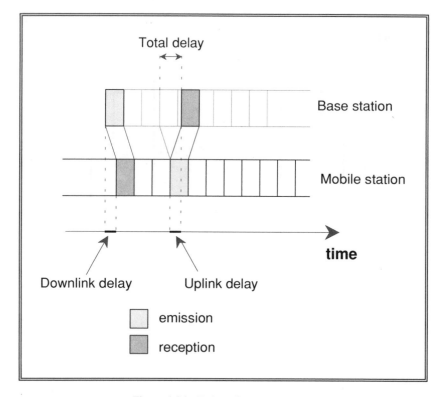

Figure 4.26 – Delay of an access burst

The time offset of an access burst as seen by the base station is equal to twice the propagation delay.

Since the propagation time between mobile station and base station is not known when such bursts are used, an access burst arrives at the base station with a time error of twice the propagation delay (see figure 4.26), compared to the reception window. The small duration of the access burst is there to compensate for this effect: the mobile station has to be very far indeed for the burst not to fit in the reception window. Mobile stations may wander up to 35 km away from the base station before they miss the target.

4.3.1.3. The S Burst

The S burst is only used in the downlink direction on the SCH. It is as long as the normal burst—142 bits—but its contents are different (see table 4.5).

Tail	Information	Training Sequence	Information	Tail
3	39	64	39	3

Table 4.5 – Contents of an S burst

A burst on the SCH bears less information than a normal burst,
but has the same length (142 bits)

As the access burst is the first burst that a base station needs to demodulate in the uplink direction, the S burst is the first burst that a mobile station needs to demodulate in the downlink direction. Therefore, its training sequence is longer than the one of a normal burst. The training sequence is unique, by necessity: the mobile station would otherwise not be able to know the sequence chosen by the base station if several sequences were defined.

4.3.1.4. The F Burst

The F burst is a very peculiar burst. It is a long burst, whose sole use is to enable mobile stations to find and demodulate an S burst of the same cell.

It is the simplest of all bursts: all of its 148 bits are set to "0". With the modulation technique used, the resulting signal is a pure sine wave which frequency is 1625/24 kHz higher than the carrier central frequency (two successive bits with equal value lead to a phase shift of $\pi/2$, as explained in the section dealing with modulation).

4.3.1.5. Preemption

As explained on page 190, it is possible to insert signalling blocks in the TCHs, when those are normally used to transfer user data. In both cases normal bursts are used, but the channel coding differs. It is therefore advisable for the receiver to know before starting decoding whether the burst includes user data or signalling. This information is included through an addressing mechanism inside the burst. The receiver is therefore able to get the information after demodulation, and thereby avoids the need to attempt decoding in two different ways.

The binary information indicating whether a coding block is user data or signalling is called the stealing flag. Two bits are used in the normal bursts for the purpose of carrying it, one bit for each half of the burst. They are the bits closest to the training sequence on each side. When a signalling block is transmitted through fast associated signalling on the TCH/F or TCH/H, it uses 8 burst halves, and the corresponding stealing flag bits are set to indicate signalling. On other channels than a TCH/F or a TCH/H, these bits are of no use. They are always set to "1" (meaning signalling).

On channels other than a TCH/F and a TCH/H, the stealing flags could be considered as an extension of the training sequence, because they are known beforehand. Unfortunately, the good autocorrelation properties of the training sequence are not extended, whatever the sequence. Curiously enough, half of the sequences bear the good properties when one more "0" is added on each side of the central peak (Training Sequences 0, 1, 6 and 7). The other cases result in asymmetric correlation functions.

4.3.2. INTERLEAVING AND CHANNEL CODING

The interleaving and coding schemes are all different for the different transmission modes. However, they are all from the same family, and we will present first the general aspects which apply to most of the cases. Then a splendid table will summarise the features of different modes, and an example, speech, will be presented in more detail.

4.3.2.1. General Principles of Interleaving

Interleaving is meant to decorrelate the relative positions of the bits respectively in the code words and in the modulated radio bursts. Why is that useful? Firstly, bit errors tend to occur consecutively rather than singly. This is true inside a burst, and comes both from error statistics on radio transmission, and from intersymbol interference introduced by the modulation. The burst structure itself is a reason for error grouping: because of the variations of reception and interference level in time (and in frequency if frequency hopping is used), a burst can have a small number of errors where the next will have a lot. Secondly, it appears that it is more difficult to design efficient codes when several adjacent bits are in error: better performance can be achieved when errors are randomised.

Interleaving consists in spreading the b bits of a code word into n bursts in order to change proximity relations between the bits. The larger the value of n, the better the transmission performance. On the other hand, the larger the value of n, the longer the transmission delay. A compromise has to be sought, and depends on the channel usage. Several interleaving schemes are therefore specified in GSM.

In order to avoid complex interleaving schemes, it is best to have simple arithmetic relations between b, n and the number of bits per burst (114 in the GSM case). For example, a code word of b=456 bits (i.e., 4×114) could be spread into, e.g.:

- 4 parts of 114 bits, each one filling up a whole burst;
- 8 parts of 57 bits, each one filling up half a burst; or
- 24 parts of 19 bits, each one using one sixth of a burst; or
- 76 parts of 6 bits, each one using one 19th of a burst.

With these examples, a burst would include contributions from different code words, respectively 1, 2, 6 and 19. The first two cases are used in GSM and are fairly easy to understand. A third case of interleaving is also used (for data services), but it is much more complex than the three listed above.

The interleaving for data at 9.6 kbit/s is derived from the last example above, and is referred to as "19 bursts interleaving". The careful reader will have deduced from the above examples that no regular interleaving scheme can be defined for spreading blocks of 456 bits onto 19 bursts: each piece would be 24 bits long, and therefore one burst has to fit exactly 4.75 such pieces! What kind of jigsaw puzzle is that? In fact, the interleaving is done on 22 bursts. A code word is cut into 22 pieces: 16 pieces of 24 bits, 2 pieces of 18 bits, 2 pieces of 12 bits and 2 pieces of 6 bits. A burst includes 4 pieces of 24 bits plus either one piece of 18 bits or two pieces of respectively 12 and 6 bits. A burst may therefore include contributions from either 5 or 6 different code words. And that works! Just take a few minutes to be convinced by figure 4.27 overleaf.

Good. Now why "19 bursts"? The easiest way to understand is to cut the 456 bits code word into 4 sub-blocks of 114 bits each. These sub-blocks should then be spread over 19 bursts in a regular manner (6 bits on each burst), shifting from 1 burst at each sub-block. The current specification has been generated this way (at some stage during the specification of the system, code words were 114 bits long), hence the "19 bursts" terminology.

More on interleaving shall be said in conjunction with the description of the coding schemes, scanning each transmission mode separately.

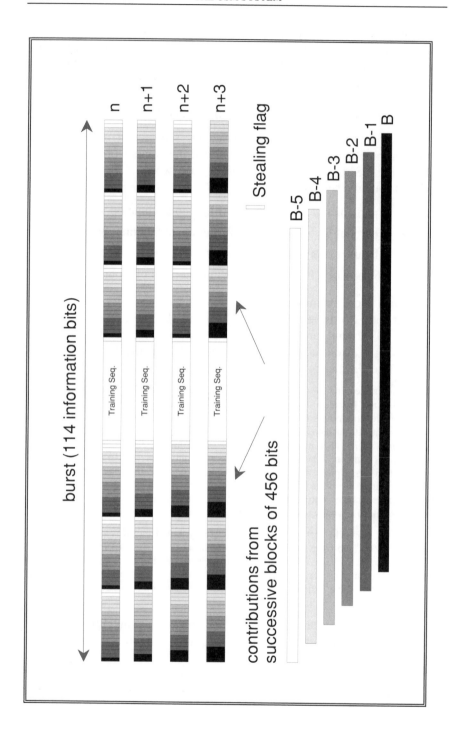

(↗) Figure 4.27 – "19 burst" Interleaving scheme for data services

There are 4 different types of bursts with regard to the contribution they contain
in terms of data blocks. Each burst may contain 6, 12, 18 or generally 24 bits
from a given block. The organisation repeats itself every 4 bursts
(burst n+5 would bear the same organisation as burst n,
but with contributions of blocks B–4, B–3, B–2, B–1, B and B+1).

4.3.2.2. General Principles of Channel Coding

Channel coding aims at improving transmission quality when the
signal encounters disturbances (significant noise when the reception level
is low, interferences, multipath propagation, Doppler shift, and so on...).
It results however in an increased number of bits.

Coding consists in adding to the source data some redundant
information calculated from this source information. Decoding makes use
of this redundancy to detect the presence of errors or estimate the most
probable emitted bits given the received ones. Errors are detected when
the transmitted redundancy is different from the one calculated with the
received data.

Depending on the transmission mode, radio path transmission uses
different codes. Moreover, several codes are "concatenated" in certain
cases, i.e., redundancy is added by applying a given code, then more
redundancy is added by applying another code.

The codes used in GSM are few:

- **block convolutional codes**; these codes are only used for
  correction purposes. They achieve tremendous efficiency when
  they are combined with likelihood estimation such as that
  coming from the demodulator;

- **a Fire code**; this code is dedicated to the detection and
  correction of "bursty" errors, i.e., errors which are grouped. It is
  used in concatenation after a block convolutional code, for
  · which residual errors often come out in groups;

- simple **parity codes** used for error detection.

After a brief description of the concepts used in these codes, a
summary of all interleaving and coding characteristics for the different
transmission modes is given (table 4.7, page 246). The example of full
rate speech is then studied in detail. Once this case is understood, the
reader should find it easy to understand the other cases described in the
Specifications.

4.3.2.3. Convolutional Codes and Block Codes

Strictly speaking, a convolutional code consists in transmitting the results of convolutions (that is to say what is obtained by adding a sequence and shifted versions of itself) of the source sequence using c different convolution formulas.

As an example, for $c=2$, two result sequences would be transmitted. The first one could be obtained by applying the "exclusive or" operation (which is the addition when speaking of bits) to respectively:

- the source sequence, the source sequence delayed by one bit period, and the source sequence delayed by 3 bit periods;

- the source sequence, the source sequence delayed by two bit periods, and the source sequence delayed by 3 bit periods.

The transmission bit rate after coding is twice the original information rate and is then redundant. Each information bit is "sent" 6 times, each time as a part of a sum with two other information bits. This multiplicity of sending will be used by the receiver to check the received flow, and possibly recover the original information flow even if transmission errors occurred.

The condensed form used to describe these recipes consists in using polynomials, here respectively D^3+D+1 and D^3+D^2+1, where "+" stands for the exclusive-or and D for a delay of 1 bit.

This method, as described, has two drawbacks: it is adapted to infinite sequences, and it results in an efficiency ratio (ratio between the number of useful bits in the source sequence and the number of bits actually transmitted) which is limited to fractions of the form $1/c$.

Two modifications of this basic scheme are used in GSM. First, the source sequence consists of finite blocks of p bits. They can be coded as described above, but with the addition of a few known bits at both ends of the block, called **tail bits**[1] . Second, an efficiency of the form p/q was needed in one case, and is obtained by **puncturing** the code, i.e., by keeping only q bits among pn, through a pre-determined rule.

For instance, if we want to apply the code described above to a block of $b=12$ bits, we could add 3 bits positioned to "0" at the start and

---

[1]   In fact, in the *Specifications,* the term *tail bits* is reserved for those added at the end; the first bits are specified to be 0 by the specification of the initial state of the encoder. However, the difference is only superficial, and the presentation here keeps the symmetry.

Source block (12 bits)	1 0 0 1 0 0 1 1 0 1 0 1
Addition of tail bits	0 0 0 1 0 0 1 0 0 1 1 0 1 0 1 0 0 0
Delay=1 bit period Delay=2 bit periods Delay=3 bit periods	0 0 0 1 0 0 1 0 0 1 1 0 1 0 1 0 0 0 0 0 0 1 0 0 1 0 0 1 1 0 1 0 1 0 0 0 0 0 0 1 0 0 1 0 0 1 1 0 1 0 1 0 0 0
1st conv. sequence (15 bits)	1 1 0 0 1 0 0 0 1 0 0 1 0 0 1
2nd conv. sequence (15 bits)	1 0 1 0 0 1 0 1 1 1 1 0 1 1 1
punctured 2nd sequence (8 bits)	1 1 0 0 1 1 1 1
transmitted block (23 bits)	1 1 1 0 1 0 1 0 0 0 0 0 1 1 0 0 1 1 0 1 0 1 1

Table 4.6 – Example of a punctured convolutional code

Every bit of the first convolutional sequence is kept,
but only every other bit of the second convolutional sequence is kept.
The convolutional sequences are obtained by applying the "exclusive or" operation
to the sequence and its delayed self, after addition of fixed bits
at both ends of the original sequence.

at the end of the sequence and therefore get $bc=30$ bits. In order to get a convolutional efficiency of, e.g., 2/3, every other bit of the second sequence could be omitted. The coded result uses 23 bits (see table 4.6).

The actual efficiency coefficient is slightly smaller than the convolutional efficiency, because of the tail bits, and amounts in that case to 12/23.

The convolutional codes used in GSM are not much more complex than the one described above: the maximum degree of the corresponding polynomials is 4. Now, convolutional codes of larger degrees have usually much better performance, at the price not of bit rate but of decoder complexity. In GSM, however, simple codes have been chosen because the performance gain is limited by the interleaving depth. The *span* of a code can be understood in broad terms as the length of the sequence of coded bits which must be analysed to decode one information bit. Now, it can be shown, with the assumption that bursts are mostly all good or all bad, that little gain is obtained if the span is greater than the interleaving depth. The GSM convolutional codes have been chosen as simple as possible, with a span just big enough (e.g., 22 for the 9.6 kbit/s data transmission mode).

4.3.2.4. Fire Code

A Fire code is a conventional linear binary block code, that is to say (in simplified terms) it consists in transmitting in addition to the information bits a number of redundancy bits computed by exclusive-or formulas applied to the information bits. Moreover Fire codes derive from codes belonging to the "cyclic" code family (to be precise, the Fire code used in GSM is a shortened cyclic code). In this case, the redundancy computation formulas can be expressed as follows. The original sequence can be used to build a polynomial (in binary arithmetic), whose coefficients are the bits of the sequence. The redundancy can be expressed as the coefficients of another polynomial obtained as the remainder of the division of the polynomial representing the sequence by a pre-determined polynomial, characteristic of the code and called the generator polynomial.

A Fire code has a generator polynomial designed to allow good detecting and/or correcting performance when errors happen in groups. The GSM Fire code uses the following generator polynomial:

$$(X^{23} + 1) (X^{17} + X^3 + 1)$$

This polynomial being of degree 40, the remainder has 40 coefficients, and then there are 40 redundancy bits. The properties of this code are such that error groups of up to 11 bits can be detected and corrected.

4.3.2.5. Parity codes

These are linear block codes, derived, like the Fire code, from cyclic codes. Three of those codes appear in the *Specifications*. For speech, a 3-bit redundancy code enables assessment of the correctness of the bits which are most sensitive to errors in the speech frame. Its detection capability is quite limited: all one-error patterns are detected, but there are patterns of two errors which are not detected.

For the RACH and SCH bursts, codes of respectively 6 and 10 bits of redundancy are aimed at detection, though they have adequate properties to be used for error correction (they have a minimum distance of 4, which means that they detect all patterns of 3 errors, and that they could be used to correct all single error cases).

For those interested, we give here their generator polynomials, and the factorisation thereof:

RACH: $X^6 + X^5 + X^3 + X^2 + X + 1 = (X+1)(X^5 + X^2 + 1)$

SCH:

$$X^{10} + X^8 + X^6 + X^5 + X^4 + X^2 + 1 = (X^4 + X^3 + X^2 + X + 1)(X^3 + X + 1)(X^3 + X^2 + 1)$$
$$= \frac{(X^5 + 1)(X^7 + 1)}{(X+1)(X+1)}$$

The latter code is a cyclic code. It is the best (35,25) code as given in the tables of *Error Correcting Codes* from Peterson (see bibliography).

4.3.2.6. Decoding

The *Specifications* only describe in detail the way in which the signal has to be transmitted. They do not impose explicit decoding methods. Still, many decoding algorithms exist, using different, more or less complex methods, or having different goals. The same code can be exploited at reception for error correction or for error detection. Usually some level of detection of the remaining errors is possible after error correction, but the performance of detection is much better if no error correction is applied.

The *Specifications* include however some constraints for the decoding process (and more generally on the reception chain); these constraints are provided as a minimum performance specification. For instance, they impose that the parity codes must be used only for detection. On the other hand, it is not so clear that using the Fire code for grouped-errors correction is needed to reach the required performance.

For convolutional codes, the required performance is quite severe. There are many decoding algorithms which can be used in general, but here the choice seems limited. A first point is that it seems necessary that the demodulator provides for each bit not only the indication whether it is more likely a "1" or a "0" ("hard-decision"), but also its opinion about the probability that it was a "1" or a "0" ("soft-decision"). There is more information to be used in the second case by the decoder. An obvious case, quite relevant in GSM, is the indication of an erased bit, that is to say a bit about which nothing can be provided by the demodulator (e.g., because the whole burst was so interfered that it was not possible to demodulate). In this case a hard-decision demodulator would provide garbage not indicated as such to the decoder, whereas a soft-decision demodulator would indicate the "garbageness" (the bit would be given as "1" or "0" fifty-fifty). It seems that some extra information beyond hard-decision is needed in GSM to reach the required performance. There are still many decoding algorithms for convolutional codes able to use likelihood information provided by the demodulator, but it appears that only a maximum likelihood decoding (i.e., finding *the* sequence of highest likelihood) fits the requirements.

4.3.2.7. Interleaving and Coding: a Summary of Transmission Modes

Table 4.7 summarises the interleaving and coding schemes used for the different transmission modes. The first column gives the channel and, where relevant, the transmission mode. The "input block" column indicates the size (in bits) of the data block before channel coding, whereas the "output block" column indicates the resulting coded block length (also in bits). In the "coding" column, codes are given in the same order as they are applied during coding. Decoding proceeds of course in the reverse order.

To avoid overburdening this book and the reader, we will detail only one case, speech. The table above gives the needed information for the other cases.

Channel and transmission mode		Input rate (kbit/s)	Input block	Coding	Output block	Interleaving
TCH/FS	Ia		50	Parity (3 bits) Convolutional 1/2		
	Ib	13	132	Convolutional 1/2	456	On 8 half-bursts
	II		78	none		
TCH/F9.6 TCH/H4.8		12 6	240	Convolutional 1/2 punctured 1 bit out of 15	456	Complex, on 22 unequal burst portions
TCH/F4.8		6	120	Addition of 32 null bits Convolutional 1/3	456	Complex, on 22 unequal burst portions
TCH/F2.4		3.6	72	Convolutional 1/6	456	On 8 half bursts
TCH/H2.4		3.6	144	Convolutional 1/3	456	Complex, on 22 unequal burst portions
SCH			25	Parity (10 bits) Convolutional 1/2	78	On 1 S burst
RACH (+ handover access)			8	Parity (6 bits) Convolutional 1/2	36	On 1 access burst
fast associated signalling on TCH/F and /H			184	Fire code 224/184	456	On 8 half bursts
TCH/8, SACCH; BCCH, PAGCH				Convolutional 1/2		On 4 full bursts

Table 4.7 – Interleaving and coding for the different transmission modes

4.3.2.8. The TCH/FS Transmission Mode

The input rate for the speech transmission mode on a full rate traffic channel is equal to 13 kbit/s, by blocks of 260 bits (i.e., one block every 20 ms). When errors occur, the received speech is disturbed in different ways depending on the role of the bits in error. This is why it was decided to protect the 260 bits of a source block in different ways:

- 182 bits are protected by a convolutional block code with a convolutional efficiency equal to 1/2;

- among these 182 bits, 50 are additionally protected by a detection code adding 3 redundancy bits (which are themselves protected by the above convolutional code). These 50 bits are the category Ia bits; the other 132 bits are category Ib bits;

- the 78 other bits are not protected.

Coding

The polynomial representing the detection code for category Ia bits is $X^3 + X + 1$. Bits of class Ia are of such importance that, if any one of them is wrong, the user will hear a—potentially loud—noise in place of a 20 ms speech slice. Detection of such errors is therefore important and allows the bad block to be replaced by something less disturbing such as an extrapolation of the preceding block.

The convolutional code consists in adding 4 bits (set to "0") to the initial 185 bit sequence, and applying two convolutions whose polynomials are respectively $D^4 + D^3 + 1$ and $D^4 + D^3 + D + 1$. No puncturing is applied, and the result is composed of twice 189 bits, i.e., 378 bits. The efficiency for class I bits is just below 0.5. The 78 non-protected bits must of course be added to these 378 bits, leading to a coded block length of 456 bits.

Interleaving

Blocks of full rate speech are interleaved on 8 bursts: the 456 bits of one block are split into 8 groups of 57 bits, each one carried by a different burst. A burst therefore contains the contribution of two successive speech blocks A and B. In order to destroy the proximity relations between successive bits, bits of block A use the even positions inside the burst and bits of block B the odd positions.

0 8 448	even bits of burst N
1 9 449	even bits of burst N+1
2 10 450	even bits of burst N+2
3 11 451	even bits of burst N+3
4 12 452	odd bits of burst N+4
5 13·. 453	odd bits of burst N+5
6 14 454	odd bits of burst N+6
7 15 455	odd bits of burst N+7
<— — — 57 columns — — —>	

Table 4.8 – Interleaving algorithm for full rate speech

A speech block is split into 8 groups of 57 bits as shown,
and each of these groups is carried in a separate burst.

The major drawback of interleaving is the corresponding delay: transmission time from the first burst to the last one in a block, taking into account one additional burst for the SACCH, is equal to $(9\times8)-7 = 65$ burst periods, i.e., about 37.5 ms.

Interleaving proceeds as shown in table 4.8. The left column indicates the position of the bits, number in order 0 to 455, in coded blocks, and the right column indicates their position in bursts.

A burst therefore carries 116 bits of coded data as follows:

- 57 bits from a given block B (odd bit positions);
- 1 bit indicating whether this half-burst contains user data or is used in fast associated signalling mode;
- 57 bits from block B+1 (even bit positions);
- 1 bit indicating whether this second half-burst contains user data or is used in fast associated signalling mode.

Signalling blocks have the same length as coded speech blocks, and a block in fast associated signalling mode matches exactly the place of a speech block.

4.3.3. CIPHERING

Among the various advantages of a digital transmission system, the easy protection of data against a non-authorised third party is an important feature for the user. Such protection has been introduced in GSM by means of transmission ciphering. An important point is that the

clear text:	0 1 0 0 1 0 1 1 1 0 0 1...
ciphering sequence:	0 0 1 0 1 1 0 0 1 1 1 0...
ciphered text:	0 1 1 0 0 1 1 1 0 1 1 1...

Table 4.9 – Ciphering and deciphering mechanism

An "exclusive or" operation is performed between the clear text
and the ciphering sequence.

ciphering method does not depend on the type of data to be transmitted (speech, user data, signalling), but is applied only to normal bursts.

Ciphering is achieved by performing an "exclusive or" operation between a pseudo-random bit sequence and 114 useful bits of a normal burst, i.e., all information bits except the two stealing flags. The pseudo-random sequence is derived from the burst number (known through synchronisation mechanisms) and a session key established previously through signalling means (see Chapter 7). Deciphering follows exactly the same operation, since "exclusive-oring" twice with the same data leads back to the original value. Table 4.9 shows an example of such an operation.

The algorithm used to generate the pseudo-random sequence is called "A5" in the *Specifications*. The cipher specialists claim that protection would be just as good if the A5 specifications were known by everybody, because of the type of algorithm and because the key can be changed from communication to communication. However, as an additional security step, the full specification is not included in the public specifications. It is distributed under certain conditions by the association of European operators which have signed the GSM MoU. The algorithm is fairly simple and can easily be implemented in a single VLSI chip by manufacturers of mobile stations and/or base stations.

4.3.4. MODULATION

We arrive now at the last of the steps, the modulation and the demodulation. We enter a realm where mathematics reign even more ruthlessly than in the previous sections. In this section we will present the modulation used on the radio interface of GSM, and an example of how demodulation can be done. For any expert in signal processing, and maybe for others, to say that the modulation is GMSK with BT = 0.3, with a modulation rate of 270 5/6 kbauds, and that a possible demodulation would use a Viterbi algorithm, is the end of the story. We will elaborate slightly, in particular for people who are not experts in the domain but who want to understand what hides behind this jargon. However, this is not so easy, and we apologise that equations appear in this section.

From a *very* general point of view, radio emission is based on the generation of a propagating electromagnetic field. This field will be detected through its effect on remote conductors. The emitter controls certain characteristics of the field (amplitude, frequency, etc.), following pre-determined rules. These characteristics will more or less be kept during propagation; and the task of the receiver is to determine the most probable characteristics imposed by the emitter by looking at the detected field.

Modulation is the function which imposes the characteristics to the electromagnetic field based on a set of rules and the data to be transmitted (which "modulates" emission). In the case of GSM it is the phase of the electromagnetic field which carries the information. Demodulation studies the received signal and tries to determine the data used by the emitter to modulate the field.

It is usual to distinguish the modulation and demodulation on one hand, and the emission and reception on the other. The first processes transform binary data to and from a low frequency modulated signal, and the second pair of processes transform this low frequency modulated signal to and from the electromagnetic field. This distinction corresponds to the use of different technologies. However, the border depends on implementation choices and there is no absolute functional meaning which would describe this border independently from the technical state of the art. The description given in this book will therefore group those two functions as a transformation between data on one side and the electromagnetic field on the other side.

4.3.4.1. The Modulated Signal

Let us start with some generalities concerning modulation. The basic band available for GSM contains two sub-bands around 900 MHz. The multiple access technique calls on frequency multiplexing, the central frequencies being only 200 kHz apart: the spectrum of the modulated signal around a given carrier determines how it interferes with signals in adjacent bands.

In general and mathematical terms, for any modulation fitting in a small bandwidth, the electric field generated at a given instant may be represented by:

$$E(t) = a(t)\cos(\omega_0 t + \varphi(t))$$

ω_0 represents the carrier angular frequency (i.e., 2π times the frequency), which depends on the channel and on time. $a(t)$ and $\varphi(t)$ are respectively the signal amplitude and the signal phase[1]. Both $a(t)$ and $\varphi(t)$ must vary fairly slowly (compared with $\omega_0 t$), in order for the resulting spectrum to occupy the allocated frequency spectrum or less.

The modulation definition must give the relation between the data to transmit and the emitted field $E(t)$. In GSM, ω_0 is not a function of data, but derives from the multiple access rules. The amplitude $a(t)$ does not depend on the data either. The amplitude during the active part of a burst depends on the emission power and may vary according to the power control algorithms, but not with the contents of data. At both ends of a burst, $a(t)$ must follow a given ramping curve in order to avoid spurious emissions due to sharp changes between emission and silence. The GSM modulation scheme is therefore a "constant envelope" modulation scheme, in so far only the central part of the burst is concerned. The contents of data only impact the signal phase $\varphi(t)$.

More precisely, the modulation chosen in GSM is *GMSK* (Gaussian Minimum Shift Keying), with BT = 0.3. The formula giving the phase at a given instant relative to an infinite bit stream ... d_{i-1}, d_i, d_{i+1} ... representing the data is (BT appears only in the computation of parameter σ):

$$\varphi(t) = \varphi_0 + \sum_i k_i \Phi(t - iT)$$

with the following definitions:

$$\begin{cases} k_i = 1 & \text{if } d_i = d_{i-1} \\ k_i = -1 & \text{if } d_i \neq d_{i-1} \end{cases}$$

$$\Phi(xT) = \frac{\pi}{2}(G(x + \frac{1}{2}) - G(x - \frac{1}{2}))$$

---

[1] The signal phase is not an absolute concept, but depends on a convention to define the origin.

$$\text{with}\begin{cases} T = 48/13\,\mu s \\[2mm] \sigma = \dfrac{\sqrt{\ln(2)}}{2\pi 0.3} = 0.441684 \\[4mm] G(x) = x\displaystyle\int_{-\infty}^{x} \dfrac{1}{\sqrt{2\pi}\sigma} e^{-\frac{t^2}{2\sigma^2}}\,dt + \dfrac{\sigma}{\sqrt{2\pi}} e^{-\frac{t^2}{2\sigma^2}} \end{cases}$$

φ_0 may take any value; it is not specified and is therefore unknown to the receiver.

To be precise, the formula of $\Phi(t)$ is obtained as the convolution of a ramp of width one bit period and of height $\pi/2$ (an MSK ramp) by a Gaussian function of parameter s, hence the name Gaussian MSK. There are many formulas to express function Φ. We have chosen this one (though it does not seem very simple) because it contains no double integral.

The function $\Phi(t)$ is shown in figure 4.28. Basically, it consists of a $\pi/2$ step, smoothed in order to have a more narrow spectrum than if the step was steeper (if the step was instantaneous, the modulation would be OQPSK—offset quadrature phase shift keying—, and the spectrum larger; the MSK corresponds to a ramp of width one bit time). This reduction of the frequency spectrum has some counter-effects, such as intersymbol interference: the signal at a given moment does not depend

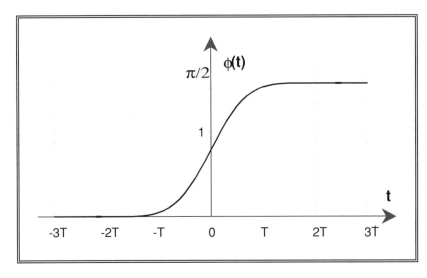

Figure 4.28 – $\Phi(t)$ in the GMSK modulation

$\Phi(t)$ is basically a "rounded" step, whose ramp is spread over 3 bit periods.

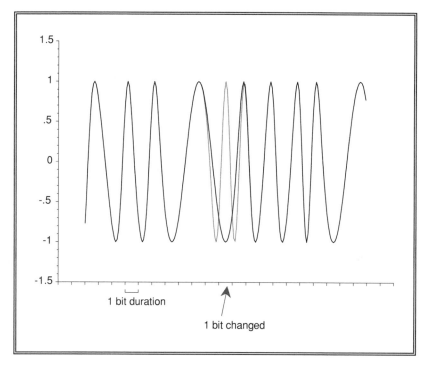

Figure 4.29 – Effect of one bit on the modulated signal

Two bit sequences differing by a single bit are used to modulate
a carrier at a frequency of 0.25 times the baud rate (one cycle every four bits),
using GMSK and the GSM step function $\Phi(t)$:
the effect of the change is negligible outside a window of 3 bit periods.

on a single data bit, or correlatively each data bit influences the signal
during a period exceeding the bit duration.

Theoretically, a modulating symbol k_i influences the signal during
an infinite period. Practically, though, this influence becomes negligible
outside a period lasting 3T, since:

$$\Phi(-\frac{3T}{2}) = \frac{\pi}{2} - \Phi(\frac{3T}{2}) = 0.003,$$

a value very small compared to $\frac{\pi}{2}$.

In order to illustrate this, figure 4.29 shows modulated signals for
two modulating sequences differing by a single bit.

This modulation is basically a frequency modulation. This can be visualised easily by taking the examples of two series of information bits d_i, one of constant value and another one of alternate bit values.

When d_i is constant (i.e., all d_i are equal to 0, or all d_i are equal to 1), all the k_i are equal to 1. A particular property of GMSK is that the phase of the resulting signal varies then linearly with time:

$$\varphi(t) = \varphi_0 + \sum_i \Phi(t - iT)$$

$$= \varphi_0 + \frac{\pi t}{2T}$$

This is a sine wave of frequency $f_1 = (\frac{\omega_0}{2\pi} + \frac{1}{4T})$.

Similarly, a sequence of alternate modulating bits (0 1 0 1 0...) results in all k_i equal to 0, and in a signal whose phase may be written as:

$$\varphi(t) = \varphi_0 - \sum_i \Phi(t - iT)$$

$$= \varphi_0 - \frac{\pi t}{2T}$$

This is a sine wave of frequency $f_2 = (\frac{\omega_0}{2\pi} - \frac{1}{4T})$.

The GMSK modulation has been chosen as a compromise between a fairly high spectrum efficiency (for radio waves...) of the order of 1 bit/Hertz, and a reasonable demodulation complexity. The theoretical modulation spectrum (calculated on an infinite random sequence of modulating bits) is shown in figure 4.30. It can be noted that the choice of a channel separation of 200 kHz results in a non-negligible overlap between the spectrum in adjacent frequency slots, quite higher than that usually tolerated in radio. This source of interference can be limited by careful frequency planning, aiming at separating geographically the usage of adjacent frequencies.

Since emission proceeds by independent bursts and not continuously, the behaviour at each end of a burst must be specified. The number of modulating information bits in a normal burst is equal to 142. By convention, the modulation scheme considers that this sequence is preceded and followed by an infinity of "1"s. The signal amplitude for a normal burst must follow the pattern shown in figure 4.23 (page 233). This amplitude falls to 0 outside the time window allocated to a burst.

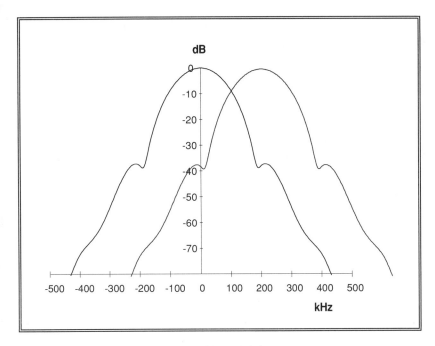

Figure 4.30 – GMSK modulation spectrum

The spectrum of the GMSK modulation used in GSM
is shown here for two adjacent central frequencies separated by 200 kHz.
The overlap is not negligible and frequency planning must take the effect
of adjacent channel interference into account.

4.3.4.2. The Modulator

A typical implementation of the modulator includes two successive
stages. The first stage, which is often referred to as the modulator by
itself, creates a signal defined as above, but with a fixed small value of ω_0
(e.g., 72 MHz). The second stage, called frequency transposition,
transforms this intermediate signal in order to transpose it at the correct
burst frequency and to adjust its power to the desired level before
delivering it to the antenna.

Frequency transposition relies on the fact that a signal may keep its
phase properties while being transposed in frequency by multiplying it by
a sinusoid function at a proper frequency and then filtering the result to
eliminate the unwanted parts. The equation may be written as:

$$\cos(2\pi f_i t + j(t)) \cdot \cos(2\pi(f_0 - f_i)t) = \cos(2\pi f_0 t + j(t)) - \cos(2\pi |f_0 - 2f_i|t + j(t))$$

The choice of frequency dictated by the multiple access algorithm is introduced during this transposition.

4.3.4.3. The Demodulator

After propagation through space, the field received by the remote antenna can be measured and an electric signal regenerated from it. However, the received field differs from the emitted field measured near the emission antenna, which itself is not quite equal to the theoretical emitted field. The differences include:

- a **variable attenuation**, both because of free space loss (depending on the distance between the emitter and the receiver through an inverse law) and also because of environmental shadowing due to the presence of "masks" along the way (buildings, hills, etc.);

- **multipath propagation**, when the signal is reflected or diffracted by different obstacles. This leads to reception of several copies of the signal, slightly shifted in time one with another;

- addition of spurious signals and **noises**, such as thermal noises, spurious emissions from other emitters, and most of all interferences from other GSM emitters using either the same frequency band at the same moment (co-channel interference) or an adjacent band (adjacent channel interference at ± 200 kHz).

The demodulator must estimate the most probable sequence of modulating data, given the received distorted signal. In order to help the demodulator to perform this task, a portion of each burst results from the modulation of a pre-determined sequence which is known by the receiver: the training sequence. This sequence allows the receiver to estimate the distortion of the signal due to propagation, in particular as far as multipath is concerned.

There exist numerous demodulation algorithms. The *Specifications* do not impose one such algorithm, but impose global performance figures measured after correction of the errors by channel decoding. Data about these minimum performances are given for a variety of conditions concerning the environment (urban, suburban, rural, ..., which determines the average nature of the obstacles), the speed, the transmission mode, and so on. One important constraint is that the algorithm must be able to cope with two multipaths of equal power received at an interval of up to 16 µs, i.e., almost four bit periods. In such a situation, the amount of intersymbol interference is dramatically increased compared to what is introduced by the modulation itself. Simple demodulation techniques cannot cope with such intersymbol interference and **equalisation** is required.

An example of a demodulation technique with equalisation, chosen among many acceptable for GSM, is given hereafter. It describes the so-called Viterbi method. One of its advantages is that it makes use of an algorithm which is also fit for maximum likelihood decoding of the convolutional codes, and so opens the possibility to implement demodulation and decoding with common parts.

Viterbi demodulation is a maximum likelihood technique, that is to say, a technique which finds the most probable emitted sequence, taking into account some assumptions on the possible signals, and on the noise statistics. It relies on the knowledge of a finite set of possible signal shapes received during one bit period. Because of intersymbol interference, the received signal depends on more than one modulating bit. If this signal depends on 3 bits, 8 different shapes have to be considered. Since the complexity of the Viterbi algorithm increases exponentially with the number of bits taken into account, this number rarely exceeds 5. A bit period in GSM lasting about 3.7 µs, this figure of 5 is enough to take into account two paths separated by 16 µs.

The demodulator has first to evaluate and to store the 32 reference signal shapes associated with the 32 combinations of 5 successive bits. This is done on a burst basis by an analysis of the part of the signal corresponding to the training sequence.

The assumption about the noise is that it is Gaussian and white. Now, let us take a candidate original sequence, and reconstruct the corresponding signal using the reference signal shapes. At a given instant, the probability that the noise explains the difference between the reconstructed signal and the received one is an exponential of minus the square of this difference. The overall probability that the noise explains the differences all along the part to demodulate is the product of the elementary probabilities. The result is that the probability of some original sequence given the received signal decreases with the integral of the squared difference between the received signal and the reconstructed signal. In GSM the demodulation is done on a burst basis, and the integral is computed on a finite time interval corresponding to n bits.

The next point is that, an integral being a sum, each segment in time, in particular each one-bit period, contributes additively to the function to minimise. Each one-bit-period segment depends by assumption on only 5 of the bits we are looking for. then we can write the value $\mathcal{S}$ to minimise as a sum

$$\mathcal{S} = \sum \eta(a_i, a_{i+1}, a_{i+2}, a_{i+3}, a_{i+4})$$

where $\eta(a_j, a_{j+1}, a_{j+2}, a_{j+3}, a_{j+4})$ represents the contribution to the integral of the one-bit-period part of the signal influenced by bits a_j to a_{j+4}, as computed from the reference elementary signals and the received signal.

The aim consists in finding the a_i values leading to the minimum of this function. Each one of these a_i appears in 5 different terms (the functions corresponding to 5 successive instants).

In order to solve that problem, a mathematical method consists in modelling it by a graph of $16n$ points whose co-ordinates are $(i, \pm 1, \pm 1, \pm 1, \pm 1)$. The "$\pm 1$" values represent the two possible states for each modulating bit. $\mathcal{S}$ is interpreted as a distance between the starting point $i=0$ and the end point $i=n-1$, and the algorithm has to find the path through the graph which minimises this distance. The distance between two points in the graph is one of the terms of $\mathcal{S}$, that is to say:

$$\eta(t, a_i, a_{i+1}, a_{i+2}, a_{i+3}, a_{i+4})$$

for possible successive points, i.e., points of the form $(i, a_{i+1}, a_{i+2}, a_{i+3})$ and $(i+1, a_{i+1}, a_{i+2}, a_{i+3}, a_{i+4})$ and is infinite for any other couples of points (since they would not correspond to a potential sequence). Looking for the shortest path is then equivalent to minimising the expression called $\mathcal{S}$ above. Finding shortest paths in a graph is a classical problem, of wider application than modulation, and efficient solutions exist. With this presentation, the Viterbi demodulation algorithm appears as an application of a classical shortest path algorithm.

SPECIFICATIONS REFERENCE

Within the *Specifications*, the 05 Series is devoted to the subject of transmission on the radio interface.

TS GSM 05.01 is a general description, introducing the major concepts.

TS GSM 05.02 describes the channels in terms of the time aspect and of the frequency hopping characteristics. It also gives the digital structure of bursts.

TS GSM 05.03 specifies the different error correction and detection codes applicable to different uses of the channels.

TS GSM 05.04 is a 3 page document specifying the GMSK modulation.

TS GSM 05.05 is related to "radio transmission and reception". The bulk of the TS is devoted to the transmission performances. It deals also with the frequency aspects, with the emission power (including the burst envelope), and with the electromagnetic compatibility problems. This is were the frequency usage is defined, including the limitations concerning the energy leaking out of the right frequency band (spurious emissions). Conversely, the specifications request receivers to accept some level of out of band energy (sensitivity). These topics are not addressed in this book (it is thick enough), but are nevertheless very important since they have impacts on the cost of the radio equipments, especially the mobile station.

TS GSM 05.08 and **TS GSM 05.10** are related a little to the topics in this chapter, but their contents are really dealt with in Chapter 6 (Radio Resource Management).

Another TS worth noting outside the 05 series is **TS GSM 04.03**, which defines the channels from a usage point of view, and tackles a little the notion of cell channel configuration.

5

SIGNALLING TRANSFER

5

SIGNALLING TRANSFER

In the two previous chapters the focus has been put on the transmission functions close to the physical medium, and we mainly studied the means to transport user information. But user information is not the only thing to be transported in a complex network such as GSM. Most functions performed by such a network are distributed over several distant machines, and information exchanges are needed to co-ordinate what is done in these machines. We will address these exchanges from two points of view. One is simply to present what these exchanges signify, why messages are sent, and what reactions they induce. This will be the subject of Chapters 6, 7 and 8. The other point of view is simply how the messages are transported from one point in the network to another. These transmission aspects call for a number of additional functions in comparison to those presented in the two previous chapters, and will be the topic of this chapter.

In all cases, the information needed for the co-operation of distinct entities, in other words the signalling information, is organised into messages. The sending of a message is triggered by some event, and its reception triggers a chain of events. A typical elementary message consists of some "message type", which indicates what reaction the message will trigger in the recipient, and some qualifying information, under the form of mandatory or optional parameters.

One of the tasks of the transmission protocols is to provide message delimitation out of bit streams. Another is to guarantee a very low level of undetected errors, since such errors can have important consequences, e.g., changing the meaning of a message into another one. These functions are part of the **link layer** functions. We will address in this chapter a variety of link layer protocols, akin to the famous HDLC, including the link protocol adapted to the GSM radio interface.

Another aspect of signalling transfer is the organisation of the message flows, and their routing. If the exchanges between machines were to be done by speech, a network such as GSM would appear as a very noisy crowd, a source of an immense cacophony. Everybody would be speaking to almost everybody else at the same time, using other people as intermediaries to overcome the problem of distance. There is then some need for organisation, independent from the actual significance of the exchanged information. Two main aspects can be identified. The first is the routing problem, i.e., how messages are passed from one point to another until they reach their final destination. The second is how to use references to handle several dialogues in parallel. These are the main aspects of the **network layer**, which will be the topic of the second part of this chapter. There we will see how messages are carried between the mobile station and the MSC. We will also visit the realm of the Signalling System Number 7, a packet data network designed for signalling exchanges.

In fact, our subject-matter is not limited exclusively to signalling information. Although most user information in GSM is of a circuit nature, there are a few user services which are basically of a non-circuit nature, such as short messages. Such services are very much akin to signalling messages in so far as they both require the same kind of transport mechanism. They will then be studied here. Another special case is the RLP, the Radio Link Protocol which was presented at the end of Chapter 3. Because of its relationship with link layer protocols, it will be presented in a bit more detail here.

Though the link and network protocols form the backbone of any system composed of co-operating communicating machines, their detailed understanding is not a pre-requisite for the understanding of the next chapters, which deal with the co-operation itself.

5.1. THE NEEDS

Both signalling messages and short messages require a packet switched system to transfer them. Packets can be stored, combined, segmented, multiplexed, and tortured in many ways along their route. They can be carried over interfaces in a variety of ways, and this is actually what happens in GSM, where virtually each segment along the transmission path bears its own stack of protocols.

One may then wonder why there should be so many different methods when all of them basically fulfil similar purposes? There is no simple answer to this question, but some sensible reasons do exist. One of them is the search for optimisation, which applies in particular for the radio interface, on which traffic load is critical. Another reason is the reuse of existing standards. For instance, the CCITT Signalling System number 7 (SS7) forms the basis for signalling exchanges between the machines of the Network and Switching Sub-system (NSS), since it would have been very uneconomical to choose anything else in a switch environment. But, beside these arguments, the historical background must also be recalled. Each interface was effectively designed by a different standardisation group or sub-group, with different antecedents, and this situation certainly contributed to the variety of protocols one can find in GSM.

When trying to understand the needs for signalling transportation, let us first remark that each entity in the system may communicate with many others. There are cases when the two communicating entities are contiguous, i.e., directly linked through a physical interface, but this is not the general rule. We will start by listing the communication needs between contiguous entities, before proceeding to the relaying cases.

5.1.1. CONTIGUOUS ENTITIES

Starting with the mobile station and unfolding the transmission chain, signalling exchanges within the scope of GSM are required on all interfaces, i.e., between:

- Mobile Station (MS) and Base Transceiver Station (BTS);

- BTS and Base Station Controller (BSC), in particular to enable the BSC to control the BTS;

- BSC and Mobile services Switching Centre (MSC), for example for co-operation in the area of communication establishment and handover;

- MSC and a point of entry into an external network, for the establishment of communications.

Looking at the internal architecture of the mobile station leads to two additional cases, between:

- Terminal Adapter (TA) and Mobile Termination (MT), in particular for the establishment of communications;

- SIM and Mobile Equipment (ME), for the exchange of subscriber related data.

In addition, exchanges must take place between NSS entities (MSC/VLR, GMSC, HLR/AuC, EIR). These entities may be physically interconnected by direct links. This may make sense for neighbouring MSCs for instance, but for an MSC to have direct lines to all HLRs in foreign networks would be very uneconomical, and administratively complex. So, most of the time, NSS entities are connected to intermediate nodes which are part of the general SS7 network. Messages between any two pairs of NSS entities are transported through this network, even when the communicating entities are in different countries. In the canonical model, we will consider that each NSS equipment has no physical interface with other NSS entities, but instead with one or several entry points in the SS7 network.

Another source of signalling needs is the Operation Sub-System (OSS). The OSS is a distributed network of its own, and as such it needs internal exchanges. This topic is outside the scope of the *Specifications* and operators have the freedom to make independent choices for the transfer of signalling inside their own OSS. However, OSS equipment must be able to give orders to the GSM transmission machines (subscription management, configuration), and to receive information from them (observations, alarm reports, charging records). Hence the BSC, MSC/VLR, GMSC and HLR/AuC have to be in contact with an entry point into the OSS. There again, operators are to a large extent free to make their own specifications in that domain, but the subject is tackled a little in the *Specifications*.

More instances of direct signalling exchanges may be found, such as the HLR-AuC interface for instance, but they are not specified by the *Specifications*.

5.1.2. RELAYING

There are a number of cases where signalling messages must be transported between distant machines, i.e., where it appears more economical not to install direct communication means, but instead to use intermediate nodes for transit. In such cases, the intervening nodes act as simple relays, which forward the messages onwards. These relays may have to adapt the messages to the transportation requirements of the different interfaces. They may also have to choose the output direction, depending on an address provided by the sender of the message: this is the routing function. All these functions pertain to the network layer of the ISO layering model.

GSM often uses the term "transparent" in this context, in a way which requires some explanation, since it might be misleading. This concept of "transparent" has already been encountered in Chapter 3, and is used in a similar way for signalling, referring to the transmission of some data by an equipment without disturbing it, and in fact without having to even understand the meaning of the transported information. It can therefore be said that an equipment is "transparent" for such data, in the same way as a transparent piece of glass will let light through without stopping it. The comparison can be pushed further: light can change its course when going through some transparent object, depending on its wavelength, and the direction of the message can be influenced based on its address. The way in which the *Specifications* may lead to confusion is when they refer to a message being transparent, rather than the equipment. In this case, what is meant is that the message is really "opaque" to the equipment! Even though the "transparent" concept and image are useful in understanding protocols, it must be treated with care because of the twist of terms in the *Specifications*.

Let us now proceed with the requirements for relaying between GSM entities. Most of the exchanges between distant GSM entities involve the mobile station at one end: MS-BSC for the management of radio resources, MS-MSC/VLR for location and communication management and MS-HLR for the management of supplementary services. The transmission chain is linear and the BTS, BSC and MSC act as relays for these signalling flows. The case of signalling exchanges between NSS entities has already been touched upon in the previous section, and there all the nodes in the SS7 network act as intermediate relays for SS7 messages.

Another area where relaying is needed is the transfer of short messages. Basically, short messages are exchanged between the mobile station and a Short Message Service Centre (SM-SC). The SM-SC is in contact with one or several MSCs in the NSS, which act as gateways. A mobile terminating short message must then transit through this gateway, the visited MSC and the BSS (BSC + BTS), before reaching the mobile station. A mobile originating short message will experience the reverse path.

In the O&M domain, the BTS is not in direct contact with OSS entities, but exchanges between BTS and OSS transit through the BSC, which therefore also acts as a relay for these messages.

In addition to this list of exchanges, all pertaining to signalling and to short message services, there is one case of circuit-type user data where communication protocols are used inside GSM above the transmission functions of the physical layer. This concerns the RLP, which has been introduced in Chapter 3 when dealing with "non-transparent" (NT) connections. RLP frames are exchanged between the Mobile Termination, inside the mobile station, and the InterWorking Function (IWF) between GSM and external networks. RLP frames therefore cross the BTS, BSC, TRAU and MSC. But these entities do not in this case act as active relays: they are only the "transparent" support of an established circuit, so that for NT data the mobile station and IWF can to some extent be considered as contiguous entities with the meaning used here, and RLP belongs indeed to the same family as other link layer protocols used on single-hop interfaces.

5.1.3. PROTOCOL INTERWORKING

The reader may wonder why logical relations between entities, such as e.g., the HLR and the SIM for maintaining consistency of their respective data, were not treated as such in the previous list of signalling exchanges. This situation corresponds to cases where no signalling message is actually transmitted by one entity (say A) to another (say B). This does not mean that no information is exchanged between A and B, but that this information is transported as components of messages, and not as complete messages. The difference seems subtle, but it is in fact important in terms of specifications. Figure 5.1 shows how this difference can be modelled. When messages are transferred without being tampered with between A and B, the intermediate entities act as carriers (they *are* transparent!): this is message relaying. In the other case, the intervening entities are involved to a much higher degree and one then speaks of protocol interworking. A first message, transmitted by A, triggers the sending by an intervening equipment of one or more messages conveying part of the information carried by the original message towards a third destination, and so on. There is some information transfer between A and B, but only at a **semantic** level. The correspondence between the content of the successive messages is then much more elusive than the identity of bit patterns. This can only be studied when the behaviour of the intermediate nodes is looked at. This will indeed be the subject of the next chapters of this book, where signalling functions will be described topic by topic and will take a complete system view.

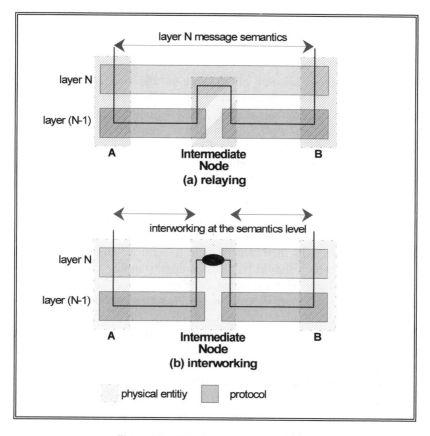

Figure 5.1 – Relaying versus interworking

In case (a), messages of layer N transit between A and B,
and the intermediate node acts as a relay,
i.e., it is not involved in the semantics of the message.
In case (b), the intermediate node is an interworking function.
It is concerned with the semantics of the layer N messages from A
and issues different messages towards B and vice versa.

For the time being, let us start by looking at the basic methods which enable messages to be transported by each individual protocol. The basic transportation means between contiguous entities will first be looked at, illustrating various implementations for the linking aspects (ISO link layer functions). Then the relaying aspects (ISO network layer functions) will be studied, both within the BSS (from the mobile station up to the MSC) and within the NSS.

5.2. LINKING

The link protocols used in GSM are not the same on all interfaces. The ones which will now be described are summarised in table 5.1. They represent the major protocols used by GSM signalling in the linking area. In the mobile station itself, the interfaces between SIM and Mobile Equipment, as well as between Terminal Equipment and Mobile Termination, should for completion be cited as well. In order not to be drawn into too many details, they will not be studied here.

These three protocols (LAPDm, LAPD and MTP 2) have very similar functionality. The presentation will therefore be done on a functional basis, along with explanations of the differences between the cases. The reader will therefore be able to gain a better knowledge of their functions, which should help in the understanding of the individual protocol's behaviour. For a detailed interface-per-interface specification, the reader is referred to the *Specifications*, as well as to relevant ETSI and CCITT recommendations. In the case of the Message Transfer Part (MTP) in particular, GSM has not introduced any new functionalities, but has used the protocol as defined in the ETSI specifications, themselves referring extensively to the CCITT recommendations of the Q series. The examples in this chapter will mostly be taken from the LAPDm protocol—and to some extent from the Link Access Protocol for the ISDN "D" channel (LAPD) protocol from which it derives—which is, among the examples shown, the link protocol with the most original features.

In all cases except for the radio interface, signalling messages are sent over plain 64 kbit/s circuits. But between MS and BTS, the physical medium is very specific to GSM. On this interface, the transport of point-to-point messages can be done in two ways, as explained in the previous chapter:

Interface	Link protocol
MS-BTS	LAPDm (GSM specific)
BTS-BSC	LAPD (adapted from ISDN)
BSC-MSC MSC/VLR/HLR-SS7 network	MTP, level 2 (SS7 protocol) "

Table 5.1 – Link protocols on GSM interfaces

Despite their apparent variety, the link protocols used on the GSM interfaces
have similar functionality.

- using the main channel, if necessary by preempting the resource (Fast Associated Signalling). This stealing method typically consists of not transmitting one block of user data (representing 20 ms), and using the corresponding resource to send a signalling message instead;
- using the Slow Associated Control Channel (SACCH).

Let us now consider how the data is structured on the link layer to be transported on these physical media.

5.2.1. STRUCTURING IN FRAMES

The prime functionality of a link layer is to structure the information to be transmitted on the channel in units bigger than a single bit. The resulting atomic units are the basic structure on which all link layer functions work. In the signalling world, such a unit is called a **frame**. The whole issue consists in including sufficient information in the bit stream so that the receiver is able to find out the beginning and the end of each frame. Both LAPD and MTP 2 are the heirs of HDLC in this area, whereas LAPDm makes use of the synchronisation scheme of the radio interface to convey information on frame limits.

In HDLC, frames start and end with an eight-bit long pattern called a flag (see figure 5.2). To prevent false starts or ends, a mechanism ("0" bit insertion) is introduced to disguise the flag pattern when it occurs inside the data. The resulting bit stream only contains the "legal" flags.

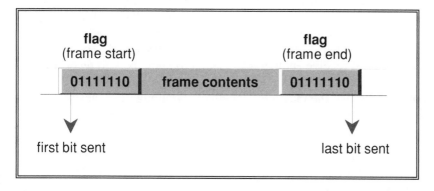

Figure 5.2 – HDLC frame flags

Frames start and stop with a defined pattern called a flag.
The figure applies both for LAPD and MTP 2.
Between transmitter and receiver, a "0" is added to each sequence of 5 consecutive
"1"s in the data, in order to disguise the flag pattern when it appears inside the frame.

The flag mechanism allows frames to be of variable lengths, without even the need to indicate the actual length value inside the frame. The same flag can be used to indicate both the end of one frame and the start of the next one.

In LAPDm, the use of flags (and the corresponding octet waste) can be avoided thanks to the ready-made blocks of the physical layer. In order to benefit from this advantage (and also for error detection reasons), the choice was made to fit each frame into a single physical block, which is 23 octets long. As a consequence, a LAPDm frame has a maximum length of 23 octets on all TCHs, and of 21 octets on the SACCHs (this difference comes from a specific usage of 2 octets per SACCH block, for timing advance and transmission power control, as will be explained in Chapter 6). But the effective information length may be smaller than this maximum value. A length indicator is therefore included in each frame. Unused octets are filled with a default value, which happens to be "00101011". This value has been chosen to minimise the probability that a frame with little information (and hence many fill octets) results in a burst similar to the Frequency Correction Channel (FCCH) bursts, which would disturb mobile stations attempting to synchronise. For historical reasons, the value "11111111" may still be used as a filling pattern for uplink transmission by mobile stations.

5.2.2. SEGMENTATION AND RE-ASSEMBLY

The maximum length of frames may be limited as a result of lower layer transmission constraints, or more generally to ease the dimensioning of buffers in the system. When the maximum length of a signalling message exceeds the maximum length allowed for frames, this message must be segmented and sent over several frames. Conversely, the message must be re-assembled at the receiver end. To do this, the receiver must receive enough information to know how to reconstruct the messages, and this causes additional overhead in the protocol. When it is not foreseen that the maximum length of signalling messages will exceed the maximum length of frames, then the segmentation and re-assembly mechanisms are usually dispensed with.

This is actually the case on the A interface. The maximum length of frames on the A interface is limited to 272 octets of information (plus 6 octets for frame control, excluding flags). This maximum value has been inherited from the SS7 world, where it replaces an older value of 62

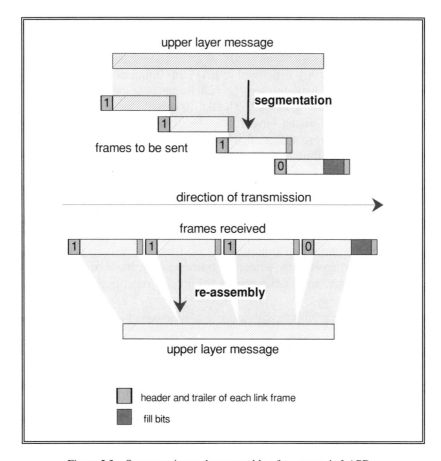

Figure 5.3 – Segmentation and re-assembly of messages in LAPDm

The "more" bit of LAPDm frames enables the receiver to reconstruct
the original message by concatenating the contents of frames
until there are no more frames for this message ("more" bit set to 0).

information octets. The value is enough to accommodate most signalling
needs, and no segmentation is foreseen at this level. As a consequence,
many GSM protocols have inherited an upper bound for its message
length which stems directly from this value.

Taking this constraint into account, there is no need to define a
segmentation and re-assembly facility on the Abis interface either. The
length of the Abis LAPD frames is simply limited to 264 octets
(excluding flags), which corresponds to 260 octets of upper layer
information.

Things are different on the radio interface, where the maximum length of a frame is either 21 or 23 octets. Such a length is clearly not sufficient for most signalling needs. As a consequence, a segmentation and re-assembly facility is defined in LAPDm. It makes use of a so-called "more" bit, which distinguishes the last frame of a message from other frames, as shown in figure 5.3. Thanks to this mechanism, there is no intrinsic limitation on the length of radio path messages, and the only constraint comes from the need to transfer such messages on the other interfaces, hence the 260 octets mentioned in the radio interface specifications.

5.2.3. ERROR DETECTION AND CORRECTION

The next important functionality of a link layer is to improve the quality of transmission, by detecting frames which have been subject to transmission error, and then possibly asking for repetition of the frames in error.

As far as error detection is concerned, both LAPD and MTP 2 use the HDLC scheme which consists in adding 16 redundancy bits (called FCS, or Frame Check Sequence in LAPD) to each frame, according to a coding scheme chosen for its error detection characteristics. In both protocols, the same generator polynomial is used for calculating the 16-bit signature: $x^{16}+x^{12}+x^5+1$. On the radio path, LAPDm can dispense with such a mechanism, thanks to the error detecting performance of the transmission coding scheme offered by the physical layer.

Error detection serves two purposes. The first one is to provide enough information on the likelihood of residual errors in a frame, so that repetition of the frame can be asked for. The second one is to monitor the quality of the link, in order to trigger the relevant alarms when the error rate exceeds some given threshold. Let us consider how monitoring is done in GSM, before describing how the repetition mechanism works.

Link quality monitoring on SS7 links is done in such a way that the link is declared out of order when the error rate exceeds some value (typically, frame error rate $\geq 4.10^{-3}$). Let us remark that a SS7 link is never inactive when in service: specific filling frames are transmitted when nothing else needs to be transmitted. Error counting is therefore reliable and watchdog timers can be defined to monitor the link behaviour. On the radio path, a similar situation (regular transmission) exists on the SACCH. This channel is then used for quality monitoring, and a counter is incremented and decremented according to the validity of each block. A link failure is reported to the upper layers when the counter (whose initial value *RADIO_LINK_TIMEOUT* is set by the operator) reaches zero.

Next comes the frame acknowledgement and repetition function, which supports a very high performance in terms of removing remaining errors. None of the three protocols studied here makes use of forward error correction capabilities (such features are usually considered to be in the physical layer). All three use HDLC-like mechanisms for backward error correction, with a choice between two modes:

- the non-acknowledged mode, in which frames are transmitted once, whatever the outcome at the receiver end;
- the acknowledged mode, ensuring correction of erroneous frames by repetition.

The two modes coexist in all three protocols. One may wonder what is the use of the non-acknowledged mode, if better error performance is available using the acknowledged mode. But the choice is not completely black-and-white. For example, for the recurrent measurement messages sent by the mobile station over the radio and Abis interfaces, the non-acknowledged mode is more adequate than the acknowledged one. Not only does a loss not cause any great damage, but furthermore, it is better for the receiver to get a new (and up-to-date) measurement report in this case rather than to receive a repeated (and obsolete) message. So measurement reports are actually transmitted using the non-acknowledged mode on the radio and Abis interfaces. A few other messages are also sent unacknowledged, such as the answer to a handover access (e.g., the RIL3-RR PHYSICAL INFORMATION messages) and the general information sent by the network to mobile stations on the SACCH (e.g., the RIL3-RR SYSTEM INFORMATION TYPE 5 AND 6 messages—all these messages come from procedures cited in Chapter 6), not to mention all the messages sent on a point-to-multipoint channel (BCCH or PAGCH). Apart from these messages, the vast majority of messages sent on dedicated channels are sent in acknowledged mode.

Acknowledgement and repetition is based on a cyclic frame numbering, which enables the recipient to detect possible repetitions and/or losses of frames, and to acknowledge specific frames. In LAPD and LAPDm, the acknowledgement is done by the receiver transmitting the number of the next expected frame to the sender, in an indicator

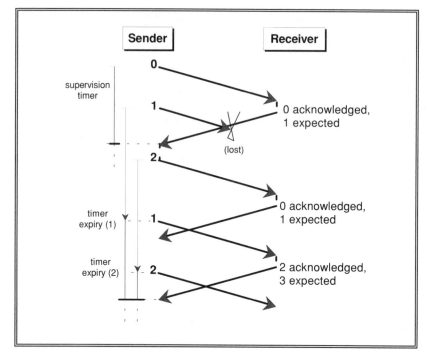

Figure 5.4 – Repetition mechanism

Repetition is triggered by the sender in two cases:
– when it receives an acknowledgement for a frame which is not the last one sent;
– when it does not receive an acknowledgement after a certain amount of time.

called N(R). The mechanism is shown in figure 5.4. If the numbering of frames is modulo 8, a receiver expecting frame 2 indicates implicitly that frames 1, 0, 7, 6, ... have been correctly received. In MTP 2, a similar goal is achieved by transmitting the number of the last frame correctly received to the sender. In all cases, the sender is then able to repeat non-acknowledged frames, if any. However, the total number of repetitions is limited in order to avoid infinite loops when a serious problem occurs.

Frames must be kept by the sender until they are acknowledged, so as to be available if repetition is needed. In order to limit the number of corresponding buffers, as well as to avoid numbering ambiguities, the concept of a window is used in LAPD and LAPDm. The size of the sending window determines the number of frames which can at any given moment be sent and not yet acknowledged. This window size K must be

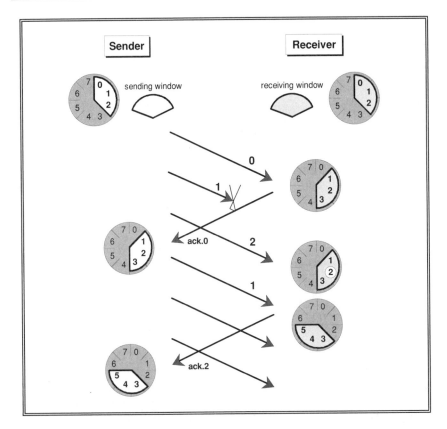

Figure 5.5 – Window mechanism for acknowledgement

Windows represent a sliding set of contiguous frames which can be either:
– sent and not yet acknowledged (sending window), or
– accepted for reception (receiving window) at a given moment.

sufficiently high to enable anticipation by the sender without waiting for acknowledgement delay. Figure 5.5 shows how the sliding window mechanism works on the sender side as well as on the receiver side.

The numbering cycle of LAPD (and MTP 2) is 128. On LAPDm, a smaller numbering cycle (8) was chosen, to reduce the size of the frame header. In LAPD, the window size can be configured. But in the case of LAPDm, the window size is set equal to 1. This choice has been justified by the expected simplification of the protocol implementation; a window size of 1 corresponds to a simple send-and-wait protocol. In the case of the small dedicated channels used for signalling (TCH/8s), performance does not suffer any degradation from this simplification, because the organisation of the channel is of a basically alternate nature, with enough time between a received frame and the next sending opportunity to build

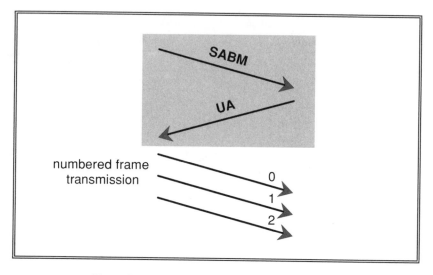

Figure 5.6 – Acknowledged mode setting procedures

The acknowledged mode is set-up on a link by an exchange of two commands,
enabling contexts to be reset on each side of the link for a given link flow.

the answering frame. As for other channels, transmission will incur additional delay when several frames are to be sent in a row because of this window size of 1. For instance, the signalling capacity of full-rate channel (TCH/F) will be at best one frame every 120 ms, i.e., no better than twice the capacity of a TCH/8. However, signalling applications suffer little from this restriction, since signalling frames usually become ready one at a time, the major exception being when a message is segmented over several frames.

In order to initialise the contexts on both sides of an interface for a transmission in acknowledged mode, a simple procedure is used in LAPD and LAPDm. It consists of two messages, as shown in figure 5.6. In LAPD, the exchange of upper layer information can only take place after such an exchange. But in the case of LAPDm, the SABM (Set Asynchronous Balanced Mode, a term inherited from HDLC jargon!) frame is able to carry a "piggybacked" message repeated in the UA answer (Unnumbered Acknowledge), in order to solve potential contention, as explained in Chapter 6.

In a similar way to the setting up of transmission in acknowledged mode, the normal release of a link is performed through a simple procedure, as shown in figure 5.7. No piggybacking is allowed on the corresponding frames in any of the link protocols.

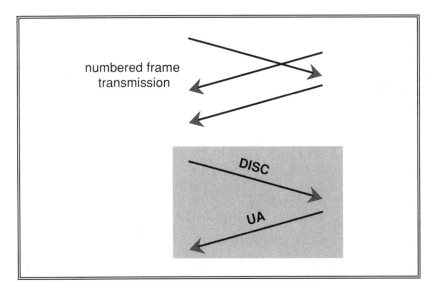

Figure 5.7 – Acknowledged mode release procedures

Acknowledged mode transmission is normally terminated
through a two-way exchange,
enabling contexts to be released on each side of the link for a given link flow.

At any time, an unacknowledged frame of information may be sent. When no frame is pending and transmission should occur, then "fill frames" are sent, which consist of an unacknowledged mode information frame ("UI" frame, for Unnumbered Information) whose information length is null.

5.2.4. MULTIPLEXING

Thus far, the flow of messages has been considered to be a single one, i.e., a succession of frames delivered one after the other by a single source. However the link layer offers the possibility of multiplexing independent flows on the same channels. These flows are independent in so far as the ordering of frames between them is not guaranteed and the window mechanism applies to each flow in a separate manner. In order to distinguish the flows from one another, an address is inserted in each frame. Such a mechanism is essential on point-to-multipoint links, such as in the ISDN LAPD, which has been designed for user installations with one line and several terminals. It has been kept in the Abis LAPD. Though the TACHs (i.e., the dedicated channels on the radio path, see

	TCH/F	TCH/8	SACCH
signalling (SAPI 0)	ack. mode	ack. mode	non-ack. mode
short messages (SAPI 3)	–	ack. mode	ack. mode

Table 5.2 – SAPIs on radio channels

All channels are not suitable to all combinations of the two SAPIs,
SAPI 0 for signalling and SAPI 3 for short messages.

Chapter 4) are point-to-point links on the radio interface, the link multiplexing mechanism is also provided in LAPDm.

On the radio interface, two independent flows can exist simultaneously. The first one is devoted to the transfer of signalling messages, and the second one to the short message services. These two flows are distinguished by a link identifier called the SAPI (Service Access Point Identifier). This term is not quite proper OSI terminology since a SAPI defines the point of entry from upper layers into the link layer on each side of the interface, and the correct OSI terminology for link identifiers is DLCI (Data Link Connection Identifier). However, in GSM, the "SAPI" refers to the link identifier transmitted in the protocol, and this usage was inherited directly from ISDN access protocols. Nothing else but history can explain the choice of such a mechanism to discriminate functional flows, since the creation of separate layer 2 links was by far not necessary. The LAPDm "SAPI" can take values 0 (signalling) and 3 (short messages). Hence the reference to SAPI 0 and SAPI 3 which appears here and there in the *Specifications*. All combinations of SAPIs on channels are not possible, and table 5.2 lists the allowed possibilities, together with an indication of the modes (acknowledged or non-acknowledged) associated with each SAPI on each channel type.

The only case in LAPDm where two real independent flows have to be transmitted simultaneously with acknowledgement and repetition is on the TCH/8. The choices were dictated both by a will to reduce the number of cases and also to prevent the use of the TCH/F for non-urgent matters, since user data might be disturbed by pre-emption. As a consequence, the transmission of short messages is fairly slow. The link rate gives a maximum throughput of roughly 80 octets per second on the TCH/8, i.e., the equivalent of a 600 bit/s circuit. We will see in the next section that upper layers further reduce this rate of transfer.

Multiplexing on the Abis interface has two facets. The first one corresponds to the distinction between different functions. It is performed

"SAPI"	Type of flow
0 62 63	Radio signalling Operation and maintenance Layer 2 management

Table 5.3 – Different flows on the Abis interface

In addition to the radio signalling procedures, the Abis interface also carries
a flow dedicated to the operation and maintenance of BTSs,
as well as a layer 2 management flow directly inherited from ISDN access.

in a similar way as on the radio interface. The "SAPI" values on this interface are listed in table 5.3. SAPI 0 is used for all messages coming from or going to the radio interface. The other application of multiplexing is to provide different links ending in different pieces of equipment inside the BTS: the TRXs (Transmitter and Receiver). This topic will be revisited in the networking section. The discrimination makes use of another field of the LAPD link layer address, the TEI (Terminal Equipment Identity). The dynamic management of the TEIs is one of the functions of the messages of SAPI 63.

No multiplexing is done at the link level (MTP 2) on the A interface.

5.2.5. FLOW CONTROL

The last issue to be dealt with as far as the link layer is concerned is flow control. When a single link is considered, the usual assumption is that the processing and buffering capability of the receiver is dimensioned to cope with the maximum throughput of the link. However, resources are often shared between different flows, with a handling capacity lower than the sum of the maximum capacity for each of the flows. One of the goals of congestion control is then to control each of the flows so that an overload on some part of the system does not cause the overall system capacity to crash to zero, but instead to try achieving the maximum throughput possible. Bottlenecks may be fairly remote from the actual sources of flows, and it is essential that congestion situations are reported backwards to control the input load, and ultimately to require the source to slow down. Flow control on each of the segments along the transmission chain is one of the tools which helps to keep the throughput under control.

Flow control is in some way provided naturally by HDLC-like protocols, simply by delaying the transmission of the acknowledgements. But this control is only marginal, because if the delay becomes too long the sender will repeat the frame, which may have counter-effects in congested situations. An additional mechanism can also be used, consisting in a "stop-and-go" control using two commands. Such a mechanism is provided in LAPD and in MTP 2, but not in LAPDm.

5.2.6. LAPD AND LAPDm FRAMES: A SUMMARY

As an application of the mechanisms presented in the previous sections, the list of frames used in LAPD and LAPDm will now be briefly presented, as well as their structure. This review is by no means detailed, since GSM has introduced few specific modifications in this domain apart from the differences between LAPDm and LAPD which have already been introduced. Table 5.4 lists the frame types identified in the two protocols, together with their respective roles.

The respective structures of a LAPD and a LAPDm frame are shown in figure 5.8. The address field contains the so-called "SAPI", and in addition, for LAPD only, the address of the destination terminal, since the interface is point-to-multipoint. The control field contains the type of

frame	meaning	role
SABM DISC UA DM UI	Set Asynch. Balanced Mode Disconnect Unnumbered Ack. Disconnected Mode Unnumbered Information	first frame to set-up acknowledged mode first frame to release ack. mode ack. to e.g., the above two frames response indicating disconnected mode information frame (non-ack. mode)
I	Information	information frame (ack. mode)
RR RNR* REJ FRMR**	Receive Ready Receive Not Ready Reject FRaMe Reject	"you may go on" (flow control), also used for acknowledgement "you should stop" (flow control) negative acknowledgement error back-reporting

Table 5.4 – LAPD and LAPDm frame types

Frames belong to three families: unnumbered frames,
information transfer frames and supervisory frames.
*The "RNR" frame (for flow control) can be ignored in LAPDm.
**The "FRMR" frame is not used in LAPDm.

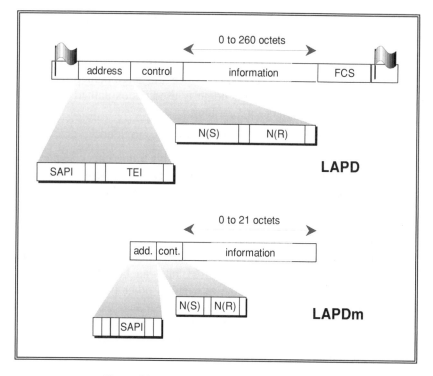

Figure 5.8 – LAPD and LAPDm frame structure

LAPDm frames are derived from LAPD frames, without the FCS,
and with shorter address and control fields.
LAPDm frames do not need flags for synchronisation.

frame, as well as the number of the frame (sending side) and the number
of the next expected frame (receiving side) for numbered information-
carrying frames.

5.2.7. RLP CHARACTERISTICS

Before moving to the next layer up, one should recall that there
exists a link protocol in GSM which is not concerned with the transport
of messages such as signalling messages or short messages. It is the RLP,
which has been introduced in Chapter 3 and which concerns the transfer
of user information between the mobile station and the IWF. The major
functions of a link protocol, as described in the previous sections, apply

for RLP as well, which is yet another HDLC-like protocol. Let us now briefly consider its characteristics.

- structuring in frames: RLP frames consist of 240 bits, without a need for frame delimitation, since the alignment provided by the radio transmission scheme is used, as explained in the two preceding chapters for LAPDm;

- segmentation and re-assembly: none is provided;

- error detection and repetition: a 16-bit frame check sequence is included in each frame. As in LAPD or LAPDm, information can be sent either in acknowledged or non-acknowledged mode. The latter is initialised through an SABM/UA exchange and released through a DISC/UA exchange. The repetition mechanism in acknowledged mode is similar to LAPD or LAPDm, using a numbering scheme modulo 64 and a window size which is usually set to 61 but can be adjusted to lower values. This window size as well as other parameters such as the maximum number of re-transmissions, the timer value for deeming a frame lost and the delay before acknowledging a frame are negotiable through a procedure consisting of an exchange of two specific "XID" frames (eXchange IDentification). In addition for the LAPDm protocol, an "SREJ" frame (Selective REJect) enables the receiver to request the repetition of a single frame;

- multiplexing: not provided;

- flow control: RR and RNR frames are used to indicate the capacity of the receiver to accept more frames or not.

A peculiarity of the RLP protocol is the possibility to piggyback information on supervisory frames, an improvement compared to HDLC. The same frames (say from A to B) can be used to carry information from A to B as well as control fields related to the B to A information flow. A partial application of this possibility has already been encountered with the N(R) indicator contained in the control field of LAPD (and LAPDm) information frames, used to acknowledge the information frames sent in the other direction. But RLP goes further, enabling information to be carried on all supervisory frames such as REJ, SREJ, RR or RNR.

The frame families listed in table 5.4 (see page 280) for LAPD and LAPDm also apply to RLP. The only difference is the addition of the "SREJ" frame and of three unnumbered frames: the XID frame (which exists in ISDN LAPD, but is not used in the Abis LAPD), the TEST frame (used back and forth to test the link) and the NULL frame (used as a filling frame).

5.3. NETWORKING

The link protocols described above enable the exchange of frames between two entities which are directly interconnected through some physical medium. Now, there are a number of application protocols which involve two entities not directly interconnected. Additional transmission functions, described in this section, are needed to provide these application protocols with end-to-end connections for the transfer of the corresponding messages. These connections make use of successive elementary links, as described in the previous section, along a route between the two extremities. Conversely, an elementary link is used for a number of network connections in parallel, with possibly different origins and ends. For example, call control messages originating from the mobile station must be routed to the MSC, whereas radio resource management messages generated by the same mobile station must terminate their course in the BSC, even though both flows of messages use the same signalling link on the radio path and on the Abis interface.

One of the networking functions is the routing, i.e., the choice of the successive segments composing the route. In this area two broad techniques are used and we will give examples of both of them. With datagrams, each message is routed following analysis on its arrival. With virtual circuits, a route is established for some time, for the use of complete dialogues: the route is established by the first message, and the following messages just follow the same route.

Another function we will see here is closely related, and consists in the possibility of having several independent connections existing in parallel between two entities. This can be used between contiguous entities as well as distant ones. The connections correspond to independent application dialogues, for instance to the management of different user communications.

A concept common to all the aspects of networking is the address, which will be a leitmotiv of this section. The network protocols add tags to the messages to discriminate between the different flows. These tags can be addresses identifying the origin or the destination, or connection references, or route references. They are used to choose a route, i.e., to forward the message onto the next appropriate segment, or to distribute it to the right application software.

Let us now consider how this issue is tackled in the different subsystems, starting with the BSS.

5.3.1. NETWORKING IN THE BSS

5.3.1.1. The Mobile Station Point of View

From the point of view of a mobile station, the origin or destination of the messages depends on the application protocol. The mobile station addresses different network functional entities, and these addresses are then used by the network to actually route the message to the appropriate equipment (BSC, MSC or HLR). Moreover, several protocols are handled between the mobile station and MSC. In addition, several user communications may have to be managed in parallel between the mobile station and the MSC, as we have seen in Chapter 1, e.g., for the indication of a new incoming call when another already exists.

The identification of the link on the radio interface gives the ability to distinguish signalling messages from short messages for any single mobile station, but this distinction is not enough to determine the application protocol a message pertains to. It must be complemented by a networking address. This function is fulfilled by the Protocol Discriminator (or PD). Several PDs are defined (see table 5.5), which were originally introduced just as a functional partition of the messages. As a consequence of the GSM functional architecture, this partition also corresponds to an entity on the infrastructure side, and this is why we can consider the protocol discriminator as an address. The case of supplementary services (SS Protocol Discriminator) raises specific problems which will be dealt with in a dedicated section.

The BTS does not appear in the list of partners on the infrastructure side (third column of table 5.5). This reflects the fact that the mobile station does not have a dialogue with the BTS for reasons other than for link management. There is however a single exception to this rule: during handover, one message is sent directly from the BTS to mobile station, for the sake of speed (see the handover execution procedure in Chapter 6).

The protocol discriminator is specified as part of the application protocols (in TS GSM 04.08), though it belongs to a sublayer common to several protocols. The PD is inserted by the originator. In the mobile station to infrastructure direction it is used by the BSC to decide whether

PD	function	Origin/destination
CC, SS	call control management and supplementary services management	MS from/to MSC (and HLR)
MM	location management security management	MS from/to MSC/VLR
RR	radio resource management	MS from/to BSC

Table 5.5 – Protocol discriminators on the radio interface

Three protocol discriminators are defined,
which correspond to an origin/destination on the infrastructure side.

it is the destination (RR) or if the message has to be forwarded to the MSC (all other cases). The PD of received messages is used by the mobile station and the MSC to distribute them to the right software module.

Now, this is not sufficient to discriminate between CC and SS messages pertaining to different user communications. The term transaction is used in this context, each transaction corresponding to a communication. In fact, transactions also exist which are used for supplementary service management. In all cases, distinct transactions can exist in parallel. Messages pertaining to different transactions are distinguished by a Transaction Identifier (TI). The TI is inserted by the originating entity (MS or MSC), and is used by the other to relate the message to the right context. The specifications related to the TI are part of the specifications of the RR protocol, which appears in this case as below the MM and CC protocols.

5.3.1.2. Abis Interface

The messages exchanged over the Abis interface signalling links have many different origins and destinations. From a functional point of view, a first split identifies the messages between BTS and BSC (for BTS control) from all other messages transmitted between the mobile station and any infrastructure entity beyond the BSC (including between the mobile station and the BSC). Going a step further, there is a need to distinguish the different mobile stations, which is done by distinguishing the different radio channels (the relationship between mobile stations and radio channels is managed by the BSC, not by the BTS).

The specifications have added another dimension to this functional view, which is the explicit discrimination on the Abis interface between sub-entities of the BTS, the TRXs (Transceiver and Receiver). The notion of the TRX has already been presented in Chapter 4, in the context of the organisation of the radio channels in a cell. It was then an abstract concept, which becomes concrete when the Abis protocols are looked at. Each TRX in a BTS corresponds to one or several signalling links, as defined in the section dealing with "linking". These links are distinguished by "TEIs" (Terminal Equipment Identities). The TRX a message pertains to is not identified by a tag in the message, but implicitly by the signalling link which conveys it. This small detail forbids any implementation where a link is shared between different TRXs, and so imposes the concrete existence of TRXs.

Still, on each given TRX-BSC link, there remains the need to distinguish the general management messages from the messages pertaining to specific radio channels, and for the latter to discriminate between the different channels a TRX manages. In order to reach these goals, each message on the Abis interface carries a "message discriminator", together with complementary data, giving enough information to know what to do with the message in the BSC (for uplink messages) or in the BTS (for downlink messages). Table 5.6 lists the different message discriminators, the complementary data, and their use. For instance, the "radio link layer management" discriminator is used for the messages which come from or are to be transmitted to mobile stations. Such messages on the Abis interface also carry a reference to a radio channel (which determines the mobile station), and to the link which is used on this channel. The channel indication contains the type of channel (TACH/F, TACH/8, BCCH, etc.), the time slot number and, when appropriate, the sub-TN of the channel. Note that this information

Message discriminator + complementary data	Communication end nodes	Use
Radio link layer management + channel reference + radio link reference	MS-BSC or beyond	Relay of radio path messages
Dedicated channel management + channel reference	BTS-BSC	Interworking for a given TACH
Common channel management + channel reference	BTS-BSC	Interworking for a given BCCH or PAGCH/RACH
TRX management	BTS-BSC	Control of TRX status

Table 5.6 – Message discriminators on the Abis interface

Four different message groups are defined between BTS and BSC, the first of which corresponds to messages transmitted or received "transparently" on the radio interface by the BTS.

From BSC to BTS	From BTS to BSC	Use
ESTABLISH REQUEST	ESTABLISH INDICATION ESTABLISH CONFIRM	link establishment
DATA REQUEST	DATA INDICATION	acknowledged info. transfer
UNIT DATA REQUEST	UNIT DATA INDICATION	non-acknowledged information transfer
RELEASE REQUEST	RELEASE INDICATION RELEASE CONFIRM	link release
	ERROR INDICATION	link error notification

Table 5.7 – Abis Radio link management messages

The messages relayed by the BTS for communication between the BSC (or beyond) and mobile station are carried on the Abis interface in an envelope containing the type of message as shown here, as well as information on the radio path channel and link.

is sufficient to determine the channel only in association with the TRX identification carried at the link level. The radio link information indicates the LAPDm link on which the message is to be sent or has been received, and discriminates between "SAPIs" 0 or 3, and between the TCH and the SACCH when applicable.

For messages for which the BTS acts as a "transparent" relay (the radio link layer management messages), the additional envelope used to carry these messages on the Abis interface is very similar to the primitives between a network layer and a link layer in the ISO model. The corresponding types of messages are shown in table 5.7. This is presented here as an example of pure relaying protocol. The BTS can be considered as a remote radio link layer entity of the BSC.

5.3.1.3. A Interface

The A interface is used for messages between BSC and MSC as well as for messages to and from the mobile station (using the CC or MM protocol discriminator as defined in table 5.5). These two flows are referred to respectively as BSSMAP (BSS Management Part) and DTAP (Direct Transfer Application Part). In addition to this distinction, BSSMAP messages which are of a general nature for the BSC must be distinguished from messages specific to a connection with a mobile station, and the latter must be identified between themselves. For this last function, a virtual circuit approach was chosen, with independent connections established and released. The choice was to use an SS7 protocol, the SCCP (Signalling Connection Control Part), which provides this function in addition to many others. The distinction between

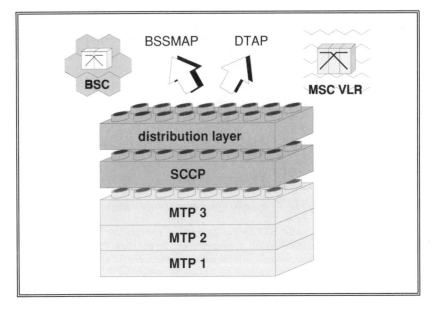

Figure 5.9 – Signalling message transport on the A interface

The A interface makes use of the SS7 MTP-SCCP stack of protocols to transfer signalling messages between BSC (or different mobile stations) and MSC.

BSSMAP and DTAP is fulfilled by a small "distribution" protocol on top, as shown in figure 5.9.

SCCP is not a GSM-specific protocol, although GSM is one of the first implementations making use of SCCP, together with intelligent networks. SCCP is part of the signalling system n°7 stack of protocols, and when used appears just above the MTP protocol. However, SCCP is not the basic network protocol in SS7. In fact, MTP includes more than the link protocol (MTP 2) described in the previous section. It is the combination of SCCP and MTP 3 (i.e., the network layer of MTP) which offers a networking service on the A interface. Let us now consider briefly the characteristics of these protocols. In a similar fashion to our earlier description of the MTP 2, details of the protocols will not be dealt with here, and the reader is referred to the SS7 specifications.

MTP 3

MTP 3 includes several aspects: one of them is the management of the SS7 network (management of traffic, channels and routes), another being the actual routing of messages inside an SS7 network. In fact, MTP is used on the A interface to transfer messages between two contiguous entities (the BSC and the MSC), and then these networking functions are of no use. But MTP 3 also fulfils some functions of local application. They are related to the possibility, which is offered on the A interface, to install several physical links rather than a single one to transport the messages (a set of such links is called a "linkset" in the SS7 terminology). The main interest in so doing is the redundancy, which enables the network to keep things going if one link breaks. MTP 3 provides some functions to manage linksets, as for instance the possibility to split the signalling load between several links (this is "load sharing"); and the possibility to send messages on a fallback link instead of the "normal" one used within a linkset, e.g., for maintenance purposes, without loosing the sequencing of messages, as well as the reverse operation.

SCCP

As an SS7 protocol, SCCP has many functions. Only a small part of them is used on the A interface, and we will see other SCCP functions when dealing with the NSS. SCCP defines several classes of services, of which only two are used on the GSM A interface: the basic connectionless mode (which in the case of the A interface equates more or less to no added function!) and a connection-oriented mode without such goodies as flow control, expedited data, or reset. These modes are referred to respectively as the class 0 and class 2 SCCP services. The class 0 mode is used on the A interface for the messages not directly related to a single mobile station, such as reset or overload indication. The class 2 mode enables separate independent connections to be set-up, a function used on the A interface for distinguishing transactions with individual mobile stations. Such SCCP connections are established when the need appears, either by the BSC (this is the general case when a signalling transaction such as a location updating or call set-up is started) or by the MSC (in the single case when handover needs to be performed to a new cell). They are released when the need disappears. The BSC manages a context per connection, where it stores the BTS and the radio channel used to communicate with the relevant mobile station. This enables the BSC to make a one-to-one correspondence between messages from/to the BTS and using a given TACH on one side, and to/from the MSC and using a given SCCP connection on the other side. A similar context is kept by the MSC, where each SCCP connection is related to the identification of a mobile station.

DTAP/BSSMAP Discrimination

Still, the use of individual per-mobile station connections does not answer all the needs of message distribution on the A interface. Both

BSSMAP messages (between BSC and MSC) and DTAP messages (between MS and MSC) may refer to a given radio connection, but their handling is different and must be distinguished. An additional "distribution function" has been introduced on top of SCCP to this avail. It consists basically in adding a small header to the application message before transferring it on a given SCCP connection. All messages on the A interface therefore bear a discrimination flag, indicating whether the message is a BSSMAP or a DTAP message. As explained earlier, this usage means that the BSC acts as a transit node between the mobile station and MSC for DTAP messages, whereas BSSMAP messages originate or end in the BSC.

In addition to the discrimination flag, DTAP messages carry information on the type of link on which they are to be (or have been) transported on the radio interface. This "Data Link Connection Identification" or DLCI matches the link identifier used on the Abis interface for the same purpose. Its main role is to distinguish what is related to the short message services from the rest.

5.3.1.4. Networking on MAP/E

The previous sections have described how the different BSS interfaces are specified in order to transport signalling application messages between MS, BTS, BSC and MSC. However, there is yet another interface which messages between mobile station and MSC must cross in some conditions: this is the interface between the **anchor MSC** and the **relay MSC**. These concepts have not yet been introduced, and will be dealt with in Chapter 6. A short explanation is then needed. When a call undergoes a handover between two cells connected to two different MSCs, the entry point of the call in GSM is not changed. The MSC in charge of the call entry point is then stable throughout the call, and is called the anchor MSC, whereas the other MSC, the one in charge of the new cell where the mobile station is now located, is the relay MSC. For the management of the call, upper layer messages must transit all the way from/to the mobile station to/from the anchor MSC, with the relay MSC acting as an intermediate node. An additional segment of transportation is therefore required in this case, and will be dealt with here.

In the *Specifications*, this transportation function is not identified as a separate protocol, but is specified as part of the MAP, more exactly of the MAP/E protocol (the MAP/E protocol is the one between neighbour MSCs, and is concerned with all aspects of inter-MSC handovers). As such, it obeys the rules for the transportation of MAP messages which will be described in the next section. However, the nature of the transportation function is very different from the other functions of MAP/E: its sole role is to carry signalling messages in a non-interfering way between the mobile station and the anchor MSC. The two MAP/E messages which are involved, PROCESS ACCESS SIGNALLING and FORWARD ACCESS SIGNALLING, are shown in figure 5.10. Their information field contains the message exchanged between the anchor MSC and the mobile station, with the same contents as when it is transported on the A interface as a DTAP message, or on the Abis interface as a Radio link layer management message, or on the radio interface itself.

The distinction between the message flows pertaining to different mobile stations is not done by using SCCP, as on the A interface. The SCCP protocol is part of the lower layers used by MAP/E, but only in connectionless mode, as explained in the next section. The discrimination function is fulfilled by the TCAP protocol, as explained in general for the NSS in the next but one section. The relay MSC has then to maintain a context for each connection with a mobile station, and to translate back and forth between SCCP references (towards the BSC) and TCAP transaction references (towards the anchor MSC).

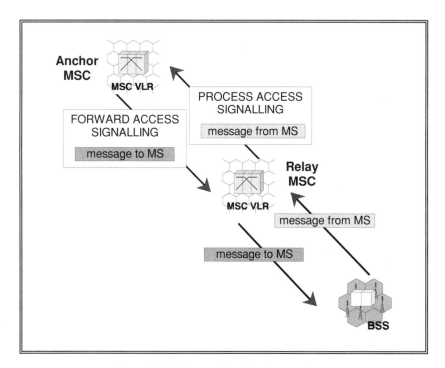

Figure 5.10 – Transport mechanism on MAP/E

Once a handover has occurred, radio messages must be relayed transparently by the relay MSC to ensure their transport between the mobile station and the anchor MSC.

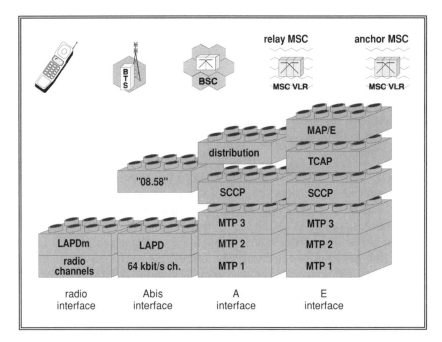

Figure 5.11 – Signalling transport structure in the BSS

The structure of protocols used between mobile station and anchor MSC
for the transport of signalling messages shows a large variety.

5.3.1.5. Connections in the BSS: a Summary

Two figures are provided as a summary of the message flow on the
BSS interfaces. Figure 5.11 shows the overall structure of the networking
protocols. Figure 5.12 (on the previous page) summarises the
organisation of the message flows from MS to MSC, the different links,
transactions, connections, ... and their identifiers, discriminators,
addresses, references, ...

Figure 5.12 – Connection identifiers in the BSS (⟋)

The structure of protocols on the BSS interfaces leads to
a hierarchical organisation of message flows shown as nested boxes,
each with its own tag. Several mobile stations are shown in this example.
The first one of them (MS a) is the only one where all the levels of detail are given:
this mobile station has two calls in progress (TI=a and b on PD=CC, on SAPI=0),
and one short message transaction (TI=a on SAPI=3).
It uses the same TRX as mobile station b, but they use different SCCP connections
on the A interface, and different TCAP contexts on the E interface.

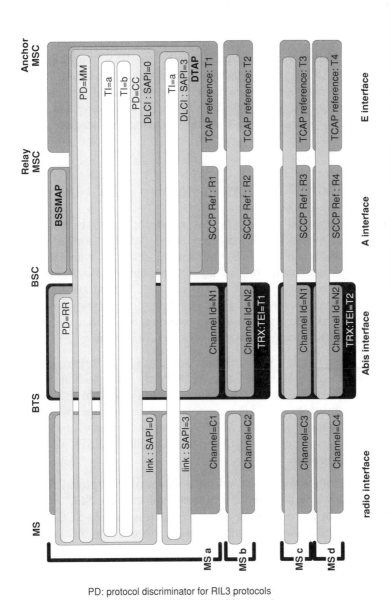

PD: protocol discriminator for RIL3 protocols
TI: transaction identifier for RIL3-CC protocol
DLCI: data link connection identifier
SAPI: service access point identifier on the radio interface
TEI: terminal equipment identifier on the Abis interface

5.3.2. Networking in the NSS

The methods of carrying messages between NSS machines are in some way much more standardised than on the BSS interfaces, since in all cases the SS7 signalling network standards are used. But roaming imposes that messages be exchanged between entities belonging to different networks, operated by different companies in different countries. The addressing and routing scheme is therefore of prime importance here.

5.3.2.1. The SS7 Network Protocols

A fundamental point to bear in mind about SS7 is that there are two network levels. The lower level, used typically to build national networks, is based on the MTP 3 networking protocol, whereas the other level, used for interconnecting all the national networks in a single global SS7 network, is based on SCCP. The rationale behind such a choice is simple: it is much easier to hold and maintain routing tables covering one national network (i.e., tables for addressing entities under the control of one operator), than many networks including foreign ones.

To these two levels correspond two levels of addressing. The MTP address, or "Signalling Point Code" (SPC) only holds relevance within a limited scope, such as one SS7 national network. Within this scope, the networking functions of MTP 3 are able to route messages with the SPC as the destination address. MTP 3 follows basically the datagram approach, that is to say that each message contains the SPC of the entity to which it is directed.

At the other level, we find the concept of the "global title". This versatile addressing scheme, at a higher level than the SPC, enables the identification of any SS7 point in the world. It is used in SCCP, where it provides in GSM the addressing capacity needed for the transport of MAP messages between NSS entities.

The addressing scheme of SCCP therefore warrants a quick description. SCCP messages can be addressed in two ways: if they are bound for a destination in the same national network, then nothing really has to be added to the information carried by the MTP. However, in this case the SPC is included as the address at the SCCP level.

Now, if the message is bound for another national network, a global title must be used. The global title may be a number with no direct relationship with SS7, such as a number typed in by a user, a PSTN number (E.164), a data number (X.121), or a GSM subscriber identity (an

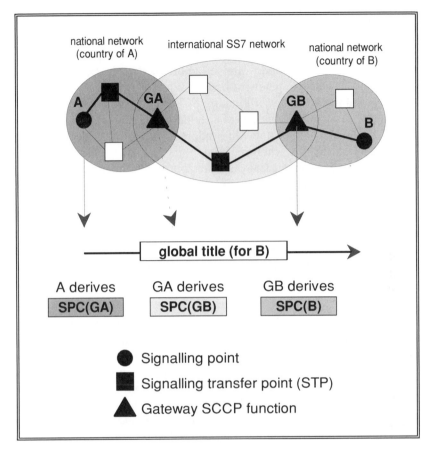

Figure 5.13 – Global title translation

When a global title is used as an address in SCCP, it must be translated
into an actual routing addresses for the MTP
within each network through which the message transits.

IMSI, specified in E.212, see Chapter 7), etc. It does not contain explicit
information on the way to route the message, and an SCCP translation
function is therefore required to derive the relevant MTP address from the
global title, at least at each network border. Figure 5.13 shows an
example of how a global title is translated into signalling point codes by
SCCP gateway functions along the way, and also by the originating node
which must identify the SPC of a relevant gateway within its own
national network. In GSM the SCCP address, whether a global title or an
SPC, includes a sub-address, called the sub-system number, which
identifies the type of the target entity (HLR, VLR, MSC, EIR).

How do the two levels interwork? In addition to the national SS7 networks there exists an international SS7 network, within which messages are transported using the MTP protocols. Typically within a national network, a few nodes are, in addition, part of the international network and can act as gateway nodes. When a message is bound for some external network, the originating SCCP entity determines from the global title the most suitable gateway node in the national network. The SPC of this node will be the MTP address of the message. When the SCCP function in the gateway node receives the message, it will determine (still from the global title) the appropriate international node to enter in the target network and will forward the message to this node, using its international SPC as the MTP address. Then, the new transit node will analyse the global title, determine the final point, and forward it using the corresponding national SPC as the MTP address. Used in its connectionless mode as for the GSM application, SCCP is a typical datagram networking protocol: each message contains an SCCP address such as a global title, and the scenario described above is repeated again for each message.

It should be noted that the SPC address can have some international meaning when used between transit nodes. In fact these nodes have two different SPCs, one for use within the national network, the other for messages with other networks. The numbering plan for international SPCs is defined in Recommendation CCITT Q.708. An international SPC consists of 14 bits, and includes a geographical zone indicator, a network indicator within this zone and a point indicator within this network. Most often, international SPCs are shown as decimal representations of these three indicators, e.g., 5-010-3 for an Australian SPC. But such addresses are only significant within the international SS7 network, and are not relevant for the entities of a GSM network (except if some of them have direct access to the international network).

The use of SCCP for the transport of MAP messages between NSS entities brings little else than the global addressing facility. In particular, the capacity of SCCP to manage independent connections, which is used on the A interface, is not used within the NSS. All messages use the connectionless mode.

5.3.2.2. TCAP

On top of the services provided by SCCP, the MAP makes use of an additional protocol, the "Transaction Capabilities Application Part" (TCAP). TCAP is not usually presented as a transmission protocol (as its name shows, it is considered as an application protocol). However, TCAP provides the means to distinguish independent message flows, and as

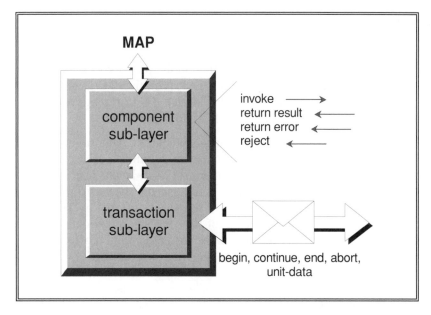

Figure 5.14 – TCAP modelling

TCAP provides the necessary correlation between individual operations,
as well as between the structured exchanges building up a transaction.

such is close in functionally to many other protocols we have seen so far. Moreover, it is part of the common platform of mandatory layers supporting all the MAP protocols.

There is no need to describe the functions of TCAP in detail to understand how GSM works. However, the MAP specification refers heavily to TCAP, so that a presentation of the vocabulary and principles is useful for any reader who will be involved with the actual MAP specification.

TCAP is best modelled in two sub-layers: a "transaction" sub-layer, on top of which sits a "component" sub-layer, as pictured in figure 5.14. The service offered by the transaction sub-layer consists in the management of transactions, or dialogues, on an end-to-end basis. This is another example of a virtual-circuit approach for distinguishing several independent flows using in parallel the same transmission means. TCAP adds a transaction indicator to each message. This indicator enables the other end to relate all the messages of the same transaction to a single context. Thanks to TCAP, the MAP protocols need not bother about the means to link together the different exchanges concerning e.g., the location updating of a given mobile station. They rely for this function entirely on the transaction capability of TCAP. In particular, there are a number of cases in MAP where important data, such as the identity of the GSM subscriber, is sent only once within a dialogue. The information is then implicit for the rest of the dialogue. It is then important when reading the specification of MAP to note in the procedures the beginning of a dialogue (a message sent with TCAP primitive TC-Begin), how it goes on (message sent with TCAP primitive TC-Continue) and its end (message sent with TCAP primitive TC-End).

In addition to the transaction management, TCAP also manages the correlation of individual commands/responses within a dialogue. These operations are managed in the "component" sub-layer. The correlation between a request issued by a MAP entity and the answer to this request is not managed by MAP, but is part of the service offered by TCAP. It is in fact the most important feature of TCAP. As a consequence, MAP often does not specify a message name for an answer to a request: this answer will simply be contained in the "RETURN RESULT" or "RETURN ERROR" linked by TCAP to the "INVOKE" component containing the initial message. Even when a message name appears in the textual description of TS GSM 09.02 (MAP), the contents of the message can be found only in relation with the operation name. For instance, the contents of the "RADIO CHANNEL ACKNOWLEDGEMENT" message cannot be found if the reader does not know that this message is an answer to the "PERFORM HANDOVER" message, and as such is carried in a "RETURN RESULT" component of the TCAP "PerformHandover" operation. This kind of knowledge is important to understand the MAP specification, and can be found in section 7.2 of TS GSM 09.02. In the rest of this book, we have chosen to use TCAP operation names to refer to the signalling exchanges carried by the MAP/X protocols. This will enable the reader to have a non-ambiguous reference to the actual information exchanged by the protocol.

TCAP provides also a grouping/degrouping function. Several operations pertaining to the same dialogue can be grouped inside a TCAP message For instance, a given TCAP message may contain the result of an operation (e.g., the positive acknowledgement of a subscriber authentication) while invoking another (e.g., requiring the start of ciphering).

A last point worth noting is that the syntax used for the description of TCAP messages and parameters (and hence used in the MAP specification) is ASN.1 ("Abstract Syntax Notation 1"), which is specified in CCITT Recommendations X.208 and X.209.

5.3.3. NETWORKING FOR SUPPLEMENTARY SERVICES MANAGEMENT

One case of signalling message transmission has been left aside so far: the transmission of supplementary services management messages between the mobile station and HLR. Let us first recall that, even though a supplementary service is in general fulfilled by the MSC/VLR, its management (e.g., activation/deactivation of the service, interrogation of its status, etc.) is performed by the HLR of the subscriber. It is the HLR which is in charge of eventually modifying the context in the MSC/VLR to keep it in line with the service state in the HLR. This choice of having a single point of control (HLR) ensures the consistency of data throughout the network.

We must consider two legs for the dialogue between the mobile station and the HLR: one from MS to MSC, which makes use of the mechanisms described in the BSS section, and another in the NSS, using MTP, SCCP, TCAP as described in the previous sections. The disparity between the two worlds made things difficult, at the conceptual level as well as for the intervening MSC/VLR. The conceptual approach in this domain is somewhat fuzzy, and the vision presented here reflects the view of the authors.

On the radio interface, supplementary services management information can be carried either in standalone messages, or be part of some call control messages. Here we find the first fuzzy conceptual point. In our view, the call control protocol can be considered as a carrier for supplementary services messages, in the same way as e.g., the messages belonging to the MM protocol between MS and MSC are carried on the Abis interface. The comparison with the Abis interface is actually interesting, because some Abis messages from the BSC are also "mixed blood" messages, in so far as they contain an order for the BTS as well as a message to be carried forward.

But the radio interface is not the only area in which supplementary services raise architectural questions. The MSC/VLR is another fuzzy area. It acts as a gateway between the MS-MSC/VLR protocol and the MAP protocol towards the HLR. As such, it must perform some analysis of the messages, be it only to know whether messages received from the HLR must be transmitted towards the mobile station or not. The corresponding information can typically be found in the message type, which is for instance the only information which provides the ability to distinguish between messages belonging to different MAP/X protocols. In

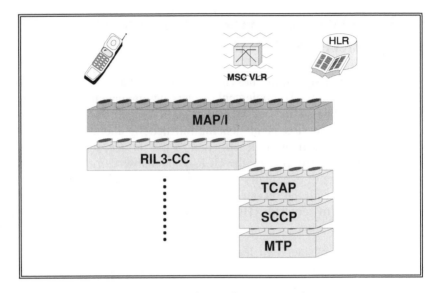

Figure 5.15 – Protocol stack for supplementary services management

The actual application protocol between MS and HLR (to be described in Chapter 8)
relies on two different transportation protocol sets,
one between MS and MSC, the other inside the NSS.

addition to this analysis for routing, the specification is such that the
MSC/VLR checks most of the syntax of the messages, and may reject
them if they do not comply with the specifications. For these reasons, it is
difficult to see the MSC/VLR as a pure transit node for the
supplementary service management messages. But it cannot be
considered either that the MSC/VLR makes use of the semantics of these
messages: it only deals with them as objects to be transported or rejected.
Taking all these points into consideration, we prefer here to present the
MSC/VLR only as a relay on the path between the mobile station and
HLR as far as supplementary services management is concerned, and
consider the application protocol (MAP/I) as an MS/HLR protocol, even
though it is not carried by SS7 all the way. This approach puts the stress
more on the functional aspects (in particular on the fact that the dialogue
is between the mobile station and the HLR), rather than on marginal
issues such as syntax analysis.

Figure 5.15 shows the overall picture: the MAP/I protocol is an
application protocol between the mobile station and the HLR. Between
the mobile station and the MSC, its messages are carried as encapsulated
messages either inside call control messages (protocol discriminator:

CC), or inside messages using the SS protocol discriminator. Between the MSC and the HLR, it makes use of all the SS7 protocol stack, and its messages are distinguished from other MAP message simply by the message type. Messages pertaining to different MSs are distinguished on the MAP leg by the use of TCAP transactions, on the A-interface by the use of SCCP connections, and finally between the BSC and the mobile station by the radio channel. Several SS management transactions can exist in parallel. They are distinguished on the BSS leg by different TIs, and on the MAP leg as different TCAP transactions. It falls on the MSC/VLR to make all the needed translations!

5.3.4. NETWORKING FOR POINT-TO-POINT SHORT MESSAGES

The short message services is the second domain where networking encompasses both the BSS and the NSS worlds. In fact it encompasses even more, as GSM interworks for these functions with external networks.

As explained in Chapter 2, the role of GSM for short messages is to transport them between the mobile station and a Short Message Service Centre (SM-SC), the latter being out of the scope of the *Specifications*, and possibly outside the control of GSM network operators. The SM-SC is connected to one or more MSCs which act as gateways between the GSM world and the SM-SC. The corresponding functions are called SMS-GMSC in the case of Mobile Terminating short messages and SMS-IWMSC (InterWorking MSC) in the case of Mobile Originating short messages. We will call them both the SMS-gateway for short. The stack of protocols between such gateways and the SM-SC is left open. What the *Specifications* specify is a set of transportation means for conveying the short messages between the mobile station and the SMS-gateway. The transportation involves potentially two domains: the MS to MSC segment, and the segment between MSC and SMS-gateway. The protocol architecture for short message transport inside GSM is shown in figure 5.16. The SM-TP ("Short Message Transport Protocol") specified between the mobile station and the SM-SC is in fact an end-to-end protocol including some features of an application protocol, and will therefore be dealt with in Chapter 8, together with other short message management functions. We will concentrate here on the relaying of the short message between GSM entities.

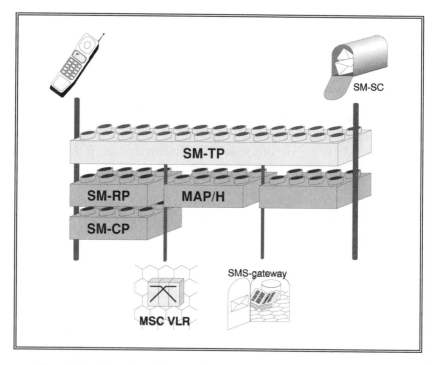

Figure 5.16 – Protocol architecture for the transport of short messages in GSM

The relay protocols co-operate to convey a short message
between the MS and the point of interconnection of the SM-SC in GSM:
MAP/H interconnects with SM-RP,
itself relying on SM-CP for transport between MS and MSC.

The stack of protocols between MSC and SMS-gateway includes MAP/H and the usual stack underlying MAP protocols, i.e., (from the bottom up) MTP, SCCP and TCAP. This interface is the most greedy in terms of message overhead, and the maximum length of short messages in GSM (140 octets, sufficient coding for 160 7-bit ASCII characters) is then a direct consequence of the MTP maximum frame length (see page 270).

We will now consider the BSS leg (from mobile station to MSC) and the NSS leg (from MSC to SMS-gateway) in turn. Since on both legs the Mobile Originating and Mobile Terminating short messages use similar transport mechanisms but in opposite directions, only one of the directions will be described in detail (the Mobile Originating short message). As a last point in the description of each leg we will identify the differences for the Mobile Terminating case.

5.3.4.1. The BSS Leg

The lower layers used to convey the messages from the mobile station to the MSC have been described in the section about linking. On the radio path, they include an acknowledged-mode SAPI 3 link on a TCH/8 or on an SACCH. This link is then relayed up to the MSC using the relay protocol on the Abis interface and DTAP on the A interface.

We now consider the short message "Control Protocol" (SM-CP), which has very little added value (if any?) compared to the service offered by the underlying layers. It is a very simple protocol, consisting of a command/answer procedure with three message types, as shown in figure 5.17, and without even a reference to correlate them since its mode of operation is basically of an alternate nature ("send-and-wait"). The SM-CP CP-DATA message is the only one carrying upper layer information, which does not necessarily include the short message itself, but could be some upper layer acknowledgement or error report. Each message of the SM-CP protocol includes a protocol discriminator specific for short messages, and a transaction identifier. This last field could enable a mobile station to manage several parallel short message transactions at the same time. This would then be the only real added value of the protocol, but the *Specifications* do not indicate clearly whether it can be used as such.

The next protocol on our list is the SM-RP (Short Message Relay protocol), whose functions include the management of references and

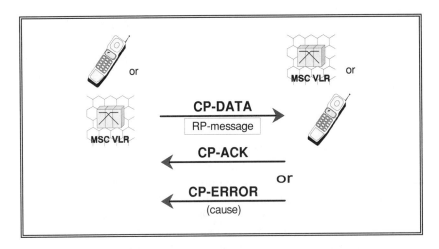

Figure 5.17 – The SM-CP protocol between MS and MSC

The SM-CP protocol forces an alternate mode of operation
(sending of the message, then waiting for the acknowledgement)
with the possibility of the message being repeated once if a timer expires.

addressing. This protocol is in fact the BSS part of the protocol for the networking functions between MS and SM-SC. The SM-RP protocol interworks with the MAP/H protocol in the MSC/VLR.

Three messages are defined in the SM-RP protocol, one to carry the message (RP-DATA), one for transporting the acknowledgement (RP-ACK) and one for indicating an error (RP-ERROR). Though the message list is similar to the one of SM-CP, the functions fulfilled are quite different. The messages are correlated together through a one-octet message reference generated by this protocol, enabling the sending or receiving of different messages in parallel at this level. Addressing is performed by including in the SM-RP RP-DATA message the destination address (for an MO message), or the origin address (for a MT message), i.e., in all cases the address of the SM-SC.

There are a few differences between the messages for Mobile Originating messages, and for Mobile Terminating messages. One of them is the inclusion in the mobile terminating SM-RP RP-DATA message of a priority indicator. However, since this priority indicator is not transported in the MAP, it is quite useless unless the MSC tampers with upper layers to find the relevant priority information.

5.3.4.2. The NSS Leg

Between the MSC and the GMSC, the transport of short messages is performed by the same means as signalling messages, i.e., using the SS7 stack of protocols supporting MAP (MTP, SCCP, TCAP), on top of which is added one of the MAP protocols, the MAP/H. Taking into account the underlying SS7 layers, the MAP/H provides the same functions as SM-RP in the BSS leg. It contains three messages, which can be mapped onto the three SM-RP messages, as indicated in figure 5.18.

5.3.4.3. The Relay in the MSC/VLR

As was the case for the protocol supporting supplementary service management, the MSC/VLR is the translator between two worlds. The key points of the translation are summarised in figure 5.18. As far as upper layers are concerned, the combination of the SM-RP protocol on the BSS leg, the MAP/H protocol on the NSS leg, and the relay in the MSC/VLR, can all be considered as a single network protocol, providing the routing between MS and GMSC (and from then on to the Short Message Service Centre), as well as the possibility to deal with several

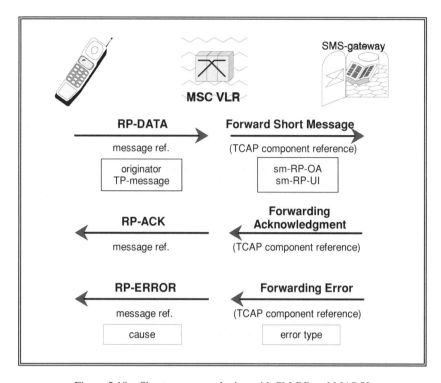

Figure 5.18 – Short message relaying with SM-RP and MAP/H

Messages from SM-RP and MAP/H map onto one another,
relaying both the short message and its acknowledgement
between the MS and the MSC to which the SM-SC is connected.

messages in parallel. This will be how these protocols will be considered in the description of the upper layer protocols for the short message services, in Chapter 8.

SPECIFICATIONS REFERENCE

The link protocols used in GSM are specified in the following specifications:

- for the radio interface, LAPDm is entirely specified in **TS GSM 04.05** (general aspects) and **TS GSM 04.06** (detailed protocol specification);

- for the Abis interface, the reference for LAPD is the CEPT T/S 46-20 recommendation, itself largely based on the CCITT Q.920 and Q.921 recommendations. **TS GSM 08.56** contains the information specific to the use of LAPD in the Abis interface context, with reference to T/S 46-20;

- for the A interface, the reference for MTP is contained in the CCITT recommendations Q.702 (physical layer), Q.703 (link layer), Q.704 (network layer) and Q.707 (test and maintenance). **TS GSM 08.06** gives a point-by-point analysis of the applicability of these recommendations to the A interface context, based on the CCITT recommendations;

- the full specification of the RLP is given in **TS GSM 04.22**.

As for BSS networking, the notion of the protocol discriminator can be found in **TS GSM 04.07 and 04.08**. There is no clear explanation of the PD as an address, though a few sentences on the subject can be found in TS GSM 08.08. **TS GSM 08.58** specifies the protocol to transfer messages on the Abis interface. The reference for SCCP is contained in the CCITT recommendations Q.711 (primitives), Q.712 (messages), Q.713 (formats and codes) and Q.714 (procedures), based on a preliminary version for the CCITT blue book. **TS GSM 08.06** specifies the simplifications which can be made in the A interface context. It also specifies the discrimination between BSSMAP and DTAP messages, and contains the complete specification of DTAP messages, as far as their transport on the A interface is concerned.

The transportation layers between NSS machines are entirely specified by recommendations outside the *Specifications*. MTP and SCCP are specified in CCITT recommendations in the Q.70x and Q.71x series as listed above, as well as in ETSI pr-ETS 300-008 (MTP) and ETSI pr-ETS 300-009 (SCCP). TCAP is specified in the CCITT recommendations Q.771 to 775, as well as in CEPT T/S 43/BB.

The special case of supplementary services management messages is treated in **TS GSM 04.80** (which also contains the full semantics of these messages) for the mobile station to MSC segment, and in **TS GSM 09.02** for the MAP part.

The general architecture of protocols for the transport of point-to-point short messages is specified in **TS GSM 03.40**, where an example of a protocol stack for the connection of an SC to a GSM MSC can also be found. **TS GSM 04.11** specifies both the SM-RP and the SM-CP protocols, which are used to carry short messages between the mobile station and the MSC. The MAP/H protocol is specified in MAP, that is to say by **TS GSM 09.02** (section 5.13).

6

RADIO RESOURCE MANAGEMENT

6

RADIO RESOURCE
MANAGEMENT

As a telecommunication system, GSM enables its users to communicate through transmission paths of various characteristics, as explained in the previous chapters. However these transmission paths are not reserved once for all between any two pairs of users. They are set up on demand, and only for the time necessary for a given communication. This requires exchanges of information, not only between the users and the network they are in direct contact with, but also between machines within the network. This and the two next chapters are devoted to the description of these information exchanges, which enable distant participants, users and machines, to act together to provide the communication services for which the networks are designed.

This technological field is known as signalling, and under this name, its reputation has not yet grown as wide as other fields such as modulation, signal processing, and other transmission techniques. Many people consider it simply as a branch of software engineering, though it is in fact at the centre of the design of complex systems, where tasks can be executed only through the co-operation of distinct machines. This is the case of telecommunications, where machines are by essence distinct and distant. The study of GSM signalling is therefore of prime importance to understand how the system operates, and the reader should not be surprised that half of the book is devoted to the signalling interchanges.

Signalling is often the juxtaposition of many simple and more or less inter-dependent procedures, and its complexity stems mainly from the number and the diversity of small issues. We have already seen in

Chapter 2, concerning the architecture, that the basic methodology to tackle such issues is "divide and conquer". Pervasive in the specifications of GSM signalling is a split in three functional domains: Radio Resource Management, Mobility Management and Communication Management. The management of the calls is the upper plane in this organisation, and deals with the establishment and the release of end-to-end transmission paths, through the GSM domain and by interworking with external networks, to support user to user communications. Communication Management, as a functional plane, relies on the Mobility Management functions for dealing with the mobility of the users and with security-related functions. The management of the radio resources groups functions specific to the radio interface.

A major difference between a radio mobile telecommunication network and a network with fixed links is the management of the access resources. In a fixed system, a dedicated communication medium exists continuously between the user terminal and the infrastructure, ready to be used when a call needs to be established. On the contrary, in a cellular system like GSM, a dedicated channel over the radio interface is provided to the mobile stations only on demand and for the duration of the call, under the control of the infrastructure. This calls for functions which bear no equivalent in ISDN for instance. Even if 64 kbit/s channels are allocated dynamically in the case of an ISDN multi-terminal installation, this resource management is rather limited compared to the one in a full GSM cell. Moreover, in ISDN, a signalling channel is always ready for use by any terminal. The matter is quite different in GSM, where the signalling capabilities offered to a mobile station in idle mode, that is to say when not allocated a radio channel to its private usage, are limited to the absolute minimum. The consequence is that a host of new procedures are needed.

Besides dynamic channel allocation, another feature of GSM (or cellular systems in general) compared with fixed networks is the **handover**. The problem consists in providing a dedicated channel from mobile station to MSC at every moment during a call, despite the movements of the user. This calls for a complex measurement and decision process to trigger the transfer of the communication at the right moment and toward the right cell. In a cellular system, the handover process is very important, since it impacts significantly the quality of the communications as perceived by the users, as well as the spectral efficiency.

This chapter will be devoted to these topics, that is to say to the functions required to co-ordinate the mobile stations and the infrastructure so as to provide the suitable transmission means over the radio interface, whatever the telecommunication service requires, and whatever the user's movements. These functions form a well defined area, which we presented in Chapter 2 as the **RR** (Radio Resource management) functional plane. They are spread among four entities in the canonical GSM architecture: the mobile station (of course!), the two base station sub-system components (BTS and BSC), and a small part in the MSC. All the higher layer functions, described in Chapters 7 and 8, are basically managed directly between mobile station and MSC, the base station sub-system (BSS) acting for these functions as a single complex transmission system. The spread of the radio resource management functions implies the existence of signalling procedures between the involved infrastructure machines; this is the purpose of the signalling protocols on the A (BSC-MSC) and Abis (BSC-BTS) interfaces, which will therefore be described in this chapter.

The chapter is basically composed of two parts. After preliminary architecture considerations needed to introduce some specific concepts, the major requirements which drove the design of the RR protocols will be looked at. Then, after a section to present the protocol architecture, the various procedures needed to fulfil these requirements will be developed.

Preliminary Architecture Considerations

A section in the middle of this chapter will be devoted to the architecture and the protocols in the Radio Resource management domain. However, some basic notions are necessary for the understanding of the first part of the chapter. It concerns mainly the notion of anchor MSC.

The major roles are played in this chapter by the components of the BSS, the BTS and the BSC. The MSC intervenes a little, to deal with handovers between cells managed by different BSCs. Some handovers may even transfer the mobile station from a cell within one MSC area to a cell in another MSC area, thus involving *two* MSCs. The roles of the two MSCs are different. In no case does the MSC in charge of the communication relinquish its control to the new MSC. This MSC is called the **anchor MSC** for the connection. This is an important design choice of GSM, with numerous consequences on the procedures. Several arguments justify this choice; a compelling one is the charging problem, since toll ticketing is much simpler when one MSC follows a call from its beginning to its end.

A consequence of this architectural choice is that after an inter-MSC handover, two MSCs (and at most two) may be involved in the connection. The transmission chain between the mobile station and the interworking point with external networks is then composed of a BTS, a

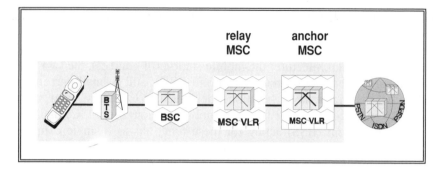

Figure 6.1 – The concepts of anchor and relay MSCs

The transmission chain may involve two (and at most two) MSCs:
the anchor MSC in charge of the communication management and
the relay MSC in charge of the BSS with which the mobile station is in contact.

BSC and either two MSCs, a **relay MSC** and the anchor MSC, or of a single MSC (see Figure 6.1). To ease the problem of terminology, a convenient approach is to consider the notions of relay MSC and of anchor MSC as functional, and to admit that when there is only one MSC, it is at the same time the relay MSC and the anchor MSC. Thus the term relay MSC will be used to refer to the MSC in direct contact with the BSC, even if it is also the anchor MSC; and the term anchor MSC will be used to refer to the MSC in charge of upper layer treatments, even if it is also the relay MSC.

6.1. RR FUNCTIONS

In this first half of this long chapter, we will study the different Radio Resource management aspects from the requirement side. While implementation issues will often be addressed, in particular for the distribution of the tasks between the involved machines, the details of the signalling procedure will be entirely in the second part.

The functions covered by the management of radio resources are centred on the management of transmission paths over the radio interface, and more exactly between the mobile station and the anchor MSC. To develop these functions, we will use the concept of RR-session, which will be presented first. After a small passage concerning the access and the paging, through which things start, we will deal with the handling of the main properties of the transmission chain, such as whether signalling, speech or data is transported, and whether ciphering is applied or not.

The next issue will be on how handovers are decided. The handover execution itself will be treated mainly in the procedural section. Addressing the handover preparation issue will take us deep into considerations about the measurements performed by the mobile station and the base station. These measurements are the basic information upon which handovers can be decided.

Next, two ancillary functions of the transmission over the radio interface will be looked at, the management of the transmission power and of the timing advance.

Finally we will deal with the management of the radio channels on the radio interface as a whole set. The two main facets are the handling of the configuration of the radio channels in each cell, and the allocation strategy of the dedicated radio channels (TACH/8s and TACH/Fs).

6.1.1. The Concept of RR-Session

Most of the functions in the Radio Resource management plane relate to the management of the transmission between the mobile station and the anchor MSC. For each mobile station engaged in a communication, there exists a transmission path, as well as a signalling path, between itself and the anchor MSC. As seen from a mobile station, such a path is set up when it enters the dedicated mode (i.e., when it leaves the idle mode), and is released when the mobile station goes back to idle mode. In the infrastructure, a transmission path exists for all this period, but can be thoroughly modified, especially by handovers. We will refer to what is managed during this period of time as an **RR-session**. As a minimum, an RR-session must include means to transmit signalling between the mobile station and the anchor MSC through the BTS, BSC and possibly a relay MSC, including a dedicated radio channel, references to manage it on both the BTS-BSC interface and the BSC-MSC interface, and means in the BSS to monitor the radio connection and take handover decisions when necessary. This minimum set suffices only in the cases where the transfer of circuit-type user data is not required, such as for location updating, short message transfer or supplementary service management. When circuit-type user data needs to

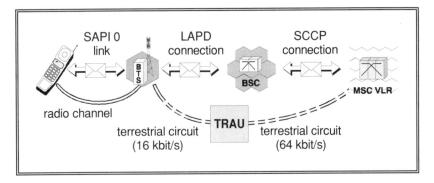

Figure 6.2 – Contents of an RR-session

An RR-session contains both the signalling resources between mobile station
and anchor MSC, including a dedicated channel on the radio path,
as well as the user data circuits if need be.

be transmitted, then a complete circuit connection between mobile station
and anchor MSC is also part of the RR-session, as shown in figure 6.2.
For instance, a speech call requires the use of a signalling connection as
well as a speech-carrying connection between mobile station and MSC.
This last connection makes use of dedicated resources such as the speech
transcoder transforming the GSM-specific speech representation into the
64 kbit/s representation used in fixed networks.

An RR-session has many different characteristics which have to be
managed by procedural means. First, two different kinds of dedicated
channels exist on the radio interface; they have been referred to as
TACH/8 and TACH/F in Chapter 4 (there will be three when "half-rate"
channels are included: TACH/8, TACH/H and TACH/F). Second, when a
circuit for user data is present, it can be used according to different
transmission modes. Finally, some other less important transmission
peculiarities characterise RR-sessions. An example is whether ciphering
is applied or not. All these characteristics may change during the lifetime
of an RR-session.

In the *Specifications*, the term "RR-connection" refers to what is
managed during the period of connection to a given BSC. A change of
BSC (e.g., at inter-BSC handover) entails a change of the RR-connection.
The *Specifications* do not have a specific term covering what is managed
for the whole period where the mobile station is in dedicated mode
between two periods of idle mode, i.e., what we call here an RR-session.
Figure 6.3 illustrates the concepts of RR-connection and RR-session, and
their relationship. The figure also shows that an RR-session can be used
for several calls in succession or in parallel, or more generally several
CM-transactions (CM for Communication Management), as will be

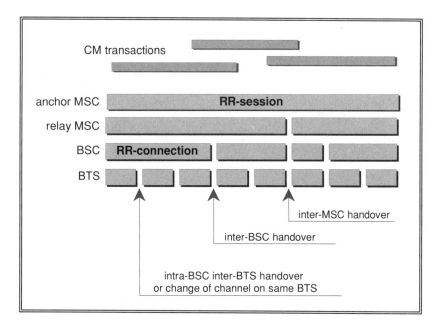

CM transactions

anchor MSC | RR-session

relay MSC

BSC | RR-connection

BTS

inter-MSC handover

inter-BSC handover

intra-BSC inter-BTS handover
or change of channel on same BTS

Figure 6.3 – The concepts of RR-session and RR-connection

From the concept of radio connection (bottom line) to the one of RR-session (top line),
different levels of transition awareness may be defined.
The *Specifications* use the concept of RR-connection, which corresponds to the BSC
view.
CM-transactions may run in parallel or in tandem during the lifetime
of an RR-session, as shown on the top of the figure.

described in Chapter 8. The beginnings and ends of CM-transactions relate to the usage of the transmission, and are completely independent from RR-connections, whose succession relates to the movement of the mobile stations.

The RR-session is the bond between the two domains of radio resource management and communication management. It represents the views of the mobile station and of the anchor MSC. An RR-session starts when the mobile station goes to dedicated mode (the access, when the initial assignment of dedicated channels is performed), and disappears when the mobile station goes back to idle mode.

The life of an RR-connection is punctuated by intra-BSC handovers and changes of radio channels, and this defines another subdivision in the lifetime of a RR-session. At the lower level, the channel connection corresponds to the continuous usage of the same radio channel by the same mobile station. A channel connection starts either through an initial assignment, a subsequent assignment (a change of channel done, e.g., because the allocated channel is no more of the

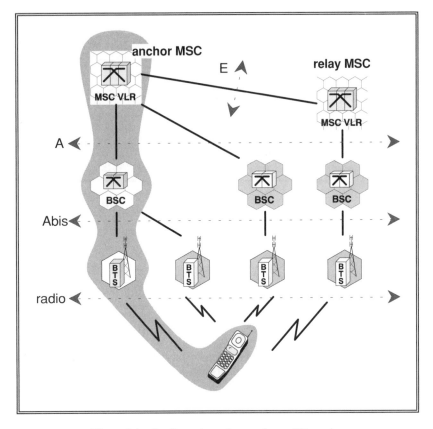

Figure 6.4 – Configurations changes for an RR-session

During its lifetime, an RR-session may go through different transmission configurations.
If the initial configuration is for example the left one, it may be changed to
any one of the others shown. The anchor MSC remains in charge of the upper layers.

needed type) or an incoming handover of any kind. It disappears either
through the release of the RR-session, the assignment of another channel
or an outgoing handover. Channel connections represent the view of
BTSs. By design choice, a BTS considers a change of radio channel
within the same cell as two independent channel connections. When a
channel connection is cleared, the related data is wiped out in the BTS,
regardless of whether the mobile station is allocated a new channel of the
same BTS or in another one.

There are very few stable characteristics of the RR-session beside
the corresponding contexts in the mobile station and in the anchor MSC,
especially when one recalls that the physical path of the transmission may
change thoroughly when a handover occurs. At a given moment in time, a
given RR-session is managed by one BTS, one BSC, an anchor MSC and

sometimes in addition a relay MSC, as shown in figure 6.4. Each of these machines maintains some context related to the RR-session. When a handover occurs, the configuration changes, some contexts must be erased and others must be created in other machines. Based on the corresponding configurations, the functional architecture of GSM distinguishes three kinds of handover. In an intra-BSC handover, only the radio channel, the context in the BTS and possibly the BTS are changed. In an (intra-MSC) inter-BSC handover, the BSC is changed in addition to the radio channel and the BTS. Finally, in an inter-MSC handover, the relay MSC is either created, replaced or suppressed. In all cases, the anchor MSC remains in place throughout all the life of the RR-session. It is indeed the only machine sure to be a constant, and the context of the RR-session in the anchor MSC is indeed the anchorage point of the session.

6.1.2. INITIALISATION

A mobile station has two widely different operating modes, the idle mode, when it is not engaged in a connection with the infrastructure, and the dedicated mode, when a full duplex channel enables actual communications to take place. The transition from idle to dedicated mode is the first step of the initialisation of an RR-session, and is called the **access**. As part of an initialisation process, it has many functional aspects, and this section will address only some aspects specific to the access issue. The full-blown detailed procedure itself will be presented much later in the chapter, after all the relevant concepts will have been introduced.

Access may be triggered either to fulfil a need expressed first on the mobile station side (e.g., a call originated by the user of the mobile station, but also a location updating), or on the infrastructure side (e.g., a call toward the GSM subscriber). In all cases the access procedure is the same, and is initiated by the mobile station. When the network desires the establishment of an RR-session, it **pages** the mobile station which constitutes a request for it to access. To this avail, when normal service is provided, the mobile station listens in idle mode to a paging sub-channel, part of the PAGCH. If a message on this sub-channel indicates that its subscriber is paged, the mobile station starts the access procedure, as it does when the user requests it. We will then deal separately with the two aspects, the access proper, and the paging.

6.1.2.1. Access

The mobile station manifests its will to access by sending a message on the random access channel (RACH), which is answered by an initial assignment message on the paging and access grant channel (PAGCH), carrying the description of the allocated dedicated channel. The mobile station provides very little information in its access request (the message has only 8 bits of contents). In particular the mobile station does not give its identity at this moment, nor the detailed reason for the access. Another particularity of interest is that the access on the RACH is not regulated. The consequence is that two mobile stations may send access requests simultaneously, which would result most often in neither being received by the BTS. Most of the complexity of the access procedure (repetition of the attempt, resolution of access to the same channel by two mobile stations) comes from fixing these problems.

The access ends with the allocation of a radio channel for the use of the mobile station. This is called the **initial channel assignment**, referred to in the *Specifications* as immediate assignment. The access procedure, though limited to the means needed for the transition between the two modes, makes exclusive use of two specific channels, the PAGCH and the RACH.

From a more general point of view, the access function is the initiation of an RR session. As such it includes the initialisation of all the contexts and all the recurrent functions which are part of the RR session. The access will therefore be revisited in many of the functional sections, such as those dealing with the management of the timing advance, with the transmission power control (which must be started during the access), and with the channel allocation. The detailed description of the access procedure, in the last part of this chapter, will provide the synthesis of these different aspects.

6.1.2.2. Paging and Discontinuous Reception

Compared to the other functions described in this chapter, paging is somewhat particular, because it is not directly linked to an RR-session. What is indeed the relationship between paging and radio resource management? Since RR-sessions are only established at the initiative of the mobile station, the network infrastructure needs some means to trigger such an establishment; this role is indeed fulfilled by the paging procedure. But there is no common reference, no clear relationship between the paging message and the ensuing RR-session establishment. The only clue for the network is the indication by the mobile station in the first dedicated message that the RR-session has been established in response to some unspecified paging indication.

Paging is in some sense closer to Mobility Management functions than to Radio Resources management functions, as it serves to locate a mobile station within a whole location area. The grouping of the paging function with RR management, which is also the one followed in the *Specifications*, reflects the relationship which exist between paging and a number of true RR functions. For instance, paging messages and initial assignment messages share the same channel (the PAGCH). This approach is also sensible from a pragmatic point of view since the main job for paging is done by the BSS, which is otherwise only concerned with RR functions.

How does paging start? When an incoming call arrives, the MSC/VLR requests the BSS to perform a paging in some of the cells of the BSS. The MSC provides the concerned BSC(s) with the identity of the mobile subscriber to be paged and the list of cells in which the paging should be issued. The BSC is in charge of managing the PAGCH, i.e., the grouping and scheduling of paging messages as well as initial assignment messages. This scheduling may be more or less optimised, depending on the manufacturer. The *Specifications* describe a framework for such an operation, but the operator/manufacturer can choose how often to repeat unanswered paging messages, whether to send initial assignment messages also on those parts of the PAGCH which correspond to the paging sub-channels, etc.

The split of functions between BTS and BSC with regard to paging also allows some flexibility and is somewhat manufacturer-dependent. Typically, the high-level tasks, such as priority decisions, rest with the BSC. Sophisticated approaches may take the system load into account for the respective priority of paging and initial assignment messages.

Another aspect of paging also tackled by the BSS concerns the concept of discontinuous reception. For the sake of battery consumption in handheld mobile stations, it is important to minimise the amount of information the mobile station has to receive, demodulate and analyse while in idle mode. To this avail, the downlink common control channel can be divided into several paging sub-channels, and all paging messages pertaining to a given subscriber are normally sent on the same sub-channel. The sub-channel organisation can vary on a cell basis, but broadcast information allows the mobile stations to determine it. Such a scheme allows mobile stations to restrict their monitoring of paging messages to their own paging sub-channel, thereby increasing significantly the lifetime of their battery, at the expense of a small increase in delay for the setting up of incoming calls. Such a feature is referred to as discontinuous reception, or DRX, and is not to be confused with discontinuous transmission (DTX) with which it bears no relationship except the similar names. Mobile subscribers are assigned to paging sub-channels in a pre-determined way taking into account the three last digits of their international mobile subscriber identity (the

IMSI), so that the knowledge of the PAGCH configuration is enough for each mobile station to determine which blocks of which CCCH carrier unit it should listen to.

The PAGCH follows a 51 × 8 BP cycle, where BP denotes a burst period, using 9 blocks per cycle for a PAGCH/F (the PAGCH of large capacity) and 3 blocks per cycle for a PAGCH/T (the PAGCH of small capacity), as described in Chapter 4. A certain (parameter-controlled) number of these blocks belong to some paging sub-channel, the others being reserved to initial assignment messages. This number may range from 2 to 9 for a PAGCH/F and from 1 to 3 for a PAGCH/T. A paging sub-channel is (almost) cyclic, with a cycle ranging from 2 to 9 times 51 × 8 BP (that is to say from 0.95 second to 4.25 seconds), here again under the control of a parameter. Hence, on a given PAGCH/F, there can be from 4 to 81 paging sub-channels (2 to 27 for a PAGCH/T). The two parameters describe the PAGCH configuration and are broadcast on the BCCH. The choice of these parameters is the operator's. The second parameter (linked to the cycle of paging sub-channels) corresponds to a compromise between the access time and the power consumption of the mobile stations. The first one was introduced solely to enable the development of very simple PAGCH scheduling algorithms which do not use the possibility to send initial assignment indications on a paging sub-channel. In such cases, the choice of the parameter is related to the ratio between the paging load and the initial assignment load. Otherwise, the parameter is set so that any PAGCH block belongs to some paging sub-channel.

A small detail is that the interval between two successive paging sub-blocks of the same sub-group is constant, except (in some combinations) once every 3.5 hours, when the numbering scheme goes through 0. This happens when the paging sub-channels "cycle" is not a divider of the numbering cycle (lasting 2048 × 26 × 51 × 8 BP), i.e., for cycles of 3, 5, 6, 7 or 9 times 51 × 8 BP.

Procedural Requirements for Paging

The procedural requirements for paging management include means for the MSC to require a given subscriber to be paged, a mechanism for the BSC to control the sending of this paging message (or alternatively to provide the BTS with the relevant data to build and schedule the paging messages) and a way to indicate the PAGCH configuration to all mobile stations. In addition, some means to configure the PAGCH are required in the operation sub-system (OSS), as part of the more general techniques for controlling the cell configuration. This is dealt with in Chapter 9.

6.1.3. Transmission Management

The life and deeds of the RR-sessions have only been sketched in this previous section. We will now see in more detail what constitutes the management of the transmission characteristics of an RR-session. These characteristics are chosen depending on the service to be provided. They are decided by the anchor MSC, but transitions are co-ordinated by the BSC. The most obvious area of co-ordination relates to the type of data which must be carried. On this depends the existence or not of terrestrial circuits, the type of the radio channel and the transmission mode (speech, or data with some data rate). Beside these, the BSS has also to manage the cipher mode, as well as the discontinuous transmission modes.

We will take each of these aspects in turn, to examine what is to be done.

6.1.3.1. Transmission Mode Management

We group under the term **transmission mode** the main transmission characteristics. More precisely, the concept of transmission mode refers to the way the GSM transmission chain is used for carrying circuit user information, from the mobile station to the point of interconnection with partner networks. The set of possible transmission modes differs depending on the type of channel used on the radio interface. The transmission modes have already been described in detail in terms of their "physical layer" characteristics in Chapters 3 and 4. Table 6.1 summarises which transmission modes exist on each radio

TACH/8	TACH/F	TACH/H
signalling only	signalling only speech data 3.6 kbit/s data 6 kbit/s data 12 kbit/s, transparent data 12 kbit/s, non-transparent	*signalling only* *speech* *data 3.6 kbit/s* *data 6 kbit/s, transparent* *data 6 kbit/s, non-transparent*

Table 6.1 – Transmission modes used on the radio channels

The only transmission mode on a TACH/8 corresponds to signalling only,
whereas the "full-rate" channels may carry 6 different modes,
and the "half-rate" channels will presumably carry the 5 modes listed.
The "transparent" and "non-transparent" modes (see Chapter 3)
use the same channel coding on the radio path,
but lead to variations in the transmission functions within the BSS.

channel type. The transmission modes which were defined on half-rate channels are not in the phase 1 *Specifications*, but are also shown in this table.

The "signalling only" mode corresponds to the non-usage of the channel to carry circuit-type user data. The transport of signalling information is a capability which exists in all transmission modes, once an RR-session has been established. There are even cases where it represents the only transmission need; for instance, at the beginning of a call (before the conversation phase), for transmission of user data other than circuit-type (short message transmission), for location updating or for supplementary service management.

The existence of a full transmission path including terrestrial circuits between the BTS and the MSC/IWF, and the inclusion in this path of a transcoder and rate adaptor unit (TRAU), depend on the mode; for instance, there are none of these in "signalling only" mode.

In general, the transmission mode is chosen by the MSC, depending on the end-to-end service. When the RR-session is initially established, the MSC does not intervene in the process before the point when it knows exactly the transmission needs to fulfil; up to this point, the *Specifications* impose the mode "signalling only". The channel can be chosen (by the BSC) to be of any type, though it is typically a TACH/8 (see page 355 for the allocation strategies which can be followed by the BSC). Later during the lifetime of the RR-session, the channel type and the transmission mode may change; these changes are decided by the MSC, in order to adapt the transmission media to the needs of the users.

Although the decision lies with the MSC, it is the BSC which chooses the exact channel (of the requested type) and is in charge of co-ordinating the different machines, including the mobile station. The only exception to this rule is the establishment of the terrestrial circuit (between BSC—or TRAU—and MSC), which is always done by the MSCs. This exception stems from no specific reason but the weight of history. Switches have always been in charge of establishing their surrounding circuits. This situation actually raises some problems in phase 2, and it would probably have been a better choice to let the BSC in total control of the whole transmission chain. The management of the terrestrial circuit will be addressed in a bit more detail in the next section.

Procedural Needs for Transmission Mode Management

The procedural organisation of transmission mode management has two facets. First the MSC must be able to indicate the need for a change in the transmission mode at any moment during an RR-session. Second, the BSC must have means to co-ordinate the mobile station, the BTS and the TRAU for fulfilling the MSC request. This last aspect is split into two cases as seen from the BSC and the mobile station, whether the type (full rate or eighth rate) of the existing radio channel fits the requirements or not. A typical case when it does not is at the beginning of an RR-session established for the purpose of establishing a call, if the existing channel is a TACH/8. In such a case, the radio channel must be changed. The procedure to change the radio channel used by an RR-session without changing the cell is called a **subsequent assignment**. If the type of the existing channel is appropriate, but the transmission mode is not, the procedure between the mobile station and the BSC is a **mode modification**.

It should be noted that the handover procedure can be used for changing the transmission mode, including the type of channel. Reciprocally, an intra-cell change of channel is often called a handover, if triggered by quality considerations and not by the adaptation to the needs of the service. A small digression on the terms is useful at this stage. This last usage of the term handover as to be understood from the point of view of *why* it is done. However, from the point of view of *what* is done as seen between the BSC and the mobile station, there is indeed no difference between a subsequent assignment and an intra-cell "handover". The accepted use of the term handover thus depends on the context.

6.1.3.2. Terrestrial Channel Management

An RR-session may or may not include a full circuit between the mobile station and the MSC to carry user information. This circuit is not always present; for instance, if the RR-connection is used for location updating, such a circuit is of no use.

When it is present, it includes a radio channel (a TACH/F, or in the future a TACH/H), and terrestrial circuits connecting the BTS with the anchor MSC via the BSC and a TRAU if applicable. The terrestrial circuit from BTS to BSC (possibly via the TRAU) is in a one-to-one relationship with the radio channel, and is then dealt with by the BSC as part of the radio channel management.

On the A interface, the circuit (from the BSC or TRAU to the MSC) is allocated to an RR-session by the relay MSC. The indication that a circuit has been allocated is given to the BSC through signalling

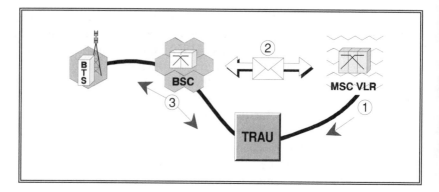

Figure 6.5 – Channel allocation on the terrestrial interfaces

When the TRAU is situated on the MSC-side of the BSC,
the MSC first chooses a circuit towards a TRAU, then a signalling exchange
takes place on the A interface, and finally the BSC controls the set-up of circuits
between BTS and TRAU.

means, enabling the BSC to then connect it to the radio part of the path through its switching matrix (see figure 6.5).

Finally, the circuit between the anchor MSC and the relay MSC, if they are different, is established in the canonical architecture using standard PSTN or ISDN methods. The establishment is initiated by the anchor MSC, using the same procedural means it would use to establish a call.

The existence of a remote transcoder and rate adaptation unit along the BTS to relay MSC path makes this picture somewhat more complex. If no BSC-controlled switching matrix exists between the TRAU and the MSC, then a one-to-one relationship should exist between a specific transmission resource in the TRAU and an MSC-TRAU circuit. Thus, it is in that case the MSC which implicitly chooses the transcoding device by choosing the terrestrial circuit. This situation bears no consequence if the TRAU devices are all equivalent; otherwise, a potential problem exists, since it is the BSS which is in charge of the TRAU, and of the consistency between what the TRAU does and what is done on the radio interface. Thus, the TRAU is in some way managed both by the MSC and the BSS, and in practice, the architectural choices of GSM make it difficult to have distinct types of TRAU.

Procedural Needs for Terrestrial Channel Management

The terrestrial part of the transmission path of an RR-session is established when a TCH/F is requested by the anchor MSC. This is a part of the subsequent assignment procedure. It is obviously modified at each handover, and is in particular entirely changed for an inter-MSC handover. The establishment of the new path, and the release of the previous one are part of the handover execution procedure.

6.1.3.3. Cipher Mode Management

The transmission over the radio path has a few characteristics which must be managed independently from the type of transported data. The first one is the cipher mode: the transmission may be ciphered or not, as desired by the MSC, according to some criteria which depend on the operator.

An RR-session is always started in "signalling only" mode, and always in clear text (i.e., not enciphered). The latter is a necessary requirement, since the RR-session setup is performed without the network knowing the subscriber identity, and therefore ciphering with a user-related key cannot be applied. Means for transition from clear text to ciphered transmission on an existing RR-session are therefore required. The *Specifications* do not provide explicitly for the reverse transition, i.e., from ciphered mode to non-ciphered mode. No need was identified for such a transition. Similarly, the *Specifications* do not cater for the change of ciphering key while in ciphered mode. However, the existing procedures could be used in the future to support these transitions, should the need arise.

Similarly to the transmission mode management case, the transition from clear text transmission to ciphered mode is decided upon by the MSC, the BSC being in charge of co-ordinating the actual change. Cipher mode management impacts the mobile station and BTS.

Procedural needs include the means for the MSC to provide the ciphering parameters (the mode, and the user ciphering key Kc if needed) to the BSC for an incoming handover, the means for the MSC to require a change of the current mode from non-ciphered to ciphered, and the means for the BSC to co-ordinate the transition in the BTS and the mobile station. This last aspect is of primary importance, since messages sent in one mode will not be understood by the peer entity if this entity is set in the other mode, resulting in an unrecoverable loss of connection. The way to synchronise mobile station and BTS will be described in the procedural section.

6.1.3.4. Discontinuous Transmission

Discontinuous transmission (DTX) has been described in Chapters 3 and 4. When discontinuous transmission is applied, actual transmission

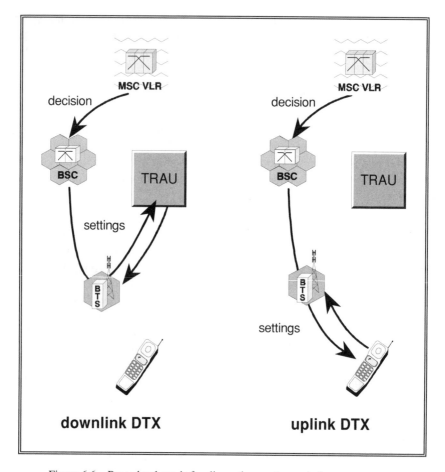

Figure 6.6 – Procedural needs for discontinuous transmission management

The use or not of discontinuous transmission needs only be notified
to the transmitting end; the receiving unit does not need to know beforehand.

on the radio path is reduced to a minimum when it is detected that the user data does not contain meaningful information (during speech silences for instance). This feature is optional and must therefore be managed. Moreover, discontinuous transmission may be applied independently to each direction, so that the control of discontinuous transmission must take into account two components: the uplink mode and the downlink mode.

Discontinuous transmission is only relevant to some of the transmission modes, speech and non-transparent data, simply because in the other cases it is difficult to assess when user data transmission can be suspended without degrading the service. The discontinuous transmission mode affects the operation of the mobile station and of the TRAU. The

BTS is obviously concerned, but derives its behaviour dynamically from data coming from the mobile station (uplink) and from the TRAU (where it exists) and MSC/IWF (downlink). Whether discontinuous transmission should be applied or not is, there again, decided upon by the MSC, and the execution is controlled by the BSC. A GSM BSS must indeed be able to cope with discontinuous transmission, whatever the strategy of the corresponding MSC.

The choice of the strategy for applying discontinuous transmission is one of the many configuration parameters on which operators may play to optimise their network. Several considerations must be taken into account in this strategy. For instance, GSM mobile to mobile calls suffer a loss in quality when discontinuous transmission is applied to both radio segments; experts refer to this phenomenon as "double clipping". The operator may therefore well choose not to apply downlink discontinuous transmission for such MS-to-MS calls, if they can be identified.

The downlink discontinuous transmission mode can be changed when the transmission mode changes; these are indeed the only instants at which the *Specifications* allow a change in the discontinuous transmission mode, since no other needs were identified.

For the uplink discontinuous transmission mode, the network can at any moment either force the mobile stations in communication to use it, forbid them to do so or leave the choice open.

Procedural needs for discontinuous transmission management include the means for the MSC to indicate whether discontinuous transmission should be applied for uplink and for downlink, and means for the BSC to configure the mobile station and the TRAU via the BTS (see figure 6.6).

6.1.4. HANDOVER PREPARATION

The possibility to change the cell during an RR-session is a very important function in a cellular system, and the major source of complexity in the Radio Resource management plane. We have already touched upon some of the aspects related to the *execution* of a handover, i.e., what has to be modified when a handover occurs. However, this represents but the tip of the iceberg. The process which precedes it, the handover preparation, may be thought of as a "behind the scenes" activity, yet it is the most important topic, which conditions both spectral efficiency and the quality of service as perceived by the users.

The decision to trigger a handover, and the choice of target cell, are based on a number of parameters, and various reasons may trigger this decision. These reasons will be studied first, then a description of the parameters affecting the choice will follow. Among these parameters, radio measurements performed by the mobile station and by the BTS are of foremost importance, and will be looked at in detail. The functional distribution of functions will be addressed last.

6.1.4.1. Handover Purposes

At first sight, the aim of handovers is to avoid loosing a call in progress when the mobile station leaves the radio coverage area of the cell in charge. Such cut-offs are very badly perceived by the users, and have an important weight in the overall perception of the quality of service by the subscribers. We shall call this type of handover "**rescue handover**", where a high probability exists that the call will be lost if the cell is not changed. An extreme form of the rescue handover is the call re-establishment, which is an attempt for the mobile station to salvage the connection *after* an effective loss of communication with the serving cell.

In other cases, it may be of interest to change the serving cell of a given mobile station even if the transmission quality is still adequate. This may happen when the global interference level would be significantly improved if the mobile station would be in contact with another cell. Computations and simulations show indeed that there is usually a "best cell" from the point of view of interference. This statement is especially true when power control is being used, since the cell corresponding to the minimum path loss will minimise the mobile station transmission power (ordered by the BTS), thereby statistically minimising the overall interference level. A handover triggered with the goal to optimise the interference level and not for the sake of the ongoing communication shall be referred to here as a "**confinement handover**". Such handovers result in a "confinement" in optimal geographical areas of the mobile stations which have a connection in a given cell, preventing them from wandering out of the optimal cellular coverage even if their connections are still of adequate quality. Confinement handovers may potentially conflict with local optimisation of the transmission quality, and should not be performed toward cells for which transmission quality is not correct.

A third kind of handover is referred to as the "**traffic handover**". It may happen that a cell is congested whereas neighbour cells are not. Such a situation happens typically when specific events lead to a very local geographical peak: fairs, sport events, and so on. Because the actual coverage of neighbour cells often overlap a lot, handing over some calls from one cell to a less congested one could temporarily improve congested situations. This kind of handover must be handled with great care, since it is obviously in conflict with the confinement criteria. Traffic handovers will necessarily perturb cell planning and increase the level of interference in the surrounding area.

The *Specifications* handle the traffic handover in a different way from the other types of handover. The concepts of rescue handover and confinement handover, on the other hand, are not developed in the *Specifications*, but the following paragraphs will show how much this distinction can help operators and manufacturers when developing smart handover algorithms.

6.1.4.2. Handover Criteria

Depending on the purpose for handover, the criteria to be taken into account differ, but they always include some information to predict what will happen with and without handover, according to the destination cell.

The main criterion for rescue handover is the quality of transmission for the ongoing connection, both uplink and downlink. The best information would be an assessment of the transmission quality as perceived by both users. With digital transmission, the transmission error rate is a good quality indicator. The propagation path loss incurred by radio transmission is also of interest. Another piece of information, although of more marginal application, is the propagation delay. Transmission on the GSM radio interface cannot usually support a high propagation delay, and a connection can be cut if it becomes too big. The case may only arise in large rural cells. In GSM, all these measurements are available to the handover decision process. Both the mobile station and the BTS measure regularly the transmission quality and the reception level, from which the path loss can be inferred. The mobile station transmits its measurements to the BTS, at the rate of once to twice per second.

The key criteria for a confinement handover are the uplink and downlink transmission quality corresponding to each neighbour cell, were the mobile station to be in connection with that cell. Since this information is quite difficult to get (it may depend on the would-be allocated channel for instance), the handover process in GSM has to make do with only the path loss between the mobile station and a number of neighbouring cells. In reality, only downlink values are measured, by the mobile station, and the assumption is made that the path loss is equivalent in both directions.

The decision process for traffic handovers requires information on the load of each BTS, and this information is known by the MSCs and

BSCs. Traffic handovers differ quite a lot from the rescue and confinement handovers, because traffic reasons dictate the number of mobile stations to be handed over, in a given cell, but do not indicate which of these should be. The choice of the favoured (?) ones usually starts with those which are closer to be handed over for other reasons. Hence, the traffic handover relies on the other criteria and the corresponding measurements.

The algorithms for the handover decision and the choice of the target cell are not imposed by the *Specifications*. An example of such an algorithm is given as an annex to TS GSM 05.08, but operators and manufacturers have complete freedom to implement more (or less) sophisticated algorithms based on available parameters. In order to summarise, a list of the parameters to be taken into account in the handover decision process, is given hereafter:

- some static data, such as the maximum transmission power of:
  ⇒ the mobile station
  ⇒ the serving BTS
  ⇒ the BTSs of neighbour cells;

- real time measurements performed by the mobile station:
  ⇒ the downlink transmission quality (raw bit error rate)
  ⇒ the downlink reception level on the current channel
  ⇒ the downlink reception levels from neighbour cells;

- real time measurements performed by the BTS:
  ⇒ the uplink transmission quality (raw bit error rate)
  ⇒ the uplink reception level on the current channel
  ⇒ the timing advance;

- traffic considerations, cell capacity and load, ...

Some of these parameters raise technical questions, for instance how does the mobile station reports its measurements, and how can a mobile station make measurements concerning neighbour cells with which it does not have a dedicated connection?

6.1.4.3. Measurements

Various aspects of the measurement process in GSM merit some explanations. One of them is how the measurements done by the mobile station and the base station must be transferred to a single point for treatment. This will be the first point we will look at. Another point is how the mobile station manages to make measurements concerning neighbouring base stations. Finally, some of the difficulties of the specifications of what is measured will be addressed.

Measurement Reporting

In order to make efficient handovers, the rate at which measurements are refreshed should be as high as possible. In GSM, the minimum rate of reporting is once per second. The mobile station must report measurements, not only for the serving cell, but also for as many neighbour cells as possible which might be candidate target cells. This is true in particular for confinement handovers, since the mobile station does not have enough knowledge of the cell planning aspects to determine in all certainty which neighbour cell would be the best target cell. Such a choice depends not only on the path loss, but also on maximum transmission power, cell size, etc. In GSM, the mobile station can report on up to 6 neighbour cells in addition to the measurements relative to the serving cell.

The measurement reporting activity of the mobile station represents a rate of roughly 130 bit/s at least. This reporting is carried by messages on the small signalling channel associated with each TCH and called the SACCH, whose maximum capacity equals twice this rate. Hence, the refreshing rate may go up to two reports per second, if the SACCH is not used for other purposes in parallel. And here lies one reason why it was chosen to have a channel separated from the main channel. If the TACH (TACH = TCH + SACCH) multiplexing had not been done on a burst basis, but through a sharing of bursts (a few bits for the SACCH, the remaining bits for the TCH), then discontinuous transmission would have been useless because of the constant requirement for measurement reporting.

Neighbour Cells Measurements

The requirement for a mobile station to measure the reception level of neighbour cells, while carrying a call in a given cell, raises a number of technical issues.

A first issue is simply when can a mobile station perform these measurements. In GSM, it is possible for the mobile station to make these measurements while maintaining a connection, and without requesting the mobile stations to have two receivers, thanks to the TDMA scheme. The mobile station indeed measures the characteristics of the neighbour cells during the interval between the transmission of an uplink burst and the reception of a downlink burst. These intervals are of various lengths, depending on the dedicated channel type. The worst case corresponds to

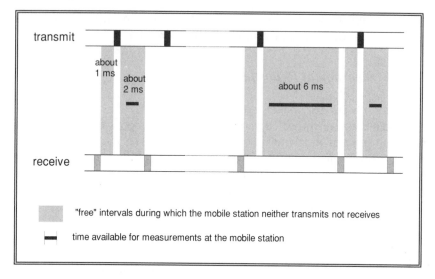

Figure 6.7 – Measurement intervals available at the mobile station

A TACH/F schedule leaves the mobile station with 26 very short intervals (2 BP – ε, # 1 ms), 24 "small" intervals (4 BP + ε, # 2 ms) and one "long" interval (12 BP + ε, # 6 ms) every 120 ms.

the TACH/F. The intervals, for each 120 ms period, are shown in figure 6.7 and are as follows:

- from end of reception to start of transmission:

  26 intervals of 2 BP –ε, where a BP (burst period) lasts for 577 µs and where ε represents the timing advance;

  these intervals are too short to be used for measurement.

- from end of transmission to start of reception:

  24 "small" intervals of 4 BP + ε and

  one "long" interval of 12 BP + ε. This interval exists thanks to the unused slot in the 26 slot cycle.

From these values, it is obvious that any efficient measurement scheme requires the mobile station to make measurements, not only during the long interval, but also during the small intervals. The measurement must be done on a frequency which is different from the one used for the preceding transmission burst or the next reception burst. This leads to a constraint on the frequency synthesis capability of the mobile station: either it is able to switch the reception/transmission frequency in less than 1 ms (thus leaving 300 µs for a measurement), or else it must have two frequency synthesisers. It can be seen that the requirement for a multi-frequency synthesis capability of the mobile station does not derive from the frequency hopping alone.

Another technical difficulty emerges from the preceding discussion, concerning the infrastructure. As explained above, the schedule of the mobile station is very tight indeed; as a consequence, the mobile stations cannot be overburdened with the requirement of listening to a specific channel in neighbour cells, which in GSM would mean find one specific burst among eight in addition to switching to the right frequency. This has led to one of the most troublesome specifications (for BTS manufacturers!) of GSM. The chosen solution requires each BTS to emit continuously (i.e., in every burst period) on one frequency, moreover at a constant power level. This must be done regardless of the status of the corresponding slots, i.e., whether they are used by some active channel or idle, and necessarily without applying discontinuous transmission or power control. This continuity of the transmission allows the mobile stations in neighbour cells to make reliable measurements whenever they can, without any other constraint than their own scheduling.

The next problem which arises is how does the mobile station determine which frequencies it must measure? First, it is necessary for the mobile station to distinguish those frequencies which are used as beacon frequencies in other cells from other potentially interfering frequencies, in order to make sure which cell to measure. Second, it is more efficient for the mobile station to limit measurements to the beacon frequencies of neighbour cells. The latter point is solved by sending to the mobiles stations a list of frequencies to be measured. The first point is somewhat more tricky. It requires the mobile station to do a bit more than just measuring a received level.

The beacon frequency is the one carrying the synchronisation and frequency correction channels (SCH and FCCH) of the cell. A way for the mobile station to check that the channel it receives is actually a beacon channel (and not another channel of another cell transmitting on the same frequency) is to find if this frequency carries a FCCH. This checking helps in fact to meet another requirement as well: **the pre-synchronisation**. Because synchronisation is necessary before transmission, as explained in Chapter 4, a mobile station must get some synchronisation information from the cell it will be handed over to, at the latest during handover execution. Since handover optimisation was felt very important by those who drafted the *Specifications*, it was decided that mobile stations should acquire synchronisation with all cells on which they report measurements. Hence, a mobile station is constantly pre-synchronised with all cells to which it may potentially be handed over.

Pre-synchronisation requires that the mobile station decodes not only the FCCH, but also the SCH of the beacon channel. When can the mobile station perform these tasks? How can it be assured to find the FCCH given the short time it is able to spend for its measurements? This is where the "long" interval comes in. The reader will recall that, in the case of a TACH/F, there is one such "long" interval (lasting about 6 ms) every 120 ms, and more of them in the case of TACH/8s. As far as the TACH/F is concerned, the "idle slot" corresponding to this "long" interval, and which happens once in every 26 slot cycle, is indeed of critical importance for the pre-synchronisation process. This *idle slot* is indeed a misnomer, since the mobile station is anything but idle during this period. Looking for a FCCH requires indeed the possibility to listen to the potential beacon channel through a sufficiently long window; the FCCH cycle recurs with a period of 51×8 BP, whereas the *idle slot* recurs every $26 \times 2 \times 8$ BP.

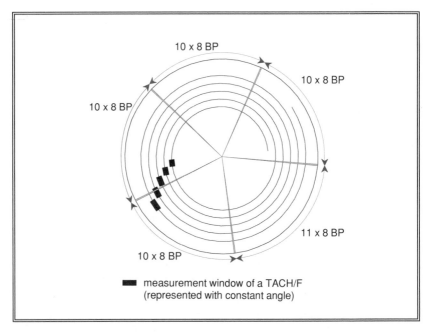

Figure 6.8 – Sliding measurement cycle

The interval between two successive FCCH bursts (including one of them)
is at most 88 BP, and lasts most often 80 BP.
Since in 11 successive long intervals the mobile station will have been able to read
more than 88 successive bursts,
it must have encountered an FCCH burst if the channel it monitors is a beacon channel.
The average time to encounter such a burst is roughly half of this value, i.e., 0.6 second.

This is where the arithmetic properties of the numbers 26 and 51 intervene: since these two numbers have no common divider, the two cycles are shifted in such a way as time passes that the *idle slot* has 100% chance of being aligned with the FCCH within 11 cycles. Figure 6.8 (previous page) shows that the FCCH cycle is shifted 8 BPs compared to the TACH/F cycle between two successive *idle slots*. Since a long interval lasts for about 12 BPs and therefore enables reception during about 9 BPs (the rest being used for frequency setting), 11 successive long intervals (i.e., about 1.2 seconds) are enough to ensure the mobile station with some opportunity to find an FCCH burst, as shown in the diagram. It is up to mobile stations to make good use of this opportunity!

In the case of the TACH/8, things are a bit different, because its recurring period is the same as the one of the FCCH. But the phasing between the two cycles can be anything. Figure 6.9 shows that there is

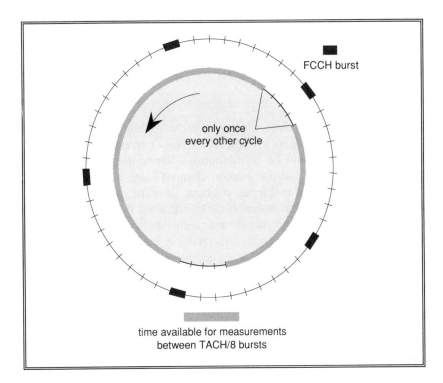

Figure 6.9 – TACH/8 cycle vs. FCCH cycle

Whatever the respective position of the TACH/8 and FCCH cycles, represented on the inner and outer circles, there is always an opportunity for an FCCH burst to fall into a measurement window of the TACH/8. The representation of the TACH/8s in this figure correspond to the scheduling of TACH/8s grouped by 8. The other case (TACH/8s grouped by 4) results in a slightly different but similar pattern.

necessarily a moment at which one of the 5 FCCH bursts in the 51×8 BP cycle can be listened to. The figure shows the case of TACH/8s not associated with the control channels, but the same reasoning applies for the TACH/8s associated with control channels.

Another question to be answered is which neighbour cell measurements does the mobile station report to the BTS among those cells measured, and how is this information packed into a small message? A measurement message can include up to 6 measurement results pertaining to neighbour cells. However, the mobile station may pre-synchronise with more than 6 neighbour cells. If this is the case, then only those measurements corresponding to the 6 cells it receives best are reported to the BTS.

Of course, this introduces a bias in the measurements collected by the network against the cells which are not among the first ones on the list. Their measurements are transmitted only when they are among the 6 best ones, and in consequence the ordering as established by the BTS by averaging several measurements may differ somewhat from the corresponding ordering by the mobile station. The measurement analysis process in the BSS ought to take into account the possibility of missing measurements; however, this potential error should not touch the best cells, which are of most interest.

The BSIC

It has been explained above how the mobile station is able to recognise that the frequency it monitors does correspond to a beacon channel. But there could be configurations where the mobile station is able to capture more than one beacon channel using a given frequency. This may happen when frequency planning must be done with very few frequencies, or at national boundaries. In order for the mobile station to be able to discriminate between the cells transmitting their beacon channels on the same frequency, a mechanism based on the BSIC (Base Station Identity Code, one of the major misnomers in GSM) has been introduced and warrants a detailed explanation.

The BSIC (Base Station Identity Code), which applies more properly to a cell, is *not* an unambiguous identification of a base station. Many cells bear the same BSIC, and moreover the normal practice is to allocate the same BSIC to neighbouring cells. So what is it? The BSIC is in fact a "colour code" (by reference to map colouring), allowing mobile stations to distinguish cells which transmit their beacon channel on the same frequency (see figure 6.10). For instance, when the radio spectrum available to a given operator is limited to, say, 2 MHz, frequency planning must cope with at most 10 frequencies. The best beacon frequency allocation scheme may not be able to avoid overlapping coverage in this case, and a mobile station will in some cases receive two beacon channels with the same frequency. A similar situation is also

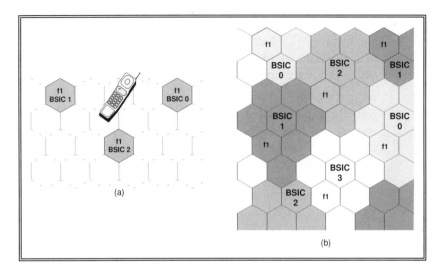

Figure 6.10 – Choice of the BSIC

The BSIC is a "colour code" allowing the mobile stations to distinguish cells
which share the same beacon frequency (a).
When a regular reuse pattern is used for beacon frequencies,
this pattern can be used as a typical basis for BSIC allocation (b).

frequent along national boundaries. Whereas inside a country the
frequency allocation of different operators are disjoint, two public land
mobile network (PLMN) operators on each side of the border will have
some frequencies in common. Frequency usage co-ordination between
operators helps, but cannot be enforced. In most cases a mobile station
will still be in a position to receive the same beacon frequency
transmitted by two base stations of different PLMNs. For all these
reasons, a scheme allowing to distinguish cells using the same beacon
frequency was deemed necessary. This is the role of the BSIC, a 6-bit
code word broadcast on the SCH of every cell.

The BSIC intervenes in different cases, all related to the distinction
between cells using the same beacon frequency:

- in order to perform neighbour cells measurements, the mobile
  station is provided with the list of frequencies to monitor. In the
  reporting message, the mobile station is required to indicate the
  BSIC for each beacon frequency on which it reports
  measurements. This implies that the mobile station has decoded
  the SCH of the beacon channels it measures, but this is not an
  additional constraint, since it must already be done for pre-
  synchronisation. The availability of the BSICs for measurement
  processing at the BSS enables it to check which cell has
  effectively been measured in the case of ambiguities.

- to prevent measurement reporting for cells toward which handover is precluded, a mechanism allows the network to indicate a subset of BSICs for which reporting should not be done. This screening mechanism is explained later.

- when a mobile station sends an access burst on the RACH of a cell, there is a risk that this random access be received by another cell than the one it was aimed at, if these cells use the same RACH frequencies and if their TDMA synchronisation do not fall too far apart. In order to avoid such spurious receptions, the RACH coded burst is "exclusive-ored" with the BSIC, so that only the right cell has a chance of decoding the burst successfully, based on the redundancy added to useful bits for checking correct decoding.

- when a mobile station in idle mode monitors neighbour cells, it regularly reads their BSICs; this provides the mobile station with a quick way to check whether the monitored cell is still the same or not. Such a check could also be achieved by decoding the broadcast messages containing the full cell identity, but at a greater cost.

The whole issue of BSIC planning consists in allocating different BSICs to cells using the same beacon frequency and between which overlapping coverage areas may exist. Inside a PLMN, BSIC planning is fairly easy; the matter is different at borders, when some co-ordination is necessary. In order to help this co-ordination, a tentative allocation of the first three bits of the BSIC has been introduced on a country basis, as shown in figure 6.11. This three-bit part has been named the "PLMN colour code" (or NCC, for Network Colour Code), a term which has been widely misunderstood. Its value is *not* a definite attribute of a PLMN, but a possibility for use near boundaries, and which can be overruled by bilateral or multi-lateral agreements between the concerned PLMNs. Indeed, nothing prevents all 64 values of the BSIC (including all 8 values of the "PLMN colour code") being used inside a PLMN when far away from any international boundary. Moreover, two PLMNs in the same country usually have disjoint frequency allocations, and can use without risk the same BSICs (and the same "PLMN colour codes"!).

The first 3-bit part of the BSIC is also the basis for screening measurement reporting. When two cells of different PLMNs using the same beacon frequency may potentially overlap near a border, it is normally best for a BSS to indicate to mobile stations that they should not report measurements concerning cells of the other PLMN, since these measurements would be worthless (since inter-PLMN handover is usually not performed), and might prevent the measurements of some cells in the

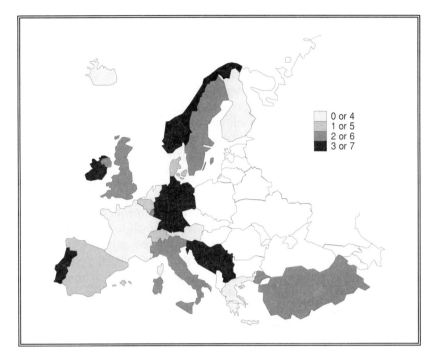

Figure 6.11 – A European PLMN code map

This tentative allocation of the "PLMN code" by country
is proposed to ease allocation of the BSIC near international boundaries;
but it does not represent a requirement.

right PLMN to be reported, since the number of reported measurements is limited to 6 neighbour cells. To that end, the information broadcast by each BTS includes an 8-bit screening indication, with one bit indicating for each of the 8 possible three-bit patterns whether cells with a BSIC starting with this pattern must be reported on. This screening indication is the *PLMN PERMITTED* indication.

The *PLMN PERMITTED* information happens to be broadcast on the BCCH. A current misunderstanding is that it influences cell selection (its name seems to carry such information indeed). This is not so: this information impacts only the measurement reporting and nothing else.

The Measurement Period

We will now look in some details to the measurement process itself. The various characteristics to be measured (reception level or reception quality) are continuous by nature, and affected by a significant level of "noise", that is to say statistical variations due to thermal or industrial noise, Rayleigh fading, interference, and also due to the change of frequency introduced by frequency hopping. To be usable for decision-making, the measurements need to be filtered (e.g., averaged). The first step of the filtering is done in the mobile station. It is quite simple: it consists in taking the average value of each measured parameter for the duration of the reporting period, which will be described in the following paragraphs. The other point worth noting is that what is averaged is the raw bit error rate for quality measurements, and the logarithm of the reception level, or, in other terms, the reception level expressed in decibels.

A measurement report pertains to a given measurement period, that is to say the period during which the measurements were done. The duration of the measurement period is always equal to the periodicity of message transmission on the SACCH (i.e., 480 ms on the TACH/F, and around 471 ms for the TACH/8). On the TACH/F, the start of the measurement period starts a fixed time before the start of transmission. But on the TACH/8, things are different. The absolute position of the measurement period is the same for all TACH/8s of the same TN (timeslot number, indicating the position in the 8 BP cycle). Its relation to the time of transmission of SACCH messages therefore varies, depending on the TN. In addition, the position of the measurement period is different for TACH/8s combined with common channels and for those which are not. In all cases however, the uplink and downlink measurement periods are simultaneous.

A small subtlety in the definition of the measurement period comes from the quality measurement. With the interleaving scheme for speech, there is one speech block spread over two successive measurement periods in the case of a TACH/F. Quality measurements for this block belong to the second period. This problem should not have arisen for the TACH/8, where the aim was to design the measurement period carefully so that blocks are all entirely within a measurement period. However, there is one exception to this rule, since the third TACH/8 of TN 4 does not comply with this aim.

A simultaneity problem arises between the measurements done by the mobile station and those done by the BTS. The second ones are known by the BTS shortly after the end of the measurement period, whereas the measurement report from the mobile station for the same period is delayed by the message transmission delay, which amounts to roughly half a second on a TACH/F. In order to provide the BSC with reports which match, the BTS must then buffer its measurements until it has received the message from the mobile station. As seen by the BSC, measurements are received roughly one second after the start of the measurement period in the case of a TACH/F, and between a bit more than half a second and close to one second for the TACH/8, depending on their position in the 102 BP cycle. The shorter delay in the case of a TACH/8 comes from the fact that each SACCH message is sent over 4 consecutive bursts, and not spread as for the TACH/F.

The Interaction with Discontinuous Transmission

The accuracy of measurements concerning the ongoing connection raises two problems, one in relation to discontinuous transmission, the other linked with power control and frequency hopping when using the beacon frequency.

When discontinuous transmission is applied, some slots belonging to a channel may not be used for effective transmission. This is indeed the goal of discontinuous transmission. But then measurements on these slots will obviously report a low reception level, and a bad quality. Even more annoying, one could imagine a measurement period, or a succession of periods, with no transmission at all, hence some difficulties for the processes which rely on these measurements. To circumvent these problems, the *Specifications* impose that at least 12 bursts are sent within each reporting period. These bursts amount to the systematic use of the SACCH for effective transmission (4 bursts constituting a coding block), and 8 bursts on the TCH itself. On a TCH/8, which corresponds to exactly 8 bursts per measurement period, this leads to the consequence that discontinuous transmission is not applied. On a TCH/F, this means that at least one block per measurement period must be sent (a block being interleaved over 8 half-bursts). For speech, this block contains a silence descriptor frame (SID frame) refreshing the comfort noise characteristics.

In addition to this minimum transmission rule, the *Specifications* require the BTS and the mobile station to report *two* sets of measurements concerning the connection: "full" measurements, done on all slots which may be used for transmission in the reporting period, and "sub" measurements, done only on the mandatorily sent bursts and blocks. On a TACH/F, the second set is less accurate than the first one when discontinuous transmission is not used, because averaging is done on a smaller set (for instance reception level is averaged on 12 bursts instead of 100 bursts). On a TACH/8, the two measurements are evidently identical, but are nevertheless both sent for uniformity. Finally, both the BTS and mobile station report for each measurement period whether discontinuous transmission was effectively used or not (or in another terms, whether all bursts were effectively transmitted), thus enabling the processes using the measurement to discard the "full" measurements provided by the other end when applicable.

The PWRC Indication

Another problem with measurements (and the last point on the topic) is in fact a consequence of a combination of different independent details of the *Specifications*. First, it is allowed that a frequency hopping TACH uses the beacon frequency as one of its frequencies (of course, not on TN 0). Second, power control can be applied on the downlink TACH. Third, the beacon frequency must be transmitted with a constant transmission power, because of the measurements performed by mobile stations of neighbour cells. The result for the channels under consideration is that power control applies only to a subset of the bursts, whereas other bursts (those using the beacon frequency) are sent with a fixed transmission power. This leads to inaccurate reception level measurements. In order to alleviate this problem, the mobile station is requested in such cases not to take into account the slots falling on the beacon frequency in the reception level estimation. This is controlled by an indicator, the *PWRC* indicator (originally called power control indicator, a misnomer), sent on a connection basis to the mobile station. This indicator should be set if the following conditions are all met: the channel hops on at least two different frequencies, one of those frequencies is the beacon frequency, and downlink transmission power control is in use.

6.1.5. POWER CONTROL AND TIMING ADVANCE

Most of the functions needed for transmission over the radio interface correspond to transformations of the signals representing the data to transmit. They were the object of Chapter 4. Two functions, the management of the transmission power and of the timing advance, lie somewhere between these pure transmission functions, and transmission management functions. As the transmission functions, they are performed continuously and relate heavily to physical features, but they share the use of signalling means with the transmission management functions. They are studied here, because of this usage of signalling means, and also because they are, as we will see, deeply entangled with the procedures managing RR-sessions. This comes from the initialisation of these processes, which must take place any time the channel is changed.

6.1.5.1. Power Control

Power control refers to the possibility to modify within some range the transmission power on the radio, both (but independently) for the mobile station and the base station. Power control shares a common goal with discontinuous transmission: improving spectral efficiency and, although to a lesser extent, battery life for the mobile station. When one side is received too well by the other, it becomes advantageous to reduce its transmission power by such an amount as to keep a similar quality level on the communication, while decreasing the interference caused on other calls in surrounding areas. In a system such as GSM, the full gain in spectral efficiency can be obtained with a small power range, and the 20 dB minimum case specified by the *Specifications* is more than sufficient for this purpose.

In GSM, both uplink and downlink power control may be applied independently one from the other; furthermore they are applied independently with each mobile station. The range specified by the *Specifications* for uplink power control lies between 20 and 30 dB, by steps of 2 dB, depending on the mobile station power class. An example is given in figure 6.12. The range used for downlink power control is manufacturer dependent and may be up to 30 dB, also by steps of 2 dB. The control of the transmission power is a network option, i.e., the operator may choose to apply it or not, in one direction or in both. All mobile stations, though, must support the feature, thereby allowing power control to be really efficient when utilised.

Power control on both directions is managed by the BSS. The transmission power of the mobile station is chosen by the BSS, and

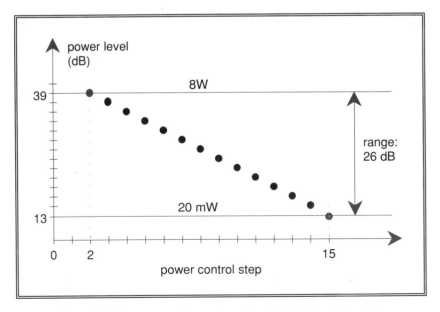

Figure 6.12 – Power control steps for a class 2 GSM900 mobile station

Power can be controlled by steps of 2 dB on a range going from 0,2 W (13 dBm) to the maximum MS power (here 8W = 39 dBm).

commands to regulate it are issued to the mobile station. The BSS computes the required MS transmission power through reception level measurements performed by the BTS, taking into account the MS maximum transmission power as well as quality measurements done by the BTS; this last parameter helps to ensure that transmission quality is kept above some acceptance threshold. For the downlink direction, the BTS transmission power is also computed by the BSS for each connection, based on the measurements performed by the mobile station and reported regularly to the BTS.

Inside the BSS, the split between BTS and BSC is basically an option for the manufacturer. The specification of the Abis interface as found in the *Specifications* is basically adapted to the implementation of power control in the BSC, but implementation in the BTS is possible. The detailed procedural means for the latter case are not specified in the *Specifications*, but a number of "place-holders" in the BTS-BSC protocol allow manufacturers to specify and implement part or all of the power control management in the BTS.

At the start of a connection, the initial value of the transmission power (both for mobile station and for BTS) is chosen by the BSC. In the case of an initial assignment, the information available to choose this power is at best very small: it consists in the reception level of a single access burst, which is necessarily of limited accuracy. Therefore, in

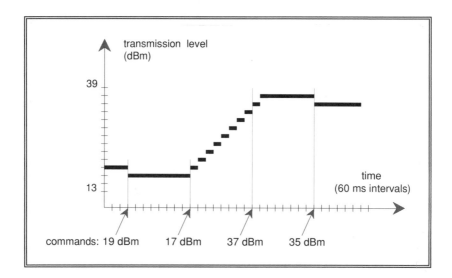

Figure 6.13 – Transmission power adaptation

The transmission power is adjusted by steps of 2 dB,
recurring not more often than every 60 ms.
A high jump in the power control commands will therefore be answered gradually.

GSM, the initial power level to be used by a mobile station for the first messages sent on the new dedicated channel is fixed on a cell-per-cell basis, and is the same level as used for sending random access bursts. The value of this level is broadcast on the BCCH, to be known by all mobile stations before any access attempt. A mobile station whose maximum power level is below the broadcast value shall simply use its maximum power level instead. For subsequent channel connections, the MS transmission power to be used when accessing the new channel is specified by the BSC, either using a default cell value (typically the case for an incoming handover between different BSCs) or based on the knowledge concerning the previous connection (this capability could be used, e.g., at subsequent assignment).

Except at the start of a channel connection, a command to change the transmission power does not trigger an immediate transition to the ordered value in the mobile station. The maximum variation speed is of 2 dB each 60 ms (see Figure 6.13). This means in particular that a high jump (more than 12 dB) will not be terminated when the next command arrives.

The basic procedural requirements for power control, as shown in figure 6.14, include the following:

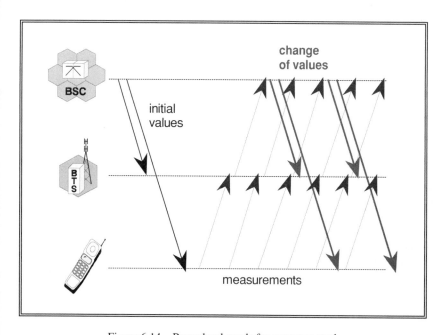

Figure 6.14 – Procedural needs for power control

Mechanisms to report radio measurements, to set-up the initial power value, and to control power by steps are included on the Abis and the radio interfaces.

- some means for the mobile station to transmit measurements (the same SACCH procedures are used for handover preparation) up to the BSC, even though some measurements may stop their journey at the BTS to be "pre-processed";

- some means for the BSC to command MS and possibly BTS transmission power;

- some means for the BSC to indicate to the mobile station the initial power level value to be used at initial assignment, as well as at each subsequent channel transition;

- some means for the BSC to indicate to the BTS the initial power level value when a channel connection is initialised.

6.1.5.2. Timing Advance

The time division multiplexing scheme used on the radio path of GSM is such that the BTS must receive signals coming from different mobile stations very close to each other. In order to reach this goal despite the propagation delay incurred by the return trip from BTS to mobile station, and taking into account that guard times between bursts have been chosen very small for spectral efficiency, a mechanism to compensate for the propagation delay is necessary. To this avail, the mobile station advances its transmission time relative to its basic schedule, which is derived from the reception of bursts, by a time indicated by the infrastructure, the timing advance.

Once a dedicated connection has been established, the BTS continuously measures the time offset between its own burst schedule and the reception schedule of mobile station bursts. Based on these measurements, it is able to provide the mobile station with the required timing advance and does that on the SACCH at the rate of twice per second.

The timing advance can take values from 0 to 233 µs, which is enough to cope with cells having a radius of up to 35 km without any other special scheme and given the speed of light. This limit comes from coding considerations (the timing advance is coded between 0 and 63, with the bit period as the unit, hence 233 µs), but there are more important hidden limitations. A first point is the guard time for access bursts, which in practice limits to about 220 µs the possibility for the initial propagation time measurement. The other point is that some minimum time is needed between the end of the reception of a downlink burst and the beginning of the transmission of the next uplink burst, in order to allow the implementation of mobile stations with the same frequency synthesiser for emission and reception.

Even in rural or low-density areas, good coverage quality will in practice require cell radii smaller than 35 km. However, there are cases when larger cells would be useful. This holds in particular for the coverage of inshore coastal areas, where high antennas (e.g., on lighthouses) could be in sight of boats more than 35 km away on sea. Such uses are indeed possible, at the expense of the number of channels per MHz. The trick consists in obtaining a huge guard time (more than 580 µs) by using only every second burst. In such cases, only the channels of even TNs can be used (since TN 0 must be used for the BCCH, odd TNs will not do). This feature requires a specific reception processing in the BTSs.

Upon establishment of a new dedicated connection, the timing advance control process must be initialised. This happens at each initial assignment, subsequent assignment or handover. Depending on the case, the mobile station and the infrastructure do not always have the same amount of information to assess the new timing advance, and the initialisation method varies accordingly. The different cases will now be examined one by one.

a) Both Mobile Station and Network can assess the Timing Advance Beforehand

This situation happens upon subsequent assignment. The mobile station simply uses on the new channel the old value of the timing advance which was ordered by the BTS on the previous channel. On the infrastructure side, the BTS device in charge of the old channel is aware of the last ordered timing advance, but there is no communication means between BTSs. The BTS is indeed not aware that the old and the new channel connection concern the same mobile station. The BSC is kept informed of the last ordered timing advance value, and it is therefore able to transmit it to the new BTS device when activating it. The timing advance control process just resumes on the new channel with the same values as on the previous channel.

There exists another case where both mobile station and network are able to assess the timing advance beforehand. This happens at handover between synchronised cells which are collocated. However, the *Specifications* treat this case as if the cells were synchronised, but not collocated (see next paragraph). A more efficient scheme could have been to give an indication to the mobile station that the old and new cells are collocated, and such an improvement may indeed be made in the future.

b) Only the Mobile Station can assess the New Timing Advance Beforehand

This case arises when handover is performed between two synchronised cells which are not necessarily at the same location. The mobile station is then able to measure the difference between arrival times of bursts coming from the two BTSs. It must indeed do so for pre-synchronisation requirements (see page 333). This arrival time offset is a combination of the transmission time offset between the two BTSs (which does not depend on the location of the mobile station), and of the two propagation times, as shown in figure 6.15. Therefore, if the mobile station is given the transmission time offset between the two BTSs (which is zero, by definition, for synchronised cells), it is able to derive

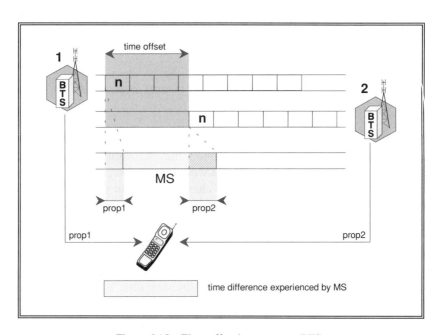

Figure 6.15 – Time offset between two BTSs

As seen by the mobile station, the arrival time difference between bursts coming from two BTSs is made up by the difference of the propagation times, plus the offset between clocks of the two BTSs.

the difference in propagation times, and therefore to calculate the new timing advance to be applied.

The indication that two cells are synchronised is therefore enough for the mobile station to assess the new timing advance, as follows (based on the notations of figure 6.15):

$$TA2 = TA1 - 2\,(prop1 - prop2)$$

On the infrastructure side, the new timing advance cannot be computed, except in the case of collocated and synchronised cells. A possibility would have been to wait for the mobile station to indicate this value (we will see later that the mobile station indicates back the value of the timing advance it uses, at least once per second). Nevertheless, to allow for a possible slightly faster initialisation on the new channel, the handover procedure between synchronised cells includes the means for the new BTS to assess the propagation time with the mobile station. To this avail, the mobile station starts transmission on the new channel with a few access bursts sent with a null timing advance, before switching to normal transmission.

c) Neither Mobile Station nor Network are able to assess the New Timing Advance Beforehand

At initial assignment, or at handover between two cells which are not synchronised, no information can be used by either side to predict the timing advance. Signalling messages are different, but the timing advance initialisation process is very similar in both cases. The mobile station is forbidden to transmit normal bursts until it knows the new timing advance to apply. Because the BTS (which is the entity which decides on the timing advance) must receive something from the mobile station in order to assess the propagation time, the mobile station is required to send access bursts to the BTS with a null timing advance. When the BTS receives such a burst, the reception instant is a measure of the double propagation time and the BTS can derive the value of the timing advance, which it sends to the mobile station in a signalling message. From the moment it receives this message, the mobile station is able to start correctly transmitting normal bursts. This exchange lengthens the duration of the handover procedure between asynchronous cells compared to the synchronised case described earlier. Moreover, the "asynchronous" handover leads to a longer communication interruption than its "synchronous" parent.

6.1.6. RADIO CHANNEL MANAGEMENT

So far, we have seen how individual RR-sessions are managed. Though RR-sessions are independent, they share the same pool of resources, in particular the radio channels. In this section we will look at the management of the radio channels in a cell as a whole, dealing with such problems as the configuration of the channels and the channel allocation strategy.

The management of the set of channels to be used in each cell includes two main aspects. On one side, the set of channels of each cell must be determined and the machines need to be configured accordingly. This "long-term" aspect is the cell channel configuration management. On the other side, the channels go through allocation/release cycles following the communication needs of the mobile stations. This "short-term" aspect is the dedicated channel allocation management. Both channel configuration management and channel allocation are the responsibility of the BSC. The MSC only intervenes to indicate which type of channel a given communication requires, whereas the BTS executes different related tasks, but always under control of the BSC. Both areas have strong impact on the procedures used over the radio interface and the Abis interface and will be described here.

6.1.6.1. Cell Channel Configuration

The channel configuration of a cell is the list of channels defined at a given time to be used in the cell. A typical cell configuration includes a set of common channels to support mobile stations in idle mode and initial mobile station access (a BCCH, a PAGCH and a RACH) and a set of traffic channels (TACHs of various rates, including what the *Specifications* call TCH/F, TCH/H and SDCCH) for carrying signalling and user data. The channel configuration of a cell may change in time. These changes may have various degrees of impact on traffic management, i.e., on the allocation and release of channels used for specific communications.

On one side, some modifications are related to the evolution of the whole network, for instance a capacity extension to cope with an increasing traffic density. Such changes are clearly within the scope of network operation. However, because network operators appreciate the possibility to handle such changes without disturbing the existing ongoing communications, mechanisms have been introduced in the traffic management area for this purpose; these mechanisms will be described here.

Configuration of the Access Channels

Depending on the full spectrum capacity of the cell, usually estimated in terms of number of frequency slots, the capacity requirement for access channels (RACH, PAGCH) will vary. The *Specifications* cater for five different access channel capacities for a given cell, which are summarised in table 6.2. These access channel capacities correspond in terms of radio consumption to a range from the equivalent of a half-rate traffic channel to the equivalent of 4 full rate traffic channels.

Since the mobile stations are assumed to be able to listen to only one unit of spectrum usage (the equivalent of a TACH/F, i.e., one slot every 8 time slots) at a given moment, they must be distributed among 1 to 4 population groups, depending on the capacity of the access channel structure. Mobile stations find the information about the applicable structure in the messages broadcast on the BCCH. These messages are sent on every carrier unit used for common control channels (of TN 0, 2, 4 and 6 as applicable).

The access channel configuration may change in time, and detailed procedures are provided in the *Specifications* to cope with such dynamic changes and ensure that the transition period between two stable configurations is as limited as possible. This impacts mainly the listening to the PAGCH in idle mode.

CCCH capacity (equiv. in TACH/F)	number of MS groups	RACH burst rate (bursts per second)	PAGCH message rate (messages per second)
1/2 (*)	1	114.7	12.7
1	1	216.7	38.2
2	2	433.4	76.5
3	3	650	114.7
4	4	866.7	152.9

Table 6.2 – Access channel capacities

Depending on the capacity to be offered by a given cell,
the access channels can be configured in several different sizes.
(*) the other half can be used only for 4 TACH/8.

The corresponding procedural aspects will be covered in the section dealing with paging procedures.

Organisation of the PAGCH

On each CCCH "unit" to which the mobile stations are able to listen, the downlink paging and access grant channel is organised in two parts:

- several "paging sub-channels", in a one-to-one relationship with sub-populations of mobile stations, on which initial assignment messages can also be sent.

- possibly a sub-channel reserved exclusively for assignment messages;

This PAGCH configuration is indicated to the mobile stations in messages broadcast on the BCCH, in order for mobile stations to determine where to listen for their own paging messages. The PAGCH configuration may change dynamically, and a mechanism is defined in the *Specifications* to enable such a change while avoiding the risk for mobile stations to lose paging messages during such changes. Both the corresponding contents of the broadcast messages and the allocation of paging messages to sub-channels are controlled by the BSC.

These procedural aspects will be described respectively in the sections dealing with general information broadcasting and with paging procedures.

Traffic Channels Configuration

Another point in the area of cell channel configuration management is the possibility to modify dynamically the set of traffic channels to meet the demand more closely. For example, the resource used at a given moment in time for a TACH/F can also be used for 8 TACH/8 at some other time. This kind of choice can be under control of the operation and maintenance sub-system (as a result of medium or long-term traffic analyses), or alternatively may be fully implemented in the BSC, so that the allocation process of the BSC would perform the conversion when needed. For instance, if a TACH/8 is needed when none is free but if a TACH/F is available, the latter could be converted into a pool of 8 TACH/8s instead of rejecting the request. The *Specifications* leave complete freedom to the operator/manufacturer to choose the implementation anywhere between these two extremes.

These functions do not impact directly the communications in progress, and are totally internal to the BSC. Hence they are not visible on the connection management protocols.

Changes in the Frequency Configuration

The previous paragraphs have dealt with changes in the functional configuration of channels, within a given pool of time/frequency resources. But that is not the end of the story; the frequency slots allocated to a cell may change dynamically in time, even though this is presumably not a very frequent event.

In the case when only single-frequency channels are used, a change in the frequency allocation of the cell will impact the communications making use of any suppressed frequencies, but each such communication can be handled independently from the others. However, when frequency hopping is employed, a frequency is used in a very tightly co-ordinated way by several connections in the cell at the same time. Any change in frequency affecting a given connection must be precisely coupled with similar changes to other connections in order to keep the non-interfering properties of the channels. These changes must happen in a synchronous way. With this objective, specific mechanisms have been introduced in the *Specifications* to enable the synchronised modification of the frequency allocation of many channels.

These mechanisms include the possibility to order a precisely timed change of the frequency parameters to the mobile stations and to the BTS for all impacted connections, as well as to have precisely timed channel assignments (whether initial assignments, subsequent assignments or handovers). The indication of the instant of change relies on the cyclic numbering scheme of TDMA slots, which has a period of about 3 and a half hours and allows an accuracy of microseconds.

As seen from the mobile station, these changes appear simply as changes of channels to be performed at a defined instant. On the BSS side, the matter is somewhat more complex. The aim is to synchronise the behaviour of several mobile stations and of the BTS, using signalling means which are by nature asynchronous and subject to losses due to transmission errors. The operation must be performed in several steps.

First, the general decision to perform a frequency change comes from the operation and maintenance sub-system, for such reasons as the setting up of new hardware, or the need for removing some for maintenance, or due to observations of unplanned interference. The decision to modify the frequency organisation of a cell is notified to the BSC, which is then in charge of co-ordinating mobile stations and BTSs to reach the new coherent configuration.

The first step for the BSC is then to determine the transition instant. This instant must not be too far away, to avoid introducing ambiguities from the cyclic numbering scheme. However, it must not be too close, in order to ensure that all concerned mobile stations have either received the command or have had their ongoing connection released. The concerned mobile stations are those engaged in transmission on a channel whose frequency parameters are affected by the change, i.e., those to which such a channel has been allocated before the actual transition time: a timed transition order must be sent to them.

There remains the case of the new allocation of channels before the actual transition time: such procedures are started, but with an indication that the mobile station will go on the new channel only at the transition time. Thus, all mobile stations involved in a connection on one of the impacted channels, as well as the BTS, perform the transition when the transition time occurs, and a normal situation is restored.

To summarise, the change of frequency configuration impacts all the assignment procedures, and require a specific procedure to deal with existing connections, called the *frequency redefinition* procedure. The corresponding details will be found in the sections dealing with each of these procedures.

6.1.6.2. Dedicated Channel Allocation

The second component of the management of the radio channels seen as a set is how the dedicated channels (TACH/8 and TACH/F) are chosen when allocated to an RR-session. As seen by the infrastructure, dedicated channels are at a given moment either allocated to the use of a mobile station, or part of a pool of idle channels from which a channel is drawn when a new need appears. To summarise the previous sections, such a new need may appear in three different conditions:

- at initial channel assignment, when a mobile station in idle mode has some communication needs, for instance because the user wants to set up a call, or because location updating must be performed;

- at subsequent assignment, when what the communication needs does not correspond any more to the type of channel it is allocated, for instance when a TACH/8 was allocated at initial assignment and a speech call needs to be connected;

- at handover, when the movements of the user or the variations of the interfering level result in a situation where the connection would be better on another channel, often through another cell.

Allocation Strategies

From the mobile station point of view, these various kinds of channel assignments are simply orders to start transmission and reception on specified channels. From the infrastructure point of view, the allocation of a dedicated radio channel involves two steps, first the choice of the channel to use, second the actual transition. The choice of the allocated channel lies entirely within the responsibility of the BSC. Sophisticated algorithms can be designed to try maximising the total amount of traffic which can be served with a given amount of resources, while maintaining a reasonable fairness level in the granting of requests.

Allocation optimisation includes several aspects. A first one is related to the relevance of the type of channel which is allocated for the effective need. This leads to a real problem in the case of initial assignment, since very little information is available at the BSC to choose the type of channel (the mobile station gives only a rough description for its reason to access in the initial access request message). A typical example concerns the setting up of a call: a TACH/F will be required to transmit user data, but a TACH/8 using 8 times less resources would be enough until the correspondents have begun conversing. Several strategies can be chosen, which can be grouped under the three following categories, as also shown in figure 6.16:

- **Very Early Assignment** consists in allocating a TACH/F at initial assignment, when it is probable that the requested connection will need such a channel;

- **Early Assignment** consists in allocating a TACH/8 initially, then subsequently allocating a TACH/F as soon as it is known for sure that this type of channel will be required;

- **Off Air Call Set Up** (OACSU) consists in allocating a TACH/8 initially, then waiting until the called party has answered before attempting the subsequent assignment of a TACH/F.

These different methods each have their pros and cons, which fomented many debates over the specification years of GSM. OACSU differs from the two other methods in that it provides the users with a different grade of service. The correct channel may be allocated a noticeable amount of time after the called party has answered, depending on the availability of channels at this instant. OACSU therefore requires

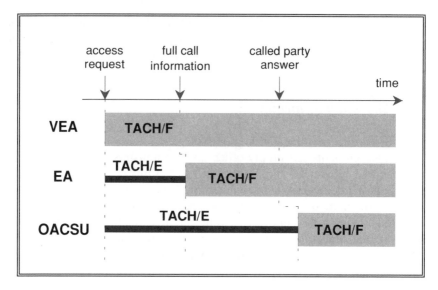

Figure 6.16 – Assignment strategies at call set-up

Very Early Assignment (VEA), Early Assignment (EA)
and Off-Air Call Set Up (OACSU) represent three different allocation strategies,
between which operators may choose.

some announcement to the called party, who may otherwise wonder why the phone rang at all! The grade of service which results is considered by many operators as unacceptable, though opinions diverge on the issue. On the other hand, OACSU is certainly the most efficient of the three schemes in terms of resource usage. The dilemma lies (as often) between efficiency and user comfort.

The main drawback of early assignment compared to very early assignment consists in an increased call set-up time, with no real gain in terms of channel usage in the case when a TACH/F is actually needed (it should be remembered that a measure of channel usage must take into account, not only the channel size (spectrum consumption), but also the usage duration, and that signalling exchanges are quicker on a TACH/F than on a TACH/8). On the other hand, very early assignment is very inefficient if a TACH/F is allocated when not necessary, e.g., for location updating. In this case, the amount of time during which the connection must be kept is mainly determined by the duration of signalling exchanges between infrastructure entities, and one cannot expect a significant reduction by using a larger channel. Therefore, very early assignment is of interest only if enough information on the use of the required connection is available to the network before initial assignment (which is not the case in the phase 1 version of GSM).

Whatever the allocation strategy, there are cases where no adequate resource is available when needed. The network may then apply one of two strategies: either the request is rejected, relying on its originator to possibly retry later, or the request is put aside to be served when a suitable channel becomes free. The latter strategy is referred to as "queuing", although there is not necessarily a queue in the strictest sense of the word.

Queuing

The interest of queuing varies with the conditions in which it is applied. At initial assignment, the repetition scheme put in place to cope with losses due to collisions or bad propagation conditions reduces the interest of queuing to nothing. Using queuing at initial assignment could even have adverse effects, since a request which is not answered in a short time will be repeated by the mobile station (the normal reason for no answer being transmission loss), and this could lead to very inefficient multiple assignments.

Queuing could of course have been made possible at initial assignment, by introducing a "please wait" acknowledgement to the mobile station request, but this has not been introduced for simplicity reasons. The WAIT INDICATION element appearing in a message answering a request for access should not be mistaken for such a scheme; it is in fact a temporary rejection preventing the mobile station from making new attempts for some time.

The major interest of queuing is indeed found in the case of subsequent assignments. The only resulting drawback is the lengthening of the call set-up time (this is perceived by the calling party when early assignment is used, and by both parties when OACSU is used). However, this is to be balanced with a rejection of the call, probably more ill-perceived by the subscribers!

For handover, the picture is not so clear. If the handover is decided to salvage a rapidly degrading situation, queuing cannot help much since the connection will most probably be lost if the channel cannot be allocated right away; but if the connection is not in immediate jeopardy, and the handover decision just stems from general optimisation reasons, queuing can be a source of improvement. Thus, a correct use of queuing at handover requires that the handover process distinguishes several cases.

Fairness between the treatment of requests is to be sought when applying queuing. When no other consideration intervenes, the order of service is usually first-come, first-served: there we have the classic queue. More sophisticated schemes can also be devised, granting requests on the basis of their estimated priorities; handovers may be considered of higher priority than re-assignments, and emergency calls are considered more important than other communications.

Even without queuing, there are ways to bias the granting of channel requests in congested situations. For instance, depending on the state of channel congestion, it could be worth rejecting the channel requests for outgoing calls to privilege those for incoming calls (a mobile user is more likely to answer than a fixed one if the mobile station is switched on, and the end-to-end circuit is almost entirely established already in the case of an incoming call). An even more drastic approach consists in forcefully terminating a connection estimated of a low importance, in order to reuse the resource for some needs deemed more important. This approach is referred to as pre-emption, and should be used with great care, given the negative impact on the preempted user.

For all the above reasons, some mechanisms have been introduced in the *Specifications* to enable priority strategies for radio channel allocation in the BSC. These mechanisms consist in conveying category, priority and pre-emption indicators, which can be used to influence the allocation decisions in cases of congestion. The *Specifications* do not describe how to use these indications, since the allocation schemes are left to the operators or manufacturers. Only their transport on interfaces is specified, and their actual use is indeed a source of difference between the products of different manufacturers.

Minimum indications relative to the purpose of the access are provided by the mobile station to the BSC at the beginning of the access procedure, to allow some priority mechanism for the initial allocation. Similarly, the MSC may provide some information when it requests a change of channel type (subsequent assignment procedure). In the case of handover the requesting BSC can also provide the MSC with some information about the reason for handover, and this reason can be carried up to the target BSC.

To summarise, a few things here and there in the procedures allow sophisticated allocation algorithms, with queuing, priorities, ..., but no constraints are put by the *Specifications* on the infrastructure equipment in this area. It is up to the manufacturers to include such functions, or for operators to request them.

Interference Considerations

The actual channel to be allocated may also be chosen with the optimisation of the transmission performance in mind. This requires the

BSC to have some knowledge beforehand about the performance of a connection for each of the free channels. Performance depends on many factors, most of them difficult to assess before the effective usage of the channel. However, there is one which is accessible: the level of interference in the uplink direction. When a channel is not allocated to any connection, the level received on the channel gives an idea of the level of interference and noise. GSM opens the possibility for the BTS to measure this reception level on all unallocated channels, and to transmit them regularly to the BSC. The BSC can then take this information into account in order to allocate a free channel of minimum uplink interference level, or to decide on an intra-cell handover if it is noticed that an active channel suffers a higher level of uplink interference level than the free one.

It should be noted that this feature is in most cases of secondary interest. The concern of an operator is to obtain a system able to support a maximum capacity; thus the situation to optimise is congestion. If all the channels of a cell are used at congestion (and this is the usual assumption), the allocation of the channels starting with the less interfered changes nothing to the eventual congestion situation. The only gain is the improvement of the average performance when far from the congestion state.

However, the assumption that all the channels are used at congestion does not always hold. One can imagine a cellular planning where the number of channels allocated to a cell is too high, in the sense that if all channels are used in all cells the overall quality of the connections is not acceptable. With such an approach congestion happens before all channels are allocated. A first consequence is that the BSCs must take care not to start additional connections when the congestion state is reached, even if channels are available. Another consequence is that taking uplink interference levels into account is then meaningful. What is obtained is a kind of automatic channel planning between the cells: the use of a channel in a cell will result in some interference level in other cells, thus preventing the latter to use interfered (and then interfering) channels. This is essentially a dynamic channel allocation method; it can be easily shown by taking the extreme assumption that all the channels are allocated to all cells. Without going to this extreme, a small over-dimensioning of the cell capacity plus this dynamic channel allocation method can be useful to cope with an abnormal distribution of the traffic between cells. If a cell is overloaded, but if the cells using interfering channels are not, the congested cell can with this approach use more channels than if all cells were equally overloaded.

Radio Channel Description

A side issue related to channel allocation is the way radio channels are described. The problem lies with frequency hopping. As explained in detail in Chapter 4, channel characteristics include time and frequency parameters. The time characteristics are not problematic; in the time domain, there are only 8 types of TACH/F and 68 types of TACH/8, depending on the type of channel and the offset relatively to the reference clock. One octet is therefore enough to code which of these 76 families of channels is being referenced (92 if half-rate channels were included).

In the frequency domain, things are somewhat more complicated. In the case of single-frequency channels, the number of different cases is 124 for GSM900 and 374 for DCS1800. In order to cope with some evolution, this frequency is coded on 10 bits, leading to a total of 18 bits to encode any single-frequency channel. However, when frequency hopping is being used, the number of combinations explodes. For GSM900 alone, a rough assessment shows that there are about

$$66,141,633,339,297,631,280,564,218,199,442,383,724,544$$

different possible hopping sequences compliant with the *Specifications* (including single frequency cases). This value needs 135 bits to be written in binary format. The matter is exponentially worse for DCS1800. This causes a problem because of the size of signalling messages, which is particularly critical on the PAGCH on which all initial assignment messages are sent. Some kind of reduction was therefore desirable. But which reduction?

First, what is needed to describe a hopping channel? The list of frequencies used by the channel, obviously. This frequency list, called the *MOBILE ALLOCATION* in the *Specifications*, is evidently the main source of length. The description also contains two other parameters used for the computation of the hopping sequence:

- the Mobile Allocation Index Offset (MAIO), of which there are as many possible values as there are frequencies in the list, hence its name since it describes the starting point for the hopping recurrence function; and
- the Hopping Sequence Number (HSN), which can assume 64 different values.

These two parameters fit on at most 13 bits.

The real coding problem lies with the coding of the frequency list. A first simplification consists in noting that only a set (i.e., an unordered

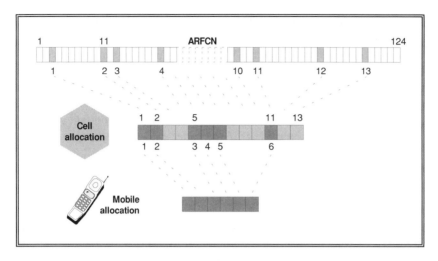

Figure 6.17 – Cell and Mobile Allocation

The cell allocation is the portion of the total resource usable in a cell,
with regard to which each mobile allocation may be defined.

list) of frequencies is needed to define a channel. The hopping sequence
generation algorithm refers to a list of frequencies, but the ordering is
implicitly defined based on the respective value of the frequencies in
Hertz. Therefore only sets are required, reducing the coding requirement
to 124 bits. This is still too much compared with the constraints of the
PAGCH.

Two approaches were possible: either restrict the number of
possibilities, or design an efficient and sophisticated signalling scheme.
Restricting the number of possibilities was inescapable for DCS1800.
The maximum number of frequencies used by any given channel varies
from 32 to 64 depending on the frequency range in which the channels
are spread. But there is still with this limitation more frequency
sequences in DCS1800 than in GSM900.

There remains the sophisticated signalling approach. All channels
in a given cell use only those frequencies which are allocated to the cell
by cellular planning. If this usually rather short list can be broadcast to
mobile stations independently from any allocation, the description of a
given channel in a known cell needs only to indicate which of these cell
frequencies are being used. This "two-step" mechanism (see figure 6.17)
leads to the concept of cell allocation, or cell channel description, which
are the terms used in the *Specifications* to designate the set of all
frequencies which are used in a given cell. (In fact, the main use of the
cell allocation is for the initial assignment, so the cell allocation can be
limited to the set of the frequencies used by channels that may be

allocated as initial channels.) The coding of the cell allocation uses 16 octets in GSM900 (where it includes a bit map for all 124 frequencies) and up to twice as much in DCS1800 (where the coding algorithm is much more sophisticated). This cell allocation is broadcast regularly on the BCCH. When the BSC sends a channel allocation message to the mobile station (whether at initial assignment, subsequent assignment or handover), it is able to encode efficiently the channel frequency list as a subset of the cell frequency list. If the latter includes n frequencies, the encoding of all possible subsets needs theoretically only n bits. The cell allocation in GSM900 is limited to less than 64 frequencies (since the bit map in the mobile allocation is limited to 64 bits). This is not really a constraint, since all frequencies cannot be used in all cells.

In the vast majority of cases, the cell allocation will not contain more than 32 frequencies; indeed, the gain brought by frequency hopping increases very little with higher numbers. Some encoding schemes have been introduced in the *Specifications* to encode efficiently the frequency set when the cell allocation consists in such a small number of frequencies. For historical reasons, two such schemes can be found: one for GSM900, one for DCS1800. The first scheme is reflected in what is called the *FREQUENCY CHANNEL SEQUENCE* element in the *Specifications*. This element allows to bypass the two-step approach described above, and is used at handover. Afterwards, another scheme was designed to cope with more frequencies, whilst maintaining compatibility with the first one. The latter scheme is indeed able to cope with 1024 different frequencies, leaving some room for extensions in the future of GSM-related systems...

6.2. ARCHITECTURE AND PROTOCOLS

The radio resource management functions are mainly dealt with by the BSS, and in particular by the BSC which acts as the orchestra conductor for these functions. It directs the mobile station as well as the infrastructure machines involved, which are more or less slave (in the RR field), i.e., the BTS (and TRAU) on one side, and the MSCs on the other. The BTS and the TRAU are indeed the main performers in the transmission chain, and they must as such be controlled by the Radio Resource functions, in fact by the BSC. The BSC is little involved in the transmission functions, but takes care of the consistency of the transmission chain, whether for its different properties (transmission mode, cipher mode, ...) or for the quality of transmission (handover preparation and execution co-ordination).

If it is the anchor MSC which decides which properties of the transmission chain are desirable to fulfil the service, the role of the MSC for radio resource management is limited to some of the handover aspects. The anchor MSC is in charge of performing subsequent inter-MSC handovers when so decided by the serving BSC. The relay MSC is in charge of the handovers when between two cells of different BSCs under its control, and of the circuits between itself and its BSCs. For all other functions, the relay MSC acts as a transit node for signalling exchanges between the anchor MSC and the BSC or the mobile station. Functionally the relay MSC functions are all within the Radio Resource management realm.

This general description of the roles of the different nodes has to be refined for the handover function. Handover preparation makes use of a number of pieces of information, described in the above sections, and which originate from different sources. The BTS is the infrastructure entry point for all measurement information (both for reports from the mobile stations and for its own measurements). The BSC is the warden of all frequency planning and cell layout data. Information about traffic is spread between the BSC and the MSCs. Thus, whatever functional split is chosen, some real-time information transfer is needed between these entities in order to obtain a coherent handover strategy.

The basic split lies between the BSS (BSC + BTS) and the MSCs. The general rule puts the BSS in charge of the management of radio resources and of the decision to perform a handover on a given RR-session. The split was not clear from the start in the standardising process, and triggered many a discussion. The main problem between the MSCs and the BSS is taking traffic into account (either to influence cell choice, or for traffic handover). One could distinguish two approaches: either to have the MSCs indicate to its BSCs the level of traffic of surrounding BSCs, or to let the MSCs intervene in the handover algorithm to ponder radio criteria coming from the BSS with traffic considerations. The second approach was finally the one chosen, even though it may not be the simplest one. The intervention of the MSCs in the handover process blurs somewhat the functional border between BSC and MSCs, and their respective role is not easy to describe, especially for traffic handovers. This is nevertheless what we will attempt to do now.

First of all, when a BSC decides that an outgoing handover is necessary, it will indicate to the relay MSC one or more target cells, possibly managed by different BSCs and MSCs. Rescue handovers might call for several target cells, whereas confinement handovers obviously call for a single choice. If several targets are proposed, the relay MSC may just try them one after the other, in the indicated order. Or it may choose among these ranked possibilities, taking into account its own traffic data. For a cell controlled by another MSC, the relay MSC forwards the request to the anchor MSC; at this level only one target cell is proposed at a time. The choice between several targets is then a function of the relay MSC not of the anchor MSC.

Another possibility for traffic handovers also exists in the *Specifications*, allowing the relay MSC to force the BSC to hand over a portion of the traffic from a cell, with the BSC being in charge of choosing which connections to hand over and to which cells; this is called the "candidate enquiry procedure". There also this possibility is not open to the anchor MSC.

In all cases, the conflict between confinement criteria and traffic criteria is obvious, and is not solved since data related to these criteria are under the control of separate nodes. In fact, there exists in the *Specifications* a means to group all data in a single place. It consists in transporting up to the relay MSC all the raw measurements coming from the mobile station and the BTS, and leave everything to the relay MSC for decision and choice. However, this possibility is not implemented by any manufacturer, and will disappear in phase 2, since the incurred load on signalling links and on the MSC processors would be formidable.

The functional split inside the BSS was also an area of long debates. The split can, there again, be done in different ways. One of the options is to group all the processing in the BSC, with all measurements being transmitted by the BTS without it performing any computation on them. The advantage of this solution is the centralisation of all data, so that handover decisions are taken based on the best data available. The drawbacks are the high signalling load on the Abis interface, since the incurred load of 2 messages per connection and per second represents by far the dominant traffic on this interface, and also an important requirement for high computational power in the BSCs. Another approach, referred to as "pre-processing" in the *Specifications*, consists in letting the BTS do an important part of the job, thereby relieving the Abis interface from the major part of its traffic and decentralising the computing load. Pre-processing is an accepted alternative in the *Specifications*, but the exact functional split between BTS and BSC in this case is not specified. Messages and information elements exist to cater for this function, but their semantics are left open to operators and manufacturers.

Pre-processing is often thought of as a way to reduce load on the Abis interface simply by making the BTS perform some averaging on the measurements. Such a simple approach could slow down the handover decision process significantly, which can be a source of inefficiency, in particular in a small-cell environment. A correct usage of pre-processing introduces sophisticated algorithms in the BTS, including some part of the decision-making process. This is the main reason why details are not specified. It would indeed require too great a workload before a scheme could be shown and accepted by the specification committees.

All these different options related to the split of handover preparation functions within the infrastructure are taken into account in the *Specifications*. More details about the relevant procedures can be found in the SACCH procedural section, as well as in the handover execution section.

Protocols

Independently of the infrastructure architecture, the implementation of the RR functions requires some kind of protocol between the mobile station and the network. On the network side, the interlocutor (or peer entity) of the mobile station for this protocol is the

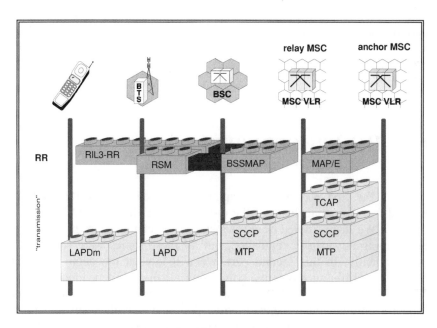

Figure 6.18 – RR protocol architecture

Protocols for RR management are needed on many interfaces,
including the A interface (BSSMAP), Abis interface (RSM)
and the MSC-MSC interface (MAP/E).
The main co-ordinator is the BSC,
which is also in relation with the mobile station (RIL3-RR)

BSC (in fact, a small part of the signalling is also handled by the BTS for efficiency reasons). This protocol will be denoted RIL3-RR.

The functional distribution between infrastructure entities calls for other protocols on terrestrial links: one between BTS and BSC, one between BSC and relay MSC, and one between relay MSC and anchor MSC. The first one, on the Abis interface, is used for the BSC to configure the transmission path and for the BTS to report measurements to the BSC. It has no official name in the *Specifications* (experts refer simply to the 08.58 protocol, from the number of the corresponding Technical Specification), and will be here referred to as RSM (Radio Subsystem Management).

The protocol between BSC and relay MSC, on the A interface, is used to carry the requests for initial connection establishment, as well as for any change in the connection attributes according to upper layer requirements. It is used also for handling handovers between the relay MSC and the BSC. This BSC-MSC protocol is called the BSSMAP protocol (BSS Management Application Part).

The last protocol, between two MSCs of adjacent coverage areas and supporting the exchanges between relay MSC and anchor MSC, is part of the MAP and will be referenced MAP/E protocol. Figure 6.18 shows the machines involved in radio resource management and the protocols between them.

6.3. RR PROCEDURES

In the first part of this chapter, we looked at the different tasks needed for the management of the radio resources from an "object" point of view. We have studied how for instance the channels are managed, or how timing advance is controlled, but all these topics were seen rather independently. In this second part, we will look at the details of what happens at different moments, combining all these independent aspects. Most procedures in the RR area are concerned with several functional aspects simultaneously, and involve, or may involve, all machines between the mobile station and the MSC. We will then revisit a number of the topics we have seen in the previous sections, but with stress on the temporal relationships.

We will start at the beginning, that is to say with the access procedure, where the RR session is created. This will be the occasion to look in detail at the use of the RACH for the initial contact from the mobile station. Next we will present the paging procedure, which precedes the access when the requirement for a connection comes from the infrastructure side. The following sections will be devoted to what happens during the life of an RR session. One aspect concerns the change of some characteristics such as the type of channel or the ciphering mode, when requested by the anchor MSC. Another aspect discussed is the execution of a handover, with all its variants, including call re-establishment, presented here as a last-resort kind of handover. The study of the main adventures of an RR-session will end with the release procedure.

We will then discuss a number of ill-assorted procedures, such as the procedural handling of the signal measurements, the timing advance and the transmission power control, which is done on the SACCH; the frequency redefinition procedures, which are rather complex mechanisms to cope with a change in the allocated frequencies in a cell when frequency hopping is used; and finally the broadcasting of various information on the BCCH.

In all cases, in this chapter and the following ones, what normally happens will be presented with some details, without going, however, into the internal structures of the signalling messages. The rarer cases, involving failures or collisions between events, will at best be hinted at, though they are maybe the most important aspects to have in mind when designing signalling protocols. But even a minimum attempt to correctly cover this subject would need a far larger text than can be included here.

6.3.1. INITIAL PROCEDURES: ACCESS AND INITIAL ASSIGNMENT

The purpose of this section is to describe the procedures enabling the transition between the two major states of a mobile station, i.e., "idle" mode (where the mobile station is everything but idle, but refrains from any active transmission towards the infrastructure), to "dedicated" mode where the mobile station is actively transmitting on a channel allocated for its own use. The transition corresponds to the establishment of an RR-session (see page 313).

The initial assignment procedure is always triggered upon the request of the mobile station, for one of three major reasons:

- to perform location updating;
- to answer to a paging; or

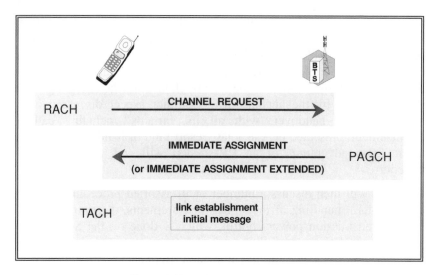

Figure 6.19 – Initial access procedure

The transition from "idle" to "dedicated" mode is always triggered by the MS, through an RIL3-RR CHANNEL REQUEST message sent on the random access channel. Only when the signalling link layer has been established and an "initial message" sent on the new dedicated channel does the network know the identity of the MS.

- as a result of a user's request, i.e., for an outgoing call, a supplementary service management request, or the sending of a short message.

In all cases, the access procedure is the same (see figure 6.19) In broad terms, this procedure starts with an RIL3-RR CHANNEL REQUEST message sent on the RACH; the answer from the network is conveyed in an RIL3-RR IMMEDIATE ASSIGNMENT (or RIL3-RR IMMEDIATE ASSIGNMENT EXTENDED) message sent on the Paging and Access Grant Channel (PAGCH), conveying the description of the channel allocated to the mobile station; finally, the mobile station establishes the link layer for the transfer of signalling on the newly allocated channel, and sends a first signalling message on this channel (the "initial message"), conveying the subscriber's identity and the reason why it requests a connection. This basic canvas appears very simple at first view; but each of the corresponding steps reveals some complexity when studied more closely.

6.3.1.1. Random Access

The channel request message is a curious animal indeed, and deserves some attention. The network has no method of knowing when mobile stations will need to communicate, and therefore this first message from the mobile station cannot be scheduled to avoid the simultaneous transmission of more than one mobile station (a collision). This is the major problem of random access schemes, and the channel name (Random Access CHannel, or RACH) indeed expresses the fact that mobile stations transmit independently from one another.

Of course, mobile stations do not transmit at any time, but follow the slotting of time imposed by the TDMA scheme (a RACH/F uses only one slot every 8 burst periods). Collisions may therefore be studied slot by slot. When two mobile stations transmit during the same slot, two things may happen: either one of the bursts is received by the BTS at a level significantly higher than the other one, allowing its correct decoding (this is called a "capture"), or none is received correctly. Collisions are therefore a source of message loss, which increase with traffic. In order to provide a satisfying rate of success for access attempts, repetitions must be used. The repetition scheme cannot be too simple, otherwise its effect on the throughput in a high load situation can be disastrous, and may indeed lead to a complete deadlock situation. This kind of problem has been thoroughly studied in the field of random access techniques, an area of interest for many networks using shared resources (Local Area Networks, Packet Radio, ...). GSM offers an example of one of the best-known (and simplest) random access schemes, with the RACH being an application of the so-called "slotted Aloha" protocol.

When a request has not been answered, the mobile station will repeat it. If two mobile stations whose attempts have collided would choose to repeat them some given constant time afterwards, their requests would collide again. Repetitions on the RACH are therefore performed after a "random" interval to avoid this phenomenon. As in all Aloha protocols, this re-transmission strategy is not enough to escape from collapse when the traffic goes over a given threshold. In practice, the offered load (in terms of number of requests, which can lead to one or more messages sent) on the RACH should not exceed about a quarter of the total sending opportunities (number of slots on the channel). In order to control this load, GSM uses three different means, corresponding to three different kinds of overload.

The collapsing threshold depends on the number of repetitions and the average time between them; one way to make the RACH robust to a higher load is to spread these repetitions further apart, and/or to reduce the number of repetitions. Of course, this may be detrimental to the quality of service, respectively in terms of delay or in terms of success probability. Such methods cannot be pushed very far and are not sufficient to control very high loads. It is nevertheless useful in cases of small and temporary overload. In GSM, both the number of repetitions

parameter	resulting value
TX-INTEGER	random scheduling of each attempt over 3 to 50 slots
MAX RETRANS	up to 1, 2, 4 or 7 repetitions allowed

Table 6.3 – RACH repetitions control parameters

Both the time interval between re transmissions of a random request
and the maximum number of such repetitions are controlled by parameters.

and the intervals between them are controlled through parameters broadcast regularly on the BCCH on a cell-per-cell basis. Since it is important to have a short delay between the moment when the BSC decides to change these parameters and the moment when the mobile stations act on them, it has been decided to send them in all BCCH messages, i.e., 4 times per second. The scheme is controlled by two broadcast parameters, the average time between repetitions (*TX-INTEGER*), and the maximum number of allowed repetitions (*MAX RETRANS*, see table 6.3).

These parameters should be controlled through a feedback loop taking into account the observed throughput. It should be noted that Aloha control is in fact not absolutely necessary: the values can be set to a constant choice representing some compromise between throughput and delay without jeopardising the system. As already mentioned, this mode of control can only cope with brief (in the order of one second) traffic peaks or sustained marginal overload.

Before studying further the other means to control the load on the RACH, it is worth noticing that this channel constitutes only the first link in a chain of resources, and is not necessarily the bottleneck of the system in congested situations. There are indeed other candidates for such a role: the PAGCH, which offers a limited capacity for carrying both initial assignment and paging messages, is one of them. Channel allocation is another, since the pool of available channels is also limited in each cell. An efficient overload control takes all these factors into account and tries to cut the traffic at the source (i.e., on the RACH) in cases of congestion. This means that overload control mechanisms must not necessarily try to maximise the RACH throughput, but must limit it to the maximum traffic the whole chain can swallow.

Having said that, the second way of "controlling" the load on the RACH consists of rejecting the requests with a message forbidding the mobile station to access the channel for some specified length of time. This mechanism prevents any further repetitions by the mobile station, either through its automatic repetition scheme (controlled as explained above), or through repeated user requests, a natural tendency of users upon failure. Obviously, this mechanism should only be used when overload threatens the overall throughput, since it significantly increases the service time. Besides, it requires the sending of a reject message (the RIL3-RR IMMEDIATE ASSIGNMENT REJECT message) from the BSC to the requesting mobile station, and is therefore inappropriate when congestion impacts the Abis interface or the PAGCH.

Last of all, a third (and most robust) line of defence exists and makes use of the concept of access class. Basically, it consists in forbidding whole populations of mobile stations to access the cell, through an indication on the BCCH. This scheme is very efficient, since it enables a cut-down in the traffic at the very source, without incurring any additional traffic towards mobile stations. In order to achieve this, subscribers are split into 10 balanced sub-populations, through a random allocation controlled by their home PLMN operator. The access class to which a given subscriber belongs is stored in the SIM and is therefore available to the mobile station. In normal load situations, all classes are allowed access. When traffic must be cut down, the BSC can decide to block 1, 2 or any number of these access classes, reducing statistically the amount of traffic by 10%, 20%, ... Mobile stations belonging to the forbidden classes refrain from accessing the network, except in specific cases (e.g., emergency calls, which are controlled by a specific indicator). In order to be fair, if the overload period lasts for a fairly long time, the BSC must take care to change the set of authorised classes regularly (though this must be handled with care, as for instance all the mobile stations waiting for a location updating will try to access at the very moment their class is authorised).

To avoid blocking special categories of users in congested situations, five more classes are defined, for "Very Important GSM Subscribers", as shown in table 6.4. Access for these classes is also controlled through indicators broadcast on the BCCH. The corresponding

"special" access class	subscriber category
11	left open to the PLMN operator
12	security services
13	public utilities
14	emergency services
15	PLMN staff

Table 6.4 – Access classes for "Very Important GSM Subscribers"

In addition to the "standard" random subscriber classes (0 to 9), the SIMs of specific users may feature one of the above "privileged" classes.

"privileged" subscribers belong both to one of the 10 standard classes and to one (or several!) of the special classes, and may access the network when at least one of their classes is allowed.

One reason for access escapes this class-dependent mechanism: emergency calls. In order to be able to control this source of traffic, the BCCH indicates instead whether emergency calls are allowed or not, and this is applied to all subscribers.

In a situation of congestion, one may wonder whether a mobile station whose access is either unauthorised or rejected in a given cell, should be allowed to try to access through other cells. There are pros and cons in this approach. When traffic is spread unevenly in an area, the overload situation of a cell can be improved if part of the traffic is redirected to neighbour cells. However, as explained in the section dealing with handover preparation, there exists one best cell for each mobile station for optimising the general interference level, and the choice of another cell contributes to a degradation in terms of spectral efficiency, and such a strategy will spread the congestion. But if the congestion situation is general, to allow the mobile station to try elsewhere would simply tend to group the mobile stations of given access classes in given cells, which would diminish if not negate the control effect, and would not be the best for interference, to say the least.

The *Specifications* distinguish different cases for this issue. When the mobile station has tried to access and has failed, it is allowed to select another cell and is indeed forbidden to choose its previously serving cell for at least 5 seconds. This enables the mobile station to attempt an access on the second best cell (at least if the user so wishes, or automatically). Because of this mechanism, some of the traffic will be diverted to neighbour cells in situations of local congestion or when the number of repetitions is reduced. In other cases, i.e., upon explicit access rejection or when the relevant access class is barred, the *Specifications* require the mobile station to stay in or leave the serving cell in the normal way, preventing an increase in congestion. This is at the expense of a "no service" period for some mobile stations, but the average service per mobile station is in fact increased through such a "drastic" mechanism. As already mentioned, fairness can be obtained by rotating the blocked classes.

6.3.1.2.   The Contents of the RIL3-RR CHANNEL REQUEST Message

The RIL3-RR CHANNEL REQUEST message, sent on the RACH, is very short indeed. Its useful signalling information consists of just 8 bits!

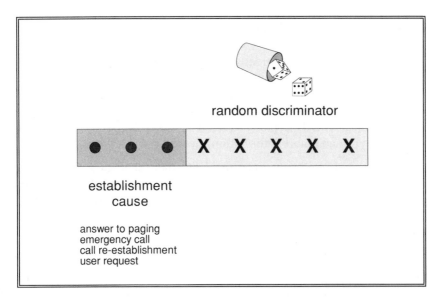

Figure 6.20 – Useful contents of an access burst on the RACH

Only 8 bits are available in this short type of burst.
Three of them indicate the reason for access, and the remaining five
serve as a random discriminator.

This capacity is obviously insufficient to carry all the information the mobile station would want to transmit, such as the subscriber's identity, the reason for requesting a channel, the characteristics of the mobile equipment, ... All of this information is in fact included in the "initial message" which will be the first information transmitted on the dedicated channel, once allocated.

But the most critical usage of a discrimination between random access attempts is not to provide information to the network. A given mobile station must be able to correlate an initial assignment from the network with its own request, with as little ambiguity as can be achieved. For this purpose, 5 bits among the total of 8 are chosen randomly by the mobile station, reducing drastically the probability that two mobile stations send identical messages during the same slot, which may in case of capture lead to an ambiguity as to which of the two requests is being granted.

Three bits remain, which are used as shown in figure 6.20 to provide a minimum indication of the reason for accessing the network. This first rough indication may be useful for discriminating rejections in case of congestion, and also to choose the best type of channel to allocate.

6.3.1.3. The Initial Channel Assignment

After the BTS has correctly decoded a channel request, it indicates it to the BSC through an RSM CHANNEL REQUIRED message, with one important piece of additional information: an estimate of the transmission delay (this indication is critical to initialise the timing advance control). A field of unspecified content (*PHYSICAL INFORMATION*) allows the manufacturer to add more information, such as the reception level.

In normal load situations, the BSC then chooses a free channel (TACH/8 or TACH/F), activates it in the BTS, and, when the BTS has acknowledged this activation, builds an initial assignment message to be sent on the PAGCH.

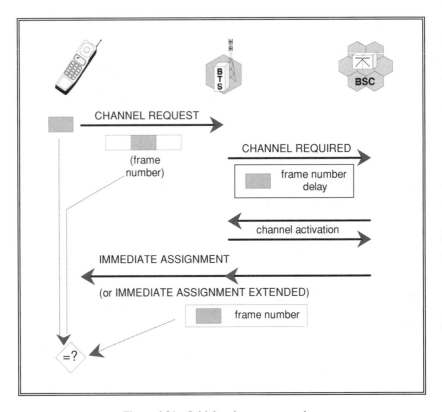

Figure 6.21 – Initial assignment procedure

After the activation handshake on the Abis interface, the BSC prepares
an initial assignment indication containing the 8 bit discriminator
as received in the correctly decoded RIL3-RR CHANNEL REQUEST message,
as well as the frame number in which it was received.
This enables the mobile station to check whether it is concerned
with the message or not.

The activation process requires the BTS to prepare for the access of the mobile station on the newly allocated channel. The timing advance is initialised based on the transmission delay estimate which the BSC indicates (back!) to the BTS. Even though this estimate has been initially calculated by the BTS, this going back and forth between BTS and BSC is necessary, since the BTS has no means to correlate the messages received on the RACH with corresponding channel assignments.

The initial assignment indication sent to the mobile station on the PAGCH contains the description of the allocated channel, the initial timing advance to be applied, the initial maximum transmission power, as well as a reference allowing all the mobile stations expecting such a message to know whether they are being addressed or not. This last point is worth some more explanation.

Addressing is done by including in the initial assignment indication the exact contents of the RIL3-RR CHANNEL REQUEST message which is being answered, plus the time reference of the slot in which it was received (such a time reference exists thanks to TDMA). This allows mobile stations to check whether they are actually concerned by each initial assignment, by comparing these values with the ones they have stored when sending the RIL3-RR CHANNEL REQUEST message, as shown in figure 6.21.

Besides, answers to RIL3-RR CHANNEL REQUEST messages can be sent in any block of the PAGCH, even on the paging sub-channels. As a consequence, once a mobile station has made an access attempt, it should monitor the whole PAGCH (of the same Timeslot Number as the RACH it used for access) for an answer from the network. Furthermore, the BCCH messages must be decoded continuously during this period, in order for the mobile station to set the RACH control parameter values in real-time. This phase is very constraining for mobile stations in terms of reception (40 bursts every 51×8 burst periods), almost comparable to TACH/F reception.

Let us examine some side issues. It may happen that the reaction of the infrastructure to an RIL3-RR CHANNEL REQUEST message is too slow to avoid a repetition from the mobile station. In such inefficient situations a given mobile station may be allocated a channel twice (or even more times), since the infrastructure has no means of knowing whether an RIL3-RR CHANNEL REQUEST is the repetition of a previous one or not. The mobile station will use the channel allocated in the first initial assignment message it decodes, and the other ones will have been blocked for a few seconds in vain. Nevertheless, the *Specifications* require that the mobile station be able to accept the network answer to any of its last three RIL3-RR CHANNEL REQUEST messages, in order to get service from such not-so-efficient BSS equipment.

In congested situations, when no channel is free for allocation, the BSC may choose not to answer to an RIL3-RR CHANNEL REQUEST message, or to send back a rejection indication. The first choice is not very efficient, since the mobile station will repeat its attempt. Explicit rejection is done through an RIL3-RR IMMEDIATE ASSIGNMENT

REJECT message, containing a time indication during which the mobile station is forbidden to make any more attempts on the RACH (the *WAIT INDICATION* parameter). If the overload situation does not concern the RACH, the value of the *WAIT INDICATION* can be null, otherwise it is a useful mechanism to help reduce RACH load (see page 370).

The PAGCH is an important potential bottleneck of the system. In order to improve its efficiency, both initial assignment indications and channel request rejection indications can be grouped together to form messages. There are two assignments in an RIL3-RR IMMEDIATE ASSIGNMENT EXTENDED message, and up to four rejections in an RIL3-RR IMMEDIATE ASSIGNMENT REJECT message. Though this point is unclear in the *Specifications*, the original intention was for the BTS to perform the grouping on the basis of the individual indications provided by the BSC. The BSC has the possibility to provide the BTS with immediate assignment indications which are not ready-to-send messages (using the RSM IMMEDIATE ASSIGN COMMAND message). The BTS must then build the corresponding RIL3-RR messages The fact is that the BSC *can* also perform the grouping and build the messages, and can provide ready-to-send messages to the BTS (this is, by the way, the only possibility for assignment rejection messages). However, nothing really precludes the BTS to un-build these messages and to build others grouping the requests differently. The same debate exists for the paging indications.

In no case does the BSC schedule the transmission of the assignment indications: this is much easier for the BTS. The counterpart is that, despite grouping, congestion may happen. This is resolved simply by the BTS, which drops messages which are in excess of the achievable throughput. While this resolves the BTS congestion problem, for the BSC it results in TACHs allocated but not "assigned". To avoid a worsening of the congestion situation through this effect, the BTS can indicate the non-sending of a message with an RSM DELETE INDICATION message. A second effect of the message is to indicate the overload to the BSC.

6.3.1.4. The Initial Message

Once it has received an initial assignment indication, the addressed mobile station modifies its reception and transmission configuration to adapt it to the frequency and time characteristics of the new channel. In phase 1, this new channel can be a TACH/8 or a TACH/F, always in "signalling only" mode (see page 321). The transmission level is set to a value broadcast on the BCCH (or to the maximum transmission level of the mobile station, whichever the smaller), and the transmission starts with the timing advance value specified by the BSC.

The first thing the mobile station does on the new channel is to transmit a link layer SABM frame for SAPI 0, i.e., the frame used to establish in acknowledged mode the link layer connection for signalling messages.

In standard HDLC protocols, an SABM frame does not carry any information other than the one necessary for the link layer level. In GSM, the SABM frame sent within the initial access procedure contains a signalling message, the "initial message". The reasons for departing from standard usage are twofold. The first reason is efficiency, though this was

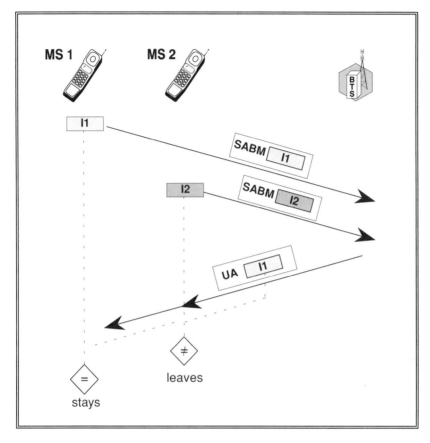

Figure 6.22 – Contention resolution at link establishment

In rare cases, more than one mobile station may find itself
on the same dedicated channel.
The transfer of a non-ambiguous initial message in the SABM-UA exchange
allows each mobile station to know whether the channel is for its own use or not.

not the leading cause. The other reason comes from the fact that the reference used in the initial assignment to address the mobile station is not fully unambiguous. It may indeed happen (though this is a rare case, around 1% in high load situations) that two mobile stations simultaneously send RIL3-RR CHANNEL REQUEST messages with exactly the same contents, and that one of them is correctly received and answered by the BSS. The ensuing channel assignment will be understood by both mobile stations as their own, and both mobile stations will access the "dedicated" channel. Therefore, until the point where mobile stations identify themselves in a non-ambiguous way, there is not a 100% guarantee that a single mobile station will access the channel.

As a consequence of this situation, potential collisions must be detected as soon as possible on the new channel, and the SABM-UA exchange provides this facility by including ("piggybacking") unequivocal information on these link layer frames. Mobile stations check the piggybacked contents of the UA frame. If one mobile station receives a UA containing something different from the contents of the SABM it sent, it must leave the channel and start the access procedure all over again, thereby enabling the "right" mobile station to stay undisturbed on its own channel (see figure 6.22).

One obvious way to obtain an unambiguous SABM content is to use an identity unique to the mobile station. This would be enough to serve the purpose of collision detection. However, the efficiency criterion was also taken into account and the SABM includes more than just this identity, and includes a full "initial message".

The "initial message" comes in four different brands, depending on the reason why the access was triggered (see table 6.5). All these messages contain an identity of the mobile station; the **classmark**, a field indicating some key characteristics of the mobile equipment, including the maximum transmission power; and complementary information making the reason for access more precise when need be. All but the first of those messages belong to the RIL3-MM protocol, and will be explained in Chapter 7. The first belongs to the RIL3-RR protocol, but could have as well been put in RIL3-MM. If there had been enough room, these messages would have been properly formatted, with a part for the RIL3-RR protocol (including the classmark), and the rest in an RIL3-MM part.

Reason for access	initial message
Response to a paging	RIL3-RR PAGING RESPONSE
Normal location updating, periodic location updating, "IMSI attach"	RIL3-MM LOCATION UPDATING REQUEST
IMSI detach	RIL3-MM IMSI DETACH
All other cases (call set-up, short message transmission, supplementary service management, ...)	RIL3-MM CM SERVICE REQUEST

Table 6.5 – Possible initial messages

Four different types of signalling messages may be used as the "initial message", depending on the reason having led to a channel request.

There is usually but one choice for this "initial message". One case of collision may arise, when, after having sent an RIL3-RR CHANNEL REQUEST message, the mobile station receives a paging indication and then the answer to its RIL3-RR CHANNEL REQUEST. What should the initial message be in such a case? Should it be consistent with the reason for sending the original RIL3-RR CHANNEL REQUEST message, or should it be the RIL3-RR PAGING RESPONSE message? It would seem fair to choose the RIL3-RR PAGING RESPONSE message if for instance the request from the mobile station concerned periodic location updating, but no priority scheme is defined in the *Specifications*, and the whole matter is left open for mobile station manufacturers.

Once an "initial message" has been received by the BTS (and sent back without any modification inside the UA frame), it is passed to the BSC in an RSM ESTABLISH INDICATION message. At this point, the mobile station classmark is stored for further use (e.g., to choose the power control loop parameters), and the BSC then sets up an SCCP connection towards the MSC (see Chapter 5). This is done through an SCCP CONNECTION REQUEST message, on which the initial message may optionally be piggybacked. Only then does the MSC become aware of the contact with the mobile station. The initial message, whether piggybacked or sent after the SCCP connection establishment, is carried in a BSSMAP COMPLETE LAYER 3 INFORMATION message, independently from its protocol (RR or MM). It contains enough information for the MSC to trigger required actions in the upper layers (MM, CC, ...), but this comes out of the scope of the access procedure.

When the access procedure ends, the RR-session is fully established with a complete signalling path between the mobile station and the MSC. With the establishment of the SCCP connection, the MSC takes control of the decisions concerning the transmission characteristics of the RR-session, and the BSS is at the ready whilst monitoring the transmission and performing handover decisions.

6.3.1.5. The Mobile Station Classmark

Mobile stations differ by many characteristics, such as their maximum transmission power and the services they may support. It is important for the infrastructure to be aware of some of these characteristics when the mobile station is engaged in a connection. Because the equipment of the user may be changed without warning the operator (subscription is linked to the SIM, not to the mobile equipment), this indication must be given at the beginning of each new connection. This is the purpose of the mobile station classmark. The full contents of the classmark are shown in table 6.6.

There exists in fact, for efficiency reasons, a subset of the full classmark, called "mobile station classmark type 1" in the *Specifications* (the full classmark being referred to as "mobile station classmark type 2"), destined to be sent in the RIL3-MM LOCATION UPDATING REQUEST message, so that this message fits in a single block (segmenting is not

parameter contained in the classmark
revision level
RF power capability
encryption algorithm
frequency capability
short message capability

Table 6.6 – Full contents of the mobile station classmark

The classmark identifies those characteristics of the mobile equipment which are needed by the infrastructure during a connection.

used when the message is piggy-backed). The shortened classmark is also used in the RIL3-MM IMSI DETACH message, for no obvious reasons.

The **revision level** is used for upward compatibility handling between successive phases of the *Specifications*. Mobile stations developed and type approved according to the phase 1 *Specifications* should set this value to 000. In the future, other values will be allocated, so that the infrastructure will know which level of upgrade is used by each mobile station.

The **RF power capability**, often referred to as the transmission power class, or even as the class, refers to the maximum power the mobile station is able to transmit. This information is used for power control and handover preparation. Power classes are not defined in the same way for GSM900 and DCS1800, as shown in table 6.7. Mobile stations of class 1 for GSM900 may possibly never be developed; indeed the typical classes in GSM900 are class 2 for portable or vehicle-mounted equipment, and class 4 for handheld. Class 5 handheld stations will most certainly experience strong limitations in coverage and will be destined for urban areas. The typical class of DCS1800 mobiles is class 1.

The **encryption algorithm** indicates which ciphering algorithm (if any) is implemented in the mobile station. In phase 1, there is but one choice: all mobile stations must implement the A5 algorithm specified by the GSM MoU. This field will enable the BSC to cope with future mobile stations with other ciphering capabilities, by being able to choose whether to cipher or not, and if yes with which algorithm.

The **frequency capability** is also present for future use. It enables the network to cope with mobile stations having different capabilities in terms of frequency bands. A phase 1 GSM900 mobile station must be able to cope with frequencies anywhere in the 2×25 MHz band allocated

Class	GSM900	DCS1800
1	20 W	1 W
2	8 W	0.25 W
3	5 W	–
4	2 W	–
5	0.8 W	–

Table 6.7 – Mobile station power classes

Five classes are defined for GSM900, and two for DCS1800,
corresponding to the maximum transmission power of the mobile station.

from the start to GSM. An extension of this band to, say, 35 MHz is under study, but all mobile stations will not be able to cope with the extension. Therefore, the classmark will enable the BSC to distinguish the different populations of mobile stations and allocate channels to each mobile station according to its own frequency capability.

Last of all, the **short message capability** indicates whether the mobile equipment is able to deal with short messages. Such an indication is not strictly necessary in the classmark, since a mobile station not equipped to deal with short messages (if any!) can always inform the network by rejecting a LAPDm SABM frame on SAPI 3. But it is more efficient to know the capability of the mobile station from the start, since it enables the network to eschew the transmission of short messages between the short message service centre and the VMSC.

We have seen that the classmark is sent by the mobile station in the initial message, at the beginning of the RR-session. The network does not store the classmark between RR-sessions, since the user can change his equipment. Now, it may happen that the classmark changes *during* the RR-session. This is not a common event, but an example is a mobile equipment composed of a handheld part and a vehicle-mounted part including an RF transmitter, with the possibility to connect and disconnect the two parts during a communication. Then the power class changes, and the new value must be provided to the network. To achieve this, a procedure is included in the RR plane. The indication from the mobile station is carried in an RIL3-RR CLASSMARK CHANGE message and the BSC can forward the indication to the relay MSC with a BSSMAP CLASSMARK UPDATE message. If the relay MSC is not the anchor MSC, the chain stops there in phase 1, since the MAP/E protocol does not support this functionality.

6.3.2. PAGING PROCEDURES

A little has to be said on the paging procedures. When a call to a subscriber reaches a MSC through which the subscriber is deemed reachable, the MSC determines the location area where the mobile station is registered and sends a BSSMAP PAGING message to all the BSCs controlling cells in this location area. The message contains the subscriber identity to page with (it could be a temporary mobile station identity, called the TMSI or the full international mobile subscriber identity, the IMSI), the IMSI to determine the paging sub-channel (to cope with discontinuous reception), and the list of cells in which the paging must be issued.

The BSC in turn sends an RSM PAGING COMMAND to the BTS device in charge of the PAGCH of suitable TN (determined by the BSC, from the IMSI and the common channel configuration), for each cell in the list. This message contains the number of the paging sub-channel, which is computed by the BSC, as well as the TN of the PAGCH.

The BTS in turn possibly packs some paging requests together and sends the resulting messages on the correct paging sub-channel. As indicated in the functional section, more sophisticated approaches are possible.

Paging messages come in three brands (called RIL3-RR PAGING REQUEST TYPE 1, TYPE 2 and TYPE 3), adapted to the size of the identity used for paging. Type 1 can carry two identities of whichever sort, whereas type 3 can carry four TMSIs, and type 2 two TMSIs and one identity of whichever sort.

A topic of interest is the repetition policy. In other areas, GSM is specified so as to provide a correct quality of service when the transmission quality is far from perfect. To be consistent, the paging indication should not be sent only once in each cell. A typical value (not specified) would be to send it three times. No repetition mechanism is described in the *Specifications*, and this leaves some obscurity as to which machine takes care of the repetition: the MSC, the BSC or the BTS? To choose the BTS has some advantages: it can optimise the use of the PAGCH, and in particular it may repeat paging requests more than the minimum required if the channel has some room left. On the other hand, if the MSC is in charge of managing the repetition process, it will wait some time before requesting repetitions, thus avoiding useless repetitions in many cells in cases of successful mobile station access. Neither the BTS nor the BSC are capable of such monitoring, since they are not able to relate the mobile station answer to the paging. This seems obvious when the mobile station has accessed another cell (the MSC does not indicate it back to the BSCs from which it requested paging, it would be too big a procedure), but the same is also true when the mobile station accesses through a cell under their control. A sensible scheme seems to provide two levels of repetitions, one in the MSC, with a long period, to cope for instance with short reception interruption (such as a change of cell, or the mobile station passing through a tunnel); and a short term repetition in the BTS, when load allows, coping with mediocre propagation conditions.

The Page Mode

The PAGCH configuration may change in time to adapt to the traffic distribution. Although this configuration is managed by the BSC, it is usually the OSS which decides on such a change, either automatically on the basis of traffic observations or through the command of an operator. The configuration of the PAGCH in a given cell must be known by all mobile stations camped on this cell, and is therefore part of the broadcast information. When the configurations changes, the BSC must co-ordinate the change of the broadcast information and the scheduling of paging messages. It is not possible to control very precisely the time at which each mobile station will decode the new broadcast parameter, and hence switch to the new configuration. The corresponding transitional period may last up to a few seconds. In order to avoid the loss of paging messages during this "fuzzy" period, a special feature has been incorporated in the *Specifications*: the page mode. The page mode indicates to mobile stations in exactly which part of the PAGCH their own paging messages may be sent.

The three values of the page mode are the following:

- the "normal" page mode corresponds to the basic scheme. Paging messages are sent only on the sub-channel as defined by the PAGCH configuration and the IMSI;

- the "full" page mode has been designed to cope with a dynamic change in the PAGCH configuration. When this mode is indicated to mobile stations of a sub-group, then it means that paging messages for the subscribers of this sub-group may be sent anywhere on the PAGCH of the same *timeslot*;

- The "next-but-one" page mode has been introduced for sophisticated scheduling algorithms. It enables the BSS to send additional paging messages for subscribers in a given sub-group in another paging sub-channel. This feature may be useful in a situation of temporary overload on some of the sub-channels, or to free a block to send an initial assignment message. Based on the fact that the definition of paging sub-channels includes an implicit numbering of these sub-channels, the "next-but-one" mode indicates that paging messages for mobile stations

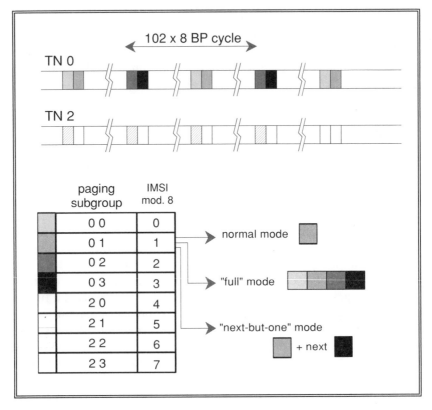

Figure 6.23 – Different page modes

In addition to the "normal" paging mode, the "full" page mode
and the "next-but-one" page mode are useful in transition situations,
or to compensate for partial congestion situations.

normally paged in sub-channel n (the sub-channel where the page mode indication was sent) will in addition be found in the next block of the paging sub-channel $n+2$, modulo the total number of paging sub-channels on the current *timeslot*. The example of figure 6.23 will help to understand this feature.

It is worth noting that the mobile station is free to listen to more than the required minimum. This may indeed represent a simplification for vehicle-mounted mobile stations, for which battery consumption is not at stake. In cases where the load on the PAGCH is very low, it might even be interesting for the BSS to send the paging messages in more blocks than those of the required paging sub-channel. This will allow a speed-up of their reception for mobile stations which listen to more than just their own paging sub-channel.

6.3.3. PROCEDURES FOR TRANSMISSION MODE AND CIPHER MODE MANAGEMENT

Transmission modes, as far as radio resource management is concerned, involve 7 different types in phase 1 and are expected to expand to 12 in phase 2 (see table 6.1, page 321). The mode is one of the properties of the transmission chain. Other variable characteristics of this chain include the cipher mode (whether transmission is ciphered or in clear text), as well as both uplink and discontinuous transmission modes.

At initial assignment, the transmission mode is chosen by the BSC. It consists necessarily of one of the "signalling only" modes, in clear text. The channel may be a TACH/8 or a TACH/F. Afterwards during the lifetime of the RR-session, the choice of the transmission mode depends on the communication needs and is done by the MSC, which can request a change of the transmission mode at any time while a connection is established. It does so through an "assignment" procedure (which does not necessarily result in a radio channel assignment). The basic procedure consists of a BSSMAP ASSIGNMENT REQUEST message, describing the transmission characteristics as desired by the MSC, and the corresponding BSSMAP ASSIGNMENT COMPLETE acknowledgement, as shown in figure 6.25. A negative answer is also possible in case of problems, and is carried in a BSSMAP ASSIGNMENT FAILURE message.

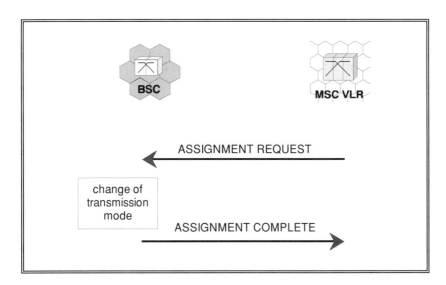

Figure 6.24 – Change of the transmission mode by the MSC

The MSC may change the transmission mode of an RR-connection at any time,
by running the "assignment" procedure towards the BSC,
which then takes charge of controlling the transmission mode change.

Another special case is when the allocation cannot proceed immediately, but the request is put in a queue; the MSC can be warned of the situation by a BSSMAP QUEUING INDICATION message (the completion or failure indication is eventually sent later). The request also includes the identity of the newly allocated terrestrial circuit between the BSC and the MSC, when required, i.e., when the mode is changed from "signalling only" to another one.

The action of the BSC when receiving a BSSMAP ASSIGNMENT REQUEST message depends on the comparison between the existing transmission mode and the required one:

- if both modes are the same, the BSC just sends back the BSSMAP ASSIGNMENT COMPLETE message back to the MSC without any other action;

- if both modes differ by the type of information to be transmitted, but use the same type of channel, the BSC performs a "mode modify" procedure before acknowledging the MSC request;

- if the new mode requires a channel of a type different from the one in use, the BSC performs a subsequent assignment procedure, i.e., it transfers the connection to a channel of the required type before acknowledging the MSC request.

The two last cases will now be studied in more detail.

6.3.3.1. The Mode Modify Procedure

As for any procedure affecting the transmission mode, the mode modify procedure includes two parts: the configuration of the transmission devices on the infrastructure side (BTS, TRAU and BSC), and the configuration of the mobile station. No means are provided to synchronise these two parts in a precise way, resulting usually in a short period of time during which the whole configuration is inconsistent. The *Specifications* do not specify in which order the two configuration steps should be managed, and this choice may have an impact on the period of inconsistency. In the following example, it has been hypothesised that the two tasks are run in parallel.

The BSC triggers the reconfiguration of the BTS and the TRAU by sending an RSM MODE MODIFY REQUEST to the BTS. Following reception of this message, the BTS modifies its coding and decoding algorithms and changes the in-band mode information in the BTS-TRAU frames. The TRAU, in turn, reacts by modifying its data processing (speech

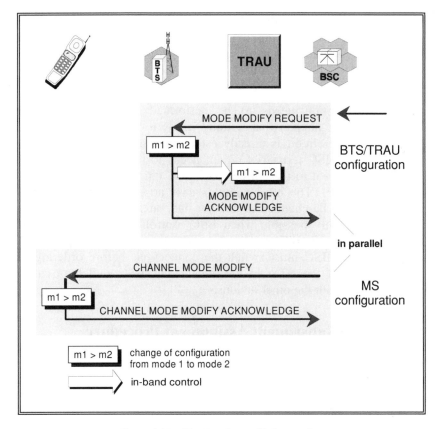

Figure 6.25 – The "mode modify" procedure

The BSC is in charge of configuring both the BTS and the mobile station,
but the order in which these steps should be performed is not specified.
The TRAU is configured through in-band information from the BTS.

coding or data rate adaptation). If the new mode is speech, then
synchronisation between TRAU and BTS is needed. When the chain is
ready (the *Specifications* do not specify whether this includes the
synchronisation with the TRAU or not), the BTS answers the BSC by
sending back an RSM MODE MODIFY ACKNOWLEDGE message.

In parallel, the BSC triggers the reconfiguration of the mobile
station by sending an RIL3-RR CHANNEL MODE MODIFY message
containing the new mode to be applied (see also figure 6.25). When the
mobile station receives the order, it modifies its channel/source
coding/decoding according to the new requirements and answers with an

RIL3-RR CHANNEL MODE MODIFY ACKNOWLEDGE message to the BSC via the BTS. It is worth noting that the connection of higher level devices (microphone and loudspeaker for speech, or terminal adapter for data) is not controlled by this procedure, but by a call control procedure (see Chapter 8).

A third action needs to be performed by the BSC: circuit switching. It may happen that nothing needs to be done, for instance when a terrestrial circuit is already established and satisfies the needs. But if the BSC-MSC terrestrial circuit needs to be reset or altered, the BSC must connect it to the correct BTS-BSC circuit (in correspondence with the TACH/F). The most complex case arises when the TRAU is remote, and must be changed to another one, adapted to the new mode. This is in fact only possible when BSC-controlled switching facilities exist between both the BTS and TRAU and between the TRAU and BSC. In this case, the BSC must switch the connections before ordering the reconfiguration of the BTS. In the other cases, BSC switching can be done in parallel with the other actions.

6.3.3.2. The Subsequent Assignment Procedure

When a change of the radio channel is required in addition to the above-described "mode modify" procedure, the procedure is somewhat more complex, because the change of channel implies a break in the signalling carrying capability between mobile station and infrastructure. Besides, let us recall that the BTS devices in charge of transmission on a given TACH are independent and do not communicate. Therefore, the whole control of the operation is centralised in the BSC, and the operation itself is quite similar to a handover.

A transfer of channel first starts with the setting up of the new path in the infrastructure. This includes the allocation of a new radio channel (with all the priority and queuing management aspects described on page 357, to cope with congestion), the activation of the corresponding BTS device, and possibly the allocation of a TRAU and the switching necessary to connect all these terrestrial segments.

The activation of the BTS is ordered by the BSC through a simple request/acknowledgement procedure, as shown in figure 6.26. The RSM CHANNEL ACTIVATION message contains all the information specifying the transmission mode, including the basic transmission mode (among those listed in table 6.1), the cipher mode and the downlink discontinuous transmission mode. An uplink discontinuous transmission mode is also

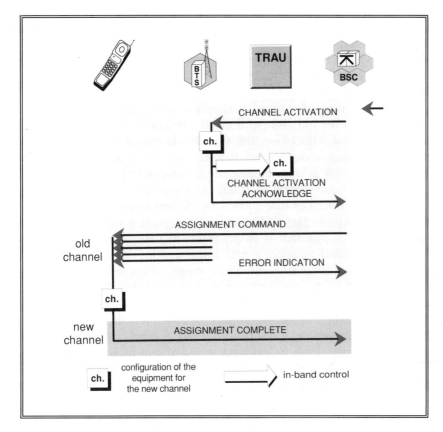

Figure 6.26 – Activation of a new channel in the BTS

After the activation handshake on the Abis interface, the BSC orders the mobile station
to change channel by an RIL3-RR ASSIGNMENT COMMAND message,
which is in general not acknowledged by the mobile station on the old channel.
The completion of the procedure is done on the new channel
after establishment of the full signalling link.

sent, though one may wonder what it may be used for, and moreover
what its meaning is, since in some cases the network leaves the choice to
the mobile station. In addition, it contains the information needed by the
mobile station for access (see the initialisation of the timing advance,
page 346) and the first power control settings. The BTS, upon reception
of this message, starts the in-band information exchanges with the
TRAU, to set the basic transmission mode and the discontinuous
transmission modes; this is the point where synchronisation with the
TRAU starts.

Once the BTS and the TRAU are activated (more exactly when the
BSC has received the RSM CHANNEL ACTIVATION ACKNOWLEDGE

message), the BSC orders the mobile station to perform the transfer of channel, through an RIL3-RR ASSIGNMENT COMMAND message. The previous path, including the signalling connection, is not released by the infrastructure at this moment. This allows the mobile station to go back to the previous channel should access fail on the new channel for any reason. In the case of a timed assignment (see page 353), the mobile station stays on the old channel until the instant of change indicated by the infrastructure. Otherwise, the mobile station performs the transfer immediately after reception of the RIL3-RR ASSIGNMENT COMMAND message, not even acknowledging the corresponding frame at layer 2. This lack of acknowledgement results in a repetition of the message by the BTS, until it decides that a link layer failure has happened, of which the BSC is advised. The BSC does not act on this indication, since it knows that an assignment is in progress. In case of return to the old channel, the mobile station starts re-transmitting on this channel by performing a link establishment, which resets all contexts irrespective of what happened in the link layer process in the BTS. The transmission mode used after return on the old channel is the one that was used on it, not the one asked for in the assignment message. Whether in the case of a successful procedure or of return to the old channel, the interruption of the link layer may result in leaving a message sent by the mobile station in a non-acknowledged state. This situation is handled by the upper layers, as will be explained below.

Once the mobile station has changed its settings to the ones corresponding to the new channel, it starts transmission and reception according to the transmission mode indicated in the RIL3-RR ASSIGNMENT COMMAND message. It must also establish a new signalling link in acknowledged mode. The *Specifications* do not say whether this link layer establishment should be done first (before the transmission of any user data) or not. Anyway, once this link is established, the mobile station sends an RIL3-RR ASSIGNMENT COMPLETE message to the BSC before any other message. Then, all the messages waiting for transmission can be sent: first the ones already sent on the previous channel but still not acknowledged, then the ones which have arisen during the procedure. The same applies in case of return to the old channel, except that the first message is an RIL3-RR ASSIGNMENT FAILURE.

A message sent by the mobile station before link interruption and not acknowledged then cannot be lost, but may be duplicated. RIL3-RR messages are considered immune to duplication, and therefore no mechanism has been introduced to cope with such duplications. This is worth remembering when new procedures are added in the future! The way in which the BSC is implemented has also some impact. A "careful" BSC will postpone the sending of an RIL3-RR ASSIGNMENT COMMAND (or an RIL3-RR HANDOVER COMMAND) while it is waiting for the answer to another RR command (e.g., for the start of ciphering!).

In the case of upper layers, e.g., call control or mobility management, message duplication could in some cases be harmful. In order to avoid problems, correction is obtained by detection and suppression of the duplicated message. Detection is obtained through the use of a simple numbering scheme, the 1-bit sequence number referred to as N(SD) in the *Specifications*. This number is included in each upper layer message, and is changed for each new message. Since messages are sent and acknowledged one at a time (window size 1), this scheme is enough for the infrastructure to detect duplication. When it receives two successive messages with the same N(SD), it discards the second one. This task is performed at the RR level, but not in the BSC. It is one of the tasks of the anchor MSC, which is the destination of upper layer messages sent by the mobile station. The reason why it cannot be performed in the BSC is that it must be done at a point which remains stable when the transmission chain changes, and the reader will recall that the only stable part of an RR-session lies in the anchor MSC.

6.3.3.3. The Change of the Cipher Mode

During the lifetime of an RR-session, the cipher mode may change on the radio interface. This procedure is not all that easy. As explained in Chapter 4, the cipher mode is applied to all transmitted information, including signalling messages. Thus, a change of the cipher mode entails a signalling break, with a possibility of message loss.

In the case of subsequent assignment and handover, the problem raised by the signalling break is solved through a full re-establishment of the signalling link. This is full-proof against call loss only because both the old and the new channels are available during the critical period, thus allowing the return to the old channel in case of problems on the new one. In the case of a cipher mode change, it would be too costly to require the BTS to perform reception in both modes (ciphered and non-ciphered) simultaneously, and a different solution was adopted. It consists of dividing the procedure into three steps instead of two:

- step 1: the BTS is configured to transmit according to the old mode, and receive according to the new mode;

- step 2: the mobile station is configured to the full (transmission

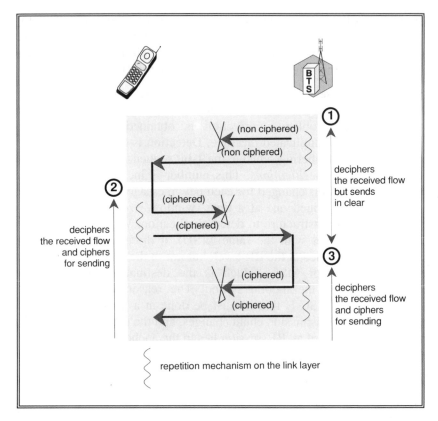

Figure 6.27 – Cipher mode change: the 3 steps

In order to cope with an interruption in the signalling link when a change
of the cipher mode occurs, the procedure is split into three steps,
so that only one direction of the transmission will be in a critical state at each transition.
The example shown here represents a transition from clear text to ciphered mode.

and reception) new mode;

- step 3: the BTS is configured to the full new mode.

It can be accepted that steps 1 and 2 be inverted, but this results in
an increased frame loss probability, since in this case both uplink and
downlink messages would be lost between step 1 and step 2, whereas
only downlink messages are lost in that phase with the given order.

The critical period is split in two with such a mechanism. From the
first to the second step, BTS to MS transmission functions correctly, but
not in the other direction. This is sufficient for a downlink message

triggering step 2 to be repeated a number of times by the infrastructure, until received. Similarly, from step 2 to step 3, only MS to BTS transmission is correct. But this is sufficient for the mobile station to repeat the uplink acknowledgement message required after step 2, to trigger step 3. In no case does a single message loss jeopardise the whole connection.

The procedure, shown in figure 6.27, relies heavily on the link layer mechanisms, namely the repetition of messages after a given timeout period, in the absence of an acknowledgement from the other side. Though not obvious, it can be shown through a careful study that, even with a window size greater than 1, and/or messages waiting to be sent on the mobile station side, the procedure terminates correctly even if the radio conditions are not perfect. Yet it should be noted at this point that the cipher mode change procedure can by itself cause the loss of a frame, and transmission is therefore marginally more sensitive to errors at the time of a cipher mode change.

Because of the strong requirement on the order of the three steps listed above, and because ciphering is implemented by the BTS on the infrastructure side, the procedure is not managed by the BSC, but by the BTS. The BSC transmits a single order to the BTS, which then runs the procedure, including its own configuration as well as the one for the mobile station. As a matter of fact, the decision to change the cipher mode is taken by the MSC, and results in a cascade of messages from MSC to BSC, then from BSC to BTS, and lastly from BTS to MS on the radio interface. The whole chain is shown in figure 6.28.

The BSSMAP CIPHER MODE COMMAND message indicates the new requested mode. After having extracted the new parameters from this message, the BSC builds up an RIL3-RR CIPHERING MODE COMMAND message targeted at the mobile station and encapsulates it in an RSM ENCRYPTION COMMAND message sent to the BTS.

The BTS then configures its reception to the new mode and sends the encapsulated RIL3-RR CIPHERING MODE COMMAND message to the mobile station using the old mode. When receiving it, the mobile station sets its configuration to the new mode and puts an RIL3-RR CIPHERING MODE COMPLETE message in the sending queue. This message is not necessarily the first one to be sent in the new mode, as another one may be there, and the RIL3-RR CIPHERING MODE COMPLETE message has no particular priority. The main reason for the existence of this acknowledging message is to ensure that at least one layer 3 message is sent at this moment, in order to trigger the switching to the new mode in

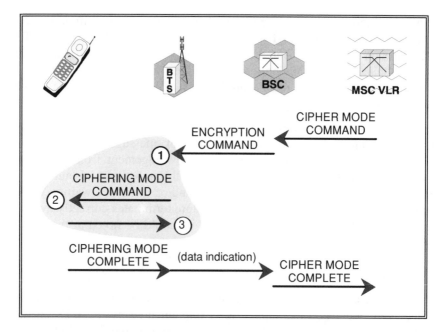

Figure 6.28 – The cipher mode setting procedure

The procedure is initiated by the MSC, but all the synchronisation management
is done by the BTS, which is in charge of ciphering/deciphering.
The grey area corresponds to this management, as detailed in figure 6.27.

the BTS. It is when receiving any correctly decoded message (in the new mode), which implies that the mobile station has indeed correctly switched to the new mode, that the BTS switches fully to the new mode. Either then or afterwards, the RIL3-RR CIPHERING MODE COMPLETE message is forwarded to the BSC, which translates it into a BSSMAP CIPHER MODE COMPLETE message to indicate to the MSC that its request has been fulfilled.

The case when the MSC requires a new mode which is already the one in place has not been dealt with. This omission is purposeful, since in this case the *Specifications* impose that the whole procedure be run anyway. This requirement comes from the second meaning of the RIL3-RR CIPHERING MODE COMMAND message, which is used to acknowledge an RIL3-MM CM SERVICE REQUEST (this will be explained in Chapter 7). The cipher mode command procedure is then strictly speaking not only spread over the RR layer and the link layer, but also on the MM layer.

It is not possible to change the cipher mode when changing the channel (assignment or handover) for unclear reasons (later phases may indeed remove this constraint). A mobile station developed according to the phase 1 *Specifications* must apply the same cipher mode on the new channel as was used on the previous one, and this situation can lead to problems in case of collision between a channel transfer and a change of the cipher mode. In order to avoid such problems, a sequential approach between these two procedures must be sought by the BSC.

6.3.3.4. Discontinuous Transmission Modes

The *Specifications* do not indicate clearly whether it is allowed to change the discontinuous transmission modes during the lifetime of an RR-connection. The need can be identified, at least for the downlink discontinuous transmission mode, since this mode may depend on the correspondent and a single RR-session can be used for several communications in succession. It should be noted as a preliminary remark that there is no need for the receiver to know beforehand whether the sender applies discontinuous transmission or not. Therefore, no specific procedure is required for the indication of the downlink discontinuous transmission mode to the mobile station or of the uplink discontinuous transmission mode to the infrastructure transmission devices. Means exist for the other cases, and will now be described.

As far as the downlink discontinuous transmission mode is concerned, it must be ordered connection by connection by the MSC to the BSC, which configures the BTS, itself configuring the TRAU. It must be recalled at this stage that, on the transmitting side, discontinuous transmission does not only result in some frames not being sent, but it also modifies the speech coding algorithm (e.g., sending of comfort noise frames is done only in discontinuous transmission mode). The initial command is issued through a downlink discontinuous transmission indicator included in the messages used for the management of the basic transmission mode from the MSC (BSSMAP ASSIGNMENT REQUEST) and for channel activation toward the BTS (RSM CHANNEL ACTIVATION). The TRAU, in turn, is configured through an in-band indicator set by the BTS. Means to change the downlink discontinuous transmission mode on an established RR-connection also exist, since the same indicator is also included in the RSM MODE MODIFY REQUEST. Besides, the BSSMAP ASSIGNMENT REQUEST message can be used by the MSC to trigger a discontinuous transmission mode change on the infrastructure side. As already mentioned, the mobile station need not be warned of such a change, and no procedure has been defined on the radio interface for this purpose.

The mobile station can be ordered to use the discontinuous transmission mode in the uplink direction as a cell option. This view is consistent with the consideration of discontinuous transmission as a means to improve spectral efficiency, but it does not take into account mobile stations or connections at an individual level. The cell options are regularly broadcast on the BCCH for mobile stations in idle mode, and they are also part of the general information sent to mobile stations on their SACCH when they are in dedicated mode. The uplink discontinuous transmission mode could thus at first sight be set on a connection basis by this slow signalling method, and changed in the same way. However, the specification of the RSM protocol requires to set the general information messages sent on the SACCH on a transmitter/receiver (TRX) basis. This state of things is even more fuzzy because of the presence of an uplink discontinuous transmission indicator (besides the downlink one) in the *CHANNEL MODE* information element, which is contained in the messages for the management of the basic transmission mode between BSC and BTS. The BTS has no need to know whether uplink discontinuous transmission is used or not; a possible justification for the presence of this indicator would be to ask the BTS to modify the RIL3-RR SYSTEM INFORMATION 6 it sends to the concerned mobile station. A point against this interpretation is that the information in the SYSTEM INFORMATION TYPE 6 message has three values (DTX must be applied, must not be applied or may be applied), whereas the information in the *CHANNEL MODE* information element is two-valued.

6.3.4. HANDOVER EXECUTION

The handover execution procedure enables the network to command a mobile station in dedicated mode to go onto another channel in another cell. The handover execution procedure is very close to the subsequent assignment procedure. The fundamental difference is the change of cell. The handover execution procedure differs mainly from the subsequent assignment procedure by the timing advance management, by the need to transmit some data specific to the new cell, and by a few limitations.

Basically the procedure was designed not to constrain the type or the mode of the new channel. As for the assignment procedure, the type and mode of channel before the handover, or the type and mode of the channel after the handover can be anything. Or almost... If the indicative value of timer T3124 (a timer used on the mobile station side for handover between non-synchronised cells) as defined in TS GSM 04.08 is applied by mobile stations in all cases, handover toward a TACH/8 does not run properly; it is therefore not part of the test requirements for type approval of phase 1 mobile stations. This will be corrected in phase 2. Yet a handover from a TACH/8 toward a TACH/F is possible. This has some application. When a connection is established, the cell has been chosen by the mobile station. While in connected mode, the cell is chosen by the network. The two cell choice algorithms are different, and hence may lead to different results in many cases. When this happens, and the initial channel is a TACH/8 whereas the needed channel is a TACH/F, a handover directly from the TACH/8 to a TACH/F in the right cell is quicker than performing a subsequent assignment first, and a handover next. This way of proceeding is called a "directed retry" (from the name of a similar action in analogue networks), and can be used also for congestion cases.

The handover procedure can be executed for different reasons, explained within the scope of the handover preparation function. But in all cases, the decision to attempt the handover of a given mobile station is taken by the BSC. Once so done, and once the new cell (or a list of candidate new cells) has been chosen, the actual transfer must be co-ordinated between the mobile station and the machines managing the old cell (BTS-old) and the new cell (BTS-new).

The handover procedure comes in several varieties, according to two main criteria.

The first criterion is related to the timing advance issue, and impacts only the "incoming" part of the radio interface procedure, between the mobile station and BTS-new. As expressed in the functional description of the timing advance management (see page 346), two cases can be distinguished:

- the mobile station is able to compute the new timing advance (to be applied with BTS-new), because the old and the new cells are synchronised (**synchronous** handover);

- the timing advance must be initialised both at the mobile station and at BTS-new during the handover procedure (**asynchronous** handover).

In a way, the subsequent assignment procedure can be considered as a third case, in which the timing advance is not changed.

The second criterion concerns the location of the switching point in the infrastructure. This location impacts heavily the procedures to be used between infrastructure entities; the procedure on the radio path is not impacted by such a consideration, except to distinguish the special case of intra-cell handover which uses the same procedure as for subsequent channel assignment.

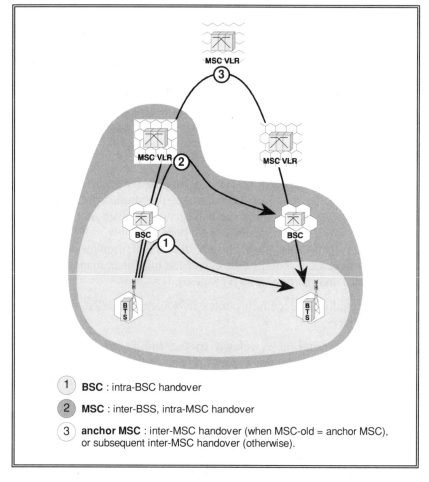

1　**BSC** : intra-BSC handover

2　**MSC** : inter-BSS, intra-MSC handover

3　**anchor MSC** : inter-MSC handover (when MSC-old = anchor MSC), or subsequent inter-MSC handover (otherwise).

Figure 6.29 – Position of the switching point at handover

The switching point depends on the relative position in the infrastructure hierarchy of the BSCs controlling the two cells. In all cases, the anchor MSC remains involved in the communication path, and there can be but only one other MSC in the path.

In order to describe the different cases, the suffix "old" shall be used to refer to all functional entities along the communication path before the handover, and "new" shall be used for the path after handover. BTS-old, BSC-old and (transit) MSC-old represent the machines in charge of the old cell, and BTS-new, BSC-new and (transit) MSC-new the machines in charge of the new cell. It may happen that BSC-old = BSC-new, or MSC-old = MSC-new. Regarding the relay MSC, it must be recalled that the anchor MSC, which originally established the RR-session, may be the same as MSC-old or MSC-new, or both, but may also be different. If the call already goes through two MSCs, the anchor MSC and MSC-old are distinct. If a handover is needed toward a cell of a new MSC different from MSC-old, the switching point is the anchor MSC. Figure 6.29 shows the most complex case, when all entities are physically distinct.

Whether synchronous or asynchronous, whether inter- or intra-MSC, and whether inter- or intra-BSC, the execution of handover is composed of two main phases:

- in a first phase, BSC-old triggers a set of events with the purpose of establishing the future communication path. Once this is done, this phase terminates with the sending of a handover command to the mobile station;

- in a second phase, the mobile station accesses the new channel. This access triggers the switch of paths in the infrastructure, and the release of the old path.

6.3.4.1. The Set-Up of the New Path

Once the decision of handing over a given communication has been taken by BSC-old, this must be indicated to the switching point. The latter must in turn establish the terrestrial resources, if need be, up to BSC-new, signal to it to allocate a radio resource and more generally provide all impacted machines with the information they need for the handover and the future management of the connection. This information includes:

- the transmission mode, used to choose and configure the transmission path in an appropriate way, including the new radio channel;

- the cipher mode;

- the identity of the origin cell, used to determine whether the handover can be done in a synchronous or an asynchronous way;

- the mobile station classmark, used for future management of the connection.

Once BSC-new is aware of all this information, it is in a position to allocate the new channel, to build the RIL3-RR HANDOVER COMMAND message and to transmit it to the switching point, which in turn will

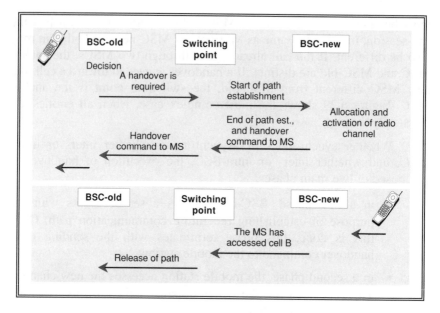

Figure 6.30 – Handover execution sequence

Once the handover decision taken, the set-up of the new path unfolds in four steps,
to prepare for mobile station access.

convey it to the mobile station along the old path, as shown in figure
6.30. Let us study these steps and the respective variations in detail.

From BSC-old to the Switching Point

The purpose of this exchange is the transmission of the information
that a handover is needed, and toward which cell (or cells). The different
cases depend on the nature of the switching point (see figure 6.31):

a. BSC-old is the switching point (BSC-old = BSC-new):

this step is internal and does not raise any problem.

b. MSC-old is the switching point (BSC-old ≠ BSC-new, MSC-
old = MSC-new):

in all cases when the target cell is not under its control, BSC-old
sends a BSSMAP HANDOVER REQUIRED message to MSC-old,
containing the identities of the target cell(s) and of the origin cell.

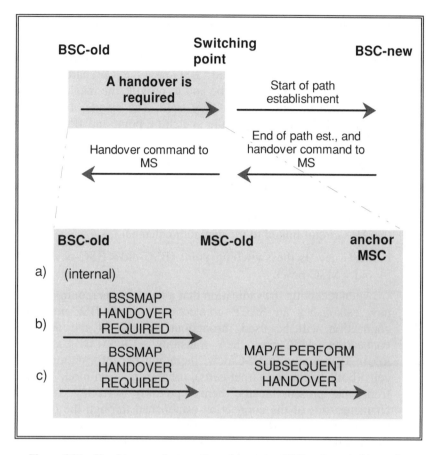

Figure 6.31 – Handover requirement from the serving BSC to the switching point

When applicable (depending on the relative position of the switching point), the BSSMAP HANDOVER REQUIRED message may be followed by a MAP/E PERFORM SUBSEQUENT HANDOVER message.

c. Anchor MSC is the switching point, and differs from MSC-old:

the behaviour of BSC-old is the same as in the previous case, but the behaviour of MSC-old when receiving the BSSMAP HANDOVER REQUIRED message is different. It translates the message in a MAP/E PERFORM SUBSEQUENT HANDOVER message towards the anchor MSC. Both messages have similar contents.

From the Switching Point to BSC-new

The purpose of this step is to establish the signalling pathway between the switching point and BSC-new, to prepare for the establishment of the circuit if need be and to provide the machines along the new path with the information they need. The events at this stage depend on the relative position of the switching point and BSC-new, as shown in figure 6.32:

a. BSC-new is the switching point (BSC-old = BSC-new):

this step is internal. The BSC is aware of all the relevant information since it already manages the current context of the connection. No terrestrial circuit needs to be allocated apart from the Abis circuit linked to the new radio channel.

b. MSC-new is the switching point (BSC-old ≠ BSC-new, MSC-old = MSC-new):

when receiving the indication that a handover is required, MSC-new establishes an SCCP connection towards BSC-new. This connection will be used throughout the life of the new RR-connection. MSC-new then transmits a BSSMAP HANDOVER REQUEST message to BSC-new, including the information on the cells (both origin and target cells), the transmission mode (derived from the present needs, hence possibly differing from the characteristics of the connection established through the old cell), the cipher mode (which *must* be the same as in the old cell, since the mobile station will assume so), the classmark and finally the reference of the terrestrial channel between MSC-new and BSC-new if need be.

c. Anchor MSC is the switching point, and differs from MSC-new:

this corresponds to the most complex case. The tasks of the anchor MSC cannot be done in a single step as in the previous case, because the new communication path between the anchor MSC and BSC-new (through MSC-new) may transit via the PSTN or the ISDN. Standard inter-switch procedures must therefore be used, which are part of TUP or ISUP (or national variants of them). These protocols do not provide the means to convey the relevant GSM information. They are therefore only used for circuit set-up and MAP/E procedures are used for the specific handover signalling needs.

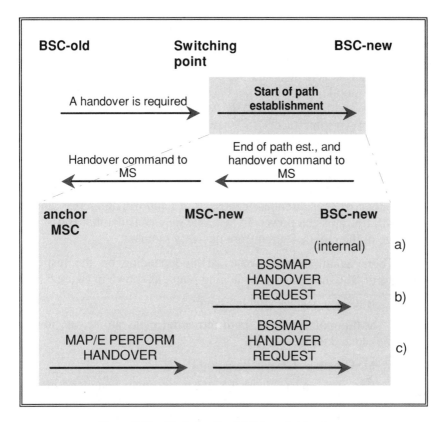

Figure 6.32 – Start of path establishment at handover

The new path is established starting if applicable
with the channel on the new A interface.
It is triggered by a message coming from the anchor MSC.
Only in the next step will the actual circuit between anchor and relay MSC
(MSC-new) be established if applicable.

The anchor MSC provides the required information to MSC-new through a MAP/E PERFORM HANDOVER message; when receiving it, MSC-new establishes an SCCP connection with BSC-new, allocates if need be a circuit on the A interface and transmits a BSSMAP HANDOVER REQUEST message to BSC-new, containing the information received in the MAP/E PERFORM HANDOVER message, as explained in case b.

From BSC-new Back to the Switching Point

At this point BSC-new must try to allocate the radio channel. This procedure results in either a positive or a negative answer. Unless otherwise specified, queuing should not be applied, because other machines are waiting for the answer, and timers are running. Negative cases will be dealt with further on. As for the "happy-ending" case, when a channel can be activated and the corresponding device in BTS-new is ready for mobile station access (an exchange of RSM CHANNEL ACTIVATION and RSM CHANNEL ACTIVATION ACKNOWLEDGE takes place on the (new) Abis interface), BSC-new builds up the RIL3-RR HANDOVER COMMAND message and transmits it to the mobile station, via the switching point and the old resources. It should be stressed that it is BSC-new which builds this message (which will be eventually sent by BSC-old), and thus decides for instance whether the handover will be synchronous or asynchronous, chooses the handover reference, and the initial MS transmission power. In fact, one may consider that BSC-new is in charge of the mobile station from this very moment.

There again, different cases arise depending on the respective positions of BSC-new and the switching point, as shown in figure 6.33.

a. BSC-new is the switching point:

at this point in time, both terrestrial paths are set-up, towards the old and the new BTSs.

b. MSC-new is the switching point:

BSC-new encapsulates the ril3-rr handover command message in a bssmap handover request acknowledge message. Nothing else needs to be done at this stage, since the terrestrial path is already completely established.

c. The anchor MSC is the switching point, and differs from MSC-new:

BSC-new acts as in case *b* above. When receiving the BSSMAP HANDOVER REQUEST ACKNOWLEDGE message, MSC-new inserts the included RIL3-RR HANDOVER COMMAND message in a new envelope, the MAP/E PERFORM HANDOVER RESULT message. This message contains a telephony-like number (provided by MSC-new) to allow the anchor MSC to set up a circuit through normal ISUP or TUP means. This **handover number** is allocated solely for the anchor MSC to establish the circuit with MSC-new, and serves as a reference for MSC-new to link the context with the incoming circuit. The same MAP/E exchange (PERFORM HANDOVER and its RESULT) serves both purposes of providing the information needed for circuit establishment and carrying the RIL3-RR HANDOVER COMMAND message back, ready to be sent to the mobile station along the old path. Upon receipt of the MAP/E PERFORM HANDOVER RESULT message, the anchor MSC is able to

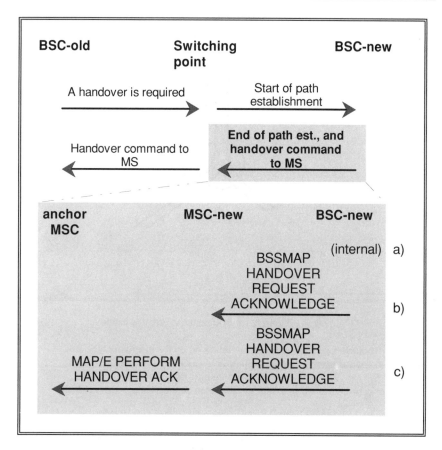

Figure 6.33 – End of path establishment at handover

Once BSC-new has established the new radio channel
(and corresponding resources in the new BTS and on the new Abis interface),
it acknowledges the BSSMAP HANDOVER REQUEST message,
which then triggers in turn and if applicable the set-up of the circuit
between the anchor MSC and the new relay MSC when
the MAP/E PERFORM HANDOVER RESULT message has been received.

set-up the communication with MSC-new, through, e.g., the IAM and ACM messages of ISUP.

Depending upon implementation choices for the switching machine, the paths can be at this stage linked in a two-way conference bridge, a one-way conference bridge or not linked at all. In the first two cases, downlink user data is transmitted to the mobile station via both

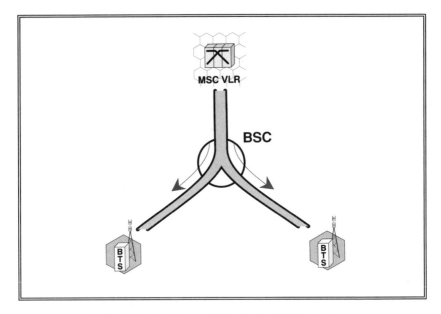

Figure 6.34 – Conference bridge in the BSC for handover

Handover performance can be improved by the insertion at the BSC
of a conference bridge, either one-way (downlink) or two-way,
so that both paths of an intra-BSC handover
may be connected in parallel to the same connection towards the MSC.

BTSs. In the first case only, the two uplink flows are combined into a single one towards the other correspondent (see figure 6.34). It is feasible to insert such a conference bridge only in some cases, for instance when the transmission mode is speech and the transcoders on the old and new paths are different (i.e., speech is carried at 64 kbit/s at the switching point).

From the Switching Point to the Mobile Station

The last step of the first phase of handover execution consists simply in sending the RIL3-RR HANDOVER COMMAND message to the mobile station, as shown in figure 6.35, according to the three following cases:

a. BSC-old is the switching point (BSC-old = BSC-new);

b. MSC-old is the switching point (MSC-old = anchor MSC);

c. the anchor MSC is the switching point, and differs from MSC-old.

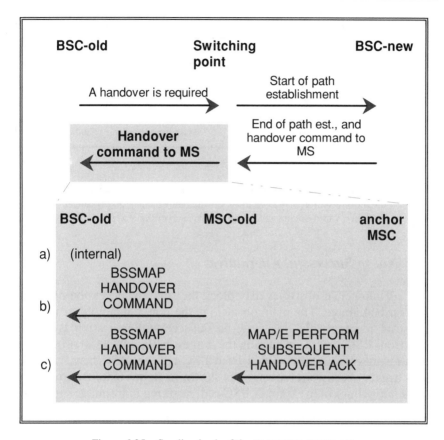

Figure 6.35 – Sending back of the HANDOVER COMMAND

As a last step in the network before mobile station access on the new channel, the RIL3-RR HANDOVER COMMAND message is sent to the mobile station.

The RIL3-RR HANDOVER COMMAND message is carried over the different interfaces in a variety of different envelopes, as shown in table 6.8, which also summarises the transfers already mentioned in the previous paragraphs.

The RIL3-RR HANDOVER COMMAND message identifies the new cell only through its beacon frequency and its BSIC. This is sufficient for the mobile station, and the full cell identity will be read later by the mobile station on the SACCH sent by BTS-new.

Interface	Encapsulating message
between BSC-new and MSC-new	BSSMAP HANDOVER REQUEST ACKNOWLEDGE
between MSC-new and anchor MSC	MAP/E PERFORM HANDOVER RESULT
between anchor MSC and MSC-old	MAP/E PERFORM SUBSEQUENT HANDOVER RESULT
between MSC-old and BSC-old	BSSMAP HANDOVER COMMAND

Table 6.8 – The transfer of the RIL3-RR HANDOVER COMMAND message

The message aimed at the mobile station contains everything
the mobile station may need to access the new channel, and is carried
unaltered in a variety of encapsulating messages over the terrestrial interfaces.

Not so Successful Alternatives

A number of obstacles may block the smooth succession of events as described above. The main obstacle is the non-availability of radio or terrestrial resources. In this case, an unsuccessful indication is carried back from BSC-new. Considering the longest path, a BSSMAP HANDOVER FAILURE message is transmitted from BSC-new to MSC-new, which in turns triggers backward messages as shown in figure 6.36. Alternatively, a watchdog timer may expire in BSC-old, resulting in a similar situation.

Two possibilities can be envisaged. Either a new handover attempt towards the same cell is performed after some time, or a handover towards another cell is attempted. In the first case, the failure indication goes all the way back to BSC-old, which will re-initiate the handover process when it decides so. All resources which have been allocated along the new path are released. The second case can also be treated this way. However, an alternative exists in the *Specifications*, whereby BSC-old may provide MSC-old with an ordered list of suitable target cells (this applies as already mentioned mainly in the case of rescue handovers, when staying in the old cell is not going to be a good alternative to the first target cell). This list is not conveyed to the anchor MSC if MSC-old differs from it and is the switching point. The failure indication, when reaching MSC-old, will therefore trigger a handover attempt towards the next cell in the list. Only when all cells have been tried in vain will BSC-old be given the failure indication (and the full control of the connection back). This possibility of multiple cell choice in the BSSMAP HANDOVER REQUIRED message is an option for the BSC. Speed requirements tend to favour the multi-cell approach, whereas the optimisation of cell allocation favours the single cell approach, since only BSC-old is in a position to change the cell list according to up-to-date measurements.

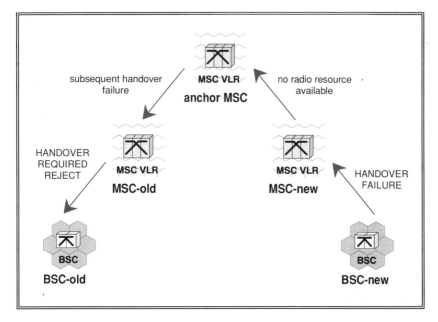

Figure 6.36 – Handover failure at the new BSC

The failure indication goes all the way back to BSC-old as shown, so that
the decision to retry or perform another action can be made.
Alternatively, MSC-old may attempt handovers toward other cells
if it is in possession of a list of candidate target cells.

6.3.4.2. Mobile Station Access and the Conclusion of the Procedure

The mobile station is completely unaware of the infrastructure processes and decisions until it receives the RIL3-RR HANDOVER COMMAND message. As already mentioned, this message contains all the information characterising transmission on the new channel (except for the cipher mode which is assumed to remain the same as on the old channel), and the data needed for the access procedure. In particular, it indicates to the mobile station whether the synchronous or asynchronous handover procedure should be followed. In both cases, thanks to pre-synchronisation, the mobile station is able to synchronise itself quickly on the new channel and starts reception immediately. It will actually receive speech or data from this point, if applicable and if the switching point uses a conference bridge.

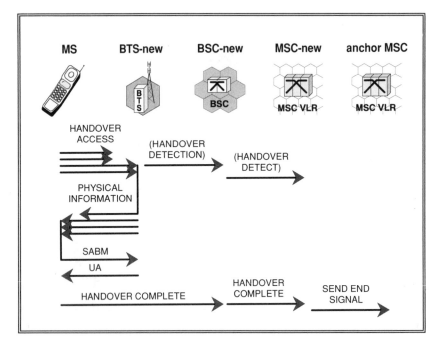

Figure 6.37 – Access in the case of an asynchronous handover

Only following reception of RIL3-RR PHYSICAL INFORMATION messages
does the mobile station switch to normal transmission mode
with the timing advance as indicated,
and sends an RIL3-RR HANDOVER COMPLETE message after having
established the SAPI 0 signalling link on the new dedicated channel.

As far as transmission from the mobile station is concerned, the type of handover intervenes. In the case of a synchronous handover, the mobile station first sends a few access bursts (the RIL3-RR HANDOVER ACCESS message), then starts normal transmission by applying the computed timing advance. If the handover is an asynchronous one (see figure 6.37), the mobile station continues to send access bursts until it has received an RIL3-RR PHYSICAL INFORMATION message from BTS-new, conveying the actual timing advance to apply. Only then does it start normal transmission. In both cases, the RIL3-RR HANDOVER ACCESS message only contains an 8-bit handover reference. This message is the only case where short access bursts are used on a dedicated channel. The handover reference (not to be confused with the handover number) is part of the data transmitted to the mobile station in the RIL3-RR HANDOVER COMMAND message and can be used by BTS-new as an additional check that the accessing mobile station is indeed the expected one.

The RIL3-RR PHYSICAL INFORMATION message is the only case in the *Specifications* where a message above the link layer is sent as an autonomous decision by the BTS. This departure from the general rule is justified by performance requirements. For efficiency reasons, the RIL3-RR PHYSICAL INFORMATION message may be sent several times in a row, until the reception of normal bursts from the mobile station makes it clear to BTS-new that it has received the message. This would not have been so easy if the RIL3-RR PHYSICAL INFORMATION message had been sent by the BSC.

The BTS may as an option indicate to the BSC that it has received adequate RIL3-RR HANDOVER ACCESS bursts on the allocated channel, through an RSM HANDOVER DETECTION message; BSC-new may in turn pass the indication on to MSC-new through a BSSMAP HANDOVER DETECT message. This mechanism allows the switching point (except in the case when it is the anchor MSC, because the information is not carried by the MAP/E protocol) to switch the communication path at this moment without waiting for the full completion of the procedure.

When it is in normal transmission mode, the mobile station sets the link layer to acknowledged mode for signalling messages by sending an SABM frame answered by a UA frame. The mobile station then sends an RIL3-RR HANDOVER COMPLETE message, which will be carried by the infrastructure up to the switching point, when applicable through a BSSMAP HANDOVER COMPLETE message from BSC-new to MSC-new and through a MAP/E SEND END SIGNAL from MSC-new to anchor MSC. The switching point will release the previous path by sending appropriate messages (MAP/E SEND END SIGNAL RESULT from anchor MSC to MSC-old, and BSSMAP CLEAR COMMAND from MSC-old to BSC-old), relayed up to BSC-old which releases the previous radio channel held up until this point. The release of resources which then takes place on the A interface, the Abis interface and at the BTS have no specific differences compared to an RR-session termination.

The sending of the handover complete indication triggers the switching of paths between the old and the new one, if this has not already been done (e.g., upon access detection on the new channel). The question of the necessity of the handover complete indication can be raised at this point: why is there a need of a two-stage mechanism? The difference between the access and the completion is that only the latter triggers the release of the previous channel. Only when it sends the RIL3-RR HANDOVER COMPLETE message does the mobile station abandon all possibility of returning to the old channel. The first stage was added to shorten the interruption time.

The return on the old channel in case of problems is similar to the subsequent assignment case; only the name of the message changes:

RIL3-RR ASSIGNMENT FAILURE in one case, RIL3-RR HANDOVER FAILURE in the other. When this unsuccessful outcome arrives, BSC-old is advised and transmits the information up to the MSC-old if applicable, through a BSSMAP HANDOVER FAILURE message. The MAP/E protocol introduces some limitations. When the anchor MSC is different from MSC-old, there is no means of passing on this information between MSC-old and the anchor MSC, since no message exists on MAP/E for this purpose. The only way for the anchor MSC to react in this case is through timer expiry upon non-reception of a message from BSC-new indicating the completion of the handover. Whatever the means by which it recognised a failure condition, the switching point releases the new path, using normal release procedures, and it is up to BSC-old to decide what action to perform, e.g., make another handover attempt.

An intra-BSC handover is usually performed autonomously by the BSC. As explained in the handover preparation section, it is a BSS implementation option not to involve the MSC (the one in charge of the BSC) at all in the decision when the best cell, as seen from the BSC, is also under the control of the same BSC. In this case, the whole handover will unfold without any knowledge of the MSC. In order to advise it that a handover has occurred successfully, a BSSMAP HANDOVER PERFORMED message is sent from the BSC to the relay MSC, possibly relayed by a MAP/E NOTE INTERNAL HANDOVER message from the relay MSC to the anchor MSC when they differ. This message may also be sent in the case of a handover internal to the relay MSC (case when MSC-new = MSC-old ≠ anchor MSC). The sending or not of this MAP/E message depends on Operation and Maintenance requirements.

6.3.5. CALL RE-ESTABLISHMENT

In a radio mobile environment, there is always some risk that a connection will be suddenly cut. This may happen because of a brutal propagation loss, due to obstacles such as bridges, tunnels, or simply buildings in the case of handhelds. But another cell could often be used to continue the communication either immediately or after a very short time (think about a short tunnel through a hill).

The handover preparation and execution are a means of limiting the occurrences of call loss, but they cannot suppress them totally. In the future, when cells become smaller and smaller, the risks will increase. The performance achieved by handover algorithms running on the network side will then lessen, in prediction accuracy as well as in reaction time, and connection loss probability will increase. The mobile station has in fact some ways to determine that a handover is needed, and may be more efficient in these cases. In some systems crafted for a microcell or pico-cell environment, all handovers are triggered by the mobile station alone. However, network handover control has many advantages when it can be applied: this stems from the obvious fact that the network has a much better understanding of the general situation than any single mobile station. Call re-establishment may then be considered as a kind of mobile station triggered handover, but limited to the extreme case of rescue handover when communication with the current cell is effectively lost. One could foresee that the importance of this feature will grow in the future. For instance, it can be imagined that in some environments the procedure will be triggered sooner, so as to improve the system performances where network triggered handover will have shown its limits.

Despite these considerations, the call re-establishment procedure is a poor relative in the GSM procedure tribe, and has serious limitations in phase 1. Let us see what there is of it.

As a general point, it should be noted that the call re-establishment procedure is not a full-blown RR procedure. We will see in this section messages from the RIL3-MM protocol. Only because of its kinship with handover do we present the re-establishment procedure in this chapter and not in Chapter 7.

Call re-establishment has two parts. The mobile station has the leading role in the first one, which is very close to an access procedure. The second part is the network's, and consists in the recovering of upper layer contexts.

The closeness of call re-establishment with initial access is quite normal, because the mobile station has to start from scratch. The differences are important, and come mainly from the requirement for speed: when a connection is lost, a timer starts ticking in the anchor MSC, and at its expiry everything related to the moribund connection is erased. As a consequence, any fraction of second lost in the call re-establishment procedure increases the risk of total loss.

The first issue is to determine the new cell. The speed requirement limits the choice to the neighbouring cells already known and with which the mobile station is pre-synchronised, because finding new cells and getting synchronisation may take seconds. The selection rule is to just select the one with the highest signal strength. The parameters to compute the different radio criteria (the $C1$'s of the possible cells, which will be described in Chapter 7) as for idle-mode cell selection are not known, and receiving them takes time. The *Specifications* nevertheless require the mobile station to check the radio criterion for the chosen cell. This constraint forces the mobile station to wait for the reception of a BCCH message containing the required parameters, and this can take up to three quarters of a second (if the first decoding attempt succeeds!). Another formality has to be checked: the chosen cell must not be barred, and call re-establishment must be allowed in it. The corresponding indications are part of all BCCH messages, so that this checking does not add any more delay.

After having received the required BCCH message, and checked the radio criteria and authorisations, the mobile station sends an access request on the RACH. Though not specified in the *Specifications* (as many other small details of the procedure), it seems that it is allowed to use the RACH on TN 0, even if there are others. The access request indicates the reason for access (i.e., call re-establishment), so that the network is aware of the critical nature of the request. The required channel is not indicated, but the network can easily play safe and allocate a TACH/F.

The initial message is an RIL3-MM CM RE-ESTABLISHMENT REQUEST. Its information contents are minimal: the subscriber identity and the classmark. The mobile station does not volunteer anything else, and the network has to use this to find out everything about the lost connection! Among the conspicuous missing data known directly or indirectly by the mobile station are the cell with which the connection was lost, the identity of the anchor MSC and the required type and mode of the channel.

In any case, the only case catered for by phase 1 protocols is when the new cell and the previous ones are managed by the same MSC. The anchor MSC is then implicitly determined. Moreover, even if the previous cell was known, it would be to no avail since no mechanisms have been included to recover the RR-session except when the new MSC is the anchor MSC. An inference is that call re-establishment is impossible when there is a relay MSC.

The fact that the required type of channel is not given by the mobile station has no clear explanation. This is a source of delay, because the BSC has to wait for the indication from the MSC to allocate the right type of channel. Unless, as already mentioned, if the BSC gambles and initially allocates a TACH/F.

The recovery of the contexts is then entirely an MSC issue, and must be done with only the subscriber identity to start with. From this the MSC must find the old context (if it still exists—it could have been erased after a timer expiry, or simply because the correspondent was not patient enough). Then the MSC performs an assignment procedure and possibly a ciphering start procedure, telling the BSC the type of channel required, the mode, and so on, and allocating the BSC-MSC terrestrial route when need be. The MSC may even choose to perform an authentication despite the incurred delay. The BSC then performs the needed procedures within the BTS and with the mobile station (subsequent assignment, ciphering start, mode modification, ...). Only then can an RIL3-MM CM SERVICE ACCEPT message be sent to the mobile station, and the end to end communication resumed. Let us note that means for the network to reject the request have been foreseen. The RIL3-MM CM SERVICE REJECT serves this purpose (in particular with the cause "call cannot be identified", which is likely to be used a lot).

A number of other fuzzy points exist in the *Specifications*. They include in particular, the recovery of several CM-transactions and the corresponding transaction identifiers (see Chapters 5 and 8), and, more important, all the cases of collision, when the connection loss happened during an ongoing procedure in any layer. Another issue is the release of the old path. As far as can be analysed, the anchor MSC can and should release it once aware of the re-establishment attempt, even if in the same BSC (or the same cell).

One wonders if call re-establishment will really be used in phase 1. This is hopefully an area in which the future phases of the *Specifications* will bring improvements.

6.3.6. RR-SESSION RELEASE

When all needs for an RR-session have disappeared, for instance because a location updating procedure has ended, because a call is terminated, or because of a failure, the mobile station must go back to idle mode and the resources must be released, in order to be free for allocation for other needs. This release mechanism is done through a so-called "normal release" procedure, which is always triggered by the anchor MSC.

If distinct from the relay MSC, the anchor MSC releases the RR-session by sending a MAP/E SEND END SIGNAL RESULT to the relay MSC on one hand, and by releasing the circuit if present, through ISUP release procedures.

The next step is the BSSMAP CLEAR COMMAND message sent from relay MSC to BSC. This message may be piggybacked on an SCCP RELEASE message releasing the SCCP connection. In this case, the BSC acknowledgement (BSSMAP CLEAR COMPLETE message) must be piggybacked on the SCCP RELEASE COMPLETE message. The clearing actions of the BSC can take place in parallel with the sending of the BSSMAP CLEAR COMPLETE message, since the *Specifications* do not

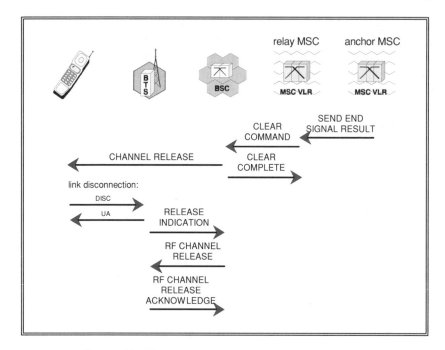

Figure 6.38 – The normal release procedure of an RR-session

Normal release is always triggered by the anchor MSC, but the BSC manages the return of the mobile station to idle mode before releasing the BSS resources.

impose any specific order between these two actions. Once the BSC has ordered the mobile station to go back to idle mode through an RIL3-RR CHANNEL RELEASE message, the mobile station disconnects the signalling link, and this event is reported by the BTS to the BSC through the RSM RELEASE INDICATION message. The *Specifications* include a number of timers and repetitions in order to ensure that whatever frame losses may occur during this period the mobile station eventually goes back to idle mode and stops using the channels. This is of prime importance to avoid allocating a channel to a mobile station when another mobile station may still transmit on this same channel. Only when the BSC is sure that the mobile station has left will it de-activate the BTS device, through the RSM RF CHANNEL RELEASE / RSM RF CHANNEL RELEASE ACKNOWLEDGE exchange. The corresponding radio channel is then considered as part of the pool of free channels by the BSC. The whole procedure is illustrated in figure 6.38.

An RR-session may also be released in other conditions, for instance when the infrastructure has lost actual contact with the mobile station. Such a situation may arise when propagation conditions are bad, or when the interference level is too high. One role of handover is to cope with such cases, but this is not always possible, e.g., when the user has gone out of coverage (in an underground car park for instance), or switched off his mobile station in the middle of an RR-session. Such cases must be detected, in order for the infrastructure to free the corresponding resources.

The mechanism specified in the *Specifications* for this purpose consists in having both the mobile station and the BTS monitor the message loss rate on the SACCH. Let us recall that messages are sent regularly (about twice per second) in both directions of the SACCH, throughout the life of a connection. A powerful error detection mechanism has been included in signalling messages, enabling the receiver to estimate message loss. This estimation is done through a counter, incremented in case of a correctly received block and decremented in the other case (see figure 6.39). When the counter reaches a minimum threshold, the link is considered as broken. On the mobile station side, this event leads to a return to idle mode (the mobile station may subsequently attempt a re-establishment, see page 412). The infrastructure is able to adjust the mobile station behaviour, in order to allow fine tuning on a cell basis, although it seems unlikely that it will be necessary to manage differences between mobile stations. The relevant parameters are sent regularly on the SACCH as well as on the BCCH. Setting them to satisfying values must be done through field experiments in operational networks. On the infrastructure side, the counter is in the BTS and the failure indication is given to the BSC in an RSM CONNECTION FAILURE INDICATION message.

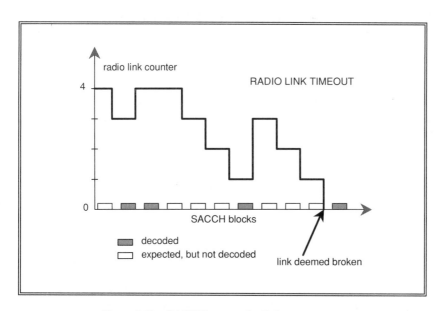

Figure 6.39 – SACCH counter for link management

A counter enables each receiver to estimate the frame loss rate on the SACCH
(downlink for the mobile station, uplink for the infrastructure).
When this counter reaches 0,
the link is deemed broken and actions are taken to release the resources.

While the detection of transmission loss simply triggers the mobile station to abandon the connection and return to idle mode, things are a little bit more complex on the infrastructure side. Since both directions of transmission may experience different qualities, reaching the threshold in the uplink direction does not necessarily imply that the mobile station also experiences a break of the link. The infrastructure must nevertheless make sure that the mobile station leaves the channel before deeming the channel to be free. The BSC therefore commands the BTS to stop transmission of downlink SACCH frames (by an RSM DEACTIVATE SACCH message), so that the mobile station counter will inexorably reach the minimum threshold after some given time. The monitoring of the uplink channel may take place either in the BTS or in the BSC. In the first case, the BTS may report the loss of the link through an RSM CONNECTION FAILURE INDICATION message to the BSC.

Once the link failure is detected and notified to the relay MSC through a BSSMAP CLEAR REQUEST message, the same A interface exchange takes place as in the case of normal connection release, i.e., the MSC sends a BSSMAP CLEAR COMMAND message, answered as expected by a BSSMAP CLEAR COMPLETE message from the BSC.

6.3.7. LOAD MANAGEMENT PROCEDURES

A few procedures in the RR plane allow the MSC and the BSC to deal with overload situations. They include means to exchange information between machines so that each one gets the information it needs about the current load situation; and means to act so as to limit the effect of the overload. Procedures dealing with load management appear in two main areas: RACH and PAGCH load, and TCH load.

6.3.7.1. Load on Common Channels

Some information concerning the load on the RACH and on the PAGCH could be inferred by the BSC from the requests it sends to, or receives from, the BTS. Still, the BTS is in a better position to assess the exact load on these channels.

A message, RSM CCCH LOAD INDICATION, has thus been introduced in the RSM protocol, to enable the BTS to send some information about the RACH and PAGCH loads to the BSC. The conditions for sending this message are set through the Operation Sub-System; it can be regularly sent, or only when the load on one of the channels is above some threshold. The message pertains to a single pair RACH/PAGCH.

The BSC may use this information to change the RACH load control parameters included in the corresponding BCCH, and to modify its assignment priority rules.

6.3.7.2. Load on Traffic Channels

The number of dedicated channels of each sort currently allocated in a cell is known by the BSC. This information may also be useful to the MSC, for instance to balance the traffic between cells. The number of TACH/F (and in phase 2 of TACH/H) currently allocated can be indicated to the MSC with a BSSMAP RESOURCE INDICATION message. The message can be sent in various cases, which are controlled by the MSC and indicated with a BSSMAP RESOURCE REQUEST message. The MSC can ask for one single immediate message; for regular sending; or for spontaneous sending when some conditions are met (these conditions are set through the OSS).

The MSC can do many things to limit useless signalling towards the BSC. However, the resource indication function was introduced mainly to support traffic handovers under the control of the MSC. In case of an overload condition local to a cell, handovers can be used to balance the traffic between cells. This is the purpose of the "handover candidate enquiry" procedure, which is used when the overloaded cell has neighbours under another BSC. Only the (relay) MSC is in a position to know which of the neighbour cells can take over a part of the load. By sending a BSSMAP HANDOVER CANDIDATE ENQUIRY message, the MSC indicates to the BSC that it would be better for traffic balance to hand over a given number of connections from one cell to other cells in a given list. When the BSC is done with the consequences of this message (precisely after having sent one BSSMAP HANDOVER REQUIRED message, with the appropriate cause, for each of a number of connections), it sends a BSSMAP CANDIDATE RESPONSE message. What the BSC must do is not specified in detail. As already mentioned, traffic balancing is contradictory with the choice of the cell for spectral efficiency reasons, and the action of the BSC must be designed with care, to avoid having the handed-over connections handed back a few seconds later because of radio criteria. It may be noted that an alternative strategy to solve traffic imbalance is to use the OSS to control handovers directly via the BSC. This uses the parameter modification procedures discussed in Chapter 9.

6.3.7.3. General Overload

In addition to the specialised procedures described above, the BSSMAP and the RSM contain a number of procedures to cope with overload in general, whether of transmission resources or computation capacity. The *Specifications* describe the messages conveying the information from MSC to BSC, from BSC to MSC (BSSMAP OVERLOAD and from BTS to BSC (RSM OVERLOAD) that the recipient must "reduce the traffic".

The purpose of overload control is to reduce the traffic as close to its source as possible. Let us examine the different cases:

- BTS to BSC: It is not clear how a BTS can be overloaded otherwise than on the common channels (and this is dealt with by specific procedures), since a BTS is normally designed to cope with the simultaneous usage of all its radio channels. The only thing it seems the BSC could do in answer to an RSM OVERLOAD message is to consider that only a portion of the channels can be used;

- BSC to MSC: The only traffic on the A interface established under the control of the MSC concerns the mobile terminating calls. A possible reaction of the MSC to the BSSMAP OVERLOAD message is then to reject a portion of the mobile terminating calls rather than to send the paging messages. On the other hand, this can be done by the BSC, and with a better correlation with the congested cells;

- MSC to BSC: An MSC overload indication can be used by the BSC to reduce the number of mobile station accesses it accepts, by using one of the various means described on page 370.

6.3.8. SACCH PROCEDURES

When the mobile station is in dedicated mode, it is always allocated a bi-directional channel of limited capacity (the SACCH), in addition to the main channel conveying the information for which the connection exists. The SACCH is used for a variety of functions. The main item, which justifies the existence of the SACCH, concerns the continuous monitoring of the connection in a mobile environment: transmission power control, timing advance control and measurement reporting. The second role of the SACCH is to convey general information to the mobile station.

6.3.8.1. Radio Transmission Control

In the downlink direction, the SACCH carries the commands related to power control and timing advance. These commands from the network are carried in the so-called *L1-header*, or layer 1 header, meaning that this information pertains to the physical layer in the radio path protocol architecture, and is therefore formatted outside the scope of the link layer, independently from the messages carried inside link layer frames. The required power control level and timing advance are then sent once per SACCH message, i.e., about twice per second. In the uplink direction, the same header exists, and contains the corresponding "acknowledgement" by the mobile station. It is coded in a similar way, and includes the values of the two parameters in use at the end of the preceding measurement period. The actual timing advance used should be equal to the one ordered, whereas the actual power level may differ from the one ordered because of the maximum variation speed.

The timing advance is managed autonomously by the BTS. On the other hand, the transmission power is basically controlled by the BSC. The BSC analyses the measurements, and uses the RSM MS POWER CONTROL and RSM BS POWER CONTROL messages to convey the requirements to the BTS.

Measurement reports are sent by the mobile station on the uplink SACCH at every possible opportunity, and at least once per second. The exact specification requires that, among any two successive messages on the uplink SACCH, at least one be an RIL3-RR MEASUREMENT REPORT message. In a basic scheme, the BTS generates a message toward the BSC (RSM MEASUREMENT RESULT) at every measurement period (about twice per second); this message indicates whether or not a measurement message was received from the mobile station, conveys its contents in the first case, and includes the result of the measurements performed by the BTS. All these data are then processed by the BSC for transmission power control and handover preparation.

The schemes described above are completely defined in the *Specifications*. However, they present a severe defect: they result in a very substantial signalling load on the Abis interface. Hooks have been introduced in the protocol for a scheme where more processing is performed in the BTS (and less in the BSC), so as to reduce the information flow in-between. This alternative scheme is only outlined in the *Specifications*, inasmuch as the messages or message elements that can be used are not completely specified (RSM PRE-PROCESS CONFIGURE, RSM PRE-PROCESSED MEASUREMENT RESULT, RSM PHYSICAL CONTEXT REQUEST and RSM PHYSICAL CONTEXT CONFIRM messages, and *MS POWER CONTROL PARAMETERS* and *BS POWER CONTROL PARAMETERS* information elements). In order for these messages to be used, complementary specifications are needed from the operator or the manufacturer. Within these constraints, transmission power control can be entirely taken in charge by the BTS, possibly with some parameters (e.g., maximum power) set up and modified by the BSC. Handover preparation is a more tricky business. The pre-processing of the measurements in the BTS can be anything from none (the basic scheme) up to include the decision that to trigger a handover is necessary or useful, barring load considerations. In the latter case, the BSC would intervene only when necessary, to check the load and channel availability aspects and to trigger effectively the handover.

6.3.8.2. General Information

A second use of the SACCH is the transmission of general information from the network to the mobile station. This information includes parameters specific to the radio connection, but not to upper layers, and which are not so important as to require the stealing of user information. Basically, the transmission of these parameters is useful only at the beginning of a channel connection (in particular after a handover) and when they change, which is very seldom. However, because there is no high load on the downlink SACCH and because these messages are not acknowledged, they are repeated continuously as a background task. They bear some relationship with the messages broadcast to all mobile stations in a cell on the BCCH (RIL3-RR SYSTEM INFORMATION TYPE 1 to 4 messages), which will be described in a later section, and their names (RIL3-RR SYSTEM INFORMATION TYPE 5 and 6) reflects this similarity. However, their information contents are different; only a part of the information broadcast on the BCCH needs to be sent to a mobile station in dedicated mode. This information includes:

- parameters for monitoring the measurement process (list of frequencies to monitor, BSIC screening, BCCH frequency indication);

- parameters for controlling the radio link failure detection (see page 417);

- requirements for the application of uplink discontinuous transmission; and

- other information which is of no direct relevance in the present state of the *Specifications* (full cell identity).

The bulk of this information concerns the measurement process. It must be noted that, after a handover, the mobile station must await the reception of this information before starting measurement reporting.

The SACCH is always used in both directions for actual transmission, to allow the other side to do reliable measurements and to detect radio link failures. This is the reason of the continuous sending of

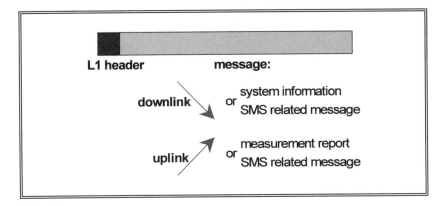

Figure 6.40 – Contents of an SACCH block

Except for messages related to the transfer of short messages,
the basic use of the SACCH concerns the reporting of measurements
from MS to BTS, and the sending of general information from BTS to MS
(the RIL3-RR SYSTEM INFORMATION TYPE 5 or 6 messages).

RIL3-RR SYSTEM INFORMATION messages downlink and RIL3-RR
MEASUREMENT REPORT messages uplink, when nothing else needs to be
sent.

A last use of the SACCH consists in carrying short messages when
the main channel is a TCH/F. This capability allows the mobile station to
either send or receive short messages while engaged in a communication
on the main channel. This situation represents in fact the only case in
phase 1 where two independent telecommunication services can be
provided in parallel.

As far as short messages are concerned, the functions of the BTS
are limited to the management of the physical layer and of the link layer
described in Chapter 5. The actual decision to send a short message is
taken by other network entities. For the sending of general information,
on the other hand, it would not be efficient to ask a machine other than
the BTS to decide on each individual sending of an RIL3-RR SYSTEM
INFORMATION message. The BTS is then in charge of the repetition of
these messages on the SACCHs. The BSC only provides information
regarding these messages when their contents are modified, using for this
purpose the RSM SACCH FILLING message. The procedure is such that the
information is not provided on a connection basis, but on a TRX basis.

Figure 6.40 summarises the different types of information which
transit on the SACCH, both uplink and downlink, and which are
scheduled in such a way as to achieve 100% use of the channel.

6.3.9. FREQUENCY REDEFINITION

The frequency redefinition procedure is used to change the frequency properties of the frequency-hopping channel used by a mobile station in synchronism with other mobile stations, in order to cope with a change of the frequencies used in the cell.

It is a very simple procedure, at first sight. It consists solely in the sending of one message to each concerned mobile station, the RIL3-RR FREQUENCY REDEFINITION message, as well as one message to each corresponding BTS device. However, there is no RSM message defined for this purpose, and the management of a frequency redefinition between BSC and BTS must use a procedure from the operation and maintenance BSC-BTS protocol for setting RF parameters in the BTS to command the configuration of the BTS. The RIL3-RR FREQUENCY REDEFINITION message includes both a starting time and the new frequency parameters (the list of frequencies as well as the MAIO and HSN, as explained on page 360), and the starting time can be set in the BTS by the operation and maintenance protocol on the Abis interface (the BTS Management protocol, BTSM). This state of affairs spoils the functional split between the two protocols, as a BTSM message is used for real-time management of connections.

6.3.10. GENERAL INFORMATION BROADCASTING

We have seen in this chapter that some information of a technical nature is needed by the mobile stations in *idle mode*, such as the configuration of the common channels. More reasons for the need of such information will be exposed in the next chapter. This data is regularly broadcast on the BCCH. It covers items of various nature which are cited here and there in this chapter and the next one. This section will act as a summary for the description of this information.

The BCCH is a low capacity channel, able to transmit one 23-octet long message every 0.235 second. It is therefore a scarce resource. The repetition rate for the different information items must be chosen as a trade-off between the use of the BCCH resource and the resulting time for the mobile station to get access to the information. As a consequence, several messages have been defined which contain different contents and which have different periodicity.

The broadcast items will now be listed together with their use and recurring rate; their order of description is based on the mobile station phase in which the information is needed.

6.3.10.1. Cell Selection Information

An important part of the broadcast information is related to the cell selection process. Some information of this kind is also directed towards mobile stations of neighbouring cells. This information will be addressed in Chapter 7, when the cell selection criteria will be described. It includes the location area identity and various parameters impacting the access choice, including the indication whether the cell is barred for access or not. The corresponding parameters in the *Specifications* are called *LAI*, *CELL SELECTION PARAMETERS* and (for the *CELL_BAR_ACCESS* flag) the *RACH CONTROL PARAMETERS*.

The transmission rate has a direct impact on the time it takes a mobile station to include the cell in its comparison list, from the moment when the BCCH is received correctly by this mobile station. This is not such a stringent constraint, but the information is nevertheless broadcast at a high rate: 2 occurrences out of 4, and even more in the case of the *CELL_BAR_ACCESS* indication which is part of every single BCCH message. It is worth noting that, even when several *time slots* are used for the common control channels, mobile stations currently located in neighbour cells only listen to the BCCH on TN 0. Therefore, the rate of transmission on other time slots need not be so frequent. For sake of simplicity, though, the specifications are the same for other time slots as for the prime BCCH.

Another constraint impacts the information used for cell selection. Mobile stations in neighbouring cells must listen to their paging sub-channel. Since recurrence on the PAGCH is a multiple of 51×8 BP, care must be taken that the relative position of a paging sub-channel and of the cell selection information in BCCH messages never results in a systematic masking of the latter. This is achieved through two mechanisms:

- the maximum rate of a paging sub-channel is half the rate of a BCCH. Therefore, a paging sub-channel may at worst mask half of the messages of any BCCH, but never all messages;

- the cell selection information is not transmitted in every other message, but according to the following pattern:
  in, in, out, out, in, in, out, out, ...

These configurations result in the masking, at worst, of every second occurrence of the cell selection information by any paging sub-channel.

6.3.10.2. Information for Idle Mode Functions

Information for idle mode functions is used by mobile stations once they have selected the cell and are staying there for some time in idle mode. They include the configuration of the common channels, the neighbour cells to monitor and the configuration for cell broadcast messages. Once settled in the cell, a mobile station will regularly check the values for changes; this checking is done at a very slow rate and the transmission rate therefore only impacts the initial settling time.

A first need concerns the common channel configuration, i.e., the number of *time slots* used for common channels (this is enough for a given mobile station to know which *time slots* are used and which one to listen to), as well as the parameters enabling the mobile station to calculate where to find its own paging sub-channel. This is found in the *CONTROL CHANNEL DESCRIPTION* parameter.

Once camped on a cell, the mobile station must also know the list of beacon frequencies to monitor. This is found in the *NEIGHBOUR CELLS DESCRIPTION* parameter. This element is also relevant for dedicated mode, but only as a default value during the short time between access and the reception of the first SACCH message containing the information. Because of this possible usage for measurement reporting in dedicated mode, this parameter contains a binary indicator (the *BA_IND* flag) enabling the list of beacon frequencies to be changed while keeping the network aware of which one was used by the mobile station in its reported measurements.

Lastly, mobile stations equipped for receiving broadcast short messages must know whether the cell provides a Cell Broadcast CHannel (CBCH), as well as where to find it when applicable. The *CBCH CHANNEL DESCRIPTION* and, if needed, *CBCH "MOBILE ALLOCATION"* information elements are therefore broadcast for this purpose.

Each of these three types of information is broadcast in every fourth message, i.e., roughly once per second.

6.3.10.3. Information Needed for Access

Sooner or later, the mobile station will want to access the cell, i.e., to obtain a bi-directional dedicated channel for its transmission needs. This may happen at the point of entry into the cell, when location updating is required or for a call re-establishment. Critical time constraints therefore apply to some of this information.

As explained in the section dealing with random access (see page 368), some control means are available for the BSC to limit access attempts, through the mechanism of "access classes". The BCCH information therefore includes the list of access classes allowed for access and the indication whether emergency calls are allowed. Another flag indicates whether call re-establishment is allowed. The BSC also controls the scheduling of access attempts and repetitions, and the corresponding parameters are broadcast. All of these parameters are part of the so-called *RACH CONTROL PARAMETERS*.

The most critical case of access to this information is call re-establishment. Any slowing down of the access procedures in this case increase the probability of losing the call. For this reason, all the access control information is broadcast in every BCCH message, i.e., 4 times per second.

Once having sent its RIL3-RR CHANNEL REQUEST burst on the RACH, the mobile station must decode the corresponding channel assignment message sent back by the network, if any. Since such a message must fit in a single block, its length may be too short in case of frequency hopping to send a complete description of the channel frequency characteristics. Therefore, as explained on page 361, part of the channel description information, applicable to *all* initial assignments, is broadcast regularly (once per second), in the *CELL CHANNEL DESCRIPTION* parameter. In most cases, this information will correspond to the list of all frequencies which might be used for dedicated channels in the cell, although it need only contain the list of all frequencies which might be used for *initial* channel assignment in the cell.

This list is critical information for access, since the other frequency parameters of the allocated channel cannot be understood by the mobile station without it. However, a mobile station could perfectly start the access procedure (i.e., send an RIL3-RR CHANNEL REQUEST message) before decoding the corresponding BCCH block, and therefore wait until it has done so after having received the network initial assignment, before actually accessing the dedicated channel. The *Specifications* are not clear on this point, and it seems a good way to improve the performance of call re-establishment.

6.3.10.4. Information for Mobile Stations in Dedicated Mode

Strangely enough, part of the information broadcast on the BCCH has no application other than after access, i.e., for mobile stations in dedicated mode. This information, given again on the SACCH, includes parameters to control the reporting of measurements, in particular the BSIC screening information included in the *PLMN PERMITTED* parameter, to which we will come back in Chapter 7. It also includes the "power control indicator" which we encountered in the section dealing with measurements, as well as the indication whether mobile stations are obliged, forbidden or permitted to use uplink discontinuous transmission. These three indications are included in a *CELL OPTIONS* parameter.

This information bears no real timing constraint; its absence would indeed little affect system performance. It is sent in every fourth message.

6.3.10.5. Cell Identity

The final piece of information found on the BCCH is the complete *CELL IDENTITY*, which is sent in every fourth message. This element has no direct usage as far as the contents of the *Specifications* can be analysed, although it seems mandatory because of general radio regulations. Beside, it may be useful for network testing purposes.

6.3.10.6. Message Scheduling and Contents

The different items described above are grouped in four different messages for GSM900, bearing the non-informative names of SYSTEM INFORMATION TYPE 1 to 4. An additional RIL3-RR SYSTEM INFORMATION TYPE 2BIS has been defined for DCS1800, in order to cope with the extra length of the list of frequencies coming from the number of available frequencies. All these messages are sent according to an $8 \times (51 \times 8)$ BP cycle, which includes 8 message occurrences for a duration of about 2 seconds. The scheduling is performed according to the description of table 6.9. RIL3-RR SYSTEM INFORMATION TYPE 1 and 2 are sent at least once every 2 seconds, and TYPE 3 and 4 are sent at least every second.

The broadcast information is filled by the BSC. Most of the parameters pertain to system configuration, and as such are set by the OSS which indicates them to the BSC as part of the general configuration

Occurrence (modulo 8)	possible messages: SYSTEM INFORMATION TYPE
0	1 if sent at all, or 2, 3, 4, or 2bis
1	2
2	3
3	4
4	1, 2, 3, 4 or 2bis
5	1, 2, 3, 4 or 2bis
6	3
7	4

Table 6.9 – Scheduling of BCCH messages

The four types of BCCH messages are broadcast using a period of 8 occurrences, corresponding to a duration of about 2 seconds.

management. Some parameters must be managed dynamically, and their values may change as the result of local observation. This is the case mainly for the RACH control parameters. As far as the access class is concerned, it can be controlled by BSC only, by OSS only or by both; this is a choice of implementation.

SPECIFICATIONS REFERENCE

The concept of radio resource management as a specific area is introduced basically in **TS GSM 04.07** and in **TS GSM 04.08** (the radio interface application protocols specification), at the start of section 3.

Functional descriptions can be found on some topics in the 03 series. **TS GSM 03.09** deals with the handover function (almost only from the MSC point of view), and **TS GSM 03.13** deals with discontinuous reception.

Power control, measurement reporting and handover preparation are described in detail in **TS GSM 05.08**. Timing advance and synchronisation aspects are dealt with in **TS GSM 05.10**, where the different kinds of handover are described.

TS GSM 04.04 describes the contents of the L1-header in the SACCH messages.

The bulk of the specifications is in fact in the interface specifications, that is to say in:

- **TS GSM 04.08**, section 3, for the RIL3-RR protocol (a part of section 4 is also relevant, since call re-establishment is dealt with in TS GSM 04.08 as part of the mobility management), section 9.5.1 for the description of the messages, and sections 10.5.1 and 10.5.2 for the coding of their information elements;

- **TS GSM 08.58** for the RSM protocol;

- **TS GSM 08.08** for the BSSMAP protocol; and

- **TS GSM 09.02** for the MAP/E protocol. The relevant sections are mainly 5.5 (which deals with handover), plus bits in section 5.7 (operation and maintenance). For those interested in the MAP/VLR interface (the "B" interface, not addressed in this chapter), section 5.15 (paging and search procedures) is also relevant.

7

MOBILITY AND SECURITY MANAGEMENT

7

MOBILITY AND SECURITY MANAGEMENT

The radio resource management plane, studied in the previous chapter, covers signalling aspects very specific to the radio interface of GSM. The communication management plane, which we will see in the next chapter, covers on the contrary functions of little GSM specificity, implemented in a way very close to their counterparts in ISDN. In between lies some other functions, not specific to the radio interface, but for the moment peculiar to cellular networks. These functions are not covered by the ISDN protocol specifications, and are grouped in an intermediate functional plane.

This modelling by exclusion results in a rather disparate set of functions in this plane. Two coherent groups stand out. The first one includes the functions rendered necessary by the movements of the subscriber. This management of the mobility has given the overall name of the functional plane. It includes two facets: how the mobile station deals with a changing environment, and how the infrastructure manages the subscriber location data, to enable efficient establishment of calls towards GSM subscribers.

If the allowance for the mobility of the subscribers and its automatic management is a fundamental service in a cellular network, it also raises some technical problems, which are amplified in GSM by the notion of roaming between networks operated independently. The study of the mobile station side will be centred on the way to choose the cell and the network when a choice exists. The management of the subscriber location data on the infrastructure side is the other facet of the problem, and is concerned with getting and updating the location information needed to route calls toward a GSM subscriber who can move between cells or even networks.

The second group relates to the management of the security features of GSM, that is to say the protective measures against fraud or eavesdropping on the radio interface.

Both groups share common aspects when the implementation is looked at. They involve the same equipments, and they interact in some procedures. In both cases, the Subscriber Identity Module (SIM) and the Home Location Register (HLR) play an important role.

Some little related functions are added to these two groups, according to the modelling of the *Specifications*. They are dealt with rapidly in the last section of this chapter.

7.1. LOCATION MANAGEMENT

The mobility of the subscribers has major technical consequences in the infrastructure, but has also the important consequence that the service provided to a given subscriber changes as he moves, because of radio propagation (he may move out of coverage); because his subscription may be limited geographically; and because he may be served by different networks, providing different services.

In idle mode, the mobile station must choose *one* cell from which it expects to receive call attempts towards the subscriber. To this avail, the mobile station listens to the Paging and Access Grant Channel (PAGCH). It is said to be **camping on** this cell. The way the mobile station should choose which cell or which network it will camp on depends a lot on these service considerations. We will then present the mobility management functions starting with an overview, including the various factors influencing the service provided to the subscribers. Some are of administrative nature, another is propagation, and still others come from the system behaviour, such as congestion control. This will allow us to present the way the mobile station chooses (or helps the user to choose) between networks and between cells, while explaining the rationale behind these design choices. Only then will we tackle the infrastructure side, that is to say how the infrastructure keeps track of the subscribers' location.

7.1.1. THE FACTORS DETERMINING THE SERVICE

In the fixed telephone system, the service as it appears to a subscriber depends on which network the subscriber's telephone line is connected to, and hence on the location. For instance, the way a called number is entered, the price of the communication, the additional services that may be available, all depend on the location. Fixed network operators are working towards harmonisation, but the process is very lengthy. The consequences of these differences are however minor in the fixed system; in a system where the users move, the situation is quite different.

GSM has been designed to enable an international coverage, for instance users of the European GSM900 will be offered a European-wide system area. A subscriber will be able to get full access to the service from many countries in Europe with a single subscription. However, in order to adapt the system to various types of users, several levels of service may be offered on a geographical basis. For example, GSM operators may offer their customers different subscription choices, ranging from a (cheap) subscription limited to part of the country (regional subscription), to a (more expensive) subscription encompassing the whole area covered by GSM900 networks. This can go even further with SIM-roaming. If suitable agreements exist, the subscription may extend to networks of other types. Since the SIM interface is common to all GSM-based systems, the subscriber of a GSM900 network may obtain service from a DCS1800 network, and reciprocally, provided he uses an adapted mobile equipment.

In order to manage this flexibility, subscription in GSM is defined around the leading concept of PLMN ("Public Land Mobile Network"). This concept will be developed together with the description of the other administrative aspects.

Subscription on the one hand, and coverage limitations on the other, impact the services a user has access to when he moves. A first rough division distinguishes three levels:

- the "normal service", where a user can be called and can call, using all the services he has subscribed to (at least those the serving network can provide);

- the "limited service", where the only possibility left to the user is making emergency calls (typically in an area within coverage but not included in the subscription entitlement); and

- the "no service" case, typically when the user is completely out of coverage of any compatible network.

Let us see in more detail what determines the level of service given to a subscriber.

7.1.1.1. Administrative Aspects

The Notion of PLMN

The development of the Technical Specifications of GSM was concurrent with the organisation of a European-wide service by the would-be GSM operators. Most of the administrative features of the system have been deeply influenced by the context of European Telecommunications, and a number of these features have left their marks in the GSM Technical Specifications.

The European GSM system is divided into a number of separate operational networks, each being operated to a large extent independently from the others. Each of these networks is called a PLMN (Public Land Mobile Network—the term is much more generic than its specific usage in the *Specifications*). One of the restrictions, probably derived from the organisation of CEPT, is that the commercial coverage area of each PLMN is confined within the borders of one country. PLMNs of different countries may nevertheless overlap a little in border areas (radio waves have no respect for political borders). Most countries have several PLMNs, whose coverage areas overlap partly or completely: competition between operators is the rule of the game. Presently, licences for operating GSM900 or DCS1800 in Europe have been granted to typically two or even three operators per country. The operator may be a private company, a public company or an administration. The total number of European operators holding a GSM licence is in the order of 25 in 1992.

Roaming

It should be noted that the grouping of several operationally independent PLMNs in a single system open to roaming, in which the users can move and keep access to the service, is only possible if some conditions are met. First, the PLMNs must communicate between themselves. This requires standardised means of communication between PLMNs. Second, a subscriber must have a piece of equipment enabling him to access the different networks. As explained in the first chapter, GSM is designed to support MS-roaming, where the piece of equipment is the mobile station itself (thanks to the standardised GSM900 or DCS1800 radio interfaces), and moreover opens the door to SIM-roaming, where the piece of equipment is the SIM only.

The air interface and the inter-PLMN interface are the only standardisation requirements which are required to provide MS-roaming. For SIM-roaming, the standardisation of the inter-PLMN interface is still needed, but the only other requirement is a common interface between the SIM and the mobile equipment (SIM-ME interface).

Subscription

A GSM customer has a subscription relationship with a single PLMN. This specific PLMN is called the **home PLMN** of the subscriber. Service can be obtained from other PLMNs, depending among other conditions on subscription. In the *Specifications*, the term of visited PLMN (or VPLMN) is sometimes used to refer to a PLMN other than the home PLMN. In order to remove ambiguities, this term will be used in this book only when it is relevant to mention explicitly that the PLMN referred to is not the home PLMN of the subscriber. In other occasions, i.e., when the relationship with subscription is not relevant, the term "PLMN", or "serving PLMN" will be used.

Subscription information includes the set of services chosen by the user, as well as regional or international entitlements. Emergency call is the only service which is available anywhere in the system, whatever the subscription conditions. In fact, this service may even (in most PLMNs) be open to anonymous calls, i.e., calls for which no subscriber identity is mentioned. In this case, the SIM is not necessary, and a mobile equipment without SIM may be used for emergency calls. Access to normal service is of course a different matter: the home PLMN, the visited PLMN and subscription entitlements all have a role to play.

PLMN Accessibility

We will now look more in detail at the different conditions governing access to a given PLMN, taking into account roaming agreements and subscription limitations. To do this, we will follow a subscriber called Alan.

Access to the Home PLMN

Depending on his subscription, Alan can access normal service in the whole area covered by his home PLMN, or only in a part of it. The last case is referred to as a regional subscription. Presently, the home PLMN is the only one in which a regionally limited access is possible, on a subscription basis. There are no technical problems in doing otherwise, although the way the mobile station selects cells in the phase 1 *Specifications* takes this restriction into account. This will be changed in phase 2, and regional subscription over several PLMNs may be offered, if the commercial interest is worth the complexity of the administrative steps.

According to the *Specifications*, the regional limits for subscription zones are constrained by the requirement that they must not include parts of VLR areas. A VLR relates to one or several MSCs, each controlling a number of cells. The coverage areas of all these cells form the VLR area. It would have been more flexible for operators to enable the management of regional subscription on a smaller basis, and indeed the restriction would be very easily alleviated by a change of the inter-location register protocol. In this case the subscription zones could consist of location areas (introduced in Chapter 1). It might even have been better on a cell per cell basis, because location areas should be designed so as to balance location updating traffic with paging traffic, and this traffic optimum would have been easier to reach without mixing in administrative aspects such as the boundaries of regional subscription. However, there are no simple means to go lower than the location area level without a large increase in the technical complexity. Because the list of cells composing a subscription area constantly evolves with network extension and reconfiguration, the mobile station cannot know this list beforehand. Only Alan's home PLMN is aware of this information. The mobile station has to learn in real time whether normal service can be provided or not in a given cell, by some enquiry means. This would be very costly in terms of signalling, unless done at the same time as location updating, and this is possible if and only if subscription area borders are also location area borders. This was the choice, and the location updating procedure has also the function of verifying subscription entitlement in the location area.

Access to PLMNs of the Same Country

The rules for roaming in PLMNs of the same country as the home PLMN are one of the aspects of the *Specifications* which had continued to evolve during the elaboration of the standard. GSM900 phase 1 and DCS1800 phase 1 offer different views, not to mention later phases.

In GSM900 phase 1, all PLMNs other than the home PLMN are treated on the same basis for selection, independently from their country. Access to them may be allowed or not depending among other conditions on subscription choices, but always on the basis of "everywhere" or "nowhere" within the PLMN coverage area.

When DCS1800 phase 1 was standardised, there was a strong will to introduce some more controlled form of competition between operators of the same country. Operators asked for a mechanism by which subscribers of other PLMNs of the same country could be tolerated in a (typically low-density) area where a single PLMN provides coverage, whereas these same subscribers could be barred in other (typically high-density) areas where several PLMNs provide coverage. Such a mechanism would allow each operator to install only part of the infrastructure needed for full coverage of low-traffic areas, whilst providing overall a full nation-wide service for customers. This mechanism was introduced in DCS1800 as early as phase 1, and is called "national roaming". It allows users to access parts of PLMNs in the same country, these parts being chosen by agreements between operators and not by subscription. Relevant areas may in fact evolve with the deployment of the networks, without the subscriber being directly aware of these changes.

As for regional subscription, national roaming is performed on a location area basis, for the same reasons. The mobile station must learn by location updating attempts which location areas are acceptable or not.

In future phases, the mechanism of national roaming will be part of both GSM900 and DCS1800.

PLMNs of Other Countries

Access to PLMNs in countries other than the home PLMN country is possible for subscribers entitled to it by subscription. If so, the access is possible in the whole PLMN area. This feature is referred to as "international roaming".

Since new PLMNs can be created any time, the mobile station does not keep a list of the subscribed-to PLMNs. It will learn if a PLMN accepts the subscriber or not by attempting location updatings.

Mobile Station Constraints

Because the service offered to Alan may depend on the PLMN used for access, an important feature of the mobile station is how it chooses, or how it helps Alan in choosing, the serving PLMN when several PLMNs are possible. Moreover, a number of operational details may change when the serving PLMN changes, such as cost and local dialling format. Therefore, Alan needs to be aware (and, if possible, in control of the choice) of the serving PLMN.

PLMN selection is one of the subjects which have been under discussion even after the freezing of phase 1. There are some differences between GSM900 and DCS1800, and there are much more important differences between phase 1 and phase 2. The whole issue arises from the conflict between two opposite aims: quick response of the mobile station to a change in the configuration of available PLMNs on one hand, and battery life on the other hand. Any solution reflects a compromise between these two ends.

Quick MS Response to PLMN Availability ...

If power consumption was not at stake, the solution would be to let the mobile station explore continuously the whole GSM900 (or DCS1800) spectrum for BCCHs. It would then detect as soon as possible any new PLMN, and take it into account in its selection algorithm.

... Versus Low Power Consumption

The monitoring of the radio environment for beacon frequencies is an operation that costs power consumption. Now battery life is a very important aspect of a handheld mobile station: no subscriber would accept such a device if refuelling was needed every second hour... Therefore a trend is to try to limit as far as possible the search for neighbouring cells done by the mobile station.

The scheme adopted for GSM900 phase 1 is an extreme example of this method: the mobile station in normal service only monitors the cells in the same PLMN and in the neighbourhood of its serving cell. To this end, the serving cell broadcasts the list of the beacon frequencies used by neighbouring cells. This scheme is very efficient for limiting power consumption, but not for finding alternative PLMNs when the user moves into an overlapping area. The resulting behaviour is not optimum as seen by the PLMN operators: they would like the mobile station to make the right choice, in particular when the home PLMN becomes available for a mobile station who was being served by another PLMN.

In the cases where the list of frequencies used by neighbouring cells is not available, or more generally when a PLMN selection needs to take place, the mobile station has to search the whole spectrum.

7.1.1.2. Radio Considerations

Administrative considerations are not, by far, the only factor determining the service a given cell may provide to a given subscriber. A very important aspect is that the transmission between the base station and the mobile station must offer a good quality. Radio propagation considerations must then be within the criterion for the choice of the serving cell.

In idle mode, the only thing the mobile station has to do is to listen to the information broadcast by the cell it camps on (including the paging requests). If we were to take only this into account, then the best cell to choose would be simply the cell which has the best reception level or quality. But the rule in GSM is that when a mobile station wants to exchange information with the network, e.g., to set-up a call at the user's request, or to answer a paging, it must do it in the cell it is camping on. It could have been different: one could imagine that the mobile station selects the cell to communicate with at the very last moment. Of course, nothing precludes the network to perform a handover very soon after receiving a call request. This mechanism, called "directed retry", is however far from systematic in GSM networks: in the general case, the call will stay for some time in the same cell as was selected by the mechanism in idle mode.

Because of this choice, the camped-on cell should also be as close as possible to the best cell in which a potential connection will be set up. As a consequence, the quality of reception by the mobile station must not be the only parameter taken into account, but also the quality of reception by the base station. This cannot be measured directly in idle mode, since the mobile station is not transmitting, but this can be derived from reception measurements and from the maximum power the mobile station can use for transmission.

A criterion used to choose a cell in idle mode combines the reception level of the mobile station on the beacon frequency, the maximum transmission power of the mobile station, and several parameters depending on the cell (and broadcast on the BCCH). The exact algorithm is described later on (see the description of the C1 criterion, page 453).

7.1.1.3. System Load Control

Congestion is a risk which exists in any telecommunication system. Mobility changes slightly the bases of the problem. The traffic variations are of bigger amplitude, since they come not only from the change of traffic per subscriber, but also from the movements of the subscribers. For instance a sport event may see a huge concentration of mobile subscribers in one small place at the same moment. Another point is that, because subscribers are not physically linked to a cell, they may "move" from a congested cell to another if the second one can provide the service. Another aspect to look at in connection with cell selection is then the way in which a network can control the traffic distribution among cells. Two mechanisms exist, one of which impacts cell selection.

The network can bar completely a cell against access by all normal subscribers. This "barred" status is indicated in the information broadcast by the cell, in the CELL_BAR_ACCESS flag described in Chapter 6. Such a mechanism is used when a BTS is unable to operate properly, e.g., for maintenance purposes. It may also prove handy when the operator sets up new cells and performs tests on these cells before opening them for normal operation. Test mobiles (which ignore the "barred" status, and do not then conform to the *Specifications*) are able to establish connections with such cells for test purposes. Another potential application of cell barring concerns cells restricted to handover access.

Cell barring is an all or nothing control mode, which must be taken into account by the mobile stations for cell selection in idle mode. GSM also includes a subtler mechanism, called the access class mechanism, which allows selective access of certain mobile stations to certain cells. Its purpose is to cope with abnormally high traffic load or emergency situations, but it does not influence cell selection. For example, a mobile station may perfectly select and camp on a cell which will not at this very moment accept a connection request from this particular mobile station (even for location updating purposes) because of a temporary high load. Allowing such a cell to be nevertheless selected avoids the congestion situation to spread in neighbour cells. This is an example of a choice in the specifications where a global optimum, evaluated on several cells, is favoured against the local improvement. The topic of access class has already been developed in the appropriate place, in the Radio Resource management chapter.

7.1.1.4. Paging and Location Areas

Finally, an important point to take into consideration for the PLMN and cell selection is that the network must be able to route calls toward the subscriber. The infrastructure must know some minimum information concerning the location of the subscriber to do so. This information can be provided only by the mobile station, and the service provided to the user depends on the consistency between the location currently assessed by the infrastructure and the cell chosen by the mobile station. It is then necessary to look in general terms at how the infrastructure deals with calls toward GSM subscribers.

In order to avoid a waste of signalling, the system is so designed that a subscriber is only looked for (paged) in a few cells of the system when a call toward him has to be established. Cells are grouped in location areas, and a mobile station is typically paged only in the cells of one location area when an incoming call arrives (see the basic concepts of location management, in Chapter 1). Therefore, the mobile station must inform the system of the location area in which the subscriber should be paged. It does so by a location updating procedure. The network, on the other hand, must store the present location area of each subscriber: this storage is inside location registers as will be detailed together with the description of the location updating procedure. Each change of location area puts an extra load, not only on the radio path, but also on the infrastructure equipments; the cell selection mechanism therefore includes some features to limit the number of location updatings.

Location Areas

For obvious technical reasons, the *Specifications* impose that each location area be a subset of the cells of a single PLMN. In fact, because of the way a mobile terminating call is routed in the network, a location area must include cells managed by a single MSC (see figure 7.1). This restriction to an MSC could have been avoided, but only at the price of complex procedures. Within these constraints, the operator has complete

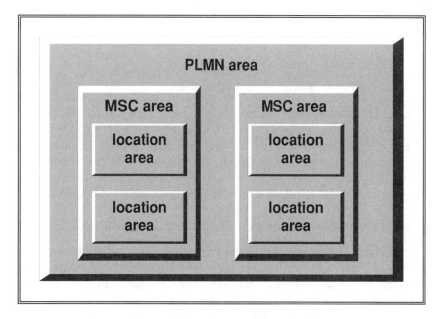

Figure 7.1 – Location area vs. MSC and PLMN areas

A location area may only contain cells of a single MSC, in a single PLMN.

freedom to allocate cells to location areas. The goal of this operation is to minimise resource consumption, taking into account the signalling load on the radio path (both from paging and location updating) as well as the processing load of the equipments.

Location Updating

If a mobile station wants to obtain normal service from a cell, and in particular to receive calls, it must make sure its subscriber (represented by the SIM) is registered in the location area of this cell. The registration state of the subscriber, except in network failure cases or after some very long inactivity period, can only be changed at the initiative of the mobile station. The outcome of the last registration attempt is stored in the SIM, as well as the identity of the location area. If this storage indicates success, normal service is assumed automatically if the mobile station camps on a cell of the same location area.

The situation is different if the mobile station is switched on in a different location area than the one where it was last successfully registered; when the mobile station moves into a place where a cell from another location area is better suited; or when the mobile station tries to get normal service in a different PLMN. In all these cases, the mobile station must attempt to register the subscriber by performing a location updating procedure prior to camping on the cell in normal service.

Status after Location Updating

Several outcomes of location updating are possible. As we have seen, the location updating procedure does more than just tell the network where the subscriber is. It also provides the network with a mechanism to tell the mobile station whether the cell can grant normal service, taking into account subscription limitations, national roaming restrictions, and so on. The best outcome of the procedure is when the procedure is run correctly and the network indicates that normal service is possible. The subscriber is then registered correctly in the network, and the mobile station is allowed to camp on the cell in normal service mode.

Several outcomes are possible when the registration is not successful. We must distinguish the cases where the procedure is run correctly (i.e., the network provides a meaningful answer, possibly denying normal service) from the others (i.e., the network does not answer or answers that it cannot give a meaningful answer). It is worth mentioning that the *Specifications* put all these cases together in an "abnormal cases" category, but there is quite a difference between no answer and a bad answer!

Meaningful negative outcomes can be of three kinds, corresponding to the different limitations presented in the administrative aspects section:

- the cell may belong to a PLMN not supported by the subscription. Mechanisms are specified ("forbidden PLMNs" list, see further on) so that cells of this PLMN will not be tried anew, except on explicit request from the user. A new PLMN (and hence a new cell) will be looked for if normal service is sought.

- the cell can be in a location area not suitable because of regional subscription. The rule is then that the mobile station must stay in the cell (and in the home PLMN), but only limited service can be provided to the user. There are some reasons for this choice: because of the implicit assumption that, when a subscriber is only entitled to regional subscription in the home PLMN, access to other PLMNs of the same country is most probably not allowed by the operator: therefore looking for other PLMNs is not essential. However, this choice leads to a peculiar behaviour of the mobile station when the subscription covers a part of the home PLMN *and* foreign PLMNs. In such cases, the mobile station will look for these other PLMNs only when leaving the coverage of the home PLMN, not when leaving the subscribed-to part of it.

- in DCS1800, the cell can be in a location area which does not accept roamers of another PLMN of the same country. As with non-subscribed-to PLMNs, mechanisms are included to prevent further attempts in the cells of the same location area. In addition, the mobile station will look immediately to see if the home PLMN is available.

An important point is that in some cases (if reject comes from the home PLMN or from a PLMN which is connected to the home PLMN), the subscriber is effectively deregistered in the HLR: it is marked in the network as not being in a state where calls to his destination can be established. Such calls are either directed to an announcement, or forwarded to another destination if the subscriber has activated call forwarding unconditional or call forwarding on not reachable. In the other cases, the subscriber keeps the same HLR registration state he had before the attempt. However he is considered as deregistered by the mobile station, since there is no indication from the network about what happened with the HLR state.

When the mobile station receives one of the negative answers listed above, it may look for other PLMNs. If none of the found PLMNs are able to provide normal service, the mobile station goes to the limited service state.

Cases of procedural failure are even more complex. Details can be found in the *Specifications* and will not be dealt with here. Basically, the mobile station will not try other cells (unless the radio criteria so determine) despite the lack of knowledge about the service that can be provided to the user: this is an important point. There is a first phase during which the mobile station will try several times to get an answer from the network. During this phase, the mobile station will behave as if it was granted the same service as before. If all attempts fail during this phase, the mobile station enters a special state, in which it does not assume any registration state in the network. It tries now and then a location updating procedure, to get out of this unsatisfying state. The mobile station does not reject a request from the user, but this request triggers the mobile station to start a location updating, and the mobile station will eventually reject the user's request if the network still does not answer positively.

7.1.2. CELL AND PLMN SELECTION

The cell and PLMN selection mechanisms are almost totally specified in the *Specifications*. We have seen in the previous section the different factors to take into account, as well as some general mechanisms. Let us see now the detailed process.

7.1.2.1. PLMN Selection

Though the PLMN selection process (as well as cell selection) affects only the mobile station, it is specified in a fair amount of detail in the *Specifications*. One of the reason is that the SIM intervenes, and the SIM-ME interface has to be specified. Other reasons are to harmonise somewhat the behaviour of the mobile station, so that it can be predictable when a user changes his equipment, and also to avoid the possibility that some implementations bias the choice of PLMN and thus the competition between operators.

However, the *Specifications* are not totally constraining, and have to be completed for some marginal cases by the mobile station manufacturers. We will mention these latitudes, with some of the possibilities. What the mobile station really has to do is to select a cell.

There is however an important difference between the selection of the serving PLMN, and the selection of the serving cell. The first is under control of the user, whereas the second is fully automatic. The PLMN selection is important for the user when the service finally obtained is normal service. In the limited service case, the choice is of little importance, and in the case where no PLMN can be found the selection of a PLMN is of no immediate application. We will look at the three cases in turn, before summarising the user's view.

The Normal Case

If already in normal service mode the mobile station only looks for the cells in the serving PLMN, independently from other PLMNs. A change of PLMN can occur at only two occasions: when the user decides so, and when the mobile station finds out that the serving PLMN can no longer provide normal service (e.g., because the mobile station is leaving the PLMN coverage area). In those cases, the mobile station will search for cells in the whole spectrum, to find which PLMNs cover the location. Access to some of these PLMNs is then tried, according to the PLMN selection method. Two methods are provided for the choice of the PLMN to try, the choice being left to the user: the **manual mode** and the **automatic mode**.

Several aspects are common to both modes. For example, the home PLMN is set as the PLMN to try at switch-on, independently from previous history (even if the mobile station knows, from the SIM, that the subscriber is currently registered in another PLMN). Another common aspect between the manual and automatic modes is the *"forbidden"* PLMNs list—the first one in a series of 3 PLMN lists to juggle with! The actual list of PLMNs accessible to a subscriber according to his subscription may change as new PLMNs are opened (or closed!) for service. The list of these PLMNs cannot be stored once and for all (for instance in the SIM). Instead, it has been chosen to build dynamically a list of PLMNs which are *not* accessible according to subscription. This list is updated according to the result of access attempts performed by the mobile station, and is stored in a non-volatile memory in the SIM (and hence is not lost when the mobile station is switched off). The list is limited by the *Specifications* to 4 entries, but nothing precludes a longer list to be stored in the ME (therefore possibly lost at switch-off).

This so-called *forbidden PLMNs* list includes PLMNs which are not subscribed to. They are not really "forbidden", since they could be used for emergency calls, and in manual mode the user is perfectly authorised to select one of them. When a mobile station attempts to update its location in a PLMN to which the user's subscription does not authorise access, the network will tell the mobile station so, and the PLMN identity will be put in the list, ejecting if needed the oldest entry. The *forbidden PLMNs* list is used for PLMN selection both in manual and automatic mode, though in different ways.

Another list, this time stored in the mobile equipment (not in the SIM) intervenes a little in the PLMN selection process in automatic mode. It contains the identities of the location areas which have rejected access before because of national roaming limitations. Cells belonging to the location areas in this list can no longer be candidates for selection, and a PLMN for which all the found cells are in this category cannot be tried in automatic mode. This list is filled as a result of location updating rejections with a suitable cause (but the PLMN is not put in the *forbidden PLMN* list), and is erased when the mobile equipment is switched off, or the SIM removed.

Manual Mode

In manual mode the list of PLMNs the mobile station has found as potential candidates for providing normal service is presented to the user, whether or not they are in the *forbidden PLMNs* list. This list of *found PLMNs* is displayed using explicit names (such as D1-Telekom or D2 Privat for the German PLMNs, DK TDK-Mobil or DK Sonofon for the Danish PLMNs, etc.) and with an explicit mention telling the user whether the PLMN is in the *forbidden PLMNs* list or not. Normally the user then chooses one of the PLMNs of the *found PLMNs* list as the selected PLMN.

The explicit names of the PLMN cannot be found in the *Specifications*, though they are part of the mobile station specifications. The list is distributed by the GSM MoU, and regularly updated. There is no requirement on the update of already manufactured (and sold) mobile stations, or on the way in which they should visualise the names of new PLMNs who have been introduced meanwhile. Typically, if the knowledge of the network commercial name is unknown, the mobile station will display the country initial together with the numerical network code, for instance "DK Network 05".

The user can choose any PLMN in the *found PLMNs* list, even if it is also part of the *forbidden PLMNs* list. This possibility might be useful after a change of subscription category: for example, if Alan requests his subscription to be changed from "home PLMN only" to "all PLMNs" while he roams abroad, he might want to force his mobile station to attempt location updating on a network previously known as "forbidden", in order to unlock the situation.

Since, at switch-on time, the mobile first looks for the home PLMN and does not (in manual mode) take into account the PLMN the subscriber is registered in, some situations can become quite irksome when the user stays in a foreign country for some time. Automatic mode is in such cases more appropriate.

Once the user has selected a PLMN, the mobile station will attempt to get normal service on this PLMN. Several outcomes are possible, as we have seen. It may well succeed: fine! The PLMN may also be indicated as not allowed by subscription, and will therefore join the *forbidden PLMNs* list. Unfortunately, the other cases are not so simple, and not always clearly specified. In manual mode, these cases can be treated generally by letting the mobile station do its best to get normal service (possibly by trying several cells in different areas). After some time, the PLMN list is presented again, including the PLMN on which the attempt failed: it is up to the user to choose another one... If everything fails, the mobile station should at some stage reach the "limited service" described below.

Automatic Mode

In automatic mode, the mobile station will choose which PLMNs to try all by itself. The automatic mode is based on the existence of another list of PLMNs, the *preferred PLMNs* list, which is stored in a non-volatile memory in the SIM. This list, capable of holding at least 8 entries, includes a number of PLMN identities in order of preference and is under the control of the user. The most preferred is usually the home PLMN, but it is nowhere specified whether the home PLMN should appear at first rank in the list or not, whether the user may choose to have a list with a different PLMN as "top of the list" and what happens in this case. The list may originally be filled in by the home PLMN operator, during the SIM personalisation process. It can afterwards be modified at will by the user through a mechanism to be specified by the mobile station manufacturer. No automatic modification of the list can take place.

When a PLMN selection takes place in automatic mode, the PLMNs are tried starting with the first PLMN in the list of *preferred PLMNs* which is in the list of *found PLMNs* and not in the *forbidden PLMNs* list. The treatment of one of the failure cases is clear: if the PLMN is indicated as not allowed by subscription, its identity is put in the *forbidden PLMNs* list and then cannot be automatically selected again, except in the (rare) case where more than 4 PLMNs are found, none being allowed by subscription: in that case, the mobile station may rewrite the *forbidden PLMNs* list in a cyclic manner! In the other failing cases, the mobile station must try the other possible PLMNs, in order of preference for those in the *preferred PLMNs* list. It is not clear what is the status of PLMNs found by the mobile station but not included in the *preferred PLMNs* list; the most literal interpretation is that they cannot be selected. This raises different problems, for instance in the case where the list is empty. It seems more logical for the mobile station to also include them in the choice process, according to some ordering rule, for instance randomly. The situation where none of the *found PLMNs* can provide normal service is not clearly specified in the *Specifications* either, but it seems logical that the mobile station should go to the limited service state in that case. The mobile station should keep some record of its attempts to avoid a deadly loop in which it will stubbornly attempt to get service from a PLMN without success when another would grant normal service.

As in manual mode, the user can force PLMN selection at any moment. The process is then the same as described above when the serving PLMN ceases to provide normal service, except that the serving PLMN is in the list of *found PLMNs*. The mobile station will then stay as the selected PLMN, except if a PLMN with higher preference rating has appeared since the last PLMN selection.

The Limited Service Case

In limited service mode, the only service available is emergency call. The theory is that all PLMNs can provide this service. The purpose of selecting a PLMN is then fairly limited. It could however prove useful for two main reasons: in border areas, the user might want to control the choice of the country where a potential emergency call would be routed; besides, the user could choose a PLMN not for immediate use, but to be the one used for normal service as soon as the mobile station can find a cell of this PLMN.

The *Specifications* specify that the choice of the cell in limited service mode is done generally independently from the PLMN or the location area. This general rule admits a noticeable exception, when the home PLMN is available for limited service, and no other PLMN is available for normal service provision.

In parallel, the mobile station monitors continuously (though possibly at a slow rate to save its battery) the 30 strongest carriers it receives for new PLMNs. The phase 1 *Specifications* leave open the behaviour of the mobile station, especially in manual mode, when a new PLMN is found. In automatic mode, the mobile station will perform a PLMN selection and try to get normal service on a newly found PLMN which is not in the *forbidden PLMNs* list (and for DCS1800 if the location area is not forbidden either). In manual mode, one behaviour would be to ask the user every time a PLMN pops up; this could however lead to annoying situations in border areas where PLMNs may appear and disappear very frequently. Another approach for the manual mode could be to stay in limited service mode as long as some PLMN chosen by the user stays unavailable, whatever other PLMNs pop up, which is not very much more satisfactory. It is hoped that mobile station manufacturers will find a satisfying compromise.

When the home PLMN is available, the limited service mode can happen only when the user has a regional subscription and is in a part of the home PLMN area where he is not entitled to normal service. In this case, the *Specifications* impose a cell selection mechanism identical to the normal service state, i.e., limited to the cells of the PLMN. The mobile station tries each new location area of the home PLMN that pops up so as to find one granting normal service, but will go on another PLMN (on which the subscriber may be entitled to service) only when leaving the home PLMN coverage area, or when asked by the user. The mobile station does not store any list of the location areas in which the user is or is not entitled to normal service. Protection against trying again and again the same location area is obtained by the hysteresis mechanism for the change of location area, as in normal service mode (see further on).

The No Service Case

If no cell can be found at all, the 124 carriers in the band (374 for DCS1800) must be monitored, not just the 30 strongest carriers found as in the other cases. Otherwise, the mobile station behaves as in the limited service case, except that it cannot accept even emergency calls.

What the Users see

The mobile station must indicate the service state to the user. Different levels of precision seem allowed. Typically the information provided by the mobile station to the user distinguishes the normal service, the limited service and the no service cases. We will see further along that in cases of signalling failure an additional state exists, but it is not necessarily indicated to the user. In addition, the mobile station must be able to indicate to the user, possibly on request, the serving PLMN when in normal service state. The PLMN identity is given in clear text when possible, including country initials and network name.

For the users, PLMN selection is first a choice between manual mode and automatic mode. Some control must be provided to allow the user to know whether the mode is set to automatic or manual, and to change the mode.

In automatic mode, the user usually does not intervene in the selection mechanism. He is however able to direct the process by two actions: he controls the *preferred PLMNs* list and he can ask for a forced PLMN selection at any time, in which case the mobile station will search the whole spectrum and select a PLMN, possibly the same as before. In order to enable the user to control the *preferred PLMNs* list, the mobile station must provide commands to display the list and to edit it. Whether a modification of this list is taken immediately into account (e.g., by a forced PLMN selection) or not is left open to mobile station manufacturers.

In manual mode, the user has a total control, but is solicited in each case, maybe too often in some situations. He is asked for a PLMN choice in several instances. This happens after switch-on (or SIM activation) if the home PLMN is not available. This happens also when the mobile station moves out of coverage of the serving PLMN. In addition, this may happen in DCS1800 when the mobile station was in a visited PLMN of the home country, and cannot find any more a cell that accepts it. In this case, a forced PLMN selection happens, and hence the user will be asked to choose a new PLMN. Finally, a prompt to select a PLMN may appear when not in normal service mode, and a new PLMN is found. As in automatic mode, the user can in addition force a PLMN selection at any time.

Whatever the triggering event, the list of all found PLMNs is presented to the user, including those in the *forbidden PLMNs* list, though with a distinguishing mark. The order of presentation was a topic raising some discussion between operators, since it was felt that this order may influence the user's choice. The issue was settled on the specification that the order should be random.

7.1.2.2. Cell Selection

As explained in the requirements, only cells can be chosen with which transmission will be a priori of at least minimum performance, and the cell choice should aim at maximising the transmission quality. The radio criteria therefore play the foremost part in cell selection. Thus, before describing the actual cell selection algorithm, we will first study them.

Radio Criteria

In order to maximise transmission quality, a criterion has been defined, which takes into account the level of the signal received by the mobile station on the beacon frequency, the maximum transmission power of the mobile station and some parameters specific to the cell. This criterion is named *C1*. (There is no *C2* in the phase 1 *Specifications*, but it will appear in phase 2.)

Description of the C1 Criterion

C1 is defined as follows:

$$C1 := (A - \text{Max.}(B, 0))$$

$$A := \text{Received Level Average} - p1$$

$$B := p2 - \text{Maximum RF power of the mobile station}$$

(all values expressed in dB)

The two parameters *p1* and *p2*—called respectively *RXLEV_ACCESS_MIN* and *MX_TXPWR_MAX_CCH* in the *Specifications*—are broadcast by the cell. The first one can take a value between −110 dBm and −48 dBm, the second between 13 dBm and 43 dBm in GSM900 (in DCS1800, the range is different). The second parameter has another independent usage: it represents the maximum transmission power a mobile station is allowed to use on the RACH. Because the range of the mobile station maximum transmission power is 29 to 43 dBm, only this sub-range of the second parameter is useful, whether for *C1* or as a maximum transmission power on the RACH.

C1 is used as follows. When looking for cells, either when looking for neighbour cells in normal service mode, or when searching PLMNs, only cells of positive *C1* (calculated from the *p1* and *p2* broadcast by each cell) are taken into account. When a choice between cells has to be made, the cell of best *C1* is chosen among those equivalent for other criteria. As a consequence, *C1* determines two things:

- the coverage limit of each cell taken in isolation, in the sense that outside the area where C1 is positive, the cell does not exist for the mobile stations;

- the boundary between two adjacent cells for selection in idle mode, determined as the locus where *C1=C1´*. The boundaries

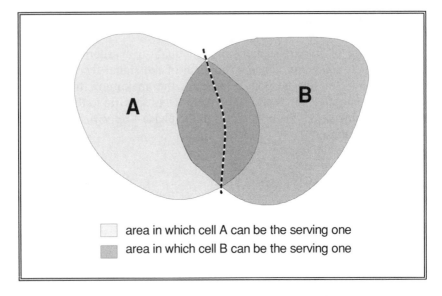

area in which cell A can be the serving one
area in which cell B can be the serving one

Figure 7.2 – Cell boundaries according to $C1$

The figure shows the boundaries of two cells A and B,
according to the values $C1_A$ and $C1_B$ of the $C1$ cell selection criterion.
The dashed line is the locus where $C1_A = C1_B$.
Since $C1$ depends on the mobile station maximum power, these boundaries
differ from one mobile station class to another.

with all adjacent cells determine a second cell limit, usually inside the area delimited by $C1=0$.

Figure 7.2 shows an example of two cells, with their $C1=0$ limits, and the line of equal $C1$'s. Two points are important to keep in mind with these limits. Because the maximum transmission power of the mobile station intervenes in $C1$, the limits are different for different mobile station classes. Second, that other cell limits exist, the ones determined by the selection of the cell for handover. It is up to the operator to choose $p1$ and $p2$ to obtain the correct compromise between cells boundaries, traffic and quality of transmission for the different classes of mobile stations, as well as consistency with the handover algorithms and parameters.

Criteria Other than C1

Because of its radio-electric nature, $C1$ varies quickly and to some extent randomly around a mean value depending on the location and on the movement of the mobile station. This means that the mobile station would often change between cells if $C1$ were the only selection criterion.

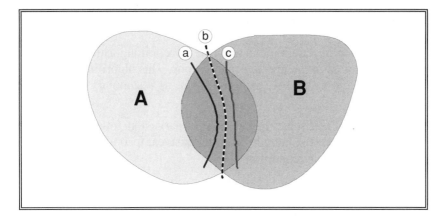

Figure 7.3 – Impact of *CELL_RESELECT_HYSTERESIS* on cell boundaries

The boundary between two cells A and B differs depending on the requirement
for a mobile station to perform location updating:
(a) A and B belong to different location areas,
and the mobile station is registered in the location area of cell B;
(b) A and B are in the same location area;
(c) A and B belong to different location areas,
and the mobile station is registered in the location area of A.

This is acceptable if the cells are equivalent, but not otherwise, for instance between two cells belonging to different location areas. This is why the comparison between $C1's$ is modified to be biased against the cells on which camping on must be preceded by a location updating. This is obtained by a handicap, added to the $C1$ of these cells. The value of this handicap is not fixed, and is broadcast by each cell. It is called *CELL_RESELECT_HYSTERESIS*, though it is not strictly speaking a hysteresis value (the real hysteresis is the sum of the values in the two cells).

A consequence of this specification is that the boundaries between adjacent cells belonging to different location areas is not the same whether the mobile station goes from one cell to the other or the contrary. The hysteresis in terms of received level is transformed by the variations of propagation into a geographical hysteresis. Taking the same cell configuration example as before, figure 7.3 shows the cell boundaries as obtained when such handicaps are used. We have three different boundaries for a given mobile station class: one for mobile stations going from cell A to cell B, one for mobile stations going from cell B to cell A, and a third for mobile stations which do not apply the bias (e.g., foreigners not entitled in the PLMN by subscription). Mobile stations located in the area between lines a) and c) can be attached to either of the cells, depending on the direction they come from.

The Cell Selection Algorithm

The different requirements the cell selection algorithm has to meet have been detailed in the previous paragraphs. This algorithm is specified in the *Specifications*, and we will present it, as a summary of the points seen so far.

The aim of cell selection can be summarised as follows: in order to get normal service, the mobile station must camp on one of the cells fulfilling the following conditions:

- a SIM must be inserted, and the corresponding subscriber registered in the location area the cell belongs to;
- criterion $C1$ for the cell must be higher than 0;
- the cell must not be barred.

Among the cells which comply to these three requirements, the chosen cell must fulfil the two additional conditions:

- the cell's $C1$ must be higher than the $C1$ of any other cell found by the mobile station in the same location area;
- the cell's $C1$ must be higher than the $C1$ of any other cell found by the mobile station in different location areas of the same PLMN, corrected by the applicable handicap factor.

As can be noted, potentially better cells in PLMNs other than the one the mobile station is registered in are not taken into account. This corresponds to the decision that PLMN selection is triggered only by the user, or when the mobile station leaves the coverage of the selected PLMN.

Cell Selection in Normal Service State

Having this goal in mind, the behaviour of the mobile station can be easily derived. Let us start with the mobile station being in normal service state (hopefully the likeliest state). In this state, the mobile station receives a list of frequencies (broadcast by the serving cell) indicating where to look for the beacon channels of the neighbouring cells of the same PLMN. The mobile station must then find these beacon channels one by one and get their synchronisation information in order to decode some of the broadcast information they carry; this information enables the mobile station to check the PLMN, to know if the cell is barred or not and to obtain the identity of the location area the cell belongs to, as well as to get the various radio parameters so as to compute $C1$. Only acceptable cells (i.e., non-barred and of positive $C1$) of the right PLMN will be taken into account. The mobile station can then compare the $C1$ of these cells with the $C1$ of the current cell. All this process takes place in parallel with the periodic reception of the paging channel on the current cell.

If the mobile station finds a better cell in the same location area, it changes to this cell and goes on with the process of listening to the paging channel (of the new cell) while monitoring the beacon channels in the new list. However, if the mobile station finds a better cell in a different location area of the same PLMN, having taken the location updating bias into account, it changes to this cell. At this very moment, the mobile station is no more—strictly speaking—in the normal service state: calls to it will in general not reach their destination. The mobile station tries immediately a location updating procedure to warn the network of its new location. Most of the time, the mobile station is granted normal service by the network, and is back to the normal service state in the new location area after a few seconds.

Cell Selection at Switch-on Time

Let us now describe other cases. An important one is the initialisation: how to get normal service in the first place after switch-on. The first PLMN to try (if found) is the home PLMN, according to the *Specifications*. The mobile station must search for non-barred and C1-positive cells within this PLMN. Without any information, the mobile station has to search the whole spectrum for beacon channels. This can be a lengthy operation, in particular in DCS1800 where 374 frequencies are supported.

The *Specifications* include a mechanism to help the mobile station in such conditions: a list of frequencies to look for can be stored in a non-volatile memory (in the SIM). The *Specifications* are however not clear about which frequencies the mobile station should put in that list. To be consistent with the imposed search for the home PLMN at switch-on time, it should be the neighbouring cell frequencies broadcast by the last cell of the home PLMN on which the mobile station camped (whether for normal service or not, and whether or not the mobile station was registered on other PLMNs meanwhile).

Whatever the case, the search results in a list of acceptable cells, with their $C1$. If this list is not empty, the mobile station chooses the cell of best $C1$. Also stored in a non-volatile memory is the identity of the location area (if any) in which the mobile station knows it is registered. If the chosen cell belongs to that particular location area, the mobile station goes immediately to normal service, possibly indicating its presence to the network (see the description of the "IMSI attach/detach" mechanism, page 474). Otherwise, if the chosen cell is in a different location area, the mobile station camps on the cell and starts a location updating procedure immediately.

If no acceptable cell of the home PLMN is found, the mobile station acts as if it were leaving the home PLMN coverage area.

Cell Selection at PLMN Change

Another special case is when the mobile station has to look for the available PLMNs, for instance because the mobile station has moved out of the coverage of the PLMN previously providing service, or because of a forced PLMN selection by the user. The process is similar to the switch-on case, except that the mobile station has no information whatsoever about which frequencies to search: it must search the whole spectrum. The mobile station proceeds in two steps: first, it searches all GSM (or DCS1800) carriers, then it selects the 30 strongest ones to obtain the information they broadcast: the PLMN to which they belong, whether they are barred or not for access, and the parameters controlling $C1$. The mobile station can then establish the list of the acceptable PLMNs, that is to say those for which it has found at least one acceptable cell. This will result in a *found PLMN* list as described some pages before. When one of the PLMNs in the list is chosen, the mobile station will access the acceptable cell of best $C1$ within those previously found in this PLMN and request a location updating.

Cell Selection in Limited Service Mode

Now we have to address the cases where normal service cannot be granted, but limited service is possible. This happens when the subscriber is not entitled to normal service in any of the found PLMNs. If the home PLMN is acceptable, i.e., if access to the home PLMN is locally prevented by subscription rather than radio propagation, the cell selection is the same as for the normal service state. Otherwise, the mobile station selects the acceptable cell of best $C1$, irrespective of the PLMN or the location area of the cells, and hence without applying the location updating bias. The mobile station searches continuously the whole spectrum for new cells, in order to find an acceptable PLMN as soon as possible. When such a PLMN is found, it may be selected and then the mobile station will try to get normal service on this PLMN.

7.1.3. ARCHITECTURE

Selecting a PLMN and a cell is but one side of the management of mobility. The goal of the network as far as location management is concerned is to prepare for the routing of calls toward the subscribers, taking account of their movements. To that end, the network must memorise for each subscriber (very precisely for each SIM) whether he is known to be in some place or not (he is said to be **registered**), and if so, in which location area. This information is retrieved when a call toward the user must be set up, as will be explained in Chapter 8. Because a location area is mandatorily in GSM wholly included in a single MSC area, the stored information is sufficient for routing the call up to the MSC which will be in charge of the communication. Furthermore, the knowledge of the precise location area (there may be several location areas within an MSC area) allows to restrict the paging to the corresponding cells.

A simple solution to the basic location management issue could consist in storing in a database the identity of each subscriber together with an indication on whether or not he is registered, and if so, where to find him.

Indeed, the canonical architecture of GSM identifies such a database, the Home Location Register (HLR). This function is separated from the routing function itself, which consists in choosing and reserving circuits to obtain a continuous connection between users that desire to be in communication. In GSM, the main actor on the mobile user side for routing and communication management functions is the Mobile services Switching Centre (the MSC). But the canonical architecture is a bit more complex, and before describing it, it is interesting to look at to the reasons why.

Every telecommunication system includes a database containing a variety of information concerning each subscriber, such as the subscription limitations, the services subscribed for, the states of the supplementary service activation, or the information needed for the management of the charging information. In GSM, the same information exists, plus some which is specific such as the information related to the confidentiality functions.

In a fixed network, each subscriber is connected to one local switch, for a long time. Every call involving this subscriber, whether an originating or a terminating call, goes through this switch. This is then the natural place for the storage of the subscriber related information. In a system dealing with moving subscribers, there is no such natural place for the storage of subscriber parameters. However, the two kinds of data to be stored (location information and subscriber data) call for a common storage solution. This is the choice made in GSM, and the HLR is the database for both sets of information.

If location information is needed only for the establishment of mobile terminating calls, the rest of the information is needed at various moments during any call. Basically, it is the visited MSC, the one in charge of a mobile subscriber engaged in a call, which needs these pieces of information. Then an important signalling load would result if the MSC had to interrogate the HLR each time it needs some piece of information. To avoid this signalling load, the data record of a subscriber is copied in a database close to the MSC while this subscriber is registered in a location area controlled by the MSC. The database, the VLR (Visitor Location Register), will be ignored for the moment as an entity separate from the MSC: we will speak of an MSC/VLR. The distinction will be dealt with later, and our approach explained.

This capacity of temporary storage in the MSC/VLR allows some distribution of the function of location information storage. The HLR needs only store information concerning the MSC/VLR in which area where the subscriber currently is. The identity of the precise location area is stored in the MSC/VLR, together with a copy of the remaining subscriber related information. This is sufficient to get the routing information needed to deal with an incoming call, and this is all the HLR needs.

This temporary storage in the MSC/VLR introduces new functions. The subscriber information has to be copied when the subscriber enters a new MSC/VLR area. Conversely, the corresponding record has to be erased in the previous MSC/VLR in which the subscriber was registered. Some mechanisms are needed for maintaining the consistency between what is stored in the HLR and what is stored in the MSC/VLRs, including the case of a failure resulting in a loss of stored information.

7.1.3.1. Functions

As defined in the *Specifications*, the HLR is basically an intelligent database used to store the location information and the subscriber related information needed for providing the telecommunication services. The HLR has no switching capability. It is connected to the other entities of the network and switching sub-system (NSS) through signalling means, as discussed in Chapter 2. The HLR is not a simple database which can accept only "store" or "retrieve" orders. In fact, the HLR completely manages the location information in the network: for instance, it has to tell the old MSC/VLR to erase a subscriber record when this subscriber is registered under a new MSC/VLR.

Functionally, we could say that there is a unique HLR function system-wise, possibly distributed on several equipments. In practice, be it only for operation reasons, one HLR function is implemented in each PLMN. There again, the HLR function for a PLMN can be implemented in a single equipment or distributed among several equipments. Both approaches are allowed, and used. Note that the usage of the term HLR may refer to the function, possibly encompassing several equipments, or to a single equipment. In most of the cases, this ambiguity causes no understanding problem.

The HLR has many different roles. What concerns us in this section is the management of the subscriber mobility. Functions of the HLR related to, e.g., the management of the confidentiality data, or the management of the supplementary services shall be described respectively further on in this chapter, and in Chapter 8.

The Visitor Location Register

While we have introduced the term VLR, the corresponding concept has been somewhat masked. Most readers will have also noted that the VLR was not even allocated an icon. We have to answer for this rather off-hand treatment.

In the canonical GSM architecture, what has been referred up to now as the MSC/VLR consists of two disjoint functional entities, the MSC itself and a database, the VLR. The MSC is defined as the switching function in charge of the management of the calls, and the VLR as the database where subscriber information are temporarily stored for those subscribers which are registered under a MSC connected to the VLR.

A VLR can manage the subscriber data for one or several MSCs, and can be an equipment physically distinct from a MSC. The reason why the two functions are split is not so much because of this possible implementation in distinct equipments, but because of the option to have a VLR for more than one MSC. The point to analyse is then why should such a choice be taken.

From an architectural point of view, the VLR can be seen from three different vantage points:

- a first approach is to consider the set of the HLR and VLR as a single distributed database. The distinction between the HLR parts and the VLR parts becomes a matter of internal architecture of this database. This would correspond to an approach where the VLR is introduced solely for signalling load distribution, and a VLR would naturally be serving several MSCs;

- the opposite point of view is to consider the VLR as fulfilling a set of ancillary tasks to the MSC, including the management of the visited subscriber database and the corresponding dialogues with the HLR. It is then naturally a part of the MSC, and there are little reason for having a VLR connected to several MSCs;

- a third approach is to consider the VLR as a truly independent entity, having tasks of its own, with added value compared to the natural roles of the HLR and the MSC.

The exact philosophy of the functional split between the VLR and the HLR on one side, and between the MSC and the VLR on the other, can be determined by looking in details of the corresponding protocols. In so doing, it becomes apparent that the cut between the MSC/VLR and the HLR can be considered as a minimum, whereas the cut between the VLR/HLR and the MSC seems not to follow strongly directive lines. If the VLR/HLR protocol can be easily presented as an MSC/HLR protocol, it is obviously impossible to do so with the MSC/VLR protocol (the proof of the first statement will be found in this book, whereas it would be too long to justify the second; the interested reader can look to the *Specifications* to make up his mind). This militates strongly for the second approach, that we have followed in this book. It is often invoked as a counter argument that the split between the VLR and the MSC is related to the Intelligent Network (IN) approach for building switch equipments. When looked at closer, though, it is obvious that the two philosophies are somewhat different: in an IN approach, the MSC would have no high level function and the VLR would be where all the complex protocols are dealt with. This is very different in the *Specifications*: the MSC deals indeed with most of the complex protocols. The VLR is neither a pure database (in which case, the VLR-MSC interface would be very close to an interface between a MSC and a HLR if temporary storage was not used), nor the true call manager controlling a rather dumb switch (this would be the IN approach). Implementations of a GSM network based on an IN approach are nevertheless possible and implemented, but the split between the "dumb" Switching Service Point and the "intelligent" Service Control Point is not based along a MSC-VLR line. While the architectural split as described in the *Specifications* is for these reasons certainly not the last say in the domain, GSM is considered as one of the first "intelligent" networks and concepts such as the interrogation of a centralised HLR are the first bricks for the construction of a full-blown intelligent network structure.

This situation is the main reason why the authors of this book decided not to describe the VLR as an entity separated from the MSC. This position is supported by the fact that up to now *all* the switch manufacturers have chosen to develop a combined MSC/VLR, and none offer the possibility to physically split them up.

Let us see rapidly some of the points of the *Specifications* among the casualties caused by our approach. If separated, MSC and VLR are connected through the SS7 signalling network, and the signalling procedures for the corresponding dialogue are specified in the MAP (they constitute the MAP/B protocol). The MAP/B protocol represents a major part of the MAP specification, at least as far as the number of pages is concerned. Because of the reasons explained above, it will not be treated here.

The Mobile Station

The mobile station holds a starring role in location management, for obvious reasons. It is at the origin of the location information, and in fact deals with problems posed by mobility almost entirely on its own in idle mode. An important point which may have been less evident up to now is the respective role of the SIM and of the rest of the mobile station, known in the *Specifications* as the mobile equipment.

The assumption in the *Specifications* is that the mobile equipment does not hold in a non-volatile memory any information specific to its user. Still, mobile station manufacturers are not prevented from doing so, but the system can work properly without such a memory. The converse point is that the SIM holds this information, and in particular many related to the mobility management, whether for handling location or security related information.

The mobile equipment contains however some information of some temporal scope, such as the list of forbidden location areas for national roaming, or lists of beacon frequencies for different PLMNs. This information is lost when the mobile equipment is switched off. The choice of what is in the SIM was a compromise between memory consumption (a scarce resource for a SIM) and keeping as much as possible potentially useful information over a switched-off period.

The SIM

The SIM has already been presented in the very first chapters. What is of interest here is the information it contains in relationship with location management. In this area, the SIM is but a passive information container, so listing the relevant fields will give us all the functional description we need:

- The update status;
- A location area identity;

These fields usually contain the result of the last location updating attempt, and the location area where it was done; their main purpose is to avoid a location updating attempt in some cases when the cell selected after switch-on is in the same location area.

- A list of beacon frequencies ("BCCH information");

We have already met this list. The *Specifications* are not clear about which PLMN it relates to. In order for this list to achieve its aim, i.e., to speed up the initialisation time after switch-on, it should pertain to the home PLMN. Another interpretation is that it should be the last list received from the serving network.

- The *forbidden* PLMNs list ("forbidden PLMNs");

The function of this list was described in the section of this chapter relative to PLMN selection. It should be noted that it is an *ordered* list, with entries sorted by order of introduction, because of the requirement to replace the oldest entry when the list is full.

- The *preferred* PLMN list.

The function of this list was also described when dealing with PLMN selection.

7.1.3.2. Protocols

The protocols belonging functionally to the Mobility Management plane are the one held between the HLR and the MSC/VLRs, the MSC/VLR and the mobile station, and between the mobile equipment and the SIM. To allow full roaming, it is of utmost importance that every HLR be able to exchange information with every MSC/VLR throughout all the PLMNs of the system (and possibly with switches from other types of networks if SIM-roaming is implemented). A HLR must also be able to dialogue with all the entities that want to get information about

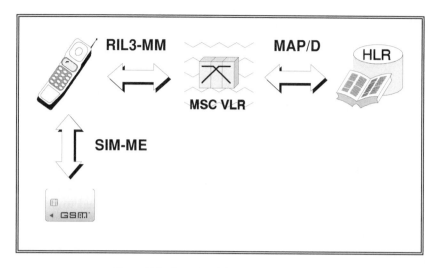

Figure 7.4 – Location Management Protocols

MM protocols involve the location registers (VLR and HLR),
which communicate between themselves and with the mobile station,
in which the SIM plays a prominent part.

the subscribers, mainly location information with the aim of setting up calls toward these subscribers. This will be dealt with in Chapter 8.

The protocol between the HLR and the MSC/VLRs is supported through the world-wide signalling network, the signalling system n°7 (SS7), as described in Chapter 5. The application protocol for the dialogues between a HLR and a MSC/VLR is part of the MAP (Mobile Application Part). In this book, we will call it the **MAP/D** protocol.

The MS-MSC protocol is called the **RIL3-MM** protocol (for Radio Interface Layer 3, Mobility Management). It uses the MS-MSC signalling connection provided by the RR layer, as seen in Chapters 5 and 6.

The **SIM-ME** protocol is limited to read and write commands as far as we are concerned here, and corresponding messages will not be cited in the text.

Figure 7.4 summarises the simple architecture of the protocols needed for location management.

7.1.4. THE LOCATION UPDATING PROCEDURES

The main procedure of interest for location management is the location updating procedure, which is triggered by a mobile station to update the location data of its subscriber. For various reasons, slightly modified versions of this procedure are used for different related purposes. These variations are also described in this section.

7.1.4.1. The Basic Procedures

The location information is stored in two different places in the GSM infrastructure, the HLR and the visited MSC/VLR. In fact the same information is known in three different places in the system, the mobile station (and more explicitly the SIM) being the third place. This information may change, and various procedures are needed to keep the consistency between the three entities.

The normal reason for a change is when the mobile station decides that the location area best fit to serve its subscriber must be changed. Then the mobile station notifies the MSC/VLR to which the new cell belongs. This MSC/VLR may be the same as before, if it controls both the previous and the new location area, or a new MSC/VLR. In the latter case, the MSC/VLR notifies in turn the HLR, which notifies the previous MSC/VLR. There are other cases where an inconsistency may appear, for instance when stored information is lost in the MSC/VLR or the HLR as a result of some hardware or software failure. Then procedures may be run to correct the failed database using information in other equipments.

In order to cover all theses cases, the following elementary procedures have been specified (see figure 7.5):

- Updating of the MSC/VLR storage at the request of the mobile station;
- Updating of the HLR storage at the request of the MSC/VLR;
- Cancellation of a subscriber record in a MSC/VLR at the request of the HLR;

The Mobile Station to MSC Location Updating Procedure

This procedure is part of the RIL3-MM protocol. It requires a radio connection, as for any dialogue between the mobile station and the network. The establishment of such a connection is a function of the Radio Resource Management functions, described in Chapter 6. This establishment has almost nothing specific to the location updating procedure.

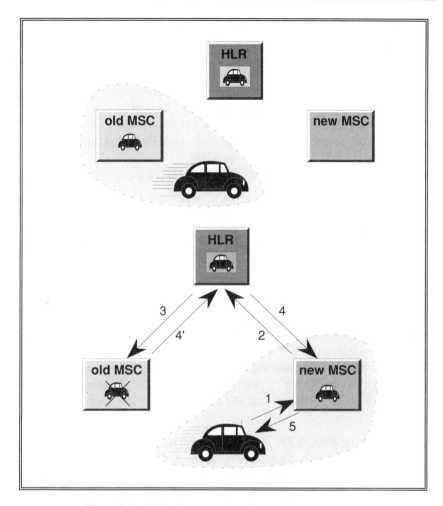

Figure 7.5 – Main elementary location updating procedures

In order to change from the old situation (shown on top) to the new situation (bottom),
the mobile station takes the initiative of location updating (1),
but the HLR, upon subsequent request from the new MSC/VLR (2),
takes care of cancelling the old record in the previous MSC/VLR (3, 4')
in parallel with confirming the updating in the new MSC/VLR (4),
which in turn acknowledges the mobile station request (5).

The MS-MSC location updating procedure is basically very simple, and consists of a request (a location updating request) and an answer. The request is carried by the RIL3-MM LOCATION UPDATING REQUEST message. This message contains mainly the information necessary to identify the subscriber.

The MSC/VLR may answer autonomously in some cases, or alternatively may have to update the HLR first, with the procedure described in the next section. There is one case when the MSC/VLR cannot do otherwise than answer on its own: when it cannot reach the HLR for lack of any roaming agreement between the two operators. This case is not addressed in the *Specifications*, though it can happen. The answer of the MSC/VLR is necessarily negative and must be chosen to ensure that the mobile station will search other PLMNs (for instance, by sending the cause "PLMN not allowed").

The usual "normal case" when the MSC/VLR can answer on its own is when the subscriber is already registered in the database of the MSC/VLR. The response in that case is usually positive, and can be negative only in case of national roaming restriction (cause "location area not allowed for national roaming", used only in DCS1800); regional subscription cannot lead to a negative MSC/VLR answer without HLR involvement, since the MAP restricts regional subscription to be offered on a per-VLR basis.

National roaming then merits some attention. The rule is that if the mobile station belongs to an unwanted PLMN, and if the requested location area is restricted, the MSC/VLR is entitled to directly answer negatively. Most of the time the MSC/VLR will not be able to contact the HLR, and this would be a particular case of the situation mentioned above, with the difference that the cause sent to the mobile station is specified.

When the MSC/VLR needs to contact the HLR of the subscriber, it must first know which HLR is concerned. Subscribers are identified for the internal business of GSM by a number, the IMSI (International Mobile Subscriber Identity). This number is provided by the mobile station anytime it accesses the network (the number is not always given directly, see the notion of TMSI, page 484). The IMSI is so specified that the MSC/VLR is able to derive the identity of the subscriber's home PLMN, and possibly more information on the HLR equipment in charge of the subscriber. Figure 7.6 shows the IMSI structure. With the help of the relevant translation tables, the MSC/VLR is then able to derive the SS7 address to which the location updating request must be sent. In practice, the HLR can usually be identified by looking at the most significant digits of the IMSI following the mobile country code and mobile network code. However, this possibility is usually only used inside the home PLMN country. PLMNs of other countries route their messages using the IMSI as a global title, towards a gateway entity in the home PLMN country. There the global title can be translated in the Signalling Point Code of the right equipment in the right PLMN, as explained in Chapter 5.

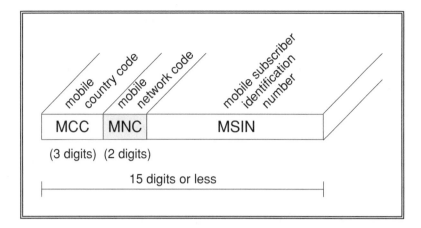

Figure 7.6 – The structure of an IMSI

An international Mobile Subscriber Identity consists of three parts:
the Mobile Country Code (MCC), identifying a country;
the Mobile Network Code (MNC), identifying a PLMN within this country; and
the Mobile Subscriber Identification Number, identifying a subscriber
within this PLMN, using no more than 10 digits.

Before giving the answer to the mobile station, in an RIL3-MM LOCATION UPDATING ACCEPT message or an RIL3-MM LOCATION UPDATING REJECT message, the MSC/VLR may proceed to some actions like authentication or ciphering setting. These procedures are detailed in the second part of this chapter.

The answer provided by the MSC/VLR, possibly after contact with the HLR, is either the RIL3-MM LOCATION UPDATING ACCEPT, thus indicating that the subscriber is effectively registered in the new location area as required, or RIL3-MM LOCATION UPDATING REJECT, with a suitable cause, such as "PLMN not allowed" or "location area not allowed". The first indicates that the subscriber has no subscription entitlement for service in the visited PLMN, whereas the second, which can be used only in the home PLMN, indicates the subscriber has no subscription entitlement for service in the location area. The HLR does not indicate which cause to use (the MAP provides only one cause "roaming not allowed"). Though not indicated in the *Specifications*, the cause in the RIL3-MM message should be "location area not allowed" if the visited MSC/VLR belongs to the home PLMN, and "PLMN not allowed" otherwise. The mobile station reacts on the receipt of a negative answer as explained in the section on cell and PLMN selection.

The MSC/VLR to HLR Location Updating Procedure

This procedure is run when a mobile station asks for registration under a new MSC/VLR. It can also be run when the HLR has suffered a failure, and request the MSC/VLR for a confirmation of the subscriber location as soon as the mobile station is in contact.

The request is conveyed in a MAP/D UPDATE LOCATION message. This message carries (among other information) the subscriber identity and enough information for the HLR to know how to find routing data for the setting up of a mobile terminating call, i.e., the SS7 address of the MSC/VLR. It does not convey the identity of the precise location area. As an option supported by the standard, the routing data can be included at this point, but we will see in Chapter 8 that this is not what is usually done.

The HLR determines whether to accept to register the mobile station in the new MSC/VLR or not by consideration of the subscription limitations of the user. If the subscriber is entitled to normal service in the requested VLR area, the answer is positive, the HLR updates its memory and triggers a location cancellation procedure with the previous MSC/VLR (if the mobile station was actually registered, and in another MSC/VLR). If the subscriber cannot be granted normal service, the HLR deregisters the subscriber: it updates its memory to indicate that the location of the mobile station is unknown, and it triggers a location cancellation procedure if applicable. The answer from the HLR is carried to the MSC/VLR in a MAP/D UPDATE LOCATION RESULT message.

If the requesting MSC/VLR receives a negative answer, it erases all information relative to the subscriber. Otherwise and if necessary, it enters the subscriber in its database. Normally, the HLR will then provide the subscriber information the MSC/VLR needs. This is done by sending a MAP/D INSERT SUBSCRIBER DATA message, which is acknowledged by the MSC/VLR through a MAP/D INSERT SUBSCRIBER DATA RESULT message. The same procedure may also be used when some change in the subscriber's information stored in the HLR has occurred, for instance at the request of the subscriber. Alternatively, depending on the nature of the modification, a procedure composed of the MAP/D DELETE SUBSCRIBER DATA and the MAP/D DELETE SUBSCRIBER DATA RESULT messages may be used.

The HLR to MSC/VLR Location Cancellation Procedure

Location cancellation from HLR to MSC/VLR is a very simple procedure, composed of a request carried by a MAP/D CANCEL LOCATION message and the corresponding acknowledgement carried in a MAP/D CANCEL LOCATION RESULT message. The HLR does not wait for this acknowledgement to confirm location updating to the other MSC/VLR who triggered the change of location.

The MSC/VLR to HLR Deregistration Procedure

A MAP/D procedure opens the possibility for the MSC/VLR to ask the HLR for the deregistration of a given subscriber. There is no identified case in GSM for this procedure, and it is indicated as not used in a laconic sentence in the first page of TS GSM 09.02.

This procedure is the simple order/acknowledge procedure. The request is carried by a MAP/D DEREGISTER MOBILE SUBSCRIBER message, and acknowledged by a MAP/D DEREGISTER MOBILE SUBSCRIBER RESULT message.

7.1.4.2. Periodic Location Updating, and Database Failure Recovery

An HLR, or an MSC/VLR, may suffer a failure such that a part of its database is damaged. These equipments are usually implemented in a secure way with back-up systems, allowing the database to be restored to some consistent state. However, there are cases when the restored database is no longer up-to-date, e.g., because updates having occurred between the last back-up and the failure have been lost.

Since such conditions may affect the possibility to set up calls correctly, a number of mechanisms are described in the *Specifications* to improve the situation. The general mechanism is as follows: in a first step, the affected equipment will mark the uncertain information as such. Then it will warn other network entities (those with which it knows to be sharing information) about the unclear status of its information contents. Consequently, these entities will also mark the corresponding records for checking. To avoid overloading the signalling system, no attempt is made to restore the consistency right away: the recovering equipment would probably not survive a high peak in signalling load! A given subscriber record (say Christian's) is corrected only when some event happens concerning this subscriber, such as a radio contact initiated by the mobile station, or a mobile terminating call attempt. If after some time nothing has happened, Christian is deregistered.

Obviously, the best place to stay informed about the subscriber position is the mobile station, and a radio contact is at the end of the day the only safe way to restore consistency. Periodic location updating has thus been introduced to ensure regular radio contact at the initiative of the mobile station.

Basically, periodic location updating refers to the requirement that the mobile station contacts regularly the network when in normal service. This is done automatically by the mobile station, and takes the form of a location updating procedure. The period is under control of the network which broadcasts its value. The possibilities range from 6 minutes (in which case the network is probably completely overwhelmed by location updating procedures!) to slightly more than 24 hours. In addition, the infinite value is included: in other words, the network can suppress completely periodic location updating if so wished. The choice of periodicity is the operator's, and is typically a trade-off between the quickness of recovery after a failure (therefore the time when it may be impossible to serve mobile terminating calls for some subscribers) and the traffic load due to periodic location updating. This clearly depends a lot on the reliability of the MSC/VLR and HLR: if the mean time between failures is very long, there is no point in incurring the load and cost of periodic updating (except if implicit detach is also used, see page 476). On the other hand, the periodicity can be increased by the operator after a failure.

Let us now see in more details how the location registers cope with losses of information after failure, and how in general location data is recovered.

MSC/VLR Failure

When a MSC/VLR suffers a database failure, it first restores its state to some previously saved state, and after this recovery marks all its records as to be checked with the mobile station and with the HLR. Then it sends a MAP/D RESET message to all HLRs for which it still has a subscriber in its tables. The MSC/VLR will eventually notice that a mobile station is missing in its records when some service must be provided to the subscriber, and at the latest when the mobile station performs a periodic location updating. In the reverse situation (i.e., a mobile station still recorded when it should not be), the correction may take longer.

In the case of a request from the mobile station (e.g., a call set-up, but not a location updating of some sort), the MSC/VLR will notice for instance that the corresponding subscriber is not in its table though the mobile station thinks so. The MSC/VLR reacts then by requesting the mobile station to perform a location updating. This is done with a rejection with cause "IMSI unknown in VLR". Because this will also trigger a location updating from MSC/VLR to HLR, consistency will be reached anew. A milder case is when the mobile station calls from a location area which is different from the one in which the MSC/VLR thinks it is registered. In this case, the MSC/VLR simply corrects its record.

In the case of a mobile terminating call, the MSC/VLR may notice a problem if it receives a request from the HLR concerning a mobile station which is not in its table though the HLR obviously thinks so. This may happen either because of an MSC/VLR failure or because of an HLR failure. In all cases, the MSC/VLR enters the subscriber in its tables, and asks the HLR for the subscriber information via a MAP/D SEND PARAMETERS message, which is answered with a MAP/D SEND PARAMETERS RESULT message from the HLR. The location area information is in this case still missing. Henceforth, and until actual contact with the mobile station, the paging is done in all cells of all location areas controlled by this MSC/VLR. Note that if the error was in the HLR, that is to say if the mobile station if effectively not under the MSC/VLR as the HLR knows, the information in the network is inconsistent: after a mobile terminating call attempt, the subscriber is registered in two MSC/VLRs (MSC/VLR–1 where it is really, and MSC/VLR-2 where the HLR imagines it is), and the HLR stores the identity of the wrong one. The error will be corrected in the HLR and in MSC/VLR-2 when a location updating procedure is performed by the mobile station. However, MSC/VLR-1 will remain in error by keeping the subscriber in its database. The cleaning-up can only be done by an internal mechanism in the MSC/VLR, erasing all records of subscribers which have been inactive for more than some long time, e.g., a month.

HLR Failure

Restoration of the HLR is a bit more complex, because the HLR is not necessarily contacted in the case of a periodic location updating or a mobile originating call. To enforce this contact, the HLR sends a MAP/D RESET message to all MSC/VLRs in which at least one of the HLR subscribers is known to be located, as indicated by the salvaged records. The MSC/VLRs will mark all corresponding records as to be checked with the HLR: the next radio contact will then trigger a location updating procedure from MSC/VLR to HLR, thereby correcting the HLR records.

In the case of a mobile terminating call, as we have already seen two paragraphs above, the interrogated visited MSC/VLR aligns its database with the (recovered and unchecked) HLR data. At least, this is all that the MAP/D protocol, as specified, enables it to do. However, HLR restoration methods described in TS GSM 03.07 include a different mechanism, by which the HLR would indicate to the MSC/VLR that the interrogation concerns a mobile station for which data is unsure: the MSC/VLR would in that case know whether it is better to take the HLR as a reference or to rely on its own state and answer accordingly. However, this mechanism has not been implemented in the phase 1 MAP protocols.

7.1.4.3. The IMSI Attach and Detach Procedures

The location updating procedure has two very close siblings: the periodic location updating procedure, which has just been described, and the *IMSI attach* procedure. On the radio path, both procedures are almost identical to a location updating procedure. They differ from it almost only by the events that trigger them. These events are such that these "location updating" procedures usually appear as a request to be registered in the location area where the subscriber is *already* registered. Hence in most of the cases, the HLR is not concerned with these procedures.

The strange names of the *IMSI attach* and *IMSI detach* procedures are due to the accidents of the standardisation, and are of little use to understand their meaning. The best way to understand these procedures is to explain what purpose they satisfy.

When a mobile station is switched off (or when the SIM is removed by the user), the calls toward the corresponding subscriber can no more be completed. Important resources are then consumed for nothing: as will be seen in Chapter 8, a circuit is established between the caller and the MSC in charge of the called mobile station, and the paging procedure is performed, all to no avail. Worse (from one point of view), the established circuit is not paid for. As will be seen when dealing with the routing of calls, the establishment of the first part of the circuit (before HLR interrogation) cannot be avoided. It is a different story with the second portion, between the point where HLR interrogation is done and the visited MSC.

To alleviate this useless load (and cost), the *IMSI detach* and *IMSI attach* mechanism has been introduced. Basically, the subscriber's record in the MSC/VLR contains a binary information indicating whether it is useful or not to try to complete a call toward this subscriber. This information makes it at least possible to economise on paging. It may also prevent the establishment of a part of the call. The *IMSI detach* procedure will set this bit to "not useful to try", whereas the *IMSI attach* procedure will do the reverse. The mobile station triggers an *IMSI detach* when it goes inactive, and either a location updating procedure (if in a new location area) or an *IMSI attach* procedure when it comes back (in the same location area).

The MAP specifications include two different ways for the management of this feature in the infrastructure: either the information is only stored in the MSC/VLR (the mobile station staying registered in this MSC/VLR as far as the HLR is concerned), or the subscriber is simply deregistered in the HLR and its record cancelled in the MSC/VLR. In fact, only the first option is allowed for GSM, since the deregistration procedure is not used.

With this first option, paging can obviously be prevented. The second part of the circuit establishment may also be prevented, but not so obviously. Moreover, even when possible, it is an option to prevent it or not. As will be seen in Chapter 8, the basic scenario of a mobile terminating call set-up attempt requires an interrogation of the visited MSC/VLR by the HLR before the latter provides the information necessary for the continuation of the routing. This phase allows the visited MSC/VLR to reject the call on the basis of the attach status before the costly set up of the traffic circuit. If it does so, call forwarding if applied can potentially be controlled by the HLR. Another possibility is that the visited MSC/VLR accepts the call, and applies the call forwarding itself if required.

To complete the option list, the support of the attach/detach feature is a network option. It is allowed that a PLMN, or a portion thereof, does not provide this facility. This is indicated to the mobile stations on a cell basis in the broadcast information. The choice between all these options and sub-options lies with the visited MSC/VLR, since the HLR does not intervene. If attach is indicated as supported in the current cell, the *IMSI detach* procedure is used by the mobile station to indicate that it (more exactly the SIM) will go inactive.

The *IMSI detach* procedure is an example of a procedure reduced to its bare bones: it consists of a single message from the mobile station to the visited MSC/VLR, the RIL3-MM IMSI DETACH message. This message is not acknowledged, simply because it has been considered that the mobile station is typically switched off, or more generally not in a position to receive an answer from the network. The *IMSI detach* procedure must use a radio connection, as any RIL3-MM procedure. This connection is either established for the purpose of the detach, or may pre-exist. The connection can be abandoned by the mobile station immediately after the sending of the RIL3-MM IMSI DETACH message. The mobile station keeps no track of having asked for a detach (for instance by storage in the SIM): the state of the attach/detach information in the network is not monitored by the mobile station.

If attach is indicated as supported in the cell the mobile station has chosen at switch-on (or SIM insertion), and if the mobile station knows the subscriber is already registered in the same location area, it starts an IMSI attach procedure, that is to say (except for a negligible detail) a location updating procedure. It should be noted that the attach procedure may happen even if there was no request for detach beforehand (because the network did not require it at the time, or simply because it was not physically possible, e.g., in case of a loss of coverage before switch-off). This is consistent with the specification that the mobile station does not monitor the attach/detach status.

These procedures would then have been better called "Subscriber deactivation" and "Subscriber reactivation". From a very slightly different point of view, these procedures are very close functionally to the call forwarding supplementary services in the case where the mobile station is not deregistered. The *IMSI detach* procedure can indeed be understood as an automatic activation of an unconditional call forwarding (toward an announcement, or a specific number), and the IMSI attach as the corresponding deactivation procedure. It can therefore be said that, from the point of view of the purpose of these procedures, of their *why*, they have little to do with the location updating, but are akin to call forwarding. From the point of view of the mechanisms they use, that is to say from their *how*, these procedures are close siblings to the location updating procedure.

7.1.4.4. Automatic Detach by the Network

As explained above, the attach status can be modified by the mobile station, through the *IMSI attach* and *IMSI detach* procedures. A past debate in the GSM committees was whether an additional mechanism may be added, consisting in putting a subscriber in the detach state, or deregistering him, if nothing has been heard of him for more than a given period. The debate ended more or less against this feature. It can be nevertheless used, since it requires only existing signalling procedures and otherwise affects only the MSC/VLR.

Thanks to periodic updating, the mobile station of a subscriber when switched-on must indeed contact the network at least once every so often. Thanks also to the non-storage of the detach status in the SIM, the mobile station will always make itself known when re-activated: there is no period after switch-on during which the mobile station cannot be called because of a forceful detach by the network.

The advantage of a network time-controlled detach is that the detaching may happen even in the cases where the mobile station is physically unable to send an RIL3-MM IMSI DETACH message.

This feature is not mentioned anywhere in the *Specifications*. However it would work without side effects, and it is known that a number of operators intend to use it.

7.2. SECURITY MANAGEMENT

Radio transmission is by nature more prone to eavesdropping and fraud than fixed wire transmission. Listening to communications is easy and does not require access to special locations. Impersonating a registered user (and therefore having him foot the bill!) can also be very easy if specific protection means are not provided. Analogue systems have indeed suffered from such problems during the 80's. GSM had to bring significant improvements in these matters.

7.2.1. THE NEEDS

The security-related functions of GSM aim at two goals: first, protecting the network against unauthorised access (and at the same time protecting the users from fraudulent impersonations); second, protecting the privacy of the users.

Preventing unauthorised accesses is achieved by means of authentication, i.e., by a secure check that the subscriber identity provided by the mobile station corresponds to the inserted SIM. From the point of view of the operator, this function is of paramount importance, in particular in conjunction with international roaming, where the visited network does not control the subscriber's record... and his ability to pay.

Preserving the privacy of the users is achieved through different means. Transmission can be ciphered to prevent eavesdropping of communications on the radio path. Most of the signalling can also be protected in the same way, preventing third parties from knowing who is being called, for instance. Finally, the replacement of the subscriber's identity by a temporary alias is another mechanism to convince third parties that listening on the radio path is useless for tracing GSM subscribers. Since most of the calls involving a GSM user go through the fixed network, the designers of GSM did not aim at a level of security much higher than that of the fixed trunk network. Mechanisms to ensure privacy have only been introduced for the radio path. Within the infrastructure, communications are transmitted in clear text, as they are in the PSTN.

It is important to note at this stage that all the security mechanisms of GSM are under sole control of the operators: the users have no possibility to affect whether authentication, encryption, etc. are applied or not. Moreover, users are not necessarily aware of what security features are used. Conversely, these security services are not usually subscribed-for. The *Specifications* leave a lot of flexibility to apply them in various conditions. Some harmonisation is however desirable, and is settled for the GSM900 operators for instance by discussions within the GSM MoU.

7.2.2. THE FUNCTIONS

7.2.2.1. Authentication

A simple authentication method is the use of a password (or a PIN code—Personal Identity Number). The level of protection achieved by such a method is very low in a radio environment, since listening once to this personal code is enough to break the protection. GSM does make use of a PIN code in conjunction with the SIM; this PIN code is checked locally by the SIM itself, without transmission on the radio interface. But in addition, GSM uses a more sophisticated method, consisting in a layman's words in asking a question that only the right subscriber equipment (in that case, the SIM) may answer. The crux in this method is that a huge number of such questions exist, and that it is therefore very unlikely that the same question would be used twice.

More precisely, the question takes the guise of a number, called *RAND* in the *Specifications*, whose value is drawn randomly between 0 and $2^{128}-1$ (something like a few millions of milliards of milliards of milliards of milliards!). The answer, called *SRES* in the *Specifications*—i.e., Signed RESult in cryptographic terminology—is obtained as the outcome of a computation involving a secret parameter specific to the user, and called *Ki* in the *Specifications* (see figure 7.7). The secrecy of *Ki* is the cornerstone on which all the security mechanisms are based. We will see that it is stored in a very protected way; for instance a subscriber cannot know his *Ki*. The algorithm describing the computation is referred to as algorithm A3 in the *Specifications*, but its specification cannot be found there. In fact, the design choices of GSM, both in the mobile station and in the infrastructure, allow A3 to be operator-dependent while allowing full

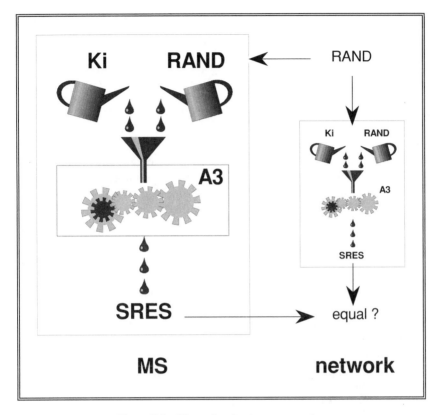

Figure 7.7 – The authentication computation

Authentication is performed by requiring the correct answer to the following riddle: what signed response *SRES* are you able to derive from the input challenge *RAND*, by applying the A3 algorithm with your personal (secret) key *Ki*?

inter-PLMN roaming. Operators can therefore choose the A3 applicable to their own subscribers independently from other operators. Such algorithms are usually kept secret (belt and braces are never too much in this domain!).

In order to obtain to desired security level, A3 should be what cryptography experts refer to as a one-way (or trap-door) function. This means that the computation of *SRES* from *Ki* and *RAND* should be easy, whereas the computation of *Ki* knowing *RAND* and *SRES* should be as complex as possible. It is indeed this level of complexity which determines which security level has been achieved. Even with the knowledge of several pairs (*RAND*, *SRES*) pertaining to the same subscriber (i.e., the same *Ki*), the computation should remain highly complex. Beyond this requirement, the only constraint imposed on A3 is the size of the input parameter (*RAND* is 128 bits long) and the size of the output parameter (*SRES* must be 32 bits long). *Ki* can indeed be of any format and length: here again the design choices leave the operator with a maximum flexibility. Only if *Ki* would be transported in the network (see page 488) would it be constrained to a maximum length of 128 bits.

At first view, it may surprise the reader to learn of the possibility for each operator to choose A3 independently, given the general specification philosophy of GSM. Special efforts were necessary to cover the case of international roaming, where the specification of a single A3 algorithm could appear an easy solution. However, several reasons justify the approach. One of them is the administrative complexity linked to the specification and distribution of cryptographic algorithms, especially when they are to cross borders. As will be explained later on, the algorithm used for ciphering in GSM is unique, and its specifications are managed in a totally different way from the other *Specifications*. The management of a single A3 algorithm would have been even more complex, since authentication is more sensitive than communication ciphering (the consequences of a "broken" algorithm are more far-reaching in the case of authentication). The management of A3 is much simpler if controlled by a single operator. Another reason is the existence of algorithms fit for authentication and already implemented in smart cards, but possibly not open for sharing. A limiting factor being the smart card memory capacity, the choice of having an operator-dependent A3 algorithm enables telecommunication operators to use a single algorithm for, e.g., GSM SIM and pay-phone access.

7.2.2.2. Encryption

Obtaining a good protection against unauthorised listening is not an easy matter with analog transmission, but digital transmission permits an excellent level of protection with relatively simple means, thanks to digital cryptography methods. This has been taken advantage of in GSM, where the position of the encryption and deciphering processes in the transmission chain allow a single method to be used for protection of all transmitted data in dedicated mode, whether user information (speech, data, ...), user-related signalling (e.g., the messages carrying the called phone numbers) or even system-related signalling (e.g., the messages carrying radio measurement results to prepare for handover). This choice is not the result of a paranoiac approach, but is justified by its simplicity. Only two cases need to be distinguished: either transmission is protected, and everything is sent enciphered, or transmission is not protected, and everything is sent in clear text. The actual procedure for changing from

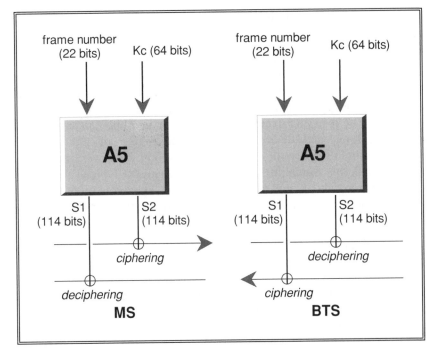

Figure 7.8 – Ciphering and deciphering

A5 derives a ciphering sequence of 114 bits for each burst independently,
taking into account the frame number and the ciphering key Kc.

ciphered to non-ciphered mode (and vice versa) belongs to the radio
resource management functions and has been described in Chapter 6.

Both ciphering and deciphering are performed by applying an
"exclusive-or" operation between the 114 "coded" bits of a radio burst
and a 114-bits ciphering sequence generated by a specific algorithm,
called A5, as described in Chapter 4. In order to derive the ciphering
sequence for each burst, A5 performs a computation with two inputs: one
is the frame number and the other is a key (named Kc) agreed between
mobile station and network (see figure 7.8). The uplink and downlink
directions use two different sequences: for each burst, one sequence is
used for ciphering in the mobile station and for deciphering in the BTS,
whereas another one is used for ciphering in the BTS and deciphering in
the mobile station.

For all types of radio channels, the frame number changes from
burst to burst, so that each burst of a given communication in the same
direction uses a different ciphering sequence. The successive values of
the frame number depends on the time organisation of each channel. The

time organisation of a TACH/E, exposed in Chapter 4, shows for example that the frame number is not always incremented by one every burst.

As far as its representation is concerned, the frame number is coded as the concatenation of three values, called respectively T1, T3 and T2 (in that order) and amounting to 22 bits all together. What these three values actually represent is in fact meaningless as far as ciphering is concerned. The resulting cycle, the hyperframe, is a bit less than 3 and a half hours long, and determines a periodic return of the ciphering sequence, should the communication last this long. However, the key Kc is controlled by signalling means and changes typically at each communication. This key is not publicised, but, since it is often changed, it does not need as strong a protection as Ki: for instance, Kc can be read freely from the SIM.

Algorithm A5 must be specified at the international level, since for achieving MS-roaming it must be implemented within every base station (as well as in any mobile equipment). Means to cater for several A5 algorithms (e.g., to cope with some regulation restrictions with regards to export outside Europe) have been introduced in the *Specifications* (partially in phase 1, and definitely for phase 2). For the time being, a single A5 algorithm has been specified for use in all countries. Its specification cannot however be found in the *Specifications*, for security reasons. This algorithm is the property of the GSM MoU, and is tightly copyright protected. Its external specifications are however public, and it can be described as a black box taking a 22-bit long parameter (the frame number) and a 64-bit long parameter (Kc) to produce two 114-bit long sequences. As for the authentication algorithm A3, the level of protection offered by A5 is determined by the complexity of the reverse calculation, i.e., the computation of Kc knowing two 114-bits ciphering sequences and the frame number.

Key Management

The key Kc must be agreed by the mobile station and the network prior to the start of encryption. The choice in GSM is to compute the key Kc independently from the effective start of encryption, during the authentication process. Kc is then stored in a non-volatile memory inside the SIM, so as to be remembered even after a switched-off phase. This "dormant" key is also stored in the visited MSC/VLR on the network side, and is ready to be used for a start of encryption. When authentication happens while the transmission is ciphered, then the active key Kc being used for ciphering/deciphering is not affected, but the new "dormant" key is stored, and is reserved for use at the next occurrence of

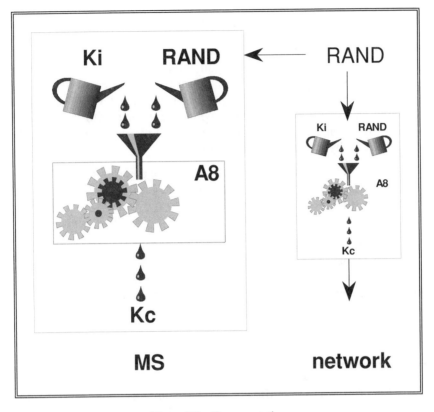

Figure 7.9 – Kc computation

Each time a mobile station is authenticated,
the mobile station and the network also compute the ciphering key Kc
by running algorithm A8 with the same inputs RAND and Ki
as for the computation of SRES through algorithm A3.

a transition between clear mode and cipher mode. Hence the terminology "dormant" key (versus "active" key) introduced in this book.

The algorithm used to compute *Kc* from *RAND* (the same one used for authentication) and *Ki* is called A8 in the *Specifications* (see figure 7.9). Similarly to A3 (the authentication algorithm computing *SRES* from *RAND* and *Ki*), A8 is not specified in the ETSI *Specifications*, but can be chosen independently by each operator. Both algorithms could in fact be implemented as a single computation. For instance, they could be implemented as a single algorithm whose output consists of 96 bits: 32 bits to form *SRES*, and 64 bits to form *Kc*. Care must be taken that the knowledge of *RAND* and *SRES* does not give away too much information about *Kc*.

It is worth noting that the length of the significant part of key Kc output by algorithm A8 is fixed by the group of signatories of the GSM MoU, and may be less than the maximum 64 bits. In that case, the significant bits are complemented with zeroes, so that the format always uses the full 64 bits. As far as A5 is concerned, all patterns of 64 bits are possible and meaningful: this mechanism allows the level of security to be increased in the future if needed, without any change to A5 (therefore without any change of the mobile equipments) by increasing the number of significant digits within the limit of 64.

Since A3 and A8 are always running together, and in most cases are implemented as a single algorithm, they will always be treated together in the rest of this chapter, and referred to as A3/A8. They are indeed so intermixed that authentication and "dormant" key management cannot be dealt with independently.

No other "Ax" algorithms are to be found in the Specs. A1, A2, A4, and so on, were place-holders during the design of the system, and disappeared eventually, but the terminology of the three survivors A3, A5 and A8 was not changed.

7.2.2.3. User Identity Protection

Encryption is very efficient for confidentiality, but cannot be used to protect every single exchange on the radio path. Ciphering with Kc applies only when the network knows the identity of the subscriber it is talking to. Obviously, ciphering cannot be applied to common channels, such as the BCCH, which is received simultaneously by all mobile stations in the cell and in neighbouring cells, or else it could be applied with a key known to all mobile stations, and therefore quite useless as a security mechanism! When a mobile station moves to a dedicated channel, there is some "bootstrap" period during which the network does not yet know the identity of the subscriber, say Charles, and therefore cannot cipher. This has a major consequence: all the signalling exchanges up to and including the first message carrying a non-ambiguous subscriber identity must be sent in clear. A third-party could at this stage listen to this identity, and know where Charles roams at this particular moment. This is considered harmful to Charles' privacy, and a specific function has been introduced in GSM to cater for such confidentiality.

Protection is obtained by using an identity alias, the TMSI (Temporary Mobile Subscriber Identity), which is used instead of the subscriber identity (the IMSI) when possible. This alias must be agreed before-hand between the mobile station and the network, during protected (ciphered) signalling procedures. Since this confidentiality feature is totally independent of the other security functions, the description of the corresponding mechanisms will be dealt with separately from authentication/ciphering later on in this chapter.

7.2.3. ARCHITECTURE AND PROTOCOLS

The actors and protocols involved in security management are almost the same as for location management, and this justifies their inclusion in the same functional plane. However, for security management, the starring roles are displaced, and must be attributed to the SIM on the mobile station side, and the Authentication Centre (AuC), which can be seen as a part of the HLR, on the network side.

The SIM and the AuC are the repositories of the key Ki of the subscriber. They do not transmit these keys, but perform the A3 and A8 computations themselves. As far as authentication and setting the key Kc are concerned, all other involved equipments are intermediaries.

The AuC is not involved in other functions than the ones just listed, concerning the GSM radio path security management. The AuC may be implemented as a separate machine or as modules of the HLR. The main reason for the distinction between AuC and HLR in the *Specifications* is to sensitise operators and manufacturers to the security issue. As mentioned earlier, all the security mechanisms described in this chapter rely on the secrecy of Ki. The AuC is a means to build an additional layer of protection around the Ki's.

The SIM takes responsibility for most of the security functions on the mobile station side. It stores Ki, it implements the operator-dependent A3/A8 and it also stores the "dormant" Kc. The existence of the SIM as a separate physical piece from the mobile equipment is indeed one of the elements enabling flexibility in the choice of A3/A8. The mobile equipment manufacturers need not be aware of the specifications of these algorithms for any operators. The SIM manufacturers, on the other hand, must implement potentially different algorithms for each of their operator-customers, but competition, mass-market production and distribution issues are totally different compared with the mobile equipment market.

The SIM protects completely Ki against reading. The smart card technology, introduced some time before GSM to produce these tiny electronic safes, was exactly fitting for this purpose. The only access to Ki happens during the initial personalisation phase of the SIM, when Ki is written in the SIM. This phase happens under the tight control of the operator. Later on, *Ki* is only accessed internally within the SIM when it has to compute SRES and Kc: a procedure on the SIM-ME interface allows the mobile equipment to send a value *RAND* and to get in return, typically a few tens of milliseconds later, the corresponding *SRES* and *Kc*. Another advantage of SIM-storage for *Ki* lies with the possibility, if security requires (e.g., as a regular measure, or if it turns out that the chosen A3/A8 is not as secure as expected) to issue a new SIM on a per-subscriber basis.

The MSC/VLR plays several small roles. It initiates authentication; it decides to switch to ciphered mode; it checks the SRES provided by the SIM (through the mobile station) with the one provided by the AuC (through the HLR); it stores the "dormant" Kc on the network side; and it manages the TMSI.

Ciphering is a transmission function, and as such it involves transmission equipments (the BTS for instance), and the radio resource management protocols (and the BSC). These aspects were tackled in the respective sections, and will not be revisited here. This chapter deals only with the decision to go to ciphered mode, as well as with the management of the needed parameters.

The security management functions at this level are supported by the same protocols (plus some others) as seen for location management. The RIL3-MM protocol supports the dialogue between the mobile station and the MSC/VLR, whereas MAP/D is used between the MSC/VLR and the HLR.

The SIM-ME protocol in the area of security management uses more than just read or write commands. Others are added to provide *RAND* and to request for an A3/A8 computation, as well as for getting back *SRES*.

The additional protocols compared to those used in the location management area include the protocol between HLR and AuC, which is only indicatively specified in the *Specifications*, as part of the general Operation and Maintenance GSM protocol. Last, a small additional protocol is introduced between MSC/VLRs to enable an MSC/VLR to ask another one for the identity and subscription data of a user, before access to the HLR. This protocol is used upon access of a mobile station identified by a TMSI with reference to another MSC/VLR. It is the MAP/G protocol and is limited to one operation.

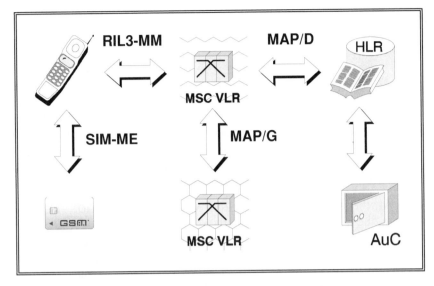

Figure 7.10 – Security management protocols

Security management is coupled with location management,
and makes use of the same protocols, with two additions:
a small protocol to transfer subscriber data between MSC/VLRs (MAP/G),
and the connection of the AuC to the HLR.

The protocol structure is summarised in Figure 7.10.

7.2.4. THE SIGNALLING MECHANISMS

7.2.4.1. Authentication and Encryption Key Management

As explained in the preceding section, the computation of the signed response for authentication and of the ciphering key Kc are performed simultaneously, based on the same inputs Ki and $RAND$, the latter changing every time. The signalling mechanisms used for managing the corresponding data in the network are strongly coupled and will therefore be described together.

The key Ki is a subscriber parameter. As such, it is stored in the HLR, more precisely in the AuC. The authentication and key setting procedure, on the other hand, is controlled by the visited MSC/VLR, which decides when to run this procedure: e.g., at call set-up, location updating, etc. There are then two different procedural aspects, distinct in time as we will see: the real-time authentication and key setting procedure between the mobile station and the MSC/VLR, and the procedure for transporting security related data between HLR/AuC and MSC/VLR.

The MS-MSC Procedure

The authentication procedure between visited MSC/VLR and mobile station consists of two messages: the RIL3-MM AUTHENTICATION REQUEST message from visited MSC/VLR to mobile station, transporting *RAND*, and the corresponding RIL3-MM AUTHENTICATION RESPONSE answer from the mobile station, giving *SRES* for checking in return.

At the reception of the RIL3-MM AUTHENTICATION REQUEST message, the mobile equipment sends to the SIM a RUN GSM ALGORITHM message, containing *RAND*, and immediately after a GET RESPONSE message whose answer contains *SRES* and *Kc*. *SRES* is sent back to the network in the RIL3-MM AUTHENTICATION RESPONSE message, whereas *Kc* is written back in the SIM at the right place, and so kept for further use.

An interesting procedural detail concerns the sequence number of *Kc*, also called CKSN (Ciphering Key Sequence Number) in the *Specifications*. To cope with possible inconsistencies between the dormant *Kc* on the infrastructure side, and the one on the mobile station side, a sequence number is associated with it. This number is provided by the MSC/VLR in the RIL3-MM AUTHENTICATION REQUEST message, and is stored together with the dormant *Kc* in the SIM and in the subscriber record in the MSC/VLR. This number is given back to the MSC/VLR in the initial message in the access procedure, so that it can be checked. If not consistent, the MSC/VLR knows that an authentication procedure is needed before ordering the ciphered mode. Value 0 of the sequence number corresponds to "no *Kc* allocated".

The MSC-HLR Procedures

The computation of *SRES* and *Kc* on the network side, requiring the knowledge of both Ki and A3/A8, must be performed so that its result is made available in the visited MSC/VLR. Two options are allowed in the *Specifications*: either the computation is done in the visited MSC/VLR, or in the AuC.

The first possibility (computation in the visited MSC/VLR) requires the visited network to cope with different A3/A8 algorithms depending on the home PLMN operator. In this scenario, the HLR has no specific role for this function, and the AuC does not exist: *Ki* is just another subscriber parameter among many others, to be stored and provided to other equipments when perchance requested. Ay! there's the rub: this scheme implies that the key *Ki* circulates through the SS7 network, introducing a weakness in the system security, because interception cannot be precluded. Moreover, international roaming requires agreement between operators as far as A3/A8 is concerned. Either a single A3/A8 algorithm is standardised on a multilateral basis, or each operator must undertake to provide the specifications of its own A3/A8 algorithm to all others, the most cumbersome aspect being the implementation of them all in each visited MSC/VLR!

The second solution overcomes both the security breach and the roaming problem, by having the computation performed in the HLR, in the AuC precisely. No need to transfer *Ki* any more, no need to divulge A3/A8 specifications either! However, signalling means must be devised to transfer the result of the computation from HLR to the visited MSC/VLR. In order to avoid such a transfer every time the visited MSC/VLR decides that it must authenticate, the computation is done in advance. For each computation, the AuC must draw a value for *RAND* and apply A3/A8. The result is a triplet of values: (*RAND*, *SRES*, *Kc*) to be sent to the visited MSC/VLR. The visited MSC/VLR stores a reserve of a few such triplets per subscriber, in which it can draw at need. This reserve is first established when the subscriber first registers in the visited MSC/VLR: it is part of the subscriber data provided by the HLR in the MAP/D INSERT SUBSCRIBER DATA message. Tight security requires that a triplet be used only once. As a consequence, when the reserve falls below some threshold, the visited MSC/VLR asks the AuC, through the HLR, for more triplets. The only accepted exception to this "throw away after use" rule is when a communication failure has occurred between the visited MSC/VLR and the HLR. The triplet replenishing procedure consists in two messages: the MAP/D SEND PARAMETERS message and its answer, the MAP/D SEND PARAMETERS RESULT message.

7.2.4.2. User Identity Protection

The Temporary Mobile Subscriber Identity (TMSI) is an alias for the subscriber identity used in order to avoid sending the IMSI in clear on the radio path. The TMSI is allocated by the network on a location area basis; at a given moment, it refers non-ambiguously to a subscriber when used in conjunction with the location area identity (LAI). Strictly speaking, the term TMSI should be used to refer to the full digit string composed of the LAI and of the digit string allocated at a given moment to a certain mobile station in this location area (which shall be called here "TMSI-code" or TIC). However, in the *Specifications*, TMSI is more

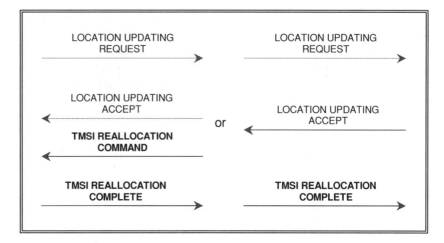

Figure 7.11 – TMSI allocation

A new TMSI can be allocated to a mobile station
through a standalone procedure consisting of two messages.
However, a more economical sequence is allowed when TMSI allocation
is performed in conjunction with a successful location updating procedure.

often used to refer to the TIC, hence with ambiguity when the context of the location area is not clear. On the radio path, most connections are set up in the location area in which the mobile station is registered. The TIC is in such cases sufficient to identify the subscriber non-ambiguously, the LAI being implicitly equal to the one the cell belongs to. The only exception to this rule is when the mobile station must perform a location updating attempt in a cell of a new location area: the full TMSI (including the "old" LAI) must then be used.

The length of the TIC is 4 octets, whereas the IMSI consists of up to 15 digits, coded in 9 octets including the length indicator. The short length of the TIC allows spectrum saving on the radio path when it can be used alone. This is especially the case for paging messages, where more than twice as many mobile stations may be paged in a single message when using the TIC, which cannot be ambiguous in these messages.

But the most interesting area of study concerns the management of TMSIs, i.e., how and when they are allocated and released. In the network, TMSIs are managed by the MSC/VLR (by the VLR in the full-blown canonical architecture). A TMSI is first allocated to a mobile station the first time it registers in the location area, and is released when the mobile station leaves it. A standalone allocation procedure uses two messages: the RIL3-MM TMSI REALLOCATION COMMAND message from MSC/VLR to mobile station and the corresponding RIL3-MM TMSI

REALLOCATION COMPLETE acknowledgement. However, when the TMSI allocation is performed just after a successful location updating (which is usually the case), the allocation message can be "combined" with the RIL3-MM LOCATION UPDATING ACCEPT message from the network: the new TMSI is then part of the RIL3-MM LOCATION UPDATING ACCEPT message and the RIL3-MM TMSI REALLOCATION COMMAND is dispensed of, but not the acknowledgement. The full sequence of 4 messages is however allowed as well. The corresponding sequences are shown in figure 7.11.

TMSI cancellation is usually implicit. In the mobile station, cancellation is automatic upon allocation of a new TMSI, or upon location updating acceptance in a new location area. Explicit cancellation can also be done, by sending the IMSI in the RIL3-MM LOCATION UPDATING ACCEPT message: this is to be understood by the mobile station as a cancellation of the previous TMSI.

Another feature of the TMSI allocation procedure enables the network to use a short version of the RIL3-MM TMSI REALLOCATION COMMAND message in order to allocate a TMSI having the same TIC part (the 4 octets meaningful within a location area) as the previous one. This is achieved by not mentioning any identity in the RIL3-MM allocation message (whether TMSI REALLOCATION COMMAND or LOCATION UPDATING ACCEPT). The gain brought by this feature seems however negligible.

Table 7.1 shows a summary of how to interpret the allocation message depending on its contents as far as the *MOBILE IDENTITY* parameter is concerned.

The TMSI is stored in the subscriber's record held by the MSC/VLR, but not in the HLR. When the record is destroyed, for example upon location cancellation from the HLR, the TMSI is then cancelled implicitly in the infrastructure. This raises some problems when database failures are considered. For instance, some situations may result

Identity used in the message	New TMSI
none	LAI (new) + TIC (old)
TIC (new)	LAI (new) + TIC (new)
IMSI	none

Table 7.1 – Rules for deriving a new TMSI upon allocation

Depending on the contents of the message sent by the network,
a mobile station can derive the new value—if any—of its TMSI.

in the mobile station using a TMSI which is not allocated to its subscriber any more, or worse which is allocated to another subscriber. In order to cope with such situations, some tools have been included in the *Specifications*.

In particular, a procedure allows the network to ask the mobile station for its full IMSI. Typical usage of such a procedure is when the TMSI by which the mobile station identifies itself is not known in the MSC/VLR, or, if known, after the failure of authentication, which may reflect a TMSI discrepancy between mobile station and network. The identification procedure consists of two messages, the RIL3-MM IDENTITY REQUEST message sent by the network and the corresponding RIL3-MM IDENTITY RESPONSE answer by the mobile station. The procedure is in fact more general than a request for the IMSI, since other types of identity may be asked for.

When a database failure has occurred in the MSC/VLR, all the TMSIs stored in the salvaged records are dubious. In this situation, when an incoming call arrives, the MSC/VLR pages the mobile station throughout its area (instead of just in one location area) using the IMSI instead of the TMSI.

A side effect of the TMSI is its usage in an RIL3-MM LOCATION UPDATING REQUEST message sent in a location area managed by an MSC/VLR other than the one which allocated the TMSI. We have seen that in this case the mobile station gives the full TMSI, and not only the TIC as for a call setup or a response to paging. While the full TMSI is unambiguous, it does not indicate the HLR or even the home PLMN as the IMSI does. But it does give the indication of the MSC/VLR which allocated the TMSI, and which then presumably knows the corresponding IMSI.

The new MSC/VLR has two possibilities. The first is to ask the mobile station for its IMSI. This is simple, but this is a breach in the protection of the subscriber identity. The other possibility is to request the IMSI from the previous MSC/VLR, as indicated by the LAI part of the TMSI. This is supported by the MAP/G procedure (the only one in this protocol), consisting of a MAP/G SEND PARAMETERS message and the answer, the MAP/G SEND PARAMETERS RESULT message. The first message includes the list of the desired data, i.e., the IMSI and optionally authentication triplets. The answering message includes the requested data. An authentication triplet allows the new MSC/VLR to perform the authentication before the HLR answer, in order to gain time.

7.3. MISCELLANEOUS MM FUNCTIONS

As explained at the beginning of the chapter, the Mobility Management realm includes more than just location management and security management, at least if all the procedures in the RIL3-MM protocol are taken into account. Even though this chapter might be a debatable location for some of these other functions, it was felt best to keep their description here, so as not to wander too far away from the *Specifications* description. It should not be forgotten that the goal of a structured protocol modelling is intended as a help for understanding more than as an implementation guideline. A protocol architecture can therefore be to a great extent subjective.

There are four MM miscellaneous functions not yet described. All of them concern only the mobile station and the MSC/VLR. In fact only two of them involve procedural exchanges (in the RIL3-MM protocol), the two remaining ones being specifications of the mobile station behaviour. These procedures or specifications are related in so far as they use a common modelling concept, the MM-connection. This should not be mixed with the concept of CM-transaction (CM for Communication Management), which refers to a transaction in the upper layer, the Call Control plane. A CM-transaction corresponds to a call transaction (RIL3-CC protocol), to a Short Message transaction, or to a Supplementary Service management session. A CM-transaction corresponds then to all the activities described in the next chapter. After the presentation of these functions, we will discuss how useful this modelling is.

Generic Mobile Originating CM-Transaction Establishment

In ISDN call set-up procedures, the first message on the originating access interface is the SETUP message. This message contains a lot of information, including the calling number. In GSM, privacy on the radio path requests that this first message be ciphered. However, because the decision to cipher lies with the infrastructure, a preliminary message from the mobile station is necessary to give enough information to the network for it to decide whether to apply ciphering (and authentication) or not. An alternative would have been to cipher systematically.

Because this preliminary message does not exist in ISDN, because of the will to keep call control as little adulterated as possible by the specific aspects of radio transmission, and because the same need exists for other kinds of services such as short message transfer or supplementary services, a generic mobile originating establishment procedure was introduced as part of the RIL3-MM protocol. When initiated by the mobile station, a call, a short message session or a supplementary service management session all must use the generic establishment procedure, even if transmission means are already established and used in ciphered mode. This allows the infrastructure to perform authentication, and/or to go to the ciphered mode before any further progress of the session.

There is no equivalent to the mobile station-originated generic establishment procedure if the session is initiated from the network side, simply because the network chooses to apply authentication and/or ciphering before starting any upper layer procedure.

The generic mobile station originated establishment procedure consists basically in a preliminary message sent by the mobile station, the RIL3-MM CM SERVICE REQUEST message, and a signalling sequence in answer from the MSC. The RIL3-MM CM SERVICE REQUEST message may be an initial message, as described in Chapter 6. The reactions of the MSC can be to start an authentication procedure (RIL3-MM AUTHENTICATION REQUEST message), or to answer positively the request. This can be done by sending an RIL3-MM CM SERVICE ACCEPT message, which seems normal, or by starting a ciphering mode setting procedure (RIL3-RR CIPHERING MODE COMMAND message), which is a curious specification. Still another possibility for the MSC is to reject the request by sending an RIL3-MM CM SERVICE REJECT message. If the answer is positive, the mobile station may start to initiate the procedure which justified the previous actions, for instance by sending an RIL3-CC SETUP message.

The only reason for "mixing up" the generic establishment procedure with the cipher mode setting procedure (from another protocol!) lies in the will to reduce the number of messages. A shortcoming of this method is however the ambiguity thus introduced: there is no means to distinguish an RIL3-RR CIPHERING MODE COMMAND message acknowledging a CM-transaction establishment and one which does not. "Collision" cases may well occur where the MSC wishes to start a ciphered session at the same time as the mobile station initiates a CM-transaction for another purpose. This situation will likely be improved in phase 2.

Another important point to note is the lack of connection reference in an RIL3-MM CM SERVICE REQUEST message and in the corresponding acknowledgement. This leads to ambiguity if two generic establishment procedures are run in parallel. Hence, this is forbidden by the *Specifications*. We will come back to these issues at the end of the chapter.

Upper Layer Synchronisation

The *Specifications* require that CM-transactions cannot be initiated while a location updating procedure is running. The requirement is even more stringent in GSM phase 1: the mobile station must go back to idle mode before setting up a new RR-connection aimed at supporting a CM-transaction.

This requirement arises from the need for a subscriber to be correctly registered with the network before accessing any service. In other words, the rationale requires that no other Mobility Management or Call control procedure be started during a location updating procedure in a location area different from the one in which the mobile station was previously registered, at least up to the reception of the RIL3-MM LOCATION UPDATING ACCEPT message. An alternative would have been to allow the mobile station to anticipate the closing of the location updating procedure and to require the MSC/VLR to store the service request until (and if) location updating is successful with the HLR. This is not allowed in the phase 1 *Specifications*. A milder position could have been to allow the mobile station to send a request just after the reception of the RIL3-MM LOCATION UPDATING ACCEPT message, or to anticipate in the case of a periodic location updating. But even this is forbidden in phase 1. The only reason for such a drastic approach was the simplicity of the MSC/VLR.

Whatever the shortcomings, the specification must be implemented as it is. The impact of this synchronisation function, modelled in the MM plane in the *Specifications*, lies only with the mobile station and does not involve any procedure.

Infrastructure Activity Monitoring

A third function modelled within the MM plane and also not involving any protocol procedure is a watchdog for MSC signalling activity. Radio channel release is indeed a privilege of the infrastructure. In case of failure, the mobile station may then find itself in a difficult situation, with an unused dedicated channel which it is not allowed to release explicitly. In such cases, the mobile station must go back to idle mode autonomously. This requires that the mobile station continuously checks if there is—to its knowledge—a CM-transaction in progress. In the opposite case, the mobile station waits some time and decides to go back to idle mode if nothing happened in-between, without sending any message to the network. The corresponding watchdog function is modelled in the *Specifications* through a timer called T3240. This timer could just as well be part of the RR plane, but is specified in the MM plane of the *Specifications*.

Re-Establishment

When a mobile station, being provided with some service, suddenly looses contact with the infrastructure, there is a possibility to resume this contact, for instance in another cell. Such cases may happen for example in configurations where the handover procedure proves to be too slow. A salvaging attempt by the mobile station has been introduced in the *Specifications*, and has been modelled in the MM plane. This is called the re-establishment procedure. All mobile stations must support this procedure, but it is optional on the network side. If it is supported, call contexts must be kept a little while after the contact loss, to allow potential re-establishment to be effective.

This feature is very close to what is called mobile station-triggered handover in other systems. It fulfils the same requirement as handover, but in a much less controlled way, though with possibly a better efficiency in configurations where propagation loss is very steep. Because of this analogy, the re-establishment procedure has been studied in Chapter 6. The initial message to request a re-establishment is the RIL3-MM CM RE-ESTABLISHMENT REQUEST. It is then worth noting that the acceptance and the rejection by the network uses the same message as for the generic CM-establishment procedure: respectively the RIL3-MM CM SERVICE ACCEPT message and the RIL3-MM CM SERVICE REJECT. The re-establishment procedure is then deeply entangled with the generic CM-transaction establishment procedure, a point which renders difficult the evolution of the re-establishment procedure.

Modelling

The four miscellaneous functions described above may seem quite ill-assorted from an architectural point of view. The two last ones would fit better in the Radio Resource plane for instance. There is however a common denominator between the four functions in the *Specifications*:

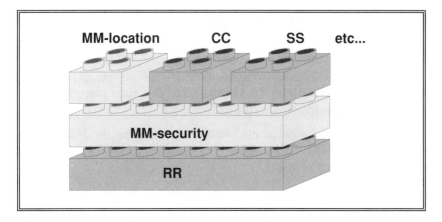

Figure 7.12 – Alternative MM modelling

GSM MM functions could be modelled in two blocks: a location management block
on the same level as call control, etc. and an intermediate layer providing service
to the location management block as well as to other upper layer blocks.

the concept of *MM-connection*. This pure modelling concept was not used here because it is not necessary and does not in fact help understanding. In fact the notion of MM-connection does not appear concretely in the protocols. There is a one-to-one correspondence between CM-transactions and MM-connections, which renders useless the insertion of a CM-connection identifier in the messages. MM-connections are explicitly established only when initiated from the mobile station side (by the generic Mobile Originating establishment presented above), but not when initiated from the MSC (the establishment is implicitly done by the establishment of the CM-transaction). Moreover it is never explicitly released.

An alternative to the modelling approach used in the *Specifications*, possibly reflecting the different roles of the MM plane more accurately, would consist in considering the MM functions in two separate groups (see figure 7.12). On one side, a location management functional block which would stand in parallel and at the same level as call control, supplementary service management, etc. A location updating connection would then be considered the same way as a CM-transaction. Below such blocks, and above the RR layer, a second functional MM block would provide security related services to the upper layers, namely authentication, ciphering key management, as well as the generic Mobile Originating establishment procedure and a synchronisation function whose sole aim is to forbid the establishment of upper layer procedures while a location management connection is in progress.

SPECIFICATIONS REFERENCE

The problems raised by inter-operator roaming are exposed from the service point of view in **TS GSM 02.11**. The main part of this short document concerns the choice of PLMN.

The technical aspects of location updating in the NSS are the subject of **TS GSM 03.12**. This specification mainly introduces the procedures included in the MAP.

It is worth noting that the general aspects of location updating, PLMN and cell selection in the mobile station will be presented in general terms in the future TS GSM 03.22, unfortunately not present in the phase 1 *Specifications*. Though requiring some caution, because of the functional differences between phase 1 and phase 2, this document can be useful.

The details of the cell choice algorithms, including measurement considerations are addressed in **TS GSM 05.08**, section 6.

The signalling aspects of the MM protocol between the mobile station and the infrastructure (dealing with all the topics addressed in this chapter, location updating as well as security management) are dealt with in **TS GSM 04.08**, section 4.

A very good synthesis of the general scheme for security management can be found in **TS GSM 03.20**.

The MAP protocols in general are the subject of **TS GSM 09.02**. Of relevance for this chapter, one can cite section 5.2 (location registration/cancellation), section 5.8 (fault recovery of location registers, a subject introduced in **TS GSM 03.07**), section 5.11 (management of security related functions), and section 5.15 (paging and search procedures).

COMMUNICATION
MANAGEMENT

COMMUNICATION
MANAGEMENT

Managing calls, that is to say principally establishing and releasing transmission paths through meshed networks, is not by far a new topic. The development of fixed networks, mainly the PSTN, and the ISDN nowadays, has seen an important evolution of signalling techniques. Public cellular networks in general, and GSM in particular, are basically access networks for these general telecommunication systems. As such, the design of their communication management signalling is very dependent on existing techniques, and offers few novelties. For instance, the signalling exchanges at the interface between GSM and the external networks are imposed by these networks. The influence goes even further, and the communication management signalling procedures defined between mobile stations and the GSM infrastructure are simplified and somewhat adapted copies of that specified for ISDN access. A part of what is described in this chapter is therefore not really specific to cellular networks, and could equally apply to fixed terminals. However, it is part of the *Specifications*, and this description is included to complete the full picture of the system.

Still, call control is not inherited totally from ISDN. Cellular systems bring some problems of their own, due to the mobility of users and to the fact that there is no fixed link between each user installation and the infrastructure. The central issue is the establishment of calls toward users that move around. One facet of the problem is the way in which the system must follow the movement of users between calls, in order to readily find them when need be. This has already been treated in Chapter 7 (Mobility Management). The other facet is the routing of the call through networks to a GSM user. This is part of call control and will be detailed in this chapter.

The technical issues related to the establishment of calls toward mobile users is the reason for a number of novel approaches for telecommunication networks. One of these important new concepts is the notion of the GMSC (Gateway MSC, an ill-coined term from the conceptual point of view, since the "GMSC" *function* has nothing to do with the MSC *function*, even if they are often grouped in the same equipment), which is the central actor in call routing. This notion is deemed to be important in the future, not only for mobile networks, but for all public telecommunication networks. Their future lies undoubtedly with the development of Personal Communication, i.e., the concept that calls are not directed to locations, as it is the case in PSTN and ISDN at the moment, but to people, who can be reached through different telecommunication means, according to their present whereabouts. Mobility is an essential point for Personal Communication, and the GSM technology provides if not a ready-for-use solution, at least a case of pioneer work in the subject.

Another area where the signalling protocols developed for ISDN do not fit all the needs of GSM is the management of the radio interface, with the specific problems raised by the complex transmission scheme over the radio path, and by the movements of the users during a communication. These issues have been dealt with in Chapter 6 (Radio Resource Management). As far as this chapter is concerned, the path between the MSC and the mobile station is considered to be a simple fixed link. This is again an application of the "divide and conquer" approach. This kind of presentation allows a more consistent description,

---

GSM

=
ISDN-like call control
+ Mobility Management
+ Radio Resource Management

Figure 8.1 – GSM functional split

The split of signalling aspects into three major domains enables a description
of a communication without taking into account
the management of user's mobility and of radio resources.

not going this way and that between different subjects (on the other hand, the complete picture of how a call is established in GSM can only be grasped by considering both points of view).

Thanks to this approach, illustrated in figure 8.1, the substance of this chapter can really focus on the notion of end-to-end communication.

The key "objects" that call control signalling deals with are communications, and we will organise the presentation in this chapter on how communications come into and out of existence, and how they change. We will first try to define what a communication is, and what are its attributes, in an attempt to identify the key functions of call management.

8.0.1. THE COMMUNICATION

Basically a communication is a temporary relationship between telecommunication users, for the purpose of exchanging information. A communication makes use of a transmission chain established through networks between users. The only communications of interest here are the ones involving at least one GSM user. A communication is by essence a "distributed" object, which exists over distances. Many of its characteristics must be managed co-operatively by the different machines which appear along the transmission path. This management concerns both static attributes, set at the beginning of the communication, and dynamic ones, which are modified during the lifetime of the call, usually at the request of the user.

8.0.1.1. The Users

The first thing to be considered in a communication link are the users, or "parties". The basic communication case involves two users, but GSM will provide (in phase 2) the possibility to have communications involving more than two users. Though not a phase 1 capability, we will address this "multi-party" possibility, at least at the functional level, principally because this allows a wider approach to the concepts of call control, which could easily be biased when only two-party communications are considered. A call is always started between only two users. As will be seen, the multi-party possibility allows the option of adding or removing users at any moment once the communication is fully established, making the attribute "users" a dynamic one.

But users are not functionally identical. One of the users has a special status: the one who originally requested the establishment of the call. This user is called the "calling party". The other one is the "called party". There is here a difficulty in terminology. Because of the possibility of call forwarding, the subscriber whose number is dialled by the calling party is not necessarily the "called party" once the call is fully established. Let us recall the example of Chapter 1, where Björn wanted to call Nina, but got Hans instead because Nina had requested her calls to be forwarded. The use of the term "called party" is therefore ambiguous, at least in the establishment phase. Unfortunately, no established terms are used to distinguish the two notions. Where need be we will use the terms "forwarded-to party", or "connected-to party". Still to make things somewhat more complex, additional users can be added by either of the existing parties. New ones are then "called parties", but the notion of "calling party" is relative to each called party. Well, it is hoped that the context will be sufficient to understand what these terms mean!

The asymmetry between users is significant for the whole duration of the communication. For instance, in fixed networks, the call will be released at once if the calling user hangs up, but will remain operational for a few seconds if the called user hangs up, enabling him, for example, to resume the call on another terminal. Another example of asymmetry is charging. But the asymmetry is obviously most important for signalling during the establishment phase. Because our purpose here is to focus on GSM users, two-party communication can be split into three categories:

- calls where the GSM user is the calling party. These are the "mobile originating calls", or more simply "MO calls";
- calls where the GSM user is the called and/or the connected-to party. These are the "mobile terminating calls", or more simply "MT calls";
- calls where the two parties are GSM users.

The last type of call bears no specific name, and being a combination of the other two, will not be considered separately.

8.0.1.2. The Partner Network

More generally, another important characteristic of the communication is the network to which the other party (other than the GSM user under consideration) belongs. This network will be referred to as the terminating network, for an MO communication, or the originating network, for an MT communication. GSM is designed to allow communications between its users and users from a variety of external networks, such as the PSTN, the ISDN, CSPDNs and PSPDNs. Establishment procedures in particular, and communication management in general will be heavily influenced by the nature of the other network. We will concentrate on the PSTN and ISDN cases, since they represent the most important part of the traffic by far.

Also important when involved in a communication link are transit networks. Many different network configurations can be imagined. There again, for practical reasons, we will address mainly simple cases, typically when two networks only are involved, one being a GSM PLMN, the other the PSTN or the ISDN.

8.0.1.3. The Type of Service

Continuing with communication characteristics, the type of service which is provided during the communication is worth some attention. The "type of service" is a concept which is concerned with the type of information exchanged between the end users (speech, fax, data, ...), but it encompasses more than that. The "service" is more generally a subscription and management concept. Its prime characteristics correspond to the capability of the transmission path which is established between the users. These "bearer capabilities" correspond to the transmission as described in Chapter 3, that is to say (from the GSM point of view) the connection type between the terminal adaptation functions inside the mobile station (the TAF) and the interworking functions at the border with external networks (the IWF). GSM can deal with many connection types, and these have been summarised in Chapter 3.

Some services call for two bearer capabilities, so that the service can alternate between them during a call. "Alternate" services make use of the speech bearer capability and a data bearer capability (other cases could be imagined, but are not to be found in GSM). The interest of these services is that the call can be started in speech mode, e.g., for the users to agree orally on some details, before the data transmission is started. Such services require the system to be able to switch between the two bearer capabilities during a communication. This is not a simple function, and this is why the network must know that the parties desire an alternate service during the call establishment phase, so that it can reserve the relevant devices. The alternate operation is an attribute of the service, and not a facility available for any speech or data communication. The procedure which enables the toggling between two bearer capabilities is the "in-call modification" procedure.

For some services, the description of the capabilities of the terminals used by the end parties is relevant. These services are the "teleservices". Along with the bearer capability (or capabilities), their description may include the "low layer compatibility" (LLC) and the "high layer compatibility" (HLC) of the terminals. The "low layer compatibility" includes parameters for the basic transmission from terminal to terminal, such as the minimum duration of the stop signal in the case of asynchronous data, whereas the "high layer compatibility" includes characteristics related to the terminal to terminal protocol, such as teletex protocols, or videotex protocols. These attributes are not specific to GSM, and the corresponding parameters are directly inherited from ISDN. Moreover, they are not usually necessary for the intervening networks, which obtain sufficient control information from the bearer capability. However, the indication of these attributes is the first phase of the negotiation between the terminals, and this allows an early compatibility check, which may result in the termination of the call if the terminals are unable to communicate together correctly. Important teleservices are speech (basic and emergency calls), fax and videotex.

Other services do not include anything concerning the terminals. They are the "bearer services", and their description as far as information exchanges are concerned is limited to the bearer capabilities. The difference between a bearer capability and a bearer service is that the latter is used in the realm of subscription whereas the former is restricted to transmission capabilities. Services must be subscribed to individually, and supplementary services are linked to the basic services (i.e., bearer or teleservices). For instance, the forwarded-to number used for speech may differ from the one used for fax.

8.0.1.4. Active and Held States

Communication attributes described so far were mostly of a static nature, except the number of called parties and the choice of bearer capability in the case of alternate services. Another dynamic aspect of a communication is the possibility for one or the other of the parties to temporarily leave the communication (typically to establish another one) and then to resume it. Such possibilities are referred to as "to put the call on hold" and then to "retrieve" it. They will be available in phase 2, and are just cited here to complete the picture. From the communication point of view, two exclusive states exist for each of the involved parties: the call can be either held or active. If, for any reason, one of the two parties is in the held state, user data is not transmitted, but all of the transmission resources are retained to allow a fast resumption of transmission when so desired by the party who asked for the held state. Of course, the charge counter is also kept running...

Because there is the possibility to establish or to answer another call while a first one is on hold, a given user can be part of several communications at the same time. However, only one communication can be active at a given moment for a given user, all others being on hold (except for short messages, which are transmitted like signalling messages and hence can be sent and received in parallel with other basic services). In the future, with half rate channels, it could be imagined that two communications are active at the same time with a single terminal, each on a different half rate channel. This is something that is foreseen, and for which some procedures are already documented in the phase 1 *Specifications*.

8.0.2. MANAGEMENT FUNCTIONS

This panorama of the attributes of communication has already hinted at most of the operations that can affect a communication. Foremost is the establishment of the communication, which must result in the setting of all the attributes, and in the establishment of the end-to-end transmission path (this description excludes the Short Messages, for which there is no establishment of a circuit-like transmission path. Short messages are addressed in a separate section).

Most **attributes of a communication** are fixed between the calling party and the infrastructure machines, more precisely with the MSC, during the establishment phase. Some of the attributes are chosen solely by the calling user, such as the directory number of the called party or the type of service. Other attributes are fixed by the infrastructure alone, such as the actual connected-to party, which depends on forwarding conditions managed by the infrastructure under the control of the called party. Finally, some attributes are negotiated between the user (or the mobile station) and the MSC. An example of such negotiation will in the future be the choice of the type of channel on the radio path, when the required service may be accommodated on either a full or a half rate channel.

Then comes **the setting of the transmission path**. Since GSM is mainly an access system, most of the job is done by external networks. For a Mobile Originating call, the MSC must analyse the called number and the requested service in order to choose the external network towards which the call will be routed. It has then to reserve a suitable link toward this network and to proceed with the relevant signalling, according to the access rules of the chosen external network. Afterwards, the MSC (and the GSM user) will just wait and react according to the signalling information available about the progress of the call.

In all standard telecommunication networks, the transmission path is set up segment per segment, from caller to called user. Each node receiving the call establishment request analyses the called number, then determines the next node, reserves a link toward this node, and passes on the ball. The path is built in this way until it reaches the node in charge of the called party. Routing tables are used in each node, giving the rules for translating the number in a link toward the next node. This simple technique works well because the called party is in a fixed location. The story is somewhat different when the called user moves around, as a GSM subscriber does. Then the called number is not sufficient to determine where to route the call, and a more sophisticated approach is necessary. **The routing of MT calls** is a complex matter that warrants a detailed study and a whole section will be devoted to this subject later.

Once the transmission path is established and the end user has accepted the call, the call is "connected", meaning that the transmission path is fully available for the transmission of data, at least between the TAF and the IWF. In the case of speech, this means that end users can hear each other, but in the case of data, further establishment steps have to be taken before user data can truly be exchanged end-to-end. We will come back to this issue.

During the "connected" phase, also called the "active" phase of the call, a small number of functions are provided. They concern mainly **the management of alternate services** and (in phase 2) the management of multiple calls. During an alternate service communication, the GSM user can request a toggling of the bearer capability (from speech to data, or vice versa). This request must be transmitted to the MSC, which then controls the changes along the transmission path (e.g., in the IWF) to meet the new requirement.

Multiple call management corresponds to multiparty calls, or the ability to resume calls or put them on hold. Here again, the MSC is the director, and signalling between the mobile station and the MSC enables a co-ordination of the operations.

In-call functions also include **the transmission of digital audio tones** (DTMF tones, for Dual Tone Multi Frequency). These tones, usually generated by pressing the keys on a modern telephone set, are used for the control of, e.g., voice mailboxes, answering machines, etc. Though DTMF transmission is not really part of the management of the communication, being more akin to a transmission facility (providing a data transmission capability of a few tens bit/s!), it has been chosen to support it in GSM by using signalling exchanges in the Call Control area. Let us recall that the speech coder in GSM has been optimised for speech signals, and DTMF tones going through a coder-decoder ordeal will not meet the quality required by external networks. This is why signalling means are used for conveying DTMF signals on the MS to MSC segment.

The last function needed for circuit call management is an elementary, but very important one, **the release of the call**. This can be initiated by any one of the users, and is controlled by signalling between the users, the mobile station and the MSC.

We have so far left to one side two functional areas, the management of Supplementary Services and the transfer of Short Messages, and these will now be introduced.

GSM enables its subscribers to inspect and modify the status or the parameters of their **supplementary services**, by using their mobile station and transmitting over the radio path. This can be done at any time, irrespective of whether a call is in progress or not. Some supplementary services can be activated or deactivated at the subscriber's request, such as for instance the call barring services. Another example is the call forwarding services, where the forwarded-to numbers can be changed at will by the subscriber. In relation to these modifications, a password can be set by the user, to improve his level of security. For instance, a password set in connection with the barring of all outgoing international calls enables the GSM subscriber to lend his SIM without having to fear that international calls will be charged to him.

All these supplementary service management functions are done between the user and his home location register (HLR), with the use of the mobile station. They call for a few signalling procedures. It should be noted that these management functions represent only one facet of the signalling related to supplementary services. The rest relates to the actual impact of the active services on call handling, introducing variations of the call establishment procedure. As such, they are described later in the section dealing with call establishment.

Short message communication differs from circuit communication in the fact that the establishment, the transmission of user data and the release are all done in one procedure (at the communication management level). This is close to a datagram technique. There is no establishment of a dedicated transmission path, since only signalling means are used. Another difference comes from a special function added for improving the quality of service when the called party cannot be reached: the undelivered messages are kept, and another attempt to send can be initiated when the destination subscriber becomes reachable. This mechanism calls for signalling exchanges between network entities to alert the short message service centre.

This completes our list of the communication management functions, and we can now look in more details to each of these issues.

8.1. CALL CONTROL

Before describing the architecture and the procedures involved in call control, we will analyse how calls to a mobile subscriber can be routed, since this aspect underlies much of the network architecture, and is of foremost importance in understanding many aspects of the system, such as, e.g., how to call a GSM user or charging.

8.1.1. THE ROUTING OF MOBILE TERMINATING CALLS

For a Mobile Terminating call, the number given by the calling party does not refer to a telephone line or a location; but points to a record in some HLR. The first digits of a GSM directory number are sufficient to indicate that the number is a GSM number, and furthermore to designate the operator with which the subscription is held. The structure of the GSM directory number, also called "MSISDN" because it is part of the same numbering plan as ISDN numbers, is defined in

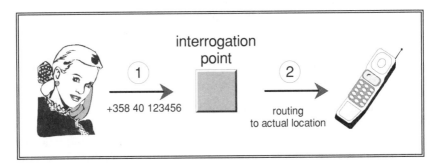

Figure 8.2 – The two parts of a mobile terminating call route

The call route consists of two parts: the first part is based on the called directory number, and continues to the point where the actual location of the called GSM subscriber is taken into account for the rest of the route.

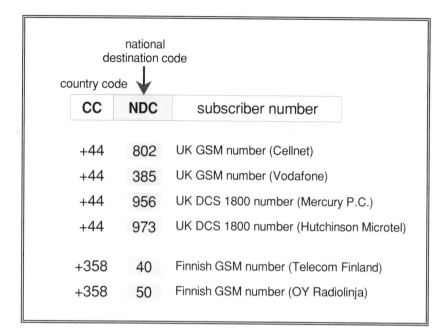

Figure 8.3 – The structure of a GSM directory number

A GSM "MSISDN" looks like a standard PSTN or ISDN number,
but the knowledge of the "National Destination Code" (NDC)
identifies an operator within a country, and not an area code.
The first digits following the NDC are used to identify
the relevant subscriber's HLR within the home PLMN.

CCITT Recommendation E.164 and is shown in figure 8.3. The HLR holding the record of the subscriber can be determined by the analysis of the first digits of this number.

The HLR record contains information necessary for finding the final destination of the call, i.e., the MSC where the GSM user is currently visiting. As a consequence, the final routing can be done only after the interrogation of the HLR. This splits the call establishment into two parts: before the interrogation, and after the interrogation. This corresponds also to a clear division of the call route into two parts: from the call originating point to the interrogation point, and the rest, as represented in figure 8.2.

What follows applies to Mobile Terminating calls requiring the establishment of a circuit. The routing of Mobile Terminating Short Messages is similar, but bears a few differences and will be described separately in the section dealing with Short Message Services. Another point is that what is specified in GSM applies mainly, if not only, to the

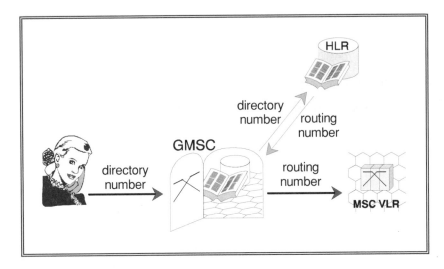

Figure 8.4 – the key role of the GMSC

A mobile terminating call is first routed towards a GMSC,
i.e., a switch able to interrogate the mobile subscriber's HLR
to know where to actually route the call.

cases where the second part of the route goes through the PSTN or the ISDN, not through a packet-switched or circuit-switched data network (PSPDN or CSPDN). An MT call coming from these latter networks must then either enter directly GSM before reaching the interrogation point, or use the PSTN or the ISDN as a transit network. We will then restrict our presentation to MT circuit calls involving only general networks, such as the PSTN, ISDN or GSM.

The first part of the routing is done only with the information that can be derived from the called number (the MSISDN) independently from the called party location. This routing is done as for any ISDN number, with tables in each of the intervening switches. The routing tables are normally set so as to reach rapidly the switch which is able to interrogate the corresponding HLR. Not only does such a switch include the software necessary for running the interrogation procedures, but it also holds a table which relates an MSISDN with the corresponding HLR. This function is referred to as the GMSC (Gateway MSC) function, and its role is shown in figure 8.4. The interrogation of the HLR is a simple request-answer procedure as seen from the GMSC. The answer contains the identity of the subscriber (for billing purposes), and the information for the next routing step. This information is basically a routing number pointing either to the called GSM subscriber in his current location, or to a third user in the case of call forwarding.

This simple description raises several interesting points. First, what is exactly a switch capable of the GMSC function? Second, where does the routing number come from? Third, who pays for what? The last question being the key to the first one, let us start with the charging issue.

8.1.1.1. Who pays What?

The charging and tariffing policies are to a large extent outside the realm of the *Specifications*. They are an operators' issue, dealt with in the GSM MoU meetings. However, the answer has some technical consequences, and so is worth some consideration. The charging principles which follow have been extracted from various public conferences of the GSM MoU operators.

In the case of a non-forwarded call, two parties are involved and may be charged: the calling party (say Woldemar, a German PSTN subscriber) and the called GSM subscriber (say Peter, who is a subscriber in the Netherlands, but who happens to be currently travelling in Spain and obtains service from the Spanish TELEFONICA GSM network). Several networks are entitled to demand some part of the call charge. These are the German PSTN, the transit PSTNs or ISDNs along the way, and the PLMN of TELEFONICA which will establish the final segment of the path. Woldemar will receive his bill from the German PSTN, and Peter, if any charge is levied on him, will be billed by his home PLMN, i.e., in the Netherlands, and in Dutch currency. The other networks are not able to bill these subscribers directly. The matter is then not simple, and accounting transfer between some of the involved networks will be necessary, as well as specific agreements between the operators. Some practical considerations must be taken into account and some rules must be fixed to simplify the combinations.

The total cost of a call depends obviously on the location of the GSM subscriber. It is also clear that the calling party would like to know in advance how much he will be charged for the call. Moreover, it seems that GSM subscribers are not willing to have anybody calling them know where they are located, even with little accuracy. These arguments lead to the principle that the charge levied on the calling party is independent from the actual location of the called party. This philosophy is in line with the similar problem of a forwarded call in the PSTN: in this case, the calling party is usually levied the same charge as if the call was not forwarded, and the party which asked for the forwarding is charged for the complement.

This principle still leaves room for different solutions. In a first extreme solution, the rate of the charge levied to the calling party can be designed so that on average it covers the cost of the call, wherever the location of the called party (and including the charge for the radio segment); and the called party never pays anything. Another extreme, and opposite, solution consists in charging the calling party with only the minimum charge (the minimum being taken on all possible locations of the called party), and have the called party pay systematically for the rest of the cost.

But levying a charge on the called party is not particularly easy, since it means raising charges on calls that the user would sometimes have preferred not to receive. In fixed networks and in most cellular networks in operation today, the entire charge is supported by the calling subscriber (except in the case of forwarding, as explained above). It seems then good practice to do the same when the subscriber is located in his home PLMN (in our example if Peter was still in the Netherlands), i.e., to charge only the originator of the call. This will still require compensation mechanisms between operators, but in a more deterministic way.

Let us consider several examples, to understand better how this principle applies. Starting with the simplest case, let us see what happens when Woldemar (using the PSTN) calls a German GSM subscriber, say Frieder, who is roaming in his home PLMN. Figure 8.5 illustrates the

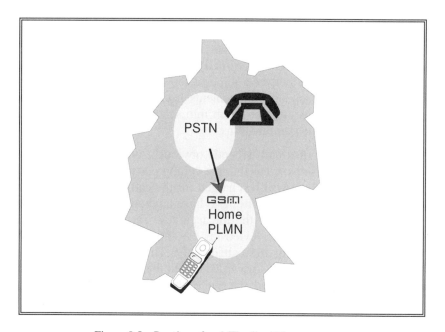

Figure 8.5 – Routing of an MT call within one country

When the called GSM user roams in his home PLMN, a call from a fixed user in the same country is simply routed to the home PLMN through, e.g., the PSTN.

case. Woldemar is the only one charged. The amount of compensation paid by the German PSTN to Frieder's home PLMN, if any, is a matter of agreement between the operators. It seems logical that this compensation should cover the costs concerning the "radio" segment (MSC to MS) at least, and possibly more depending on the point of entry in the GSM network. However, since for Mobile Originating calls the situation is reversed (the charge is recovered by the GSM operator, who should give some money to the German PSTN for routing the call toward the called user), there is not necessarily a positive flow of money at the end of the month from the PSTN operator to the GSM operator! In any case, the German telecommunication networks taken as a whole cannot obtain any more money for a Mobile Terminating call than what is levied on the calling user by his PSTN.

Let us now offer a trip to Frieder, say in Portugal. Woldemar still dials the same number to call Frieder (i.e., a German number), and the interrogation point (the GMSC) automatically re-routes the call from Germany to Portugal, without Woldemar being aware of it (principle of location confidentiality). He should then pay the same charges as in the previous case. But more networks have been involved in the call, which now includes an international leg, as shown in figure 8.6. The standard compensation mechanisms in use between fixed networks for

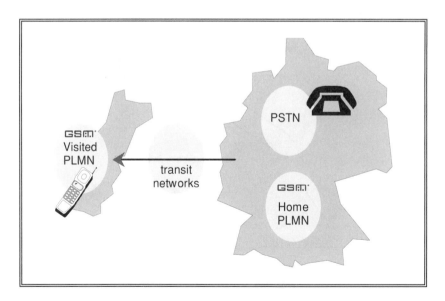

Figure 8.6 – MT call toward a roaming subscriber

The calling party is unaware of the actual location of the called subscriber
(for him, everything is the same as in figure 8.5).
The cost of the international leg can be recovered
by levying a charge on the called GSM user.

international routing will apply to the German PSTN, so that the transit networks get their share. In such cases, an additional charge to cover the international leg will have to be raised, and it can be levied on nobody but Frieder if the principle of location confidentiality is retained.

Let us now take the most complex case, where the originating network, the home PLMN of the called user and the visited PLMN are all potentially in different countries. This corresponds to our original example, when Woldemar calls Peter (GSM subscriber of the Netherlands), himself roaming in Spain. In this case, Woldemar is dialling an international number starting with 31 (the country code for the Netherlands), and as such is prepared to pay the cost of an international call towards the Netherlands, and possibly a little more if he knows that the number he is calling is the one of a mobile subscriber (and if the German PSTN also knows this information, which is not by far an obvious matter). Depending on the location of the interrogation point (the GMSC able to interrogate Peter's HLR in the Netherlands to know where he is located), the call may experience one or several international legs, as shown in figure 8.7. Besides the German PSTN, which gets the money

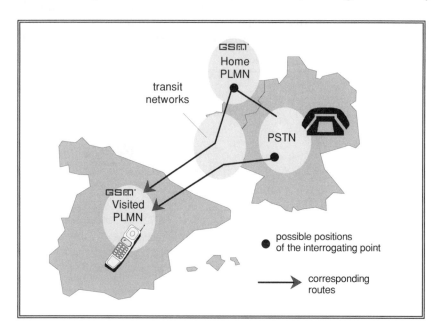

Figure 8.7 – MT call involving three countries

Depending on the location of the interrogating point, the call may be routed through one or several international legs, without the calling user being aware of it.

from Woldemar, Peter's home PLMN may charge him for some amount linked with his current location, but it is the GMSC which holds the relevant information on the routing cost of the second segment of the call (from GMSC to VMSC). As for the other networks through which the call is routed (including the visited PLMN), they need to get some compensation for the costs incurred.

All this leads obviously to complex transfer mechanisms. The GMSC appears as an important participant, and the network to which it belongs heavily influences the complexity of the problem. In fact things can be simplified if some choices are made concerning the position of the GMSC. Let us examine this issue in more detail.

8.1.1.2. The GMSC Function

GMSC

The GMSC function requires only a switching capacity, and special software (including the ability to establish a toll ticket for the second branch of the call). It must hold a table linking MSISDNs with HLRs. There is no reason why the table should be complete. Indeed, a given GMSC may fulfil its function only for the subscribers of some Home PLMNs, or even for a single Home PLMN. As will be seen, a simplification is to have the GMSCs specific to one PLMN.

By nature, the GMSC function is independent from the radio access function provided by a PLMN, and can be implemented as a part of any network through which the call is routed, typically the PSTN or the ISDN. As specified in the *Specifications*, the GMSC function can be implemented in any switch of the PSTN, the ISDN, or directly connected to those networks. From a service point of view, the greater the density of GMSCs the better. Indeed, the closer the interrogation point to the calling party, the more efficient is the routing. The epitome of what should be avoided is what is called in technical jargon a "trombone", that is to say the case where the second branch begins by backtracking the path of the first branch. This is obviously not efficient, and is prevented if GMSCs are everywhere.

This ideal view is not so ideal when taking into account the charging considerations. The billing record established by the GMSC covers the second segment of the call (from GMSC to called party). This record must be transferred towards the home PLMN of the called subscriber, in order to bill him the amount not covered by the fixed charge of the calling subscriber. A first problem is that this transfer of billing records must be organised with all Home PLMNs for which the GMSC is able to interrogate the HLR. A second problem is that the charging mechanism for the second branch depends on the position of the GMSC. If the GMSC is not in the home country of the called party, a part of the charge for the second branch may have to go to the calling party if we want to follow the principles developed in the previous section. An important simplification for charging management is obtained if the first leg is always charged to the calling party and the second branch always charged to the called party. Now, applying this principle is in contradiction with the search for an efficient routing. As explained in the previous paragraphs, in a network configuration optimised for routing, there should always be a GMSC close to the calling user, and he would not pay for, e.g., the international path if the called user is a foreigner. To solve this dilemma, the choice was made in favour of simplifying the accounting procedure, and to forget routing efficiency. To achieve this aim the interrogation function for the subscribers of a given PLMN is performed only by switches of this PLMN. The GMSC is always in the home PLMN (and hence in the home PLMN country), and the two segments of the call can be considered as "from calling party to home PLMN", and "from home PLMN to visited PLMN", and each charged to one end of the route. The resulting principles appear in figure 8.8.

Let us consider the consequences of this approach. If the called party is in his home PLMN (Peter is in the Netherlands), the second branch of the call is reduced to a minimum, and this is consistent with not levying the called user in this case (Woldemar pays the whole amount, and Peter is not charged). Conversely, when the called party is roaming abroad (Peter is in Spain), he will pay for the international leg between his home PLMN to his actual location. This also does not raise an acceptance problem, because this rule is simple to understand, and the charge is easy to predict (it consists in the price of an international call between the home PLMN country and the visited country). Moreover, a GSM subscriber has the possibility to ask that incoming calls are not completed when he is roaming abroad; to do that, he must activate the supplementary service "Barring of Incoming Calls when roaming outside the home PLMN country" (BIC-roam).

Call charge transfer is much simplified. The same network (the home PLMN of the called subscriber) generates the charging information concerning the second portion of the route, collects the corresponding charges if any from its subscriber, and receives the invoice of the other intervening networks further down the call route, for compensation.

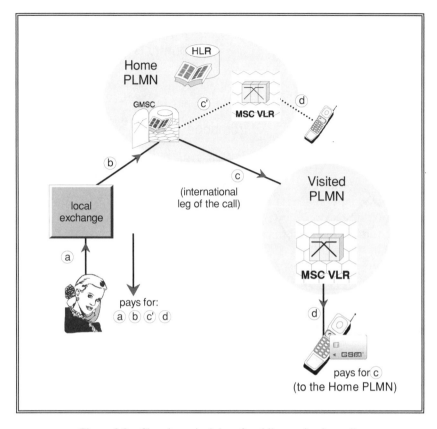

Figure 8.8 – Charging principles of mobile terminating calls

The charging principles retained by GSM MoU operators
are based on the following principle: the originator of the call is charged
as if the called GSM subscriber was in his home PLMN.
The international leg, if any, is paid by the called GSM user. Conversely
the GSM subscriber is not charged for incoming calls when located in his home PLMN.
The originator is unaware of the actual location of the GSM subscriber,
and will always know how much she will pay.

The only problem with this approach is that "tromboning" is not suppressed. Routing optimisation can still be achieved on a national basis, if the density of GMSCs is high, but this is not possible when the calling user and the home PLMN are in different countries. The lack of efficiency is obvious when the called GSM subscriber is roaming in a foreign country A and is called by somebody from country A, possibly a few meters away. Then the call is routed to the home PLMN country and

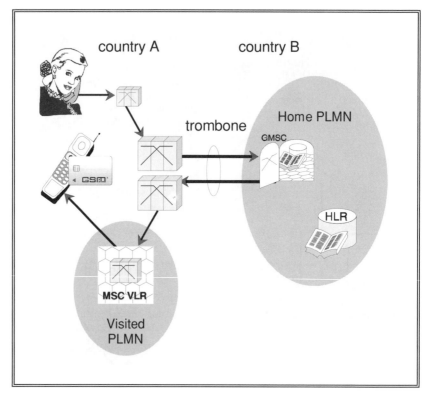

Figure 8.9 – The "tromboning" effect

When the originator of the call and the mobile subscriber happen to be
geographically close to each other, a "trombone" may appear in the routing
of the call when the home PLMN of the mobile subscriber is in another country.
It consists in routing the call back and forth between two countries,
leading to two international legs instead of a national (or even local) call.

back, as shown in figure 8.9. Two international calls are then established
where a local route would have been sufficient (this is the problem of
"tromboning"). The two international calls are being paid for (one by the
calling party, the other by the called party), and only in a small number of
cases will the calling user realise the actual situation, since he remains
unaware of the location of the called user if the latter does not tell him
explicitly.

Of course, the example shown represents an extreme and fairly
marginal case. If both users realise the situation, they can both save on
their bills by having the mobile user immediately end the call, and restart

it from his end, since all the problems of subscriber location do not exist from the mobile user to fixed user. More generally speaking, the lack of efficiency (and therefore the additional cost) in certain situations such as the one described represents the price to pay for location confidentiality in an international environment where billing and accounting is a complex and constraining issue.

To conclude this long discussion, let us just recall the result: the GMSC function, though designed to be used in a versatile way, will be at least in the first years of GSM always co-located with an MSC, and limited to the interrogation of the HLRs inside the same PLMN.

8.1.1.3. Where does the Routing Information come from?

The answer to the question of where the routing information comes from is strongly linked to how subscribers are identified in GSM, as well as in ISDN and PSTN, for routing purposes. We have already encountered the MSISDN, which is the "directory number" used to call GSM subscribers. The MSISDN is part of the E.164 numbering plan. Also part of this numbering plan is the MSRN, or Mobile Station Roaming Number, which is the routing number (see page 512) used on the second leg of an incoming call between the GMSC to visited MSC. The MSRN is not visible to GSM users or to calling parties, but is used solely between infrastructure machines. It is not allocated permanently to a subscriber, and it is geographically integrated into the numbering plan of the fixed networks, since its purpose is for routing towards the visited MSC.

A third type of identity, introduced in Chapter 7, is the IMSI, which is used as the main subscriber key in the GSM location databases. Both MSISDN and IMSI contain an identification of the country and of the network within this country. Table 8.1 (overleaf) gives a few examples of the correspondence between the "country code" (CC) of the international telephone (or ISDN) numbers and the "mobile country code" (MCC) of the IMSI.

The problem now is how the GMSC gets the MSRN to point to the MSC where the subscriber is located. The method by which the system originally determined the location of each subscriber has been studied in detail in Chapter 7. To sum up, the HLR stores within each subscriber record some location information, including at least an address of the visited MSC/VLR which can be used by the SS7 signalling. The HLR record may also include an MSRN usable for the routing of the second branch of the call, if the visited MSC/VLR has provided such a roaming number when updating the location information in the HLR. In this case, the answer to the interrogation can be readily given, but when the record

MSISDN   **IMSI**

| CC | NDC | | | MCC | MNC | |

Country	CC	MCC
Denmark	45	238
Finland	358	244
France	33	208
Germany	49	262
Italy	39	222
The Netherlands	31	204
Norway	47	242
Portugal	351	268
Spain	34	214
Sweden	46	240
Switzerland	41	228
UK	44	234

Table 8.1 – Correspondence between CC and MCC for a few European countries

The country code (CC) known to subscribers in the telephone numbering plan
is defined in Recommendation CCITT E.164, whereas
the mobile country code (MCC) used in the first few digits of the IMSI
is defined in Recommendation CCITT E.212.

is limited to the visited MSC/VLR address, the HLR has first to interrogate (using this address) the visited MSC/VLR to get the routing information. The flow of information is shown in figure 8.10. When receiving the message, the visited MSC/VLR chooses the roaming number from a pool of free numbers, and links it temporarily with the IMSI. When the call eventually reaches the visited MSC, using the roaming number as the address, the MSC can retrieve the IMSI from its records, and can go ahead with the establishment of the call towards the mobile station. The roaming number can then be freed as soon as the call is fully established.

The two scenarios for the provision of the MSRN (available continuously at the HLR or provided by interrogation at each incoming call) triggered a long debate in the specification committees. The issue is now settled in favour of the provision of the MSRN on a call per call basis, though both solutions still appear in the phase 1 *Specifications*.

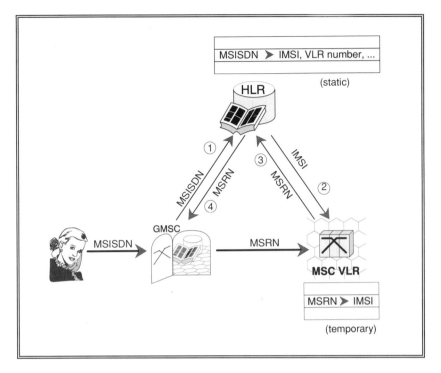

Figure 8.10 – The provision of the MSRN

The HLR requires the visited location register to provide an MSRN
which will be used to route an incoming call towards the correct MSC.

One of the fundamental arguments is the consumption of directory numbers. If the MSRN is allocated to a subscriber for the whole duration of its stay in the VMSC, the quantity of numbers reserved for this purpose in each MSC must be about equal to the number of subscribers who may be registered at the same time under this MSC. By comparison, a call per call allocation requires a quantity of numbers which are about equal to the number of simultaneous call establishments by the MSC, a much smaller value. This advantage is amplified when the multi service problem is taken into account (see next section). In the following, only the call per call allocation of the MSRN is considered.

8.1.1.4. The Problem of Multi Service

GSM is designed to provide different telecommunication services. Most of them, such as speech, fax, or data bearer services, can be provided through the PSTN or the ISDN. A GSM mobile station, or an

ISDN terminal, is designed so that the user can indicate which service it requires amongst the supported ones. This is not so easy for a calling party using a PSTN access line. The problem is for instance how to make the network understand that a fax call is requested instead of a speech call? In addition, even if the information is provided by a GSM or ISDN calling party, an intervening PSTN is unable to carry it. The problem exists if any part of the route is via the PSTN.

A solution exists if the first branch does not include any analogue PSTN segment. This applies if the call is issued from GSM or from ISDN, and if the GMSC can be reached via ISDN using SS7. Then the information can be transferred up to the GMSC, and from then on to the HLR, and thence to the VMSC (in substance bypassing a possible passage of the second branch through the PSTN).

The problem remains when the only information received by the GMSC is the called number (the MSISDN). Two solutions of general application have been put on the table, and were the source of intensive debates. The first solution consists in letting the service be chosen by the called party. The message setting the call from the network to the mobile station does not specify the service, and the mobile station indicates it in return. This solution imposes the requirement that the service is set by the user in the mobile station *before* the actual start of the communication. A typical scenario to send a fax to a GSM subscriber is then first to phone him (speech communication), asking him to set the mobile station so that the next call will be treated as a fax call; then hang up and re-dial to establish the fax call.

This solution has minimal impact on the network, but is not very convenient for the users. This is why an alternative solution was proposed, consisting in providing a GSM subscriber with as many MSISDNs as services for which he wishes to receive incoming calls (for instance a speech number and a fax number). The service can then be chosen by the calling party, by using the right number. The relationship between numbers and services is held in the HLR. The next issue is to convey the information to the VMSC. This is simply done in the procedure used to get the roaming number, which allows once again to bypass the networks intervening between the GMSC and the MSC/VLR. The HLR sends the reference of the requested service to the MSC, which stores it against the provided roaming number. When eventually a call with this roaming number reaches the VMSC, it proceeds with the call establishment for the requested service. This scheme is shown in figure 8.11. Its only drawback is the consumption of directory numbers for the MSISDNs, which can be unacceptable in some countries. On the other hand, it enables the provision of a service no worse than that which PSTN subscribers are familiar with.

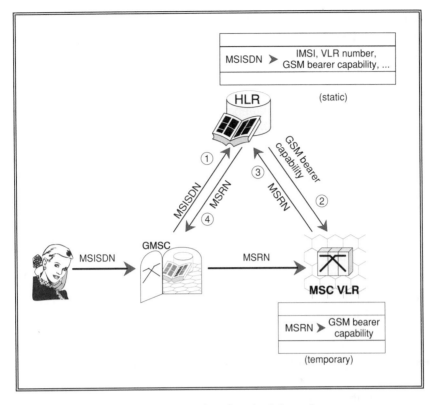

Figure 8.11 – Routing of service information

When an incoming call arrives from the PSTN, the only way to convey the information on the service (based on the dialled directory number) up to the visited MSC/VLR is to carry a service profile (bearer capability) between the HLR and visited MSC/VLR. This service profile is stored in the HLR in association with each MSISDN.

The two schemes are provided for and both must be supported by all MSC/VLRs. The choice of the scheme is done by the Home PLMN of the called party. When no service information is provided by the HLR, the MSC/VLR can either choose the service or let the mobile station decide. The first approach is acceptable only in the case when the subscription is limited to one service.

8.1.1.5. Data Services

Calls originating from a PSPDN or a CSPDN and directed towards a GSM subscriber raise specific problems. The most important issue is related to numbering. While PSTN, ISDN and GSM use E.164 numbers, public data networks (PDNs) use a different numbering plan, specified by

CCITT Rec. X.121. GSM and ISDN allow for mobile originating calls to be addressed by X.121 numbers, but the support of E.164 numbers by PDNs will become mandatory in a few years (this is planned for in the beginning of 1997). Several cases can be envisaged.

The GSM subscriber may be given an X.121 number, to be used to call him from PDNs. This number must be allocated so that the PDNs will route calls towards GSM. This can be obtained by defining a GSM PLMN as a "PDN", in substance by allocating a data network code to the GSM PLMN used as the first part of the X.121 number of its subscribers. The call is routed through PDNs directly to an interworking function of the target PLMN, which acts as a GMSC. Roaming may be dealt with by using ISDN/PSTN routes. It should be noted however that if the called GSM user is served by his home PLMN a circuit can be set up without audio modems, whereas such modems may be necessary if the roaming requirements force the call to go via the PSTN.

A variant consists in defining in the PDNs the PLMN as part of ISDN. The data network code then refers to the ISDN, and the calls are routed first to the ISDN (and more precisely to a packet handler in the case of a call from a PSPDN), which has then to translate the X.121 number into an E.164 number for routing the call to any GMSC for the target PLMN and then to the suitable MSC/IWF. Roaming is dealt with normally. Whether audio modems are used or not is determined by the ISDN. This determination may be difficult when the user is roaming, and the default solution is to use audio modems.

Another case is when the PDN does support E.164 numbers. The routing depends on the capacity of the PDN to determine that the number refers to a GSM user. The basic assumption is that it does not, and the call is routed to ISDN, which will then route it to a suitable GMSC. If the PDN can further analyse the number, the route can be routed through PDNs directly to an IWF which may act as a GMSC.

It is not foreseen to have an HLR interrogating function in the middle of the PDN leg. This means that tromboning routes will not be avoided. This is not too much of a problem, as we have seen that this would save only a part of the route within the home country.

8.1.1.6. The Impact of Call Forwarding

The category of supplementary services which have the greatest influence on call routing are the forwarding facilities. There are various reasons which can lead the infrastructure to forward a call; and the process differs accordingly. There are two main cases:

- either the HLR is able to decide to forward the call, in which case it sends back the forwarded-to number to the GMSC as a result of the interrogation;

- or forwarding can only be decided by the visited MSC/VLR in charge of the subscriber.

The first case is obviously more efficient for routing, and should be favoured as much as possible. Let us now study the reasons to forward a call and see how they fit into these two broad categories.

In the case of an unconditional call forwarding (CFU), the HLR, knowing the status of the supplementary services, has no difficulty to do it by itself. However, a call forwarding done because the end party is found to be busy (Call Forwarding on Busy, CFB) can only be performed by the VMSC/VLR in the present state of the *Specifications*. The route is first established toward the VMSC/VLR, and only then to the forwarded-to party.

Two types of conditional call forwarding exist: Call Forwarding on Not Reachable (CFNRc) and Call Forwarding on No Reply (CFNRy). The first case can be treated by the network in several ways, according to the method by which knowledge that the mobile station is not reachable is obtained. In some cases, for instance because the mobile station last tried to register from a geographical area where the subscriber is not entitled service by subscription, the HLR knows of the situation (see Chapter 7), and performs the forwarding itself. In other cases the situation may be known only after an effective (and unsuccessful) attempt to contact the mobile station over the radio path. This can be done only by the MSC/VLR. Moreover this is done after the route from the GMSC to the MSC/VLR has been established, and then the forwarding is done by the MSC/VLR.

Another case for Call Forwarding on Not Reachable happens when the mobile station has been able to indicate that it was going to be switched off (this is the "IMSI detach" procedure, described in Chapter 7), or if the subscriber has been inactive for a long time and "implicit detach" is implemented by the visited PLMN. The "unreachable" indication is stored in the VLR, but not in the HLR. It will become possible in phase 2 to send back this information to the HLR during the interrogation procedure, but this is not the case in phase 1 and forwarding must also in this case be carried out by the MSC/VLR.

Call Forwarding on No Reply is triggered when the mobile station has been successfully reached on the radio path, but when no answer from the user has been received after some time. CFNRy is therefore always triggered by the VMSC.

Whether the forwarding is done by the HLR or by the visited MSC/VLR is of no importance for the calling party, whose bill is not affected by the actual routing of the call beyond the GMSC, but it can make a big difference for the called party, who is billed for the remainder of the route. The most serious case is when the visited MSC of the called party is in a country other than the home country, and when the forwarded-to user is in the home country. The difference is then between a national, or even local call, and two international calls! Users who want to save on their bills will probably prefer to activate Unconditional Call Forwarding in this case.

8.1.2. ARCHITECTURE

So far we have presented in general terms the functions for call management. In the following we will describe more precisely the signalling procedures, and for this objective it is necessary to indicate first the different entities and protocols which are involved in these procedures.

The main actors of Call Control are the user and the mobile station on one side, and the Network and Switching Sub-system (NSS) and the external network on the other. The functional entities of the NSS which are involved in Call Control include the MSC/VLR and IWF, the GMSC and the HLR. The main one is the MSC/VLR, which deals with all the communications, mobile terminating or originating, of which one end is a GSM mobile station within its radio coverage. The MSC is the connection point between the mobile stations on one side, through the BSS, and the external networks on the other. As such the MSC is the function where the call control procedures held with the mobile station interwork with those held with the external network. In addition, for Mobile Terminating calls, the HLR and the GMSC intervene. They communicate with each other and with the VLR/MSC through the SS7 network.

The signalling protocols on the Call Control scene are shown in Figure 8.12. They include the following:

- the MS-MSC protocol enables user requests to be conveyed between the mobile station and the network and service provision to be co-ordinated between them; this protocol will be referred to as **RIL3-CC**. It acts as a relay of the man-machine protocol held between the user and the mobile station;

- the GMSC-HLR protocol enables the interrogation of the HLR by the GMSC to get routing information for incoming call establishment; this protocol is referred to as **MAP/C**;

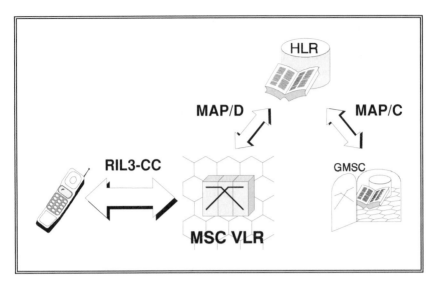

Figure 8.12 – Protocols in the CC domain

The key infrastructure entity for Call Control is the MSC,
which is in contact with the mobile station.
But most of the complexity comes from the routing of calls toward mobile stations,
involving two additional protocols between the Gateway MSC and HLR, as well as
between the HLR and MSC (the latter appears also in the MM domain).

- the **MAP/D** protocol between the HLR and the MSC/VLR,
  which is mainly a protocol for Mobility Management (and as
  such has already been cited in Chapter 7), also serves to convey
  call-related information for incoming calls.

The protocols with the external world must be added to these
internal GSM protocols: the user-MS protocol on one side, protocols on
the interface with external networks on the other side.

On the mobile station side, the man-machine interface (MMI)
between the user and the mobile station has already been mentioned. It is
to a large extent not specified by the *Specifications*, leaving each mobile
station manufacturer free to develop user-friendly means for this purpose.

On the MSC side, the protocols with external networks are specific
to each of these networks. The functionalities are always more or less the
same, but variations exist according to the type of network, and also to
the specific implementation adopted by each country. The basic examples
are the TUP (Telephone User Part) and ISUP (ISDN User Part), or

national variations thereof, which are the standard circuit call management protocols used in an SS7 environment. The TUP or ISUP is in direct interworking with the RIL3-CC protocol in the MSC. Similarly, the GMSC interfaces with the PSTN or ISDN for which it performs its gateway role, through the call management protocol specific to the relevant external network. In the description of procedures which follows, the ISUP will be taken as an example for interworking.

A last area where relevant signalling protocols exist is between the mobile station and the IWF or the other terminal. For most data services, the establishment of the transmission circuit is but the first phase of the call establishment. Additional negotiations and synchronisation are performed "in-band", that is to say using the establishment circuit to convey the messages necessary for the dialogue. This phase is of little concern to the intervening entities, and to GSM, but some aspects have an impact on the IWF, so we will briefly address these protocols.

8.1.3. The Mobile Originating Call Establishment Procedure

As seen by the calling user, the establishment of a communication follows a number of steps, which he perceives through displayed or audio information. The rough sequencing for basic telephony is well known by anybody. An originating call starts by the user lifting the receiver, an action which the network answers with some tone. Then the called number is keyed in. Once the last digit is entered, there happens in some cases nothing for some time; in other cases, a waiting tone can be heard. Eventually, an answer comes. It can be a busy tone, indicating that the called party is already engaged in a communication, an announcement stating for which reason the network was unable to fulfil the request (e.g., congestion, or non-existent number), or, sometimes, a tone indicating that the called party is being alerted. Then, in the last case, it may happen after a while that the other party lifts the receiver, and this is perceived by the speech path being connected, thus allowing the person to person discussion to take place.

Cellular telephony does not introduce many modifications to this basic scheme. The main distinction is that the number is keyed in **before** establishing the contact with the network. It is in addition displayed, and a very useful consequence is that it can be modified if need be before transmission. Another difference, applicable to multi-service networks such as GSM, is that some additional information may be exchanged between the user and the network at the beginning of the call, such as the type of service or in the future the type of channel. In most of the cases, these issues will be dealt with automatically by the mobile station, with default or implicit values.

For all these interactions, the mobile station intervenes between the user, with which it exchanges messages according to a man-machine protocol, and the infrastructure, with which it exchanges information by electronic means. As far as Call Control is concerned, these two flows of information are for the important parts in a one to one correspondence with each other. The mobile station acts as a protocol translator. It receives orders from the user in the form of key pressing for instance, and translates them into signalling messages for the network according to the RIL3-CC protocol. In the other direction, the MSC provides its answers or indication to the mobile station in the form of RIL3-CC messages, and the mobile station translates them into signals understandable by the user, such as audio signals (tones) or visual signals (lights, alphanumerical display, ...).

This modern approach, which is also the one of the ISDN (from which the RIL3-CC protocol has been derived), is different from the one still used in the PSTN. In the PSTN, all the information provided to the user are in the form of audio signals (tones, announcements), and are generated by the network. There is no digital message to be found coming from the network to the telephone set. Inside the PSTN, the transmission of the information ultimately destined to a user may be done in the guise of messages (digitally encoded), in which case the translation in tones or announcements is done at the end, by the switch in charge of the user line. Yet there are still old machines not supporting this scheme, and then the tones and announcements can come from switches far away from the user, the role of the local switch being limited to their transmission, as during the actual communication phase. The consequence for GSM (and for ISDN) is that an escape mechanism ("progress" information) must be provided so that the dialogue between the user and the network can be done in the old fashioned way.

There is yet another approach, which is specified (besides the "functional" protocol described above) in the ISDN terminal to network protocol for simple terminals, and is called the "stimulus" mode. In this mode, the network has the full detailed control of the signals (display or audio) provided to the terminal, and instructs it through simple messages, directly giving the signals to be provided to the user. This approach is not used in GSM, but some traces of it can be found, in the guise of the

SIGNAL information element. It should be understood that this is **not** a full stimulus protocol, but a modified usage of the information element to qualify a message.

We will describe first the modern scheme, using messages, and then address the escape mechanisms.

The RIL3-CC protocol between the mobile station and the MSC is entirely specified in the *Specifications*, but the man-machine interface between the user and the mobile station is left very much open. Similarly, the interworking between the RIL3-CC protocol and the man-machine protocol is not specified in detail, though the functions to be fulfilled are often described in the RIL3-CC specifications. The description which follows corresponds to an invented man-machine protocol, derived from the knowledge concerning existing terminals. It does not cover, by far, all the possibilities. This area is one of the most important for the marketing of GSM terminals (and GSM), and it is desirable that terminal manufacturers have many ideas about the man-machine interface and protocol, to help and guide the user through the tortuous route of GSM. Eventually sophisticated methods will appear, such as voice recognition, enabling the user to give his commands orally, menu driven protocols, and other yet to be imagined features making the use of technique more convivial for human beings.

8.1.3.1. The Signalling Flow for a Mobile Originating Call

Let us take the example of a GSM user (Ansgar) calling a fixed user (Rémi). In the first part of the call establishment, Ansgar indicates, through the mobile station, what he wants (directory number of Rémi, type of service requested—speech, fax, ...), and the mobile station will pass the information to the MSC. In GSM, as in ISDN, the user has to provide this basic data before the mobile station contacts the network. This can be done in different ways, depending on the terminal, but also on the type of service. For instance, for emergency calls, this can be reduced to the pressing of one or two keys (typically an "emergency" key and then a "send" key). Some effort was made to prevent the keyed information depending too much on the place where the user is. One of the issues is the format of the called number, and we will come back later to this topic. Explicitly giving the characteristics of the service will often be skipped. For example, for a speech only terminal, speech is obviously the default service, and then has not to be explicitly requested. Typically,

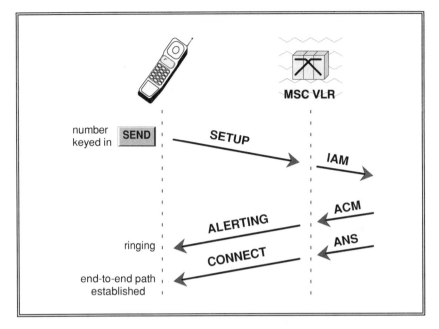

Figure 8.13 – Mobile originating call establishment

A successful MO call establishment leads to a signalling sequence as shown,
each step being related to the following events:
– the man-machine interface between user and mobile station;
– the protocol between MSC and external networks (ISUP for example).

Ansgar enters first the called number, which is shown on a display, may correct it, and then presses a "send" key. Only when the "send" key is pressed does the mobile station trigger the effective establishment phase with the network, shown in figure 8.13.

Then the mobile station runs the procedure needed for call establishment. This starts by the access procedure (RR level), and the MM connection establishment. These steps have been dealt with in Chapters 6 and 7 respectively, and their effect as far as Call Control is concerned here is to establish a suitable signalling connection between the mobile station and the MSC. Then the mobile station sends to the MSC an RIL3-CC SETUP message or an RIL3-CC EMERGENCY SETUP message, which is the translation of the Ansgar's request. This message contains Rémi's number and a description (which can be long and detailed in the case of data services) of the required service. It may contain additional information, in relation with the supplementary services.

When the MSC receives the setup message, it analyses the request, and checks whether it can accept it. Whether it is accepted depends on the capacity of the MSC/VLR to provide this service (in a compatible way with the mobile station capacity), on Ansgar's subscription characteristics (this is determined locally thanks to the subscriber information record sent by the HLR during location updating, and stored in the MSC/VLR), and on the availability of resources (interworking devices, a free circuit with the external network, etc.). If some of these checks fail, the call establishment is aborted by sending an RIL3-CC RELEASE COMPLETE message to the mobile station, before the release of the lower layer connections returning the mobile station to the idle mode. If everything is all right, the MSC on one hand starts the establishment through the network (for instance by sending an ISUP INITIAL ADDRESS message (IAM) in the case of a communication with the ISDN), and on the other hand sends an RIL3-CC CALL PROCEEDING message to the mobile station, which indicates simply that the request has passed the MSC tests, and that the MSC is proceeding with it. More pragmatically, it tells Ansgar (if so indicated by some signal on the man-machine interface) to be patient.

Sooner or later, the MSC will receive from the external world a report of the requested call establishment, as seen by the switch in charge of the called party. Such a report may indicate that Rémi is being alerted (alerting indication), or that the call establishment is aborted (call release indication) because it failed for some reason (congestion, or because Rémi is already busy, or not reachable, etc.). In the case of ISUP, the report of a successful alerting takes the form of an ISUP ADDRESS COMPLETE message (ACM), and a failure is indicated by an ISUP RELEASE message. The MSC reacts respectively by passing the information to Ansgar (RIL3-CC ALERTING message, translated for instance to an alerting tone by the mobile station), or by aborting the call establishment. If an RIL3-CC CALL PROCEEDING message has been sent before, the abortion is done by sending an RIL3-CC DISCONNECT message, which will be answered by the mobile station with an RIL3-CC RELEASE message, itself acknowledged by the MSC with an RIL3-CC RELEASE COMPLETE message (explanations on these exchanges will be found in the section concerned with call release). Only then can the lower layer connections be released, if not used for some other context.

Some time can then elapse before the end-to-end answer, i.e., before the acceptance of Rémi (lifting the receiver in the case of plain telephony). At this stage the mobile station (and Ansgar) is still waiting for the result of the request. The ball is on Rémi's side, who may answer, or not. In the case of a no answer, the call is aborted by the network after some time (say 3 minutes), provided Ansgar has not decided to abort it himself, usually by pressing an "end" key. The acceptance of the call by Rémi leads to an ISUP ANSWER message being received by the MSC. When this happens, the call is connected-through, that is to say that the transmission path is completed between the two end users, and this is indicated to Ansgar through an RIL3-CC CONNECT message. The mobile station reacts first by stopping the alerting indication if any, secondly by answering the network with an RIL3-CC CONNECT ACKNOWLEDGE message, and thirdly by connecting the circuit transmission on the radio path with the suitable terminal. In the case of speech, this consists of completing the speech path to the microphone and to the loudspeaker. The call then enters the connected phase ("is connected" in short), charging starts and effective bi-directional transmission is provided between the two users in the case of speech, or between the terminals, or the terminal on the mobile side and the IWF in the cases of data calls. In the last case, the establishment procedure goes on in-band.

Let us now see some of the variants from this basic scenario.

8.1.3.2. Automatic Answering

When the called party is a machine, which is often the case for data calls, some of the steps can be merged. For instance, the connect-through can happen without any alerting indication. In some even more expedient cases, the connect-through can happen as an immediate answer to the request. In the latter case, the RIL3-CC CONNECT message is sent directly as an answer to the RIL3-CC SETUP message. In the former case, it directly follows the RIL3-CC CALL PROCEEDING message.

8.1.3.3. Entered Number

One of the problems specific to an international mobile communication network like GSM, where each user can establish calls from different countries, is the format of the called number. As PSTN users, we usually know three basic types of formats: a local format; which refers to one destination only in some regional area; an interurban or national format, used when the destination is in the same country but in a different region, and an international format used to call abroad. The

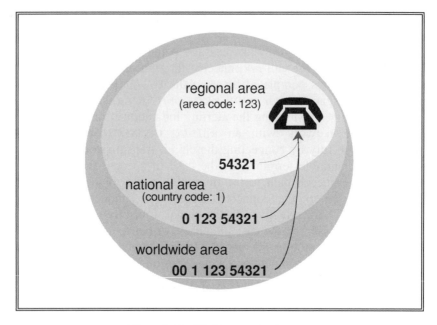

Figure 8.14 – Dialling number formats

Depending on the relative positions of the calling and called users,
PSTN numbers can usually be entered in three different ways,
which are distinguished by different prefixes.

distinction is done by the use of prefixes. Figure 8.14 shows the three
formats in an example using the standardised (but not yet universally
applied) prefixes 0 and 00.

The way in which a number is dialled in the PSTN then depends on
the location of the calling party. This is quite unfortunate in a mobile
environment. To avoid the difficult task of requiring the user to know in
which telephone area he currently is, and how numbers are to be entered,
the issue must be treated differently in a cellular network. The ISDN
protocol, and RIL3-CC, support many different number formats, by
qualifying the digit string by two indicators, the "type of number" (TON)
and the "Numbering Plan Identification" (NPI). However, the
Specifications are not very informative on the subject of what the *user*
has to enter. There is however one mechanism that must mandatorily be
supported, concerning the international format. In order not to ask the
user to know all the prefixes in all the countries where he wishes to roam,
an international number can be entered by pressing the "+" key followed
by the country code, etc. Moreover, a switch must be able to correctly
treat a number presented as international, even if it addresses a
destination in the same country. This behaviour is by no means obvious:

for instance in the French PSTN, keying an international call with country code 33 (France) from a PSTN telephone set results in a call to the international directory enquiries for the country whose code followed 33. A different treatment is required from the MSC. This minimum specification enables the user to enter the number of a specific destination totally independently from his location: it is therefore advisable for GSM subscribers who travel abroad to always store their abbreviated dialling numbers in this format (e.g., +44 701 1234...), instead of a national format.

As already mentioned, this format is only one among many that can be supported by the RIL3-CC protocol. The translation between the number in the man-machine protocol (the number keyed by the user), and the number sent in the RIL3-CC message is not specified by the *Specifications*, nor is the way the MSC must interpret the numbers it receives. This is unfortunate, because what the user has to key does depend on the terminal as well as on the PLMN operator. Most probably, some standardisation will appear. Though other schemes cannot at this date be precluded, the most widespread format for user dialling (besides the "+" key format) seems to be the national format of the visited country, usually without prefixes, to be used for numbers within the country.

8.1.3.4. Off Air Call Setup

In the basic call set-up scenario, the transmission path between the MSC and the MS is fully operational for the requested service before the MSC is aware that both parties are ready to converse. The call can be connected-through immediately. A variant is possible, called Off Air Call Setup (OACSU), where the allocation of the suitable radio resource is delayed as much as possible, in an attempt to save radio resources (see Chapter 6). In this case, the MSC has first to request the BSS to establish the circuit connection with the right transmission mode, before connecting the communication. For the called party, the instant of connection is not delayed, but the connection may be temporarily routed to an announcement machine until the actual connect-through happens.

8.1.3.5. Progress

The basic set-up scenario assumes that signalling messages are received from the originating network. Now, this is not always guaranteed, since an important proportion of the PSTN equipment are not using the advanced protocols. It may then happen that the indications concerning the progress of the call come as audio tones or announcements. This case is foreseen and coped with in the ISUP protocol, and the MSC will know when the call establishment must go on in the old fashioned way. The MSC reacts by performing a kind of one-way through-connection in advance to the normal point, so that the speech path from the called party to the GSM user is established and the tones and announcements can be heard. (Note that this is not easy with data services except as alternate speech/data services.) This early connection is commanded by way of the RIL3-CC PROGRESS message, or the *PROGRESS* information element included in messages such as RIL3-CC CALL PROCEEDING or ALERTING, depending on the moment it is known that the call is not ISDN-compatible all the way. In addition, if the RIL3-CC PROGRESS message is used, there are no more messages for call establishment, except the RIL3-CC CONNECT message used to indicate the two-way through-connection.

8.1.3.6. Additional In-Band Establishment for Data Services

In the case of data calls, the connect-through is not sufficient to allow user-to-user information exchange, and not even terminal-to-terminal exchanges with some cases of interworking. The connect-through can be understood as the establishment of the lower protocols in the transmission plane, but other data specific protocols must also be initiated.

The next step is the confirmation of the connect-through between the terminal adaptation functions in the mobile station (the TAF) and the IWF. The *Specifications* do not state clearly when the V.110 synchronisation pattern is established between the rate adaptor (TRAU) and the IWF. It can be imagined that it is when the transmission path is established between the IWF and the mobile station, that is to say before the connect messages exchange. After this exchange a small procedure takes place between the TAF and IWF, as follows. First one end sends data and auxiliary information all set to "1's", and then the other answers in the same way. When this is completed, the transmission path is used by higher layer protocols, for instance to establish a link layer in the case of an X.25 connection. This procedure raises some problems. There is no clear indication in the phase 1 *Specifications* on what is transported before the connect-through, that is to say just before the above mentioned example. Unfortunately it can well be an all "1's" pattern! The second point is that this detail in the *Specifications* ruins the possibility of a "null" interworking function between GSM and ISDN for T connections (the synchronisation procedure described above does not exist in ISDN).

As explained in Chapter 3, the possibility of such a "null" IWF with ISDN was an argument for the peculiar management of the E1, E2 and E3 bits in the V.110 frame.

Once the existence of a V.110-like connection is confirmed, the RLP protocol must be initialised in the NT cases. This is done by an exchange of SABM and UA frames, as explained in Chapter 5. The next step involves the modem control signals, and is done in the same way as between a terminal and a modem.

The next, and last steps, depend on the service and are normally carried between the terminals themselves. They correspond to the establishment of the upper layer protocols for transmission, or even to further call establishment as with the double-numbering cases we have seen for PSPDN or CSPDN access. They do not generally affect GSM, but there are some exceptions. For instance, the IWF is involved in the procedure in the case of facsimile, which consists mainly in a negotiation concerning various characteristics of the transmission, and in particular the modem speed. Another example is for dedicated PAD or packet access, where the establishment procedure through the PSPDN is taken care of by the IWF.

It should be noted that part of the procedures presented in the previous paragraphs duplicate the basic call establishment, or could be supported by it. The trend is to limit them, at least for the user, as exemplified by the introduction of the dedicated PAD and packet access services.

8.1.4. THE MOBILE TERMINATING CALL ESTABLISHMENT PROCEDURE

Let us now consider the case of Mobile Terminating calls in the same way as we just did for Mobile Originating calls, i.e., from the point of view of the user to network interface. At this stage, the call has just reached the Visited MSC and we are concerned with what happens from this moment on.

In usual telephony, the interaction between a called user and his handset is very simple: the latter rings, the former then lifts the receiver and communication ensues. In cellular telephony, the behaviour is basically the same. Some sophistication appears if additional facilities are invoked. For instance the directory number of the calling party may be displayed, as in ISDN, allowing the called party the option of not responding to unwanted calls.

From a signalling point of view, a Mobile Terminating call reaches the visited MSC through one of its interfaces with external networks. If ISUP is used on this interface, this event corresponds to the reception of an ISUP INITIAL ADDRESS message (IAM). From the contents of this message, and from the data stored previously (during the interrogation phase) and which can be linked with the incoming call through the roaming number, the MSC/VLR can derive all the information it needs, such as the IMSI, the requested type of service, ... Let us see what most often happens afterwards, and then study the variants of a call from an ISDN user called Carlo to a GSM subscriber called Jan. The signalling exchange corresponding to the basic successful mobile terminating call establishment is shown in figure 8.15.

If Jan is not known to be already engaged in a communication, the next step consists in "paging" the mobile station, that is to say in a few words, to find whether the mobile station is actively in coverage, and to ask it to establish a signalling connection with the MSC. When this and other ancillary tasks are done (see Chapters 6 and 7), an RIL3-CC SETUP message is sent to the mobile station, indicating many details concerning the call, which include the type of service required and, if applicable, Carlo's directory number. The mobile station checks if it can deal with the type of service, and if not rejects the setup by an RIL3-CC RELEASE COMPLETE message. Otherwise, the mobile station answers with an RIL3-CC CALL CONFIRMED message, and starts to alert Jan, by a display or a ringing tone.

The RIL3-CC CALL CONFIRMED message is also the vehicle for the choice of parameters which can be decided upon by the mobile station. Two cases are of interest. First there is the future case of mobile stations able to deal with both half rate or full rate channels for the same service: the message then conveys the choice of the mobile station. The other case is when it is the called party which chooses the type of service. We have seen that this may be the case with a call originating from the PSTN, and when a single directory number is allocated to a mobile user who desires access to different services (speech and fax for instance). Then the RIL3-CC SETUP message does not contain any service indication (the MSC knows by the subscription record that the subscriber may receive calls for different services), and this information is provided by the mobile station in the RIL3-CC CALL CONFIRMED message.

The following step is the start of user alerting. There is however a problem here. The allocated radio channel at this stage is not necessarily suitable for the communication. If the mobile station is alerted, Jan may answer (this is the very purpose of alerting!), and if at this moment the

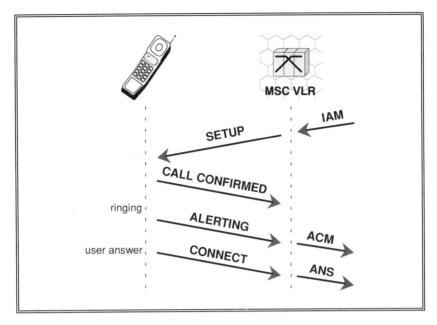

Figure 8.15 – Basic MT Call Establishment sequence

After the negotiation phase (SETUP, CALL CONFIRM),
the mobile station sends back the ALERTING message while ringing the user.
The sending of the CONNECT message happens only once the called user
has answered the call, triggering full connect-through.

channel is still not suitable, the communication cannot be established readily and Jan will not be happy. On the other hand, it is useful, for saving spectrum, to delay the allocation of the full rate traffic channel (TACH/F) as much as possible: this is the notion of OACSU, a channel allocation strategy which we already encountered with mobile originating call establishment. Two schemes are possible, and are left for choice by the operator. Either the start of alerting is postponed until a suitable channel is allocated (early assignment), or alerting is started immediately so as to delay the allocation of a suitable channel (off-air call set-up).

The mobile station must know which of the two schemes is used, in order to decide whether to start user alerting or not. This information is conveyed by an indication in the RIL3-CC SETUP message (a SIGNAL information element, hijacked from its customary usage in ISDN, where it is used for the operation of the stimulus mode). When the SIGNAL information element is absent, the mobile station only alerts the user once a suitable channel is allocated. Conversely, the presence of this information element indicates an off-air call set-up, where the suitable transmission path will only be established after the user has accepted the call (and hence after his being alerted).

Once the appropriate alerting signal has been started, the mobile station sends an RIL3-CC ALERTING message. At the MSC, the alerting stage is reflected towards the network from which the call comes, in the case of ISUP with an ISUP ADDRESS COMPLETE message (ACM). When off-air call set-up is used, this message is sent as soon as the RIL3-CC CALL CONFIRMED message is received from the mobile station; otherwise, it is sent when the RIL3-CC ALERTING message is received from the mobile station.

The next step is the acceptance of the call by the GSM user, which happens in the case of speech when Jan lifts the receiver or presses some key. This action is translated into an RIL3-CC CONNECT message. At the receipt of this message, the MSC joins the network transmission path and the access path, and the end-to-end transmission may start.

Let us see now some variants from this general scheme, walking through the procedure from the start onwards.

8.1.4.1. Call Waiting

When the MSC receives an incoming call from the external world, it checks whether by ill chance the called user is already engaged in another communication. If the answer is yes and the subscriber has not activated the Call Waiting facility (CW), or if for any other reason (e.g., too many calls on hold already) call waiting cannot be applied, the request is rejected or forwarded. Conversely, if Call Waiting is active, the MSC proceeds directly with the setting up of the call, using the signalling means which already exist between the MSC and the MS. The paging phase and the establishment of the lower layers between the mobile station and the MSC can then be skipped, and the procedure begins with the RIL3-CC SETUP message. As explained in Chapter 5, distinct call control transactions are identified by a Transaction Identifier (TI) included in each RIL3-CC message. This number enables the mobile station to distinguish the messages pertaining to different communications. Once having received the RIL3-CC SETUP message for this new call, the mobile station behaves at the call control level exactly as for the first call, and it is up to the user to release one of the calls or juggle between them if Call Hold is available.

8.1.4.2. Automatic Answering

In the case of a data call involving a separate Terminal Equipment (TE) connected to the Mobile Termination (MT), the RIL3-CC SETUP message is echoed on the interface between the MT and the TE. This interface is not specified within GSM, but uses existing standards such as the V.25bis or the X.21 access methods. The RIL3-CC messages are then relayed by signals on this interface. Typically, in this case, the incoming call signal is answered immediately by the TE, requesting the connection of the line. This is an example of automatic answering, which can also of course be implemented in an integrated mobile station. In these situations, the mobile station is allowed to skip the RIL3-CC ALERTING message and it directly sends the RIL3-CC CONNECT message to the network after RIL3-CC CALL CONFIRMED message.

8.1.4.3. Other Points

As for a mobile originating call set-up, it may happen that the originating network does not indicate the progress of the call establishment through messages, but by in-band tones. The same escape mechanisms we have seen for Mobile Originating calls can be used for Mobile Terminating calls. In this case, the *PROGRESS* information element is included in the RIL3-CC SETUP message. This commands an early connect-through, but does not impact otherwise the running of the call establishment procedures.

Data calls need additional in-band establishment procedures, as for mobile originating calls, and there are little differences, apart from the point that most procedures are initiated from the network side rather than from the mobile station side.

8.1.5. THE INTERROGATION PROCEDURES

As studied in detail in previous sections of this chapter, a call toward a GSM subscriber must go through a complex routing procedure, involving the interrogation of the HLR, before it reaches the visited MSC. There is no such equivalent for mobile originating calls. We have studied the general issue of MT call routing. Let us now describe in detail the procedures involved when a call reaches a GMSC, taking as usual the

ISUP message names as examples for the signalling interworking. The basic flow is shown in figure 8.16.

The incoming ISUP INITIAL ADDRESS message contains at least the MSISDN of the called GSM subscriber, and it can, in addition, include the type of required service if the call was issued in the ISDN. The GMSC derives from the MSISDN an SS7 identification of the corresponding HLR, and sends it a MAP/C SEND ROUTING INFORMATION message, containing the MSISDN and the service indication if available.

At the receipt of this message, the HLR examines the data record of the subscriber, and takes different courses of action, depending on what it finds in the record. It may answer the GMSC directly in the following cases:

- the call cannot be routed to a destination; this situation may happen when, e.g., the subscriber has not paid his bill and is temporarily suspended (this is called "operator determined barring"), or when the subscriber is known not to be reachable but has not activated forwarding;

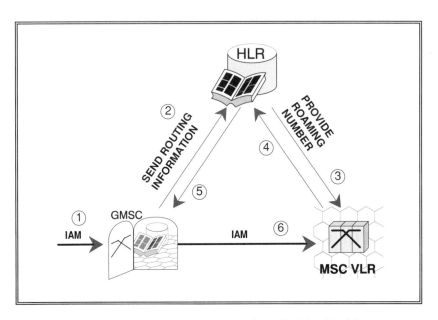

Figure 8.16 – Routing of an MT call from GMSC to VMSC

The interrogation of the HLR by the GMSC,
which usually triggers the request of a roaming number from HLR to VLR,
provides the GMSC with a routing number used in future IAM messages.
The flow of identities in these messages is shown in figure 8.10, and the exchange
also enables the transfer of service information, as shown in figure 8.11,
when the originating or transit networks are not able to generate or transport it.

- the call cannot be delivered because the subscriber has activated the Barring of All Incoming Calls (BAIC), or the Barring of Incoming Calls when Roaming (BIC-roam) and he is known to be roaming abroad;

- the HLR knows that the call must be forwarded (e.g., when Call Forwarding Unconditional is active);

- a re-routing number (MSRN) is readily available at the HLR; as explained earlier, this possibility was left in some *Specifications* but is not used in practice, and will disappear in phase 2.

A negative answer from the HLR comes back to the GMSC with an error message to the MAP/C SEND ROUTING INFORMATION message, whilst a positive answer includes the forwarded-to number or the MSRN in a MAP/C SEND ROUTING INFORMATION RESULT message.

In the alternative and more usual scenario, the HLR knows only some part of the identification (SS7 address or global title) of the MSC/VLR visited by the subscriber. To get the MSRN, the HLR sends a MAP/D PROVIDE ROAMING NUMBER message to the MSC/VLR. This message contains a variety of information, including the IMSI of the subscriber, and in the case of multi-numbering, the GSM bearer capability needed for the required service. If the original IAM message contained service information, the high layer part of it (relevant for end terminals) can also be transported to the GMSC through the MAP/D PROVIDE ROAMING NUMBER message. The answer of the VMSC, if positive, may contain a forwarded-to number (if the MSC/VLR knows already that the subscriber cannot be reached), but more often an MSRN. This is included in a MAP/D PROVIDE ROAMING NUMBER RESULT message. The receipt of this message by the HLR, is translated into a MAP/C SEND ROUTING INFORMATION RESULT message sent towards the GMSC, which can then proceed with the call establishment towards the actual destination.

8.1.6. CALL RELEASE

We have seen how communications are routed and established. Means to end them must be provided. We will now study the signalling means associated with these call releases, which are not specific to GSM.

A communication can be stopped at any moment by one of the two users. A user will indicate that he wishes to terminate the call by replacing the telephone receiver, or by pressing an "end" key, etc. In the case of a GSM mobile station, such an action is translated by the mobile station into an RIL3-CC DISCONNECT message (see figure 8.17). The result is that the call is "disconnected", in the sense that the other party is notified of the complete termination of the call (the MSC sends an ISUP RELEASE message to achieve this in the case of an ISDN call) and the end-to-end connection is terminated. The call is however not fully released at this stage. The local context between the MSC and the mobile station is kept, enabling the completion of side tasks, such as charge indication. When the MSC determines that the call has no more reason to exist, it sends an RIL3-CC RELEASE message to the mobile station, which answers back with an RIL3-CC RELEASE COMPLETE message. Only then are the lower connections released (unless they are used for something else in parallel) and the mobile station returns to the idle mode.

In fact, the phase 1 procedure described in the *Specifications* requires the MSC to send the RIL3-CC RELEASE message straight away after reception of the RIL3-CC DISCONNECT message. This is consistent with the fact that there exists in phase 1 no possibility for the user to exchange signalling data during the period between disconnection and release. The phase 1 release procedure could then have been designed in a simpler way, e.g., using two messages only. The choice of a more complex three-way handshake procedure was done in order not to block future possibilities.

The case of a release triggered by the other user is done in a similar way: the MSC receives an ISUP RELEASE message, which causes the sending of an RIL3-CC DISCONNECT message toward the mobile station,

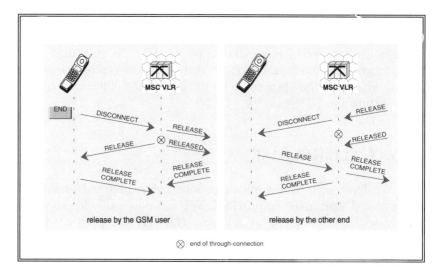

Figure 8.17 – Call release

The release procedure is a three-way handshake procedure,
whichever end initiated the release.

with a cause field giving some more explanation to cope with abnormal termination. The mobile station should answer by an RIL3-CC RELEASE message, acknowledged by an RIL3-CC RELEASE COMPLETE message from the network. Between MS and MSC, the procedure is symmetrical with the one used when the release is initiated from the mobile station side. As in the case of call establishment, this procedure is not suitable for the situation when the PSTN is involved. Explanations for the disconnection are then given if needed as tones or announcements. To cope with this case, the RIL3-CC DISCONNECT message may include a *PROGRESS* indicator, in which case the mobile station may keep the audio connection, and the release will be completed afterwards at the network initiative. In this case the DISCONNECT and the RELEASE messages both come from the network.

8.1.7. IN-CALL FUNCTIONS

We have seen so far how calls are routed, established and released. Between the establishment and the release phase, the call is deemed "active". During this period, several events may happen which require some additional functions on the Call Control level: for instance, the type of service may change within the same call, the user may want to juggle between several calls or may want to press some keys to send control signals to the other party (e.g., his voice mailbox). Let us consider these events in turn and describe the corresponding Call Control procedures.

8.1.7.1. Alternate Services

A few of the services proposed in GSM allow the user to toggle between two transmission modes, speech and a data mode. An important example (the only phase 1 case, other alternate services being part of phase 2) is the alternate speech/fax. Some signalling means are introduced in the *Specifications* to co-ordinate the transition of all machines along the transmission chain from one mode to another. The corresponding procedures will be fully used mainly in phase 2 implementations, where they will be corrected and enhanced. We will therefore only briefly describe them, looking at the case of a mobile to mobile fax call.

For the fax application, the toggling of transmission modes is always initiated by the user, through some man-machine command, and applies only to his end of the call. This is translated by the mobile station in an RIL3-CC MODIFY message, aimed at the MSC. At this stage the mobile station does not take the initiative to change its own transmission

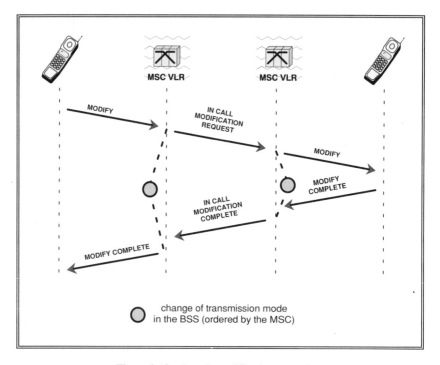

Figure 8.18 – In-call modification procedure

The change of mode within a call is controlled by a call control procedure,
which in turns triggers the relevant actions on the radio resource level,
to change the connection type to the appropriate mode.

mode. When the MSC serving the initiating mobile station receives the
message, it changes if applicable the interworking function, and then it
orders the change of mode to the BSC, which is in charge of the
transmission aspects on the MS to MSC segment. It is the BSC which
will order the effective transmission mode change to the mobile station
(this is outside the scope of Call Control, and has been described in
Chapter 6) and the BTS. When the change of transmission mode is
successfully indicated back to the MSC, it sends to the mobile station an
RIL3-CC MODIFY COMPLETE message. Then, when the new mode is data,
the further re-connection procedures needed by the service
(synchronisation, modem commands, ...) are run.

At the other end, the same procedure should take place, triggered
by the other user. The important point is that the intervening system does
not take care of the synchronisation of the two changes. In some cases,
the other end will be aware of the change from fax to speech by hearing
or detecting the start of the modem tone exchange, and will only then ask
for the mode modification on his/its side.

Why was it so specified? Alternate fax/speech service is by nature designed for interworking with the PSTN. Clearly, indicating a mode change to another extremity in a PSTN, or even simply through the PSTN, is not possible. This comes simply from the PSTN not being equipped for this kind of signalling. There is then in these cases no alternative to letting the users cope with the co-ordination themselves. Things can be different in the case of a full ISDN compatible call (through ISDN, and terminating for instance in the ISDN or in GSM): ISUP supports this function, and the information indicating that one end has asked for changing the transmission mode is carried by an ISUP IN-CALL MODIFICATION REQUEST message. Figure 8.18 illustrates this case. The receipt of this message at the other end triggers the sending, by the MSC, of an RIL3-CC MODIFY message, and an order to the BSC to change the transmission mode for this MSC-MS segment. Eventually, the mobile station will answer with an RIL3-CC MODIFY COMPLETE message, the BSC will indicate the correct completion of the command, and then the MSC can report this successful outcome to the other end, with an ISUP IN-CALL MODIFICATION COMPLETE message. There is necessarily a (short) period during which the end-to-end transmission path is not consistent, before transmission can resume in the new mode. Let us recall that the full procedure as described here will be in use only in phase 2, though it is for its major part specified already.

8.1.7.2. Multiple Call Handling

As a preliminary to this section, we would like to remind the reader that all cases of multiple calls are also phase 2 features. Their specifications were not frozen at the time of writing. We will therefore not go to any level of detail, but briefly describe the procedures which enable the users to juggle between several calls in parallel.

We have already met one case of multiple calls, which is call waiting, when an RIL3-CC SETUP message is sent to a mobile station fully engaged in another communication. As seen in that example, and true in all other cases, dealing with multiple calls involves only the mobile station and the serving MSC. Beyond this point in the network, the calls are all independent, and managed as such. The related functions are then essentially local to GSM.

The basic issue raised by the need to exchange signalling messages pertaining to several calls in parallel is solved, as already mentioned, by the notion of parallel Call Control transactions, the messages of which are distinguished by a Transaction Identifier. This mechanism enables signalling activities to take place for several calls, and this can be done even when one of them is fully connected (through the fast associated

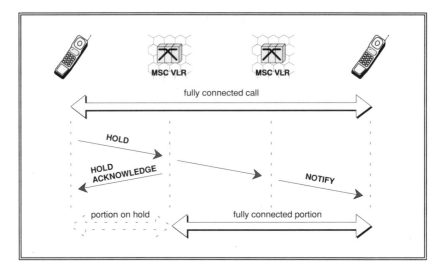

Figure 8.19 – Call hold

Either end of a call can be put on hold independently from the other.
Here, user A puts his end of a mobile to mobile call on hold.

signalling scheme). However, the only cases where this currently is allowed is the rejection of a newly arrived call (call waiting) or the release of another one on hold. In all other cases the user can intervene in a call only if all others are on hold.

Putting calls on hold, and retrieving them, represent the two major actions in the domain of multiple calls. Putting his own end of the communication on hold is done at the request of the user. This action triggers the procedure shown in figure 8.19. First, the mobile station sends an RIL3-CC HOLD message to the MSC. At the receipt of the message, the MSC acknowledges the new state with an RIL3-CC HOLD ACKNOWLEDGE message, and if possible warns the other party. In an ISDN environment, this is done with an ISUP SUSPEND message, which would be, in the case of a GSM user at the other end, translated into an RIL3-CC NOTIFY message. These messages are only indicative, and do not entail any state change at the other end. It should be noted again that the hold condition pertains to a single end of the communication, so that a call may be in one of four states, depending on whether none, one or both extremities are on hold.

The converse procedure consists in resuming full connection on the end of the communication which was put on hold. It is called the "retrieve" procedure, and can be done only if no other call is connected. It consists in the exchange of an RIL3-CC RETRIEVE message and the corresponding RIL3-CC RETRIEVE ACKNOWLEDGE message. There again

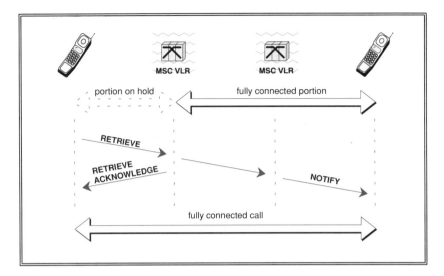

Figure 8.20 – Retrieving a call on hold

The user may resume end-to-end communication on a call previously put on hold, using for this purpose the retrieve procedure.

the other extremity may be notified in a full ISDN/GSM environment, using in the GSM radio part, an RIL3-CC NOTIFY message. The procedure is shown in figure 8.20.

Once all existing calls are on hold, the user can retrieve any one of them, release any one of them, establish a new one (if the allowed maximum number of communications in parallel is not reached), or answer a pending Mobile Terminating call. This last procedure is not done by a retrieve procedure, but by sending an RIL3-CC CONNECT message, as described in the Mobile Terminating call case.

Another action which can be done in the connected state is to merge calls into a multi-party conference. More exactly, a user can ask for the conferencing of his calls. At the other ends, the calls have nothing specific from a management point of view. The calls to merge must exist when the conferencing is asked for. The first step will then be to merge two calls, the active one, and a held one (there can be no more than one held call if another is active). Other calls can be added afterwards at the same end, by putting the multi-party on hold, establishing a new call, and conferencing the new call and the multi-party. Calls in multi-party can be disconnected temporarily, or released independently. The multiparty state disappears when reduced to one call.

8.1.7.3. Transmission of Audio Tones

In the PSTN, dual tone multi-frequency refers to the transmission of dual-frequency tones generated on the speech path by pressing the number keys of the telephone set (or by a separate tone generator with old-fashioned sets!). The specification of the GSM mobile station requests that DTMF tones are not generated by the mobile station, but instead by the MSC, to avoid going through the 13 kbit/s speech encoding. Instead signalling messages are sent as the result of pressing keys on the mobile station side, and tones are generated by the MSC at the reception of such messages. Note that this applies only for transmission from the mobile station. Nothing precludes sending DTMF tones to a mobile station, and they will go through the 13 kbit/s encoding. As a consequence, there is no guarantee that a DTMF receiver listening to the loudspeaker of the mobile station will recognise the tones.

The procedure is not that simple, because the period during which a tone is generated is under the control of the user, and because it was considered necessary that the MSC acknowledges these messages (though the probability of an RIL3-CC message being lost without the call being cut is very small, thanks to the lower layer protocols). As the result, the sending of one tone requires four RIL3-CC messages, and this does not include the layer 2 acknowledgements. The generation of a tone is started by an RIL3-CC START DTMF message, which is acknowledged by the MSC with an RIL3-CC START DTMF ACKNOWLEDGE message (or alternately rejected with an RIL3-CC START DTMF REJECT message). It is stopped by the sending of an RIL3-CC STOP DTMF message, acknowledged predictably by an RIL3-CC STOP DTMF ACKNOWLEDGE message. This is repeated for each tone, resulting in a total of 40 messages for the sending of a ten-digit number.

8.2. SUPPLEMENTARY SERVICES MANAGEMENT

In the previous sections, a number of supplementary services have been described as far as their interaction with call handling is concerned (e.g., call forwarding, call waiting or call hold). Most of these facilities can be activated (allowing their use or their effect) and deactivated (preventing their use or stopping their effect) by the subscriber, and some have parameters which have to be set, or may be changed, such as, e.g., the forwarded-to number. Signalling means are provided to support their modification, or to check their value, by the subscriber and using the mobile station to access the GSM network. They will now be described.

8.2.1. ARCHITECTURE

The signalling requirements for supplementary services management actively involve only two entities: the mobile station and the HLR, as shown in figure 8.21. The signalling exchanges are grouped in a special protocol, dubbed here the MAP/I, which differs from the other MAP protocol in that one of the two protagonists of the protocol is the mobile station, which is not directly connected to the CCITT signalling system number 7 network (SS7 network). Between the mobile station and MSC/VLR, the MAP/I messages are then not carried by SS7, but are piggybacked on RIL3-CC messages encapsulated in a *FACILITY* information element, either as stand-alone information (in RIL3-CC FACILITY messages, which may be carried in fact under a special protocol discriminator, SS), or as information carried in some other RIL3-CC messages such as SETUP or ALERTING. The exact meaning of a FACILITY message or of a *FACILITY* information element coming from the mobile station is determined by information inside the message or the element. They are only carriers, of limited semantic value. In some cases (as for the management of a multi-party conference) the meaning of the facility is local, and relates to the call. In this case the MSC/VLR treats these messages by itself. In the other cases, the message is forwarded by the

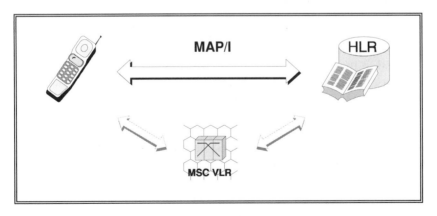

Figure 8.21 – Protocols for supplementary services management

Users can change and check the status of their supplementary services through GSM;
the MSC/VLR only acts as a relay for the corresponding MAP/I protocol,
which is handled between the mobile station and the HLR.

Supplementary service	Sequence of keys
Call forwarding unconditional applied to telephony	**21*[forwarded-to-number]*11#[SEND]
Call forwarding on Busy applied to all basic services	**67*[forwarded-to-number]#[SEND]
Barring of All Outgoing Calls applied to all basic services	*33*[password]#[SEND]
General deactivation of all call barring services	*330*[password]#[SEND]
Change of password for all supplementary services requiring a password	**03**[old password]* [new password]*[new password]#[SEND]

Table 8.2 – Generic man-machine interface commands

As a fall-back solution, in particular for public GSM phones,
a generic method of entering supplementary services control orders
on the keyboard of a mobile station has been standardised.
But most mobile stations offer more user-friendly means.

MSC/VLR to the HLR. In the other direction things are simpler for the MSC/VLR, since all MAP/I messages are forwarded to the mobile station.

8.2.2. PROCEDURES

Facility management can be done at any moment, at the request of the user. A generic, but not very user-friendly, man-machine protocol can be found in the *Specifications*. It specifies how commands can be issued by the user on his mobile station keyboard. This protocol only uses the 12 basic telephony keys (the 10 digits, plus * and #), and examples of it are given in table 8.2. The main advantage of this generic method is that it can be used for new facilities in the future without having to modify the mobile stations. However, it is likely that mobile station manufacturers will add more friendly methods, to help the users to control efficiently the different facilities, which can, once mastered, bring a real improvement in user comfort.

Whatever the method used by the subscriber to issue the control commands, the mobile station will generate signalling messages to be sent to the network. If a signalling connection already exists between the MS and MSC, it is used, otherwise one is created. The procedures can be

Supplementary services	Type of activation message	Parameters for activation
Call Forwarding Unconditional, on Busy, on Not Reachable	REGISTER SS	forwarded-to-number, basic service
Call Forwarding on No Reply	REGISTER SS	forwarded-to-number, basic service, condition time
Call Barring	ACTIVATE SS	basic service

Table 8.3 – Activation of supplementary services

For both call forwarding and call barring, a single "activation" action is defined, which requires specific parameters depending on the service.

done in parallel with any other process, including a communication which is fully operational. Messages pertaining to a call independent facility management are then distinguished from call related messages by a separate Protocol Discriminator, and among themselves by different Transaction Identifiers if several transactions exist.

Facility messages pertain to different kind of operations, which can be divided into several classes. First we find the messages enabling the activation or deactivation of facilities. There are several of them, with marginal differences. Activation (as a general term) makes use of the MAP/I ACTIVATE SS, REGISTER SS or INVOKE SS messages. Reciprocally, deactivation (still as a general term) makes use of the MAP/I DEACTIVATE SS or ERASE SS messages. The choice of the message to use depends on the facility. Such a message contains a reference to the facility to activate or deactivate, a reference to the basic service (speech, fax, Short Message, ...) to which it is linked, and, for activation, various parameters depending on the facility. Table 8.3 indicates, for some of the major supplementary services including the phase 1 services, what type of activation parameters are relevant.

A second group of messages enables the user (let us call him Finn) to enquire about the status of his facilities. It consists of the MAP/I INTERROGATE SS message and its answer. An interrogation message pertains to a single facility, which is referred to in the message. The answer contains the value of the different parameters as set in the HLR for this facility. As an exception to the general rule that MAP/I messages are handled between the mobile station and HLR, the *Specifications* consider that the MAP/I INTERROGATE SS message is handled between mobile station and MSC/VLR when it pertains to conditional call forwarding.

Finally we consider the handling of passwords. A password can indeed be associated with each facility, as an additional protection for the subscriber against a modification of its status or parameters by a third person. The MAP/I REGISTER PASSWORD message enables Finn to set, change or suppress a password in relation with a given facility. Conversely, the MAP/I GET PASSWORD message and its acknowledgement convey respectively a request from the HLR to provide the password and the password itself, as given by the user. This is of course invoked by the HLR when a request to tamper with the concerned facility is received.

All the messages seen so far relate to the cases where the mobile station understands exactly what is going on. If facility management commands are entered as a sequence of digits, # and *, a MAP/I PROCESS UNSTRUCTURED SS DATA message is sent by the mobile station. If a password is needed, it has to be included with the request. Status checking is not possible this way. This mechanism enables operators to introduce new services between the user and the home network, without the need for existing mobile stations or other networks to handle them in a specific way. As such, it contributes to the upward compatibility of the system.

The MAP/I FORWARD SS NOTIFICATION message is used when the HLR detects that an activation of a supplementary service requires deactivation of another service in order to avoid conflicts in the operation of the two services.

The last MAP/I message, FORWARD CHECK SS INDICATION, is used at the initiative of the HLR, in cases when some failure may have resulted in an erroneous state of the facilities for the subscriber. The subscriber is in essence asked to check for himself if all is correct.

8.3. SHORT MESSAGES

All communications of a circuit nature, such as speech or data transfer, are established, released and generally managed by the procedures described in the previous sections. Let us now see how communications of another nature are treated in GSM. The only GSM services not requiring the end-to-end establishment of a traffic path are the Short message services. As a consequence, short message transmission may take place even if the mobile station is already in full circuit communication.

A short message communication is limited to one message, or in other words the transmission of one message is a communication all by itself. The service is then asymmetric, and Mobile Originating Short Message transmission is considered as a different service from Mobile Terminating Short Message transmission. This does not prevent a real dialogue, but the different messages are considered to be independent by the system. The transmission of a message is always relayed by a Short Message Service Centre (SM-SC), considered to be outside GSM[1] . The consequence is that the transfer of a short message always takes place between a mobile station and some SM-SC from the point of view of the GSM infrastructure. However, for the user, the message has also an ultimate destination or origin, identified by some field in the message, but relevant only for the user and the SM-SC not for the GSM infrastructure.

8.3.1. ARCHITECTURE

As explained in Chapter 2, the point-to-point short message services defined in GSM enable the transfer of short messages between the mobile station and a short message service centre which is in contact with GSM networks through specific MSCs called SMS-GMSC (for Mobile Terminating Short Messages) or SMS-IWMSC (for Mobile Originating Short Messages), referred hereafter, as in Chapter 5, both as "SMS-gateway". The protocols involved in SMS management are shown in figure 8.22. They include the following:

- the mobile station to SM-SC protocol enables the transport of short messages, whether from or to the mobile station. This protocol is referred to as **SM-TP** (Short Message Transport Protocol); this protocol relies on underlying protocols which have been described in Chapter 5;

- the protocol between the SMS-gateway and HLR enables the SMS-gateway to interrogate the HLR in search of the address of the subscriber when reachable; it is part of the **MAP/C** protocol already mentioned for the interrogation of the HLR by a GMSC within the standard Call Control procedure;

---

1   The detailed functions of the SM-SC, and its interfaces are out of the scope of the *Specifications*. This does not however preclude a GSM operator to operate an SM-SC, and thus to include it in its network.

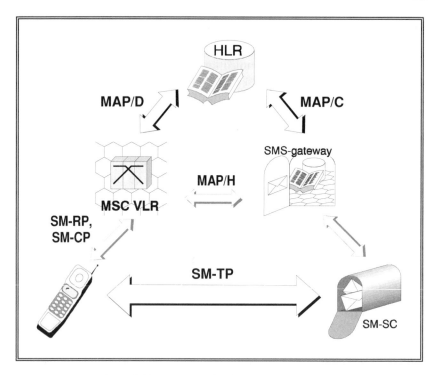

Figure 8.22 – Protocols for Short Message Transfer

The lower layer protocols (shown here as arrows of lighter shade) and the SM-TP
enable the delivery of short messages between the mobile station and SM-SC
either in real-time or as soon as the user becomes reachable,
through the help of information stored in the HLR.

- the protocol between MSC and HLR, as well as the protocol
  between HLR and SMS-gateway, enable the alerting of the SM-
  SC when a mobile station has missed a message while it was
  out of reach but has subsequently become reachable. This
  function must also be supported on the interface between the
  SMS-gateway and the SM-SC, but the protocols on this
  interface are not defined in the *Specifications*.

Of course, application protocols are needed on top of the SM-TP
(to format the user information for instance), but these protocols are there
again left to the choice of operators, and are out of the scope of the
Specifications.

We will now describe the functions of these protocols in turn,
starting with the case of Mobile Originating short messages.

8.3.2. MOBILE ORIGINATING SHORT MESSAGES

When a GSM user wants to send a short message, he must as a minimum type in its contents, the identification of the ultimate destination, and the directory number of the service centre which must deal with the message. Then, by some man-machine command, he can request the transfer of the message.

It is worth noting that there is no specification concerning the man-machine aspects of Short Message handling. Quite certainly, various solutions will be presented by the manufacturers. The simpler mobile stations will necessarily include a small display and a more or less sophisticated keyboard, which may be sufficient to support Short Message entering, e.g., with escape mechanisms or function keys to emulate a full alphabetical keyboard. The syntax to be applied to the message contents is irrelevant to GSM, but may depend on the service centre for specific applications and in this case will need to be known by the user.

Short Message transmission requires the setting up of a signalling connection between the mobile station and the MSC if none currently exists. This is done as for any other communication. The transfer of the message itself requires the establishment of a special link layer connection on the radio path, the SAPI 3 connection (see Chapter 5), and the use of specific message transfer protocols. On top of the specified protocol stack is the so called Transport Layer protocol, which consists in the case of a Mobile Originating message of a single message, the SM-TP SMS-SUBMIT message. Lower layers deal with the delivery acknowledge, which indicates only that the SM-SC has received the message. There is no support at this level of an automatic acknowledge to indicate when the message has reached its ultimate destination. If the application in the SM-SC supports this kind of service, the end-to-end acknowledgement will probably be sent in an independent short message. In any case, what the SM-SC does with the message is outside the scope of the *Specifications*, and no literature is available on the subject at the time of writing. What the user can do as offered by the SM-TP protocol is to set a period of validity for the message, after which the service centre will not try to deliver the message any more but destroy it. There are also some provisions in the format of the destination address to ask for the transfer from the SM-SC to a variety of directions, such as fax machines, teletex machines, other message handling facilities, ..., and GSM subscribers.

8.3.3. MOBILE TERMINATING SHORT MESSAGES

A short message addressed to a GSM subscriber must first be routed from the sender to a Short Message Service Centre, and from then be routed to the actual destination. The way the message first reaches the SM-SC is once again out of scope of the *Specifications*. There also a variety of solutions can be imagined to enable PSTN users to send messages towards GSM users, using human operators or inter-working with other services such as videotex.

When the SM-SC has a message to send to some GSM subscriber, it builds a SM-TP SMS-DELIVER message, containing various pieces of information for the benefit of the recipient. This information includes in particular the user content, the identification of the original sender, and a time-stamp indicating when the message was received by the SM-SC. Similarly with the Mobile Originating case, the SM-TP SMS-DELIVER message will be transferred on various interfaces, using the capabilities of lower layer protocols described in Chapter 5, in particular to convey the acknowledgement back to the SM-SC.

Before the SM-TP SMS-DELIVER message can reach its destination (the mobile station), its actual routing must be derived using the interrogation functions of MAP/C. This is achieved in the following manner. The SM-SC conveys the short message to an SMS-gateway to which the service centre is connected, which it chooses depending on the subscriber it wants to reach, since most often a gateway will be able to deal with only some of the subscribers (for instance those of some country, or some operator). The subscriber is then identified by his directory number (the same MSISDN as for telephony, typically), entered originally by the originator of the message. This enables the SMS-gateway to identify the relevant HLR and interrogate it. The interrogation is done by sending a special message, the MAP/C SEND ROUTING INFO FOR SHORT MESSAGE message. This is answered by either the corresponding MAP/C SEND ROUTING INFO FOR SHORT MESSAGE RESULT message, which contains an SS7 address pertaining to the MSC/VLR where the subscriber is visiting, or by a rejection message if the subscriber is known not to be reachable at this instant. There is no need for a specific roaming number, as for circuit calls, since the short message uses only SS7 signalling means to be transported to the visited MSC.

The SMS-gateway makes use of the SS7 address to forward the message to the relevant MSC, which delivers it to the mobile station after setting if need be a signalling connection, as for the mobile originating case. The delivery to the mobile station does not involve the user. The message can be stored until the user decides to discard it after reading. More precisely, it is stored in the SIM, and then can be kept in storage even after the mobile station has been switched off, or even read on another mobile station. The limited memory capacity of the SIM, however, raises a small problem when for any reason the memory is full. In phase 1, a message delivered to a SIM with no free memory could be lost. In phase 2, a mechanism has been specified to enable a crude sort of flow control by the mobile station, which will then be able to indicate to the network when the memory is full, or conversely when it is back to a state where messages can be accepted again.

An important variation from this basic scenario corresponds to cases when the mobile station cannot be reached. To provide a satisfying quality of service, the message is not lost in these cases, and steps are taken so that this message, and possibly following ones, are kept and delivered to the subscriber as soon as possible. Since the message, if not acknowledged, is still stored for some time in the SM-SC, it can be sent anew as soon as the subscriber resumes contact with the network. This requires the GSM network to store the lack of delivery condition and the address of the SM-SC, and to start a procedure to "alert" the SM-SC when the subscriber pops up again. The HLR is obviously the "focal point" in such mechanisms. Let us describe how they work, by looking at the different situations in which delivery fails.

Three different kinds of non-reachability can be identified, similar to what we have seen with circuit calls. The HLR can know beforehand that the subscriber is not reachable for the moment, the VLR can know it, but not the HLR, and finally it can be discovered after failure of the effective attempt to deliver the message by the MSC/VLR. When interrogated by an SMS-gateway, the HLR may know immediately that delivery cannot take place, because it already holds a non-empty list of service centres which have not succeeded in transmitting messages, and waiting to be alerted. It then adds if possible the new SM-SC identity to this list. In such situations, it will usually indicate the problem to the SMS-gateway with a negative answer to the MAP/C SEND ROUTING INFO FOR SHORT MESSAGE message. The same course is taken when the HLR knows that the subscriber is not reachable, for instance because the subscriber is not entitled to get service in the geographical area where he is currently located. There is however one exception to this rejection at the HLR level. A priority indication is linked with each message, and it is used to bypass the straightforward rejection of the HLR when some messages are still undelivered. The HLR will answer positively if the message is of high priority, and the potential delivery problem, if it still exists, will be detected by the MSC/VLR.

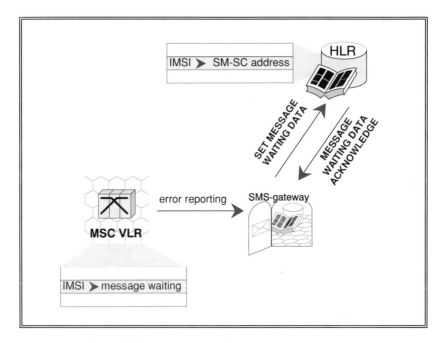

Figure 8.23 – Short message failed delivery management

When the MSC is unable to deliver a short message to the mobile station,
this state of things is stored in both MSC/VLR and HLR, for triggering retries
when the mobile station reappears.

In the cases where the MSC/VLR is given the message but is not able to deliver it, a failure indication is first sent to the SMS-gateway, as an answer to the MAP/H FORWARD SHORT MESSAGE message. The gateway then sends on one hand a negative report to the SM-SC, and on the other a MAP/C SET MESSAGE WAITING DATA message to the HLR, which acknowledges the updating of its table by a MAP/C SET MESSAGE WAITING DATA RESULT message. This state of affairs is stored by both the MSC/VLR and the HLR in the subscriber record. In addition, as already mentioned, the HLR maintains for each subscriber a list of addresses for the SM-SC holding messages in wait. The sequence of events is represented in figure 8.23.

Eventually, the subscriber surfaces again. This may be known for instance by a contact with the MSC/VLR where the subscriber was located (e.g., a mobile originating call attempt). When such an event happens, thanks to the stored indication of a previous delivery failure, the MSC/VLR notifies the HLR with a MAP/D NOTE MS PRESENT message.

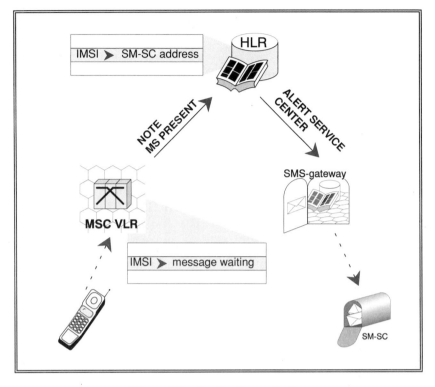

Figure 8.24 – Service Centre alerting

When a subscriber becomes reachable, the HLR alerts all the service centres
which are known to hold messages not delivered to the given subscriber.

The mobile station may also reappear within coverage of another MSC,
in which case the HLR will be directly aware of this state of things,
thanks to the mobility management procedures which have been
described in Chapter 7. In any case, the HLR then sends an indication of
the subscriber's reappearance to all the SM-SCs whose identities are
stored as holding a message for this subscriber. This is achieved by
sending a MAP/C ALERT SERVICE CENTRE message to the suitable SMS-
gateway for each service centre. The whole sequence as described above
is shown in figure 8.24. The SMS-gateway will convey the relevant
information to the service centre, to trigger a new transfer attempt.

The alerting mechanism must be supported by all MSC/VLRs, but
it is an operator's option to store the list of SCs in the HLR and to alert
them.

SPECIFICATIONS REFERENCE

The only GSM TS that can be cited for a general presentation of the call management mechanisms is **TS GSM 03.01**, *Network functions*, but the subject is only partially treated.

For more details the reader must jump immediately to the detailed interface specifications.

The man-machine interface between the mobile station and the user is dealt with in the 02 Series. **TS GSM 02.30** is devoted to the subject, and pieces can also be found in **TS GSM 02.07**, dealing with mobile station features.

All the *Specifications* cited hereafter are deeply technical, and contain little general presentation to aid an understanding of the subject.

The interfaces for call management between a mobile station and a Terminal equipment are referred to in **TSs GSM 07.01, 07.02 and 07.03**, of which the main part is devoted to transmission aspects on the interface.

Generalities concerning the modelling aspects of the RIL3 protocols can be found in **TS GSM 04.07**. The RIL3 protocols themselves are specified in **TS GSM 04.08**, where section 5 is devoted to the CC protocol. Section 6 deals with call control for packet mode data, but is of no application (it is even removed for phase 2). The supplementary services aspects of Call Control are not included in TS GSM 04.08, but are dealt with in **TS GSM 04.10** and in the *Specifications* of the 04.8x series, including the coding of messages in **TS GSM 04.80**. In phase 1, only the call barring and call forwarding facilities are fully included, and they are specified respectively in **TS GSM 03.82** and **03.88** (for the general technical aspects) and in **TS GSM 04.82** and **04.88** (for the detailed specification of respectively the call forwarding and the call barring service implementation). More generally in phase 2, the implementation of the facilities described on the service level in TS GSM 02.8x (x ranging from 1 to 8) is described in the corresponding TS GSM 03.8x and 04.8x.

The details of the MAP protocols are to be found in **TS GSM 09.02**, more particularly in section 5.4 (retrieval of subscriber parameters during call set-up), section 5.3 (handling of supplementary services) and section 5.13 (support of short message services).

The interface between MSCs or GMSCs and external networks are referred to in TSs GSM 09.03 to 09.07, for the various categories of external networks. Of particular relevance are **TS GSM 09.03** and **TS GSM 09.07** which deal with the signalling aspects of the interface with the PSTN or the ISDN; and **TS GSM 09.11**, which deals with the signalling interworking for supplementary services.

Short message services are dealt with in **TS GSM 03.40**, which covers all the higher layer aspects of the service.

NETWORK MANAGEMENT

9

NETWORK MANAGEMENT

The picture of the GSM communications system as presented so far is a functional one. We have seen the different actions performed by the traffic handling machines in order to provide communications services to the users. We have described in a quite abstract way machines, transmission lines and users. Now, for a network operator, users and machines are more than abstractions. A real network contains many machines, whose location and capacity must be chosen. Machines must be procured, installed, linked to one another, and configured so as to be part of a consistent and cost-effective network. Machines and links can suffer failures, and must be repaired. Subscribers must be sought, corresponding data must be entered in the system and money must be recovered from these subscribers. These different activities are an integral part of telecommunications systems. Because of the ever-increasing complexity of telecommunications networks and because of the search for cost-effectiveness, these tasks are supported more and more by machines introduced in the system for this purpose.

The focus in this chapter is given to the point of view of the operator. The different operation, maintenance and subscriber administration tasks will be developed, with a stress on the aspects done by electronic machines, whether by the traffic handling machines or by other machines devoted to these tasks. The latter constitute the OSS, the Operation Sub-System. Most of the concepts presented here are of general application to telecommunications systems. In each case, GSM will be taken as an example, and some importance will be given to its particularities, which are often those of cellular mobile communications systems.

The first area presented in this chapter, subscriber management, encompasses the handling of the subscriber data needed in particular by the traffic handling machines, and the charging aspects. The study of this area will lead us to present the relationship between customers and operators, and the concept of service providers. We will also look at the part of the Operation Sub-System related to all these aspects.

The second area is maintenance, that is to say the functions aiming at maintaining a satisfactory level of functioning in the system despite the unavoidable failures. We have included in this domain the management of the mobile stations (with the related machines and procedures), and their type approval, which can in a way be presented as preventive maintenance.

The third, last, and longest part is devoted to operation. This covers the "piloting" of the network by the operator personnel. This includes the engineering choices to build, deploy and improve the network; the means of knowing if the level of performance is as expected; and the means to modify the network and its parameters. The description of the GSM system engineering part is developed using a number of examples. In particular cellular planning will be addressed, and this will be the opportunity to tackle topics such as radio propagation and spectral efficiency.

The last part of the chapter will present the architecture of the Operation Sub-System, with a focus on the part devoted to infrastructure equipment maintenance and operation. This will be the opportunity to address the concept of TMN (Telecommunications Management Network), both for its architectural aspects, and for its protocol design approach, an application of which can be found in the GSM *Specifications*.

9.1. SUBSCRIBER MANAGEMENT

The relation of an operator with its subscribers has two main facets. First a commercial dialogue must be initiated with them which will lead to a subscription relationship being established. A subscription includes an entitlement to obtain service from the network, and this requires actions on the network equipment so that the subscribers are recognised as such.

The second facet is billing and accounting. The operator must calculate the call charges, and—a vital issue for operators—must recover the money, by, among others means, billing its own subscribers. The subscriber mobility, and even more the roaming capability, somewhat complicate the issue, compared with wireline systems. With roaming it is possible that some service is provided entirely by one PLMN, and paid for by the subscriber to another. Inter-PLMN billing and accounting is then a very important topic, which determines, more than the technical aspects, the possibility to provide roaming to subscribers. Some stress will then be put on the transfer of charging information, between machines, and between networks.

9.1.1. SUBSCRIPTION ADMINISTRATION

Access to GSM services is conditioned by the user being known to the system, which is the purpose of subscription. From the traffic handling point of view, a subscription is materialised by a subscriber identity module (SIM) and corresponding entries in databases (in the HLR and in the authentication centre, AuC), and is identified by an IMSI (International Mobile Subscriber Identity). Means are needed in the system to create, upgrade and cancel subscription data.

Some additional data related to the subscriber, such as his name and address, are needed for commercial aspects, and in particular for recovering the charges. The *Specifications* do not cover these aspects, and the implementation will differ from one operator to another. The HLR may, in addition to its canonical functions, be the repository of this data, or they may be kept in another database, with typically the IMSI serving as the common reference. This choice depends in fact less on technical considerations than on the commercial organisation set up by the operator.

A commercial structure which is becoming more and more common is based on the concept of service providers. Service providers are companies, usually distinct from the operator, who take the responsibility for the commercial contacts with the customers. This includes typically the establishment of the subscription, the establishment and the dispatch of the bills towards the subscribers, and the recovery of the money. A subscriber who goes through a service provider has no direct contact with the operator. Usually, the operator bills globally the service provider for all the charges related to the subscribers managed by the latter, and provides the service provider with toll ticket information necessary to produce the individual bills.

Looking to practical cases, such as existing cellular networks or the organisation of the GSM operators, there are in general many more service providers than operators, and a service provider may deal with several competing operators. Some operators retail exclusively through service providers. At the other extreme, some operators deal with all the commercial aspects themselves. Mixed cases can also be found, where both a direct sales network belonging to the operator and a number of service providers coexist. Figure 9.1 pictures the relationships between the operator, the service provider and the customer.

The intervention of service providers splits the work between a network operating company, operating a telecommunication equipment, and companies in charge of the commercial network. The subscriber data is then spread over, for instance, a database held by the service provider (including name and billing address in particular) and a database held by the operator. The latter is typically implemented in a database separate from the HLR, and it holds all the subscriber information not necessary for the provision of the telecommunication service, in particular information such as "how to bill" the subscriber (for instance from which service provider each subscriber depends).

In the following we will assume that there is a subdivision of tasks between operator and service provider. This is to be understood as covering also the case when the operator manages all commercial relations by itself, i.e., the service provider is the operating company itself.

Creating a subscription entails different tasks. One is the contact with the would-be customer, to establish with him or her the subscription contract. The other is the initialisation of the different databases and of the SIM. Because customers like to subscribe in a one-stop shopping approach, leaving the sales agency with their SIM, it is useful for an operator to prepare as many things as can be in advance of the actual subscription. In a typical scenario, the operator prepares the AuC and the data in the SIMs before the actual contact with the future subscriber, and may even prepare incomplete corresponding records in the HLR. The service provider have then at their disposal a number of SIMs which already contain a stored IMSI, but which cannot be used at this stage for getting service (by lack of a proper initialisation in the HLR). When the subscription is created, the service provider enters the data related to the customer in its own database and ascribes an IMSI and a SIM to the new subscriber. In this scenario, the service provider need not perform any action on the "ready-to-use" SIM, except possibly printing the subscriber name on it. In some sales networks, the customer may choose his directory number (the MSISDN) among a proposed list of numbers.

Another scenario, although less attractive to the customer, consists in completing the SIM personalisation process (including the storage of the IMSI and secret key Ki) once the subscriber is known. In such a scenario, the subscriber must wait a few days to receive a SIM, e.g., by mail. The main advantage of this scenario is to remove the need for the operator to have a real-time subscription management network.

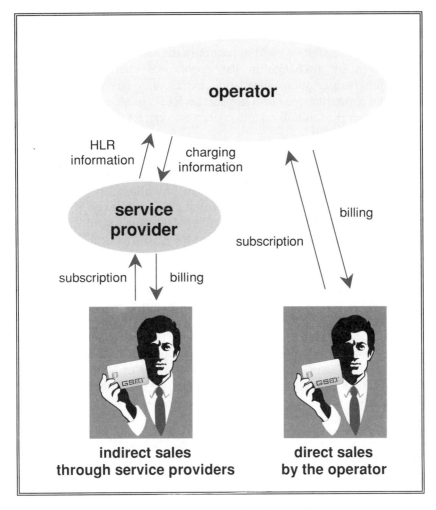

Figure 9.1 – Operators and Service providers

The operator may choose to deal with customers directly,
or alternatively to entrust service providers with customer care.
Many operators choose a mixture of both solutions.

In both cases, some steps must be taken to properly initialise the HLR record. This includes the description of the subscribed-to services, the allocation of one or several MSISDNs, and the enabling of the IMSI for actual service. Finally some tests have to be conducted to check all the operations were done correctly, so that the subscriber can actually access the services he or she asked for. The service provider must have means, whether in real-time or not, to trigger the initialisation of the HLR record, usually under the control of the operator.

There are other aspects to subscription management. Subscriber data may be modified after the initial subscription. Some modifications (e.g., a change of address) will impact only the service provider database. Others, such as a change in the operational characteristics of the subscription (subscription cancellation, extension to new services, change of regional limitations, ...) also impact the HLR. In all cases, if a service provider intervenes, it will take in charge the contact with the subscribers for these actions also.

The role of the service provider for customer care is not limited to filing new subscriptions and changing them. Beside billing (which will be described in the next section), the service provider must perform tasks such as the after-sales assistance to customers. For instance, the report of lost SIMs must be handled quickly and replacements issued with another IMSI, while the IMSI of the potentially stolen SIM is deactivated in the HLR. A first level of protection against misuse of a stolen SIM consists in blocking the SIM itself, i.e., forbidding any access to the system when a wrong PIN code has been entered three times in a row. Of course, the absent-minded but rightful owner of the SIM might occasionally fall in the trap himself. To cope with such cases, the *Specifications* define a PIN Unblocking Key (PUK), different for every SIM and which can be used to unblock the SIM. Depending on the operator's choice, the PIN unblocking key can be provided to the subscriber or alternatively made available only to the service provider. In the last (and more secure) case, the subscriber will need some assistance from his service provider to unblock his SIM.

The commercial activity requires a number of points of sales spread all over the covered area. A technical approach to support the operations described so far, with a high level of automation, is to equip the points of sales with computer terminals able to read SIMs. These terminals are connected through some data network to machines which are able on one hand to update the service provider database, and on the other to provide contact with the HLR machines.

9.1.2. BILLING AND ACCOUNTING

A subscription relationship enables access to service. The counterpart is the payment of some periodic charge, the charges for calls and other services, and possibly a fixed initial subscription fee. The collection of the corresponding revenue is vital for the operator, as well as the accuracy of the services charges. This requires recording mechanisms in the traffic handling part of the system, so that sufficient data is recorded for each chargeable service. Calls are obviously the first source of traffic-dependent charges, but supplementary services

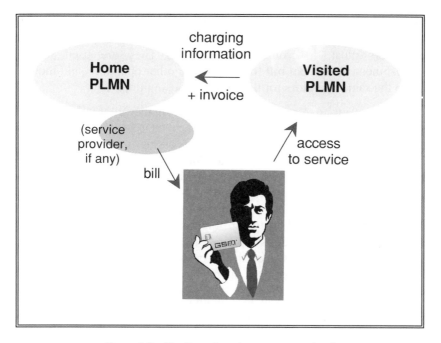

Figure 9.2 – The "transferred account procedure"

The body responsible for paying the visited network is the operating company
with which the customer holds his subscription. This way, a subscriber
receives a single bill for all his calls including those made when roaming.

management transactions may also be billed for instance, depending on the operator's policy. The GSM MoU operators have decided so far not to charge for location updating. The tariffing principles for the different transactions (calls, short messages, supplementary services, ...) are being discussed at the international level, but the tariff levels are left open for every operator (or even in some countries for every service provider) to decide upon.

We have seen in Chapter 8 that toll tickets are created by the MSC/VLRs and by the GMSCs. Toll tickets are individual records generated for each call and containing all the information necessary to calculate the call charges. Because of roaming, the MSC/VLR tickets pertain to subscribers which may belong to networks other than the one corresponding to the issuing machine. Each ticket includes the IMSI of the to-be-billed subscriber. The IMSI is the basis for forwarding the ticket to the right place. In the GSM MoU, the rule is that the charges are collected from a subscriber by the operator he holds a subscription with. The responsibility for paying the visited network operator lies with the home network operator of the subscriber: this is the concept of the "transferred account procedure", illustrated in figure 9.2. A first task for the visited network is then to sort the toll tickets and forward them to the correct subscription networks. The tickets pertaining to roaming subscribers must be grouped by network. They are used for the establishment of a global bill for each of the other operators, and they are sent to the same operators for their internal accounting.

The tickets for the home subscribers, whether internally generated or coming from other operators, must then be sorted on a subscriber basis. The operator is in charge of the processing up to this point. In a scenario with a service provider, a global bill is sent to the service provider, for all the calls of the corresponding subscribers, and the tickets are sent to the service providers to allow them to establish the individual bills. Variants of this scenario may be envisaged, where the operator establishes the individual customer bills for the service provider: this can happen when, e.g., the tariffs are fixed by some national authority. The actual dispatch of the invoices to the customers is however always the task of the service provider, as is the recovery of the charges.

All these transfers can be automated. Machines (in the operator part of the system) can collect and dispatch the toll tickets, and possibly establish the bills for the other operators or for the service providers. Then, means are required to transfer data between the switching offices and these machines, between these machines and those of the service providers, and between such machines for different operators. Often at the beginning the data transfers will be done by magnetic tapes, written by one machine and transported to be read by another. Real time is not really a constraint in this field (except for a real time advice of charge to the subscriber: the difficulty in case of roaming is the reason why this service will not be provided initially). Electronic data interchange will shortly replace tape transfer, especially between those networks where roaming leads to substantial traffic.

The *Specifications* provide a method for the real-time transfer of toll tickets between networks held by different operators. The MAP/C includes messages for carrying toll tickets on a call basis between an MSC/VLR and a HLR. This involves two messages, the MAP/C REGISTER CHARGING INFORMATION and the corresponding acknowledgement. The contents of the message gives an example of what can be a toll ticket, and we will come back to this topic in a few paragraphs. It should be noted that there is no corresponding procedure between GMSC and HLR. This is in line with the approach developed in Chapter 8, that a GMSC deals only with calls toward subscribers of the network to which the GMSC belongs.

The use of the MAP/C facility to send toll tickets back to the home network is not mandatory. It will be little used in practice, since the task of valorising the toll tickets is typically performed by specialised

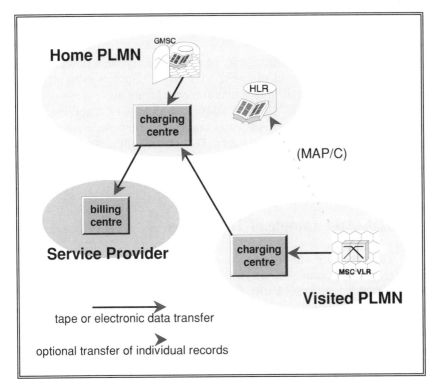

Figure 9.3 – Transfer of billing data

The transfer of billing data between networks for roaming subscribers,
and between operator and service provider, calls for an architecture
of interconnected machines.

charging centres separate from the MSCs and GMSCs. The inter-PLMN
transfers of charging data are then performed between such centres, by
tape or electronic transfer through means other than the SS7 network.

Figure 9.3 presents a general view of how the different machines
interact in the administration of subscriber charges, with the assumption
that there is a split between a network operator and a service provider.

We have described so far how GSM operators are able to exchange
toll tickets between them. The transfer of money involves more than just
GSM operators, and the transfer of the billing data is but the start of the
operation. The files transferred between two operators include the
individual call records for a given period of time (e.g., a few days), and
enable the home operator to charge its own subscribers. They indicate the
applicable exchange rate for each call and the inclusion of value-added
tax or not depending on national regulations, and they give the total

amount of all call charges they contain. These files serve as a basis for establishing invoices between operators.

The payment of invoices may be done directly, on a bilateral basis. However, since this would give rise to a lot of transfers even when the resulting balance is small, many operators prefer to settle these accounts through an international "clearing house", in charge of receiving all the agreed amounts to be transferred and combining them so as to minimise the number of actual money transfers. This concept of a clearing house is already applied within CEPT, where operators of the wireline networks settle their accounts in this way.

9.1.2.1. An Example of Toll Ticket

The structure of the MAP/C REGISTER CHARGING INFORMATION message is an example of the information which may compose a toll ticket issued by an MSC/VLR. Though this message may in most cases not be used, the contents of the files exchanged between operators will contain charge records which have fairly similar contents. Since the specification of the MAP message is public, it can be reviewed here as an example, to illustrate many different aspects seen in this chapter and in Chapter 8. Though the MAP procedure does not cover the transfer of accounting from the GMSC, the toll ticket structure covers all cases, including the re-routed part of mobile terminating calls from GMSC onwards.

First, the message contains identities: the IMSI (to identify the subscriber) and the identity of the MSC that generated the ticket. Then we find a call reference, which will enable cross referencing.

Next is included the "type" of the call. The following cases, correspond to tickets produced by a visited MSC/VLR:

- a mobile originating call;
- a mobile terminating call forwarded by the MSC/VLR; the toll ticket then concerns the portion from the MSC/VLR toward the forwarded-to destination.
- a non-forwarded mobile terminating call; this is not foreseen to be charged in networks of GSM MoU operators, but is included to keep the flexibility to do so, as well as for other purposes such as statistics, etc.

In addition, two cases correspond to tickets issued by a GMSC:

- a mobile terminating call forwarded by the GMSC; the toll ticket then concerns the portion from the GMSC toward the forwarded-to destination.

- the re-routed leg of a mobile terminating call, from the GMSC to the visited MSC/VLR.

Then we find the "status" of the call, that is to say whether it was successful or not, and in the second case the reason for failure. While toll tickets for successful calls are vital to the operator, the establishment of call records for unsuccessful calls usually serves no other purpose than for statistical analysis, and is not provided by all machines.

The next part of the information is the main data for the computation of the charge, in addition to the MSC identity which provides the ability to determine one end of the call. This includes the date and time of the beginning of the call (to cope with tariffs depending on time of the day), and its duration; the nature of the service provided (speech, data, short messages, supplementary service management transaction, with, in the last cases, more details on the nature of the transaction); and the called number (pertaining to the other end of the concerned segment of the call). To cope with packet services, a field includes the volume of transmitted data.

So as to enable the subscription operator to provide detailed billing, the calling party number (for mobile terminating calls) or the called party number (for mobile originating calls) may be included.

Finally, the record includes the charge which will be collected by the originating operator on the subscription operator. The usual unit is the "Special Drawing Right" (SDR), a "currency" used between telecommunication operators for inter-network accountings.

It should be noted that this structure has been designed only for billing and accounting purposes, and covers only the information needed for transfer between networks. Recording user activity is needed for other reasons within the network, such as network statistics. Since it would be inefficient to have a double recording mechanism, records created by MSCs or GMSCs and dealt with in the OSS usually contain more information. In particular other activities such as location updatings may be recorded, and the records may include information such as the mobile equipment identity (IMEI) or the identity of the cell where the transaction was initialised.

9.2. MAINTENANCE

Very rare are the systems which never fail, and telecommunication systems like GSM are not among them. A very important task for an operator is therefore to maintain the system in a state where the quality of service offered to the subscribers is acceptable. Maintenance includes the techniques aiming at minimising the loss of service quality incurred by a failure, the means to detect such failures and to report them accurately to the right person, and finally the means to restore the state of the network.

Failure causes are numerous. Electronic components have a finite—though long—lifetime. The hardware of any piece of equipment include thousands of components and a whole PLMN will include tens of thousands of such pieces. Even with a component lifetime of 10 000 years, several failures will occur per *day* in such a network! Other problems may come from the external world: the power supply may fail, communication lines may be cut by accident, thunder strikes regularly... Errors are also an important source of dysfunctioning. Zero-error software is long and costly to develop and in complex areas such as GSM not obviously achievable. The coverage of tests never represents one hundred percent of the running conditions, since the most marginal cases are often skipped. Failures tend to occur for instance in overload situations, where the system is pushed to its limits, and such situations are difficult to mimic on a test bed. Maintenance is faced with this large variety of failure sources, and the increase of system complexity makes automation and specific software necessary to assist the maintenance teams in their task.

Observing what happens in connection with a failure hints at the different facets of maintenance. It will serve as our guideline to describe them. What happens *before* the failure is the subject of preventive, or pro-active, maintenance. The aim is to minimise failure occurrences. In fact, this step merges with the next one, since preventive tests (such as test loops, for instance) may also serve to detect that a failure has occurred. There are other ways to detect failures, such as watch-dogs, that is to say independent devices which monitor some aspect of the activity of a processor, or temperature detectors, ... Once a failure is detected, a number of immediate automatic measures are taken locally by machines directly connected to the one suffering the failure. The aim is to limit the immediate effects of the failure. Examples of such emergency measures include software reset, the taking over of another piece of equipment in a redundant design or the stopping of the machine if material danger exists. These first-aid measures do not preclude repair, and to this avail the failure must be identified and located in a precise manner. The actual replacement may involve only the faulty module, or imply a modification of all similar modules in the system. Let us look at some of these points in more detail.

9.2.0.1. Minimising Failure Occurrences

Obviously the best method to limit maintenance costs is to use reliable equipment. During the procurement activity, an operator will consider different manufacturers and choose the best one according to specific criteria. This criteria may include the way in which quality insurance is performed by the manufacturer. In addition, the operator may also submit the pieces of equipment to its own set of tests before formally accepting them. Usually, a minimum set of tests is also passed on each single piece of equipment during the installation phase, before going into commercial operation.

Cellular networks experience a difficulty in this area. When for instance a new base station is introduced in a network, a normal practice would be to test it in a fully operational configuration while preventing commercial traffic until the machine has been fully tested. But radio waves cannot be restricted to the needs of the test, and the new cell impacts on the neighbour cells already in operation. The tested cell must be *barred*, to prevent the mobile stations of the customers to camp on it, and handovers towards this cell must be forbidden for these customers. Testing the new cell requires the operator to have special mobile stations, which (in addition to tracing facilities) will be able to camp on a barred cell. Modifications in the existing part of the network are necessary to test the new cell in a fully operational configuration, since handover parameters in neighbour cells are then different whether or not the new cell has to be taken into account. If the handover configuration had to be tested, means would be needed for BSCs and MSCs to distinguish test mobile stations from others. No such means exist for the moment, and, in short, the very final test will be done by the users themselves!

9.2.0.2. Minimising the Effect of a Failure

Failures can have very different effects on the service perceived by the users. The failure of some component involved in the transmission of a single radio channel or terrestrial circuit will only marginally degrade the capacity and hence the quality of service in terms of blocking probability. On the other hand, the failure of a switching matrix handling hundreds or thousands of circuits (which can be on a single chip) may result in the suppression of service over a whole geographical area.

The key method to limit such catastrophic events is redundancy. Where n interchangeable devices are used in parallel, $n+p$ are installed. For instance, if a cell needs 4 Transmitter/Receiver (TRXs) to accommodate the expected traffic, 5 will typically be installed. Another example of redundancy concerns the central chain of switching machines such as BSCs and MSCs, which are usually doubled.

The use of the redundant parts belongs to the domain of immediate defence, and is triggered locally by failure detection means. Several techniques exist to minimise the impact of the transition on existing traffic, and we will now consider a few examples.

In signalling system n°7, a signalling relationship between two adjacent points can consist of several links, and redundancy ensures that the required traffic load can be supported when one of the links is missing. In normal operation, the traffic is spread on all links. If acknowledgement and repetition protocols detect that one link is faulty, the corresponding load is reported on the other links. There is no loss of ongoing traffic, and capacity is maintained.

Another example is the hot-standby method, used typically for central processors. Two similar machines (A and B) are installed instead of one, and they run in parallel. Only one of them (say A) is actually in charge at one given instant, and provides its results to the outside world. When a failure is detected, the roles of the processors are swapped, and the outputs of the system are now provided by B, which then becomes the leading processor. At its highest level of sophistication, the hot-standby technique enables a transition without hiatus, and damages are limited to what occurred between the failure and the swap.

Several degrees of standby exist, from cold to hot. At the bottom of the scale, the "spare" device is not running, and is started only when the fault is identified. In this case, the swap involves a loss of context (e.g., all ongoing communications are aborted), but traffic capacity can be maintained.

A similar case exists when one of a set of ($n+1$) TRXs of a BTS fails. Typically, the ongoing communications using this TRX are lost, and a reconfiguration at the BSC level is required to restore the capacity to nominal values by using the spare TRX.

A last example in this domain is the case of A interface circuits between the BSC and the MSC. When such a circuit fails, the communication using it, if any, is typically lost. In order to circumscribe the effect of the failure, the circuit is then marked as "blocked" until repaired, i.e., tagged as not fit for allocation in the MSC. If the problem has been detected by the BSC, it indicates it to the MSC by signalling means.

Fault Detection

So far, we have briefly discussed the measures taken once a fault has been detected. How failures are detected is an important area of maintenance, and we will illustrate the matter by a few examples.

A number of detectors are incorporated in machines for the sole purpose of detecting failures. This is the case of temperature detectors, which may indicate faulty components consuming too much power, or failed fans. The power supply may be monitored by voltage or current measurement devices, triggering alarms if some thresholds are exceeded. Contacts may be used to detect board removals, door openings, and so on. In all these cases, the fault is detected locally.

In the case of hot-standby, a discrepancy between the results provided by the two parallel modules is the symptom of a failure. The problem is to decide which is the faulty one! Usually, a separate monitoring device will ask the modules to perform specific tasks and will compare the results to an expected pattern. A majority decision can also be applied if there are more than two parallel processors (e.g., if the modules are tripled).

Faults can also be detected remotely. For instance, when two machines communicate, they monitor each other. The lack of respect of some protocol rules, detected by the receiving end, may be the symptom of a failure or of a software fault in the other entity. Similarly, transmission circuits can be monitored if one end loops back what it receives when the circuit is not engaged. The other end can thus verify the continuity and consistency of the connection.

Regular and automatic testing is an important fault detection method. An advanced example of such tests is the use of a test mobile equipment in a BTS. Such a test mobile may be able to perform all the normal activities of a mobile station, initiating calls and so on. But in addition, it will analyse the way in which the network reacts, and check whether this behaviour conforms to some pre-established rules. If such a mobile station uses an antenna, it will also detect radio problems such as a decrease in the BTS transmission power, a fault difficult to detect otherwise. There is quite a variety of tests which can be performed. They are done in some cases routinely, in particular if running them does not take up resources which could be used for traffic; in other cases, they are controlled through the OSS.

A source of information which must not be neglected is the users themselves. User complaints, once analysed, often reveal faults that have escaped internal checks. Most operators have a free-phone number dedicated to the reception of customer complaints. In addition, a panel of voluntary subscribers or operator staff may regularly perform a postman's round and report strange events. A last example of failure detection is the analysis of global observations on the system. For instance, abnormally low traffic in some area may reveal a fault. Such high level detection can sometimes be helpful to detect failures affecting other failure detection mechanisms!

Generating Alarms

Fault detection can be done in a large number of different ways, including local and remote testing as well as analysis of various observations. However, failures of a similar nature can be detected by very different means, and reciprocally some analysis has to be performed to dispatch the failure information to the right place. In all cases, the original problems are translated into a general currency: the alarm. An alarm is a message sent to some central facility for analysis. It is generated at some point in the system to indicate a failure, and it reports the local analysis of the failure after immediate defence actions have been taken.

Locating a Failure

Alarms are but symptoms, which do not necessarily point to the exact illness. The more global the symptom, the less precise the identification of the failure location. For instance, a lack of coverage in some area can be caused by a number of problems, including failures in BTSs, BSCs and MSCs. A fair assumption is to consider that the failure happened at a single point, and can be pinpointed to a single piece of equipment. But a failure can also be an inconsistency between two co-operating modules or machines. The aim of the diagnosis will then be to determine the site of the faulty component, and if possible the module to replace.

Several techniques can be used for this purpose, and many of them can be done cost-effectively through a management network. The alarms themselves may carry sufficient information. In sophisticated implementations the analysis may be done with the help of a database where similar faults which happened in the past history of the system can be found, and even with the help of an expert system. If the first analysis is insufficient, complementary information can be obtained remotely by performing active tests on the machine.

Repair

Once the illness has been identified, a cure must be prescribed. It may be a hardware repair. In this case, a maintenance engineer will go on site with a replacement unit, run necessary checks, replace the (presumed) faulty module by the new one and run again some tests to check that the symptoms have disappeared, before putting the machine back to full operation. Ideally, the replacement and tests can be done without interrupting the handling of traffic through the concerned part of the network. It is useful for the maintenance engineer to communicate with the management network to order some tests or to re-enable operation. Supporting such communications may indeed be one of the functions of the data network interconnecting traffic handling machines with the management network.

In order for such replacements to be safe and efficient, consistency between hardware and software versions must be maintained. To this purpose, a database of the actual versions of all modules must be kept updated, and each module must carry an easy-to-get reference to its version. This is once again a function where management machines may improve efficiency.

Other kinds of corrective measures may apply after a failure. If the problem is a software error, the real correction requires new—corrected—software to be loaded. In the meantime, software patches may be required to reduce the effect of the failure. Alternatively, modifications of the configuration are sometimes used to avoid going through the dubious routines. Similarly, errors imputed to inconsistencies between co-operating machines can be corrected by changing the configuration.

In all cases a precise location of the fault is of foremost importance to minimise maintenance costs and the time during which service is either degraded or critically sensitive to a second failure.

9.2.0.3. Maintenance and the Management Network

The maintenance of a complex system such as GSM can represent an important part of the running costs. A computerised system communicating with all traffic handling machines contributes to lower these costs. Although fault detection is mainly a function spread among the traffic handling equipment, the network management system can fulfil a number of useful functions such as alarm centralisation, remote testing, alarm analysis to determine the nature of the fault, failure database, trend analysis, etc.

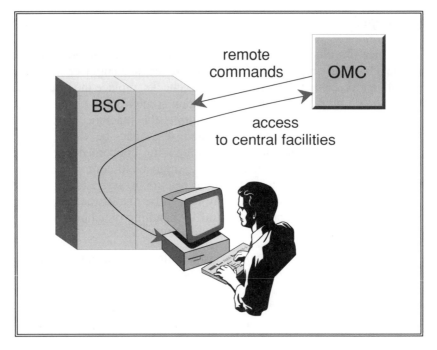

Figure 9.4 - Remote access to central maintenance facilities

It is useful for maintenance staff to be able to be in contact with the central
maintenance facilities (e.g., OMCs) while on site,
and have for instance remote tests ordered by the OMCs to help localise failures.

An integrated system to support maintenance activities will then include central facilities from where the whole network is monitored. Each of the machines in the network is linked to one of these facilities. Through these links, alarms can be forwarded, and remote tests to localise failures can be commanded. Another important feature is to allow communications between the personnel in charge of repairs on the site and a central maintenance facility. This facilitates the administration, provides the ability to command tests involving more machines than the ones on the site, permits access to the failure database to assist the diagnosis, and so on. Figure 9.4 shows an example of remote access to central facilities from a site. The maintenance functions are usually implemented together with a number of operation functions. For instance, failure localisation may require a detailed knowledge of the network structure. The implementation aspects of the maintenance support functions will then be looked at after the operation functional aspects have been developed.

9.3. MOBILE STATION MANAGEMENT

The mobile stations are often not the property of the network operator, but they intervene heavily in the quality of the service as perceived by the users. A badly designed or damaged mobile station may not only degrade the quality of the service for its user, but also degrade the quality of service for other users, for instance by being the source of unacceptable interference. The maintenance activity must take the mobile stations into account, since a symptom detected within the infrastructure may come from the failure of a mobile station and not of the infrastructure equipment. The correct functioning of the mobile stations is then a concern for the operators. However, the fact that the mobile stations are owned by the users and do not have any direct link with the subscription, and the requirement for a free mobile stations market, significantly limit the actions the operators can take.

A first approach is curative, and consists of detecting failed mobile stations, to indicate the problem to the subscribers using them, and to list them so as to bar service. The second approach is preventive: the type approval procedure gives some control to the operators, through the regulation authorities, on the mobile stations which may be used on their networks.

9.3.0.1. Type Approval

In many countries, the connection of a telecommunications terminal to a public network requires that this terminal be type-approved by a regulatory body. Until recently, such procedures were almost uniquely a national matter. For GSM, things are different. Full MS-roaming, made possible by the technical specifications, can only be achieved in the real world if common regulations are used by all countries in the domains of type approval, marking, free circulation and use of mobile stations.

Well aware of this goal, the European Community has produced directives concerning the mutual recognition of type approval of telecommunications terminal equipment. The first step in this process was a directive produced in 1986, whereby test reports produced by accredited test laboratories are mutually recognised, so as to avoid the need for each country to reproduce the tests for granting type approval. A further step was achieved in 1991 with directive 91/263 introducing the mutual recognition, not only of test reports, but of type approval itself.

(a) user safety, [...];

(b) safety of employees of public telecommunications networks operators, [...];

(c) electromagnetic compatibility requirements in so far as they are specific to terminal equipment;

(d) protection of the public telecommunications network from harm;

(e) effective use of the radio frequency spectrum, where appropriate;

(f) interworking of terminal equipment with public telecommunications network equipment for the purpose of establishing, modifying, charging for, holding and clearing real or virtual connection;

(g) interworking of terminal equipment via the public telecommunications network, in justified cases.

Table 9.1 – Essential requirements to be met for type approval

The Directive 91/263 of the European Communities states a list
of essential requirements which terminal equipment shall satisfy.

This directive defines essential requirements which must be met by the equipment, as reproduced verbatim in table 9.1.

All these requirements are not enough to ensure that the equipment is functioning correctly for the service expected by the person using it, but it certifies to a reasonable degree that the use of the equipment in a public network will not disturb the service provided to other users. In addition, services recognised as of prime importance may lead to additional tests of the ability to reliably perform end-to-end interconnection; they are referred to as "justified cases" in the directive.

Within the scope of directive 91/263, "Common Technical Regulations" (CTRs) are to be produced to describe the relevant checks. These CTRs, whose approval lies mainly with CEPT and the European Community (for member states), include both technical and regulatory aspects. The technical contents, i.e., the purpose and description of the tests, is referred to as a TBR (or "Technical Basis for Regulation"); the TBRs are to be produced by the European Telecommunication Standard Institute.

The existence of the CTR procedure being recent, the first type approvals granted to GSM mobile stations are still based on a former mechanism, known as a NET (Normes Européennes de Télécommunications). The NET involved an approval at the CEPT level, without any involvement of the European Community. The NET relevant to GSM is known as NET 10.

In the GSM area, NET 10 is replaced by two CTRs, CTRs 5 and 9. CTR 5 is concerned with the interconnection of the mobile station to a GSM network and the non-disturbance of the latter by the mobile station. CTR 9 is specific to telephony and tests the end-to-end interworking capabilities of the mobile station. More CTRs related to the end-to-end interconnection for other services may be introduced if these services become "justified cases".

The coverage of the tests may differ when going from the NET regime to the CTR regime, but in both cases the main work is to produce a technical specification of the tests, and to develop test equipment capable of performing the required tests. A test specification for GSM mobile stations is included in the *Specifications*, and this specification (TS GSM 11.10) represents the technical foundation for producing NET 10 and the TBRs.

TS GSM 11.10 covers most areas of the *Specifications* affecting the mobile stations. The bulk of it concerns the signalling protocols. It must be stressed that in this area, in particular, there is no way to guarantee with a reasonable test time that a machine as complex as a GSM mobile station will behave correctly for *every* possible sequence of events. TS GSM 11.10 has been designed to guarantee to a *reasonable* level that the mobile station conforms to the set of GSM Specifications.

The type approval tests must be conducted in a non-disputable way, by accredited laboratories who are proven to be independent from operating and manufacturing companies. Although the actual connection of a mobile station under test with operational GSM networks may help to identify potential problems, such tests cannot be included in the type approval process, because no network can be considered as a neutral reference.

The need for a reference test tool, recognised early in the process, led to the development of a single "system simulator" capable of performing the whole set of tests included in TS GSM 11.10. However, commercial pressures to start GSM operation proved incompatible with the delay for the system simulator delivery. Hence, a reduced set of tests was defined as a first priority, and a less sophisticated test equipment,

Figure 9.5 – The test equipment for interim type approval
(by courtesy of Rohde & Schwarz)

This equipment is used to type approve mobile stations on an interim basis;
before the full system simulator and full type approval procedure become available.
It performs some 140 official ITA tests,
mainly in the areas of radio and protocol conformance,
and more test sequences can be programmed on the machine.

shown in figure 9.5, was developed to cover the initial needs, before the
full system simulator becomes operational. The resulting "Interim Type
Approval" (ITA) has only limited validity in time. Mobile stations which
have been granted interim type approval will be subject to pass the full
type approval tests as soon as suitable test equipment and software
becomes available.

The granting of type approval to a mobile station type results in the
allocation of a 6-digit Type Approval Code (TAC), which is part of a
mobile equipment identity number (the IMEI). The IMEI has the

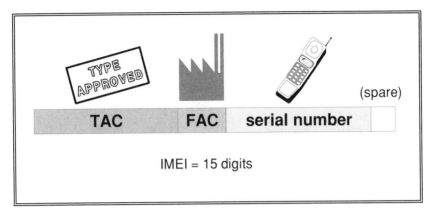

Figure 9.6 – Structure of the IMEI

The IMEI, ranges of which are allocated upon the granting of type approval,
references the mobile equipment type approval and the final assembly plant,
and includes a serial number which is unique for each unit of a given type.

structure shown in figure 9.6. The Final Assembly Code (FAC) is
intended to be used as an identification of the final assembly plant. Serial
numbers are allocated by ranges to the manufacturer for inclusion in the
produced mobile stations. Operators, through the permanent secretariat of
the GSM MoU, are notified of the valid IMEIs of type approved mobile
stations they can expect on their networks.

9.3.0.2. Mobile Equipment Management

When maintenance activities detect a problem which are attributed
to a mobile station, the network must determine the IMEI of this mobile
station. The procedures on the radio interface are such that the mobile
station does not volunteer its IMEI. The network must ask for it, through
the RIL3-MM IDENTITY REQUEST message. Depending on the
implementation, this request can be systematic, at the beginning of each
RR-session (but this creates a lot of additional signalling traffic), or can
be done at certain occasions only, e.g., at each location updating and for a
sample of call attempts. The knowledge by the operator of the IMEI of
the (presumed faulty) mobile station will help to produce statistics and to
determine the origin of the fault, in order to take corrective measures. In
the worst case, all mobile stations of a given type and series might have
to be removed from the market, if the defaults are important enough to
justify a retrofit. An advisory group consisting of GSM MoU members
has been set up to exchange information on problems identified during
commercial operation; these problems may reflect a need for new tests in

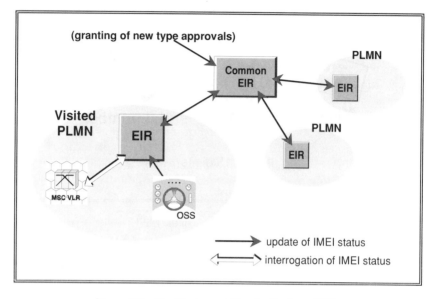

Figure 9.7 – The Equipment Identity Register (EIR)

The status of IMEIs is stored in EIRs, to be checked by the MSCs at any time.
Operators may interconnect their EIRs to update the status of IMEIs.

the type approval process, which would then be discussed with the
appropriate authorities.

The next problem for an operator is to bar mobile stations which
do not operate properly, i.e., to refuse to grant them service (except
possibly for emergency calls). This can also be applied to stolen mobile
equipment (subscription barring is insufficient in this case, since the
equipment can be used with another, legitimate, SIM). To achieve this,
IMEIs for which problems have been detected must be recorded in a
database. This database, referred to as the Equipment Identity Register
(EIR) in the *Specifications*, is updated though the OSS. It can be accessed
by the MSC/VLRs to check the status of a particular IMEI. This access is
supported in the *Specifications* by MAP procedures. Figure 9.7 shows the
position of the EIR in the system architecture. As an example of a
realistic configuration, it shows two levels of EIRs, one at the individual
network level, and one common to all networks. The interconnection of
EIRs between themselves is not specified in the phase 1 *Specifications*,
and is a matter for operators to agree on.

The information stored in the EIR can be to a large extent operator
dependent. However, the control, of, e.g., stolen mobile stations, will

only become effective if operators with a roaming agreement have an agreed IMEI checking policy. This is obtained for instance by an harmonisation at the level of the GSM MoU. Within this group, operators have agreed to use three levels for the status of the IMEI, and the *Specifications* define three corresponding "lists" to be stored in EIRs. The *white list* includes the ranges of IMEIs allocated to type approved mobile equipment. Consequently, an IMEI not in the white ·list does not correspond to a valid type-approved mobile station. The *black list* includes the list of the IMEIs for stations which need to be barred, either because they have been stolen or because of severe malfunctions. The *grey list*, as its name implies, is intermediate between the white and black lists, and includes the IMEIs of faulty stations whose fault is not important enough to justify plain barring. The grey list can also be used as a temporary buffer before authorities confirm or impose black-listing. The use of a grey listed mobile station is reported to the maintenance system, together with the IMSI of the subscriber which used it. This can be used to notify the subscriber, or to trace malfunctions in the network.

To detect *persona non grata* mobile stations the network has first to know the IMEI and then to check it in the EIR. A thorough IMEI checking is costly in signalling, because it must be applied at every establishment of an RR-session, since the user may change his equipment at any moment. The traffic within the network can be minimised by storing in the visited MSC/VLR the IMEI of the equipment in use with a given SIM, in order to interrogate the EIR less often.

The MAP protocol to support the exchanges between the EIR and the MSC/VLRs is called in this book the MAP/F protocol. This protocol consists in fact of a single request-response procedure, using the MAP/F CHECK IMEI message and its result.

9.4. SYSTEM ENGINEERING AND OPERATION

The architecture of GSM as presented in all the previous chapters is an abstract functional architecture. Now, a network in operation is composed of actual machines. The operator must choose how many of each machine type to order, with which capacity, where to install them, to which other equipments to connect them, etc. Similarly, the telecommunication links between machines must be dimensioned and provided for. In addition, a number of parameters must be set, as for instance the radio channel configuration in each cell (including the frequencies used and the number of channels for each channel type), handover parameters, priority levels, and so on.

All these aspects are covered under the umbrella term "system engineering". Not only does system engineering aim at having a consistent and operational system, but also to obtain a cost-effective system providing an acceptable quality of service to subscribers.

The main issue of cost effectiveness is to adapt the network configuration to the traffic. The system must be dimensioned to support the expected traffic with the required quality of service (probability of call loss, of call establishment failure, etc.), but over-dimensioning must be avoided to avoid wasted cost. The right choice of equipment location, of equipment dimensioning and of link dimensioning are all interdependent. System engineering appears then as a gigantic optimisation problem, aiming at minimising costs under traffic and quality of service constraints.

The aim of the first of the following sections is to set down some basics for the system engineering of GSM. We will concentrate on two main issues. One of them is general to all telecommunications systems, and concerns switching equipments and links. The problem is of a topological nature. It consists in choosing sites, switching capacities and link capacities to support the traffic with a minimum cost. This issue will be dealt with in the next but one section, and not in a great level of detail. More emphasis will be put on the other facet of system engineering, **cellular planning**, which is specific to cellular radio, and concerns the radio coverage. This area deals with the choice of radio sites, with the attribution of frequencies in each cell and with many other parameters for cell selection and handover. The concept of spectral efficiency plays a key role in this domain, and we will present some of the engineering points which impact the spectral efficiency of GSM.

Yet even the system engineering view presented above is static, and a bit narrow. Another dimension must be added, which is time. For any system, and in particular during the initial ramp up, the traffic does not remain constant. An important subscriber growth at the start of the system leads to a steep increase in the traffic during the first years of operation. One aspect of cost is investment, and installing from the start a large capacity suitable for long-term traffic is not cost-effective. The problem is then to optimise the deployment of the system across time, in conjunction with the marketing policy. The aim is to minimise the overall cost of the system (including investment costs and recurrent charges), while maintaining a balance between the traffic handling capacity at a given level of service quality and the number of subscribers. When all the uncertainties of market forecast are borne to mind, in particular when several operators are in competition, it will be agreed that this is no simple task!

A way out is to refine the system engineering while in operation. This requires first the means to monitor the system, to know if the level of performance is adequate. This is the **observation** function. The engineering of the system can then be re-assessed with the gathered data, and decisions to modify the network can be made accordingly. An important topic is then how to modify a network in operation. The is the subject of **network change control**. Observation and network change control involve actions of the traffic handling machines, and are to a large extent performed automatically. These implementation aspects will be dealt with together with the maintenance implementation, in the architecture and protocols section. The integration of the network engineering tasks with other operation tasks is not usual in telecommunications systems at the date of writing: many operating companies use separate teams and tools for these tasks. However, the complexity of these tasks in cellular networks will be an incentive for this integration, and we will see in the near future management networks handling the complete operation chain, from observation gathering to network reconfiguration through observation analysis.

9.4.1. CELLULAR PLANNING

GSM is aimed at cellular networks of wide but dense coverage, and operators typically design their networks to ultimately reach a continuous coverage over all inhabited areas of a country. The goal of cellular planning is to choose the cell sites and a number of system parameters such as the frequency allocation and the capacity of the cells, in order to provide economically such a continuous coverage, supporting the required traffic density. We will first examine more closely what is the problem, and then look at the different parameters to optimise.

9.4.1.1. Goals

In a cellular environment, the traffic to support is best presented as a function of the plane giving the local traffic density at each point. As a first approximation, this function matches the population density. An important factor is the population penetration, which gives the ratio of the number of subscribers over the total population in an area. This value varies, being higher in big towns than in the countryside. The other intervening factor is the traffic per subscriber. Taking an average penetration factor of 5%, and a traffic of 0.05 Erlang per subscriber, the large scale traffic density varies between 0 (in non-populated areas) to, say, 100 Erlang per square kilometre. At smaller scales, the traffic density can reach much higher values. If we go too far, we can obtain a traffic density of 50000 Erlang per square kilometre, taking a square meter with one subscriber in it, whereas in the next square meter the density is 0 because there is nobody in it. The scale at which to measure the traffic density must be adapted to the size of the cells, and we assume here that with GSM technology, cells are not too small when, say, they are at least 0.2 square kilometre.

In reality, to complicate things even further, this traffic density varies with time. Even if we do not consider its long-term evolution, the traffic pattern experiences short-term variations, according to a daily and weekly cycle. Usually, telecommunication operators dimension their networks according to the "peak hour" traffic, when the system is closest to congestion. In a first approximation, a network supporting the peak hour traffic will support the traffic at any time. This is in fact not always true, since the geographical distribution of traffic may be substantially different from one hour to another. As an example, the traffic around sport fields may be much larger during week-ends than during the week, but this happens when the total traffic of the whole city is much smaller than during the rush hour of week days.

The first goal—or constraint—of cellular engineering is that each cell is able to support (for most of the time) the traffic from the corresponding geographical zone. Now, there are a lot of solutions to such a problem, and an obvious one consists in installing a very large number of cells. The other goal of cellular engineering makes the picture less simple, and asks that the cost of the radio infrastructure be minimised for a given traffic load.

This cost has two aspects, and it is important to differentiate them. Part of the cost depends directly on the traffic. For instance, the total number of installed speech channels can in a first approximation be derived directly from the total expected traffic, whatever the distribution over different sites. This part of the cost can then be considered as a fixed one for a given total traffic, and cannot be decreased otherwise than by reducing the cost of each equipment unit. Using the GSM terminology, it means for example that the cost of a TRX can be neglected in the first step towards optimisation of the system parameters. The number of traffic channels (and of TRXs) depends at second order on cell planning because of the Erlang law. More precisely, more TRXs must be installed than just the minimum number to support the expected **average** traffic, allowing for normal statistical peaks. The analysis of this problem, embodied in the Erlang law, shows that the ratio between the number of installed resources and the average traffic they can support decreases when the number of resources increases. In practice, this means that the total number of traffic channels installed in a network is the sum of the number of channels supporting the average traffic, plus some quantity which increases when the average capacity per site decreases (i.e., when the number of sites increases). The inclusion of this factor in the optimisation process directs the choice towards less cells of larger capacity.

The second aspect of cost is not linked directly to the amount of traffic, but to the number of sites. Each individual site is a source of cost for the operator, independently from the traffic it supports. Not only does each site include equipments which are to a large extent independent from the traffic (e.g., antennas), but real estate costs and installation costs (e.g., air conditioning, power supplies) are far from negligible compared with the costs of the electronic parts. Cellular planning will then attempt to minimise the total number of sites, given the traffic distribution constraints.

9.4.1.2. Radio Constraints

The cost optimisation problem is not an easy one. The cost considerations alone lead to having as few sites as possible, ideally a single (enormous) cell. This is however not feasible, for two reasons. The first one is the amount of available spectrum. The provision of a number of traffic channels in a cell requires some minimum amount of frequency, which depends on the system. In GSM, 25 kHz per simultaneous conversation is required with full-rate speech, for instance. The support of 3 million GSM subscribers, each of them generating 25 milli-Erlang at the peak hour, would require a spectrum allocation of 1855 MHz, whereas 25 MHz are available for GSM900! Multiple cells are unavoidable for any big radio network.

But spectrum scarceness is only one of the reasons for cellular networks. As explained in Chapter 1, the radio transmission range is limited by the maximum transmission power of mobile stations and by noise. Maximum cell size, and hence the total number of radio sites in the system, is therefore affected by two factors: the transmission range and the interference introduced by the reuse of a radio resource too small to accommodate the total traffic.

Range

The emission power of mobile stations is limited to 20 W for GSM900 vehicle-mounted stations, to 2 W for GSM900 handheld stations, and to 1 W for DCS1800 handheld stations. On the other side, the sensitivity of base stations is limited because of noise factors, and in particular because of the thermal noise introduced in the transmission chain. Because of this combination of an upper limit for the emission power and a lower limit for the reception level, the propagation of radio waves intervenes to limit the coverage.

Radio waves have different means to go from one point to another. The first one is direct "line of sight". The power loss due to propagation varies in this case with d^{-2} (d being the distance between transmitter and receiver), provided no obstacles are too close from the line of sight. Other possibilities include reflection and diffraction, as illustrated in figure 9.8. At 900 MHz and even more so at 1800 MHz, a reflection or a diffraction leads to an important propagation loss. On the other hand, since both the base station and the mobile station are fairly close to the ground, a lot of obstacles usually hinder the direct line of sight between them, and they can communicate only through reflected or diffracted paths. As a consequence, the mean loss increases rapidly with the distance in uneven areas such as the centre of towns. The power loss is modelled typically as an exponential of the distance $d^{-\alpha}$, with typical values of α in the range 3 to 4, depending on the town. The matter is a little better in suburban areas, and even more so in flat rural areas. In the latter case, however, the cells are large because of traffic, and the roundness of the earth also intervenes to encumber the direct line of sight, and even more so if antennas are close to the ground.

So far, we have only considered the average propagation loss. Another factor makes things worse. Users want to use their mobile stations everywhere, including inside buildings, or even in underground car parks or deep basements. Even at short distance from the base station site, there are places such as these where the propagation loss is high enough to prevent communication. More generally, for a given distance, the propagation loss follows some statistical distribution. By definition, only in 50% of the cases will the propagation loss be less than or equal to the median propagation loss at a given distance from the base station. This value does not hold much interest for an operator, whose aim is to guarantee a 90% or 95% coverage, measured over the places where

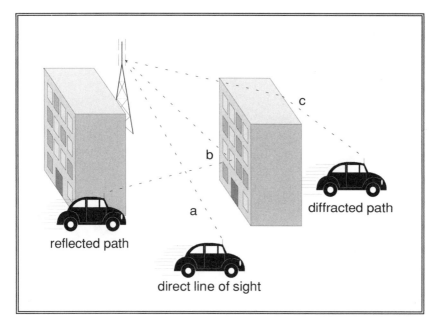

Figure 9.8 - Propagation paths

In most cases, propagation for land mobile radio systems does not follow
a direct line of sight (a), but encounters many obstacles which
reflect the radio waves (b) or diffract them (c),
leading in both cases to losses much higher than with direct line transmission.

customers are likely to want service. As a result, an operator must take into account an important margin from the mean propagation loss.

The spread of the propagation loss around its median value depends on geography, on the height of the base station antenna and on the places where customers are expecting traffic. This last point is very tricky, since it depends on the subjective point of view of the users, influenced in particular by the indications provided by the operator or the service provider. If it is clear for users that they cannot expect coverage in underground trains, these places can be taken out from statistics; if users know that they may have to move by a few meters when using their terminal indoor, the indoor part of the statistics has to be corrected accordingly. A few years ago, with only vehicle-mounted mobile equipment, things were simpler, and measurements in the streets were sufficient to have an idea of the propagation loss statistics. From this period we inherited the classical log-normal model, where the distribution of the propagation loss for a given distance is modelled as a Gaussian distribution when expressed in decibel. This is the model assumed in

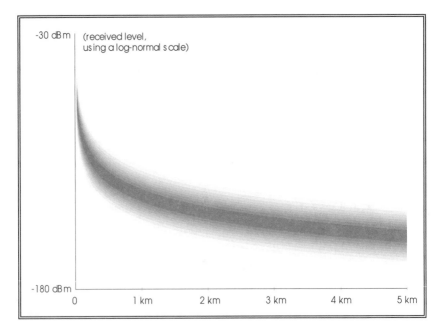

Figure 9.9 – The propagation statistical model

The grey level is proportional to the probability,
at a given distance from the emitting site, of a given propagation loss.
The attenuation with distance is $d^{-3.5}$, the log-normal standard deviation
is 8 dB for all distances, and the median reception level at 1 km is −92 dBm.

figure 9.9, and the one we will assume in the rest of the chapter, though it can be debated at length if it is adapted to handheld dominated traffic.

With the log-normal model, the spread can be measured by a single value, the standard deviation, usually expressed in decibels. As defined it may vary with the distance; however, the measurements (in the streets) show that it can be assumed constant, at least for distances over a kilometre. The standard deviation σ is greater in towns than in rural areas, because of obstacles. A typical value of the spread in urban areas is $\sigma = 8$ dB. With such a value, a 95% coverage of the cell can be reached with a propagation loss around the median reception level value at the cell radius.

When all these factors are taken into account, the cell range in GSM900 is limited to a few tens of kilometres in rural areas, depending on the terrain and on the base station antenna height. In urban areas, this limit comes down to 5 to 10 km, except for handheld stations, for which the limit is a few kilometres at best.

Interference

Interference precludes the use of the same GSM radio channel for two simultaneous communications in the same cell. Furthermore, because the signal level does not become completely null until a long way off, interference exists between any simultaneous use of the same radio resource, even if from distant places. In a cellular system, since the spectrum resources are by necessity reused in several places, the goal is to ensure that the interference caused by this reuse is negligible, or at least statistically acceptable.

The impediments of propagation as explained in the previous sections are clearly an asset for reaching this goal. Cellular systems would indeed not be so easy if propagation did not imply a loss varying significantly quicker than d^{-2} ! An obvious way to reduce interference is to increase the distance between machines using the same radio resources. As a negative consequence, the area in which a given radio resource may be used only once is larger than a cell. It extends to neighbouring cells as well. The minimum number of cells able to support a given amount of traffic T is then not equal to the ratio T/S (where S represents the traffic supported by a single cell using all the available spectrum), but to T/S divided by some value R called the average reuse factor. Correspondingly, the maximum number of resources in a cell is on average the total number of resources divided by R. As an example, an operator using 12.5 MHz (half the primary band of GSM900, 62×8 full-rate speech channels) and a frequency plan based on a 7 reuse factor will handle at most 70 simultaneous calls per cell on average.

If the traffic density is high, this capacity limit may lead the operator to build smaller cells than made necessary because of the range problem alone. For instance, a traffic density of 20 E/km^2 (a rather high value, but not excessive for large cities) would lead the same operator as above to limit the cell size to 3 km^2 in order to offer an acceptable quality of service. This corresponds to a cell radius of about 1 km.

These elements show that a cell is limited in size, either because of the range or because of interference. If range is the limiting factor, more frequencies are available than necessary to guarantee both a far-enough distance for reuse and the required quality of service in a cell of maximum range. In the other case, on the contrary, the minimum reuse factor must be used and cell size must be reduced to accept the planned traffic. For a given system, it is the traffic density and the available amount of spectrum which dictate the applicable case for each cell. Typically, a mature network will include both types of cell, the first in rural areas and the second in urban areas.

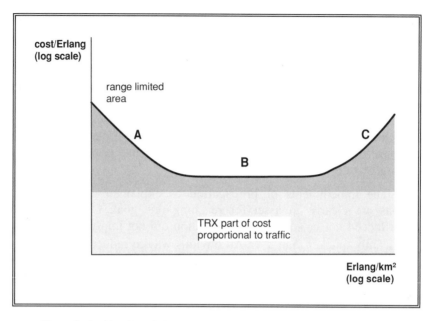

Figure 9.10 – Number of sites per Erlang as a function of the traffic density

The evolution of the cost (per Erlang) with the traffic density
identifies three types of cell configuration:
– in zone A, cells are limited by the range;
– in zone B, cells use the maximum capacity for a given reuse factor;
–in zone C, cells become too small and the cost efficiency decreases.

It is interesting to measure the investment as a function of traffic. It becomes even more interesting when noting that the operator's income is basically proportional to the traffic. Part of the investment is also proportional to the traffic. This part contributes to a fixed portion of the cost per Erlang independently from cellular planning choices. However, another part of the investment depends on the traffic density, and the resulting cost in terms of number of sites per Erlang is shown in figure 9.10.

Zone B in figure 9.10 corresponds to areas where cells use the maximum capacity. The number of cells in relation to the traffic is fixed and is dictated by the average reuse factor of the system. In zone A, the size of cells is dictated by the range only, and the traffic per cell is proportional to the traffic density. As a consequence, the number of sites per Erlang is inversely proportional to the traffic density. Zone C corresponds to cells which become too small. When the cell size has the same order of magnitude as the size of obstacles to propagation, some problems appear (which will be dealt with later). A higher reuse factor is then required compared with zone B, and the cost efficiency decreases.

The analysis so far has shown that the two key factors for cost efficiency are the possible range, and the reuse factor in zone B. Little can be done for the range at the level of cellular planning, since this range is dictated by propagation conditions, mobile station emission power and BTS sensitivity. Conversely, the reuse factor is an important parameter, and it will take us into the domain of spectral efficiency.

9.4.1.3. Spectral Efficiency

There are many definitions of spectral efficiency. When the problem is considered from the cost point of view, the key factors are on one hand the cost of BTS equipment, and on the other the number of sites to install, which depends mainly on the capacity of one cell when it is limited by interference (and not by range), as in zone B of figure 9.10. While the main thing to optimise is the cost, which combines the two factors, the spectral efficiency we will study here is the second term, that is to say the number of sites to install with a given number of MHz and a given traffic density to support. The right unit to compare the spectral efficiency of different networks is then the number of communications per cell and per MHz. The figure combines two aspects:

- the number of communications per MHz if the network was composed of a single cell: for GSM900, this figure is less than 40 with the full-rate speech coder and less than 80 with the half-rate speech coder;

- the **reuse factor**, which depends on the cellular planning of each operator, and which is constrained by interference limitations.

The reader may wonder at this stage what is the reuse factor of GSM? Until now, we have been careful not to mention any value, except as examples. Why? Simply because there is no ready-made answer. The reuse factor depends on a variety of operational choices, and its evaluation with given assumptions is still not easy. In the past, a value of 9 circulated among the GSM arena. But this value does not take into account a number of features which lead to substantial improvements of this figure. Our goal is not here to give a definite value, but to analyse the influencing factors.

Relative Interference Levels (C/I ratios)

For a cell which is part of an operational system, it is possible to measure the relative interference levels caused by the environment during a communication. These levels are usually expressed as the ratio of the received signal level from the wanted source (carrier level: C) to the interference received level (interference level: I), or C/I, and are expressed in dB. C/I ratios follow some statistics, which are best visualised by showing their cumulative distribution, as in the examples of figure 9.11.

The distribution in a given cell depends on the locations of the mobile stations in communication with it, and on the locations of the interfering sources; hence it depends on cellular planning and on frequency reuse. In order to ensure an acceptable quality of service to subscribers, some objective must be put on the minimum C/I ratio. This objective can be expressed as forbidden areas in the cumulative distribution graph. For instance, a criterion can be that at least 90% of the communications have a quality above some given threshold C/I_{90}, as represented in the figure, where we have chosen for this threshold a value of 7 dB, which corresponds to a transmission quality close to the maximum with GSM full rate speech.

Other criteria can be thought of (e.g., a C/I_{50}), but usually a "worst case" criterion is the only one taken into account. There is in fact no real incentive to provide a better quality than the minimum acceptable, because of the consequential costs. The actual goal of a telecommunication network is not to maximise quality, but in fact to **minimise** the cost, (and hence the quality!), whilst keeping the quality above some threshold.

The quality criterion provides the ability to translate a C/I distribution into a minimum reuse factor. The propagation model we presented in a previous section has the particularity that an homothetic modification of emitter positions, of ratio r relative to the reception point, changes the reception levels by a simple multiplication factor of $r^{-\alpha}$ (α being the value introduced earlier as the exponent with which the level decreases with the distance). Then we can take the assumption that the cumulative distribution of I (in dB) is simply shifted along the I axis when the reuse distance varies. With $\alpha = 3.5$, the shift is of 2.5 dB for a multiplication of the reuse distance by 2. The minimum reuse distance can then be derived from the I distribution determined for another reuse distance. For instance, if the C/I cumulative distribution is as shown in figure 9.11, case b), the reuse distance is too small and must be increased to compensate 6.5 dB, i.e., by a factor of 1.5 if propagation varies with $d^{-3.5}$ (this corresponds to an increase of the reuse factor of 2.3).

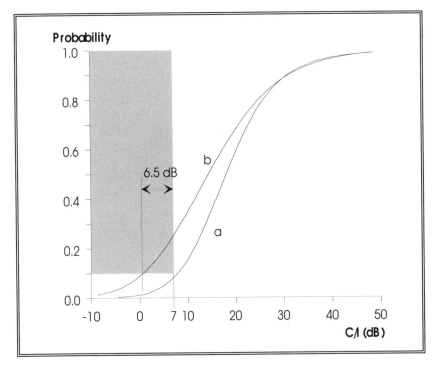

Figure 9.11 – C/I cumulative distribution

The cumulative distribution of the C/I statistics depends on cellular planning.
Quality of service criteria lead to forbidden areas of the graph; for instance,
the criterion "90% of the calls must experience a C/I better than C/I_{90}"
(here C/I_{90} is equal to 7 dB) is a "worst case" constraint
represented by the forbidden grey area.
(a) The C/I distribution is outside the area and respects the constraint,
whereas, (b) the C/I distribution is not compatible with the quality of service criterion.
In order to achieve the required quality, the reuse distance must be increased to shift the
distribution. In the example shown, 6.5 dB must be gained for the 10% worst cases.

The statistical distribution of C/I could be derived from the statistics of C and from the statistics of I, if these two distributions were independent. They are on the uplink (from mobile station to base), not on the downlink, but some insight can be obtained by studying each of them. C varies with propagation fluctuations and with the distance between mobile station and base station. I depends in particular on the distance between interfering cells, and hence on the reuse factor. These two distributions will now be studied in more detail, especially with regard to the features which influence them.

The two directions (uplink, mobile to base and downlink, base to mobile) are not equivalent as far as the interference statistics are concerned. This state of things comes in particular from the fact that a

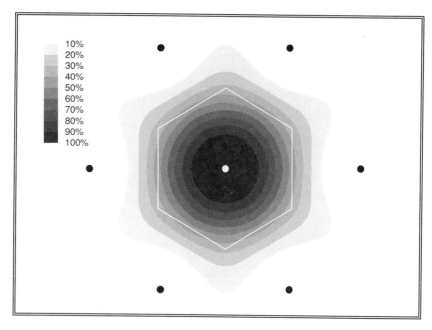

Figure 9.12 – An omnidirectional cell (using an antenna emitting evenly in all horizontal
directions) viewed as a probabilistic function

A cell "area" is best considered as a function of the location giving
the probability with which a mobile station at this location
will communicate with the relevant base station.
The black dots indicate the location of the neighbouring base stations.

mobile station receives interference from a small number of fixed sites
(the base stations), whereas a base station is being interfered by a
potentially great number of mobile stations moving around inside the
interfering cells. As a first approximation, however, we will ignore these
differences and look how the features of the system influence the C and
the I distribution.

The C distribution is determined firstly by propagation statistics
and secondly by the area in which the mobile stations are located while in
communication with a given base station. The second point is interesting,
and is not that obvious. What is exactly the geographical area of a cell?
We have been careful not to mention the boundary of a cell as a
measurable quantity, since there is no such simple border (not to mention
the fractal dimension of borders in general). There are places where a
mobile station in communication can be in one or several cells depending
on its past history, in particular depending on the direction in which it has
moved to get there. Such overlapping areas are fundamental for the
operation of any cellular system. The correct approach to describe a cell
"area" is to consider the probability of being in communication with such

a cell as a function of the location. Using this concept, a typical cell is shown in figure 9.12. At the cell centre, all communications are held within the cell, but this probability fades away when the distance increases, the calls then being supported by neighbouring cells.

What determines the cell area is not only propagation with the given base station (this is true only for isolated cells), but also propagation with the neighbouring base stations and handover criteria. Handover moves a communication from one cell to a neighbouring one, and thus influences the cell function as described above. Different handover algorithms result in different C distributions, and have a small impact on the I distribution.

Impact of Handover

In order to best visualise the impact of handover on the traffic distribution between cells, figure 9.13 shows the cumulative distribution of C in three different configurations. The first curve represents the case

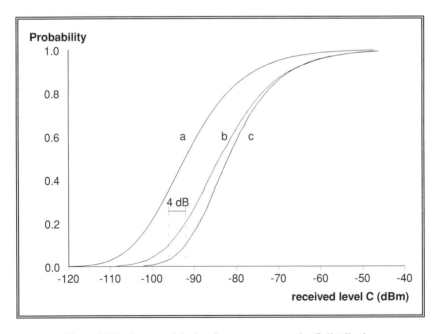

Figure 9.13 – Impact of the handover strategy on the C distribution

The cumulative distribution of the carrier level is shown here in three different cases:
a) no handover (isolated cell);
b) distance-controlled handover (cell with defined boundaries);
c) handover based on the comparison of received levels
(basic propagation model, reuse factor 4).

of a typical urban cell, if it was isolated (no neighbouring cells). For simplicity, the distribution is limited to a distance of twice the size of the cell as defined in the next case. The second curve corresponds to the same cell assuming that cell boundaries were geographically perfect; this case could be approached with a handover criterion based on precise distance measurements. The third curve represents a more realistic situation, where the cell chosen by the mobile station is the one it receives best, with an hysteresis of 6 dB. The curve shows that, in this case, the 10% C value is improved by 4 dB compared to the value computed when geographical allocation is applied.

The handover strategy corresponding to the third curve consists in comparing received levels. This is an excellent handover strategy, referred to by some as "mobile station assisted handover" and supported by GSM. The example of a handover algorithm provided in TS GSM 05.08 uses this principle. Since the choice of the handover algorithm is left open to operators and developers, we have here a first case where operational choices influence spectral efficiency.

Mobile station assisted handover has little influence on the distribution of the interference level, when no power control is used. Things are different when the two features are combined, and this will be considered after some study of the impact of power control itself.

Impact of Power Control

Handover strategy is not the only factor which influences the statistics of C. Another one is power control. This feature is provided in GSM with a dynamic range of 20 dB or more, depending on the maximum emission power of the mobile station (the mobile station power class). Its impact on the C distribution is due to the reduction of the transmission power in some cases. The reduction of the transmission power depends on the level which would be received if there was no power control. Typical cumulative distributions of C are shown in figure 9.14. Not only do they show the impact of using power control compared with not applying it, but they also show the impact of different power control strategies. Two parameters have to be chosen by the operator. One determines the reception level threshold above which power control is applied, and the second is the maximum accepted emission level, which can be chosen to be lower than the maximum emission level of the mobile stations. Two choices for these parameters are represented in curves b) and c) of figure 9.14. The parameter values should be chosen as part of the general cellular planning optimisation.

Let us turn to the interference distribution. The interference level received on a given channel is the sum of the contributions from several

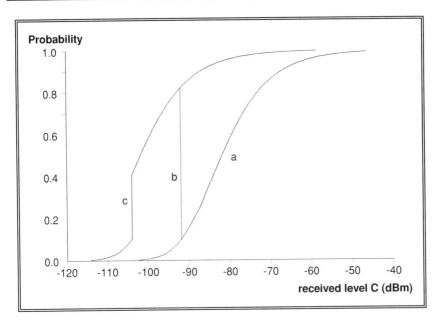

Figure 9.14 – Impact of power control on the C distribution

The cumulative distribution of the carrier level is shown here for three different cases:
a) no power control, maximum emission power;
b) power control applied to all the population except the 10% worst received,
for which emission at their maximum power is tolerated;
c) as b), but the maximum authorised power is lower.

sources. Before looking at this sum, we will study the statistics for one contribution, which is the same for the uplink and downlink. Figure 9.15, curve a), shows the cumulative distribution of the interference level from one communication using the same frequency in a cell a 3-reuse-factor away, with no use of any special feature. Having seen how power control influences the C distribution, we must now consider its effect on the I distribution, which is even more important. With power control, a large number of mobile stations in dedicated mode transmit with a power level below their maximum power capability. This leads to an important shift of the I cumulative distribution toward lower interference levels, as shown in figure 9.15, curve b).

When power control is used in conjunction with "mobile station assisted handover", the improvement on the I distribution is even greater, when compared with distance-controlled handover using the same value of the power control parameters. This comes from the fact that a choice of cells based on the reception level will favour high reception levels, favouring situations where the decrease of power due to power control is

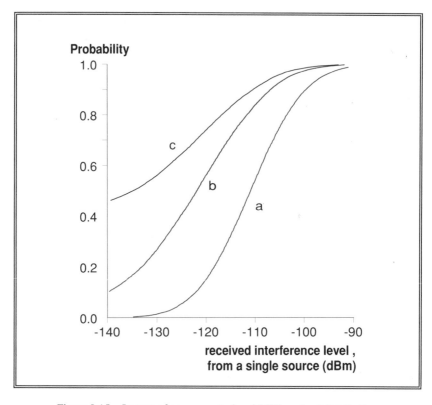

Figure 9.15 – Impact of power control and DTX on the *I* distribution

The cumulative distribution of the interference level is shifted
towards lower interference values when power control or DTX is used:
(a) without power control or discontinuous transmission;
(b) with power control, without discontinuous transmission;
(c) with power control, and transmission is effective 60% of the time.

the largest. What has been called "confinement handover" in Chapter 6 improves greatly the *C/I* statistics when used together with power control.

Impact of Discontinuous Transmission (DTX)

Another feature which has a great influence on spectral efficiency is discontinuous transmission (DTX), which has been presented in Chapter 4. DTX is an option controlled by the operator, and which may be used independently in the mobile to base and in the base to mobile directions.

Discontinuous transmission provides the ability to reduce transmission to a low activity cycle when the user is not effectively generating a traffic flow. For speech communications, this applies for typically 40% of the time in each direction. The use of DTX in such conditions changes the interference distribution from curve b) to curve c) of figure 9.15.

Impact of Frequency Hopping

Let us now consider the last influencing factor, frequency hopping. Usually several sources contribute to the interference level experienced by a given connection. When frequency hopping is not used, the number of sources is fairly small (typically between 2 and 6 interferers). With such a small number of interferers, it requires only one to be sufficiently high for the connection to suffer a bad quality. With frequency hopping, things become quite different, at least in the mobile to base direction. It is

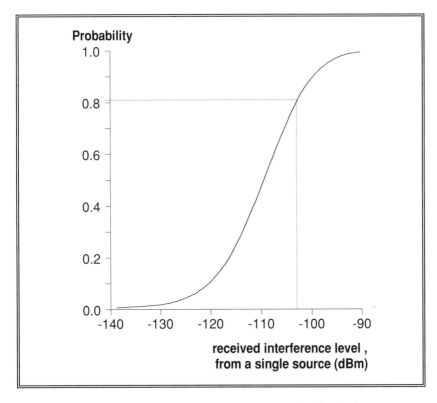

Figure 9.16 - Impact of frequency hopping on the I distribution

Because of the diversity of interferers which is generally assumed
when frequency hopping is applied, a cumulative distribution of I as shown
applies to the bursts of any given connection.

possible to choose the hopping sequences in a way such that each mobile station in a cell interferes a little bit with many communications in other cells. This is called "interferer diversity" and the corresponding choice of suitable hopping sequences is explained in Chapter 4. With this property, the probability to be interfered with by a strong interferer is not at the connection level, but at the burst level.

Let us make some simplistic assumptions to outline what happens in GSM, assuming that a burst is all good or all bad (this is obviously simplistic). The coding scheme used for speech is such that a speech frame is correct if at least 5 bursts are received correctly among the 8 bursts carrying information for this frame. Let us decide that the quality level aimed at for 90% of the cases is a frame erasure rate of 5%; the corresponding acceptable burst erasure rate is then 19%. Figure 9.16 shows the I distribution when six interfering cells are taken into account. The corresponding C and C/I distributions are those in figure 9.14, case b, and figure 9.11 respectively, and the distribution for a single interferer is the one given in figure 9.15, case b. Assuming frequency hopping with a huge number of different interferers the I curve represents the distribution of I for the bursts of any connection. On this curve, a percentage of 19% burst erasure is obtained for I equal to −103 dB, thus requiring C to be above −96 dB in 90% of the cases (if we assume that the threshold between bad and good bursts is 7 dB). This is 4 dB less than what is required without frequency hopping (−92 dB, as derived from figure 9.14). Of course, these computations cannot totally reflect the reality and the gain is somewhat smaller because the bursts are not all good or all bad, and because the number of interferers is not so large.

Sectorisation

We have seen before in this chapter that the investment cost depends on the number of sites to be installed for supporting a given traffic. A way to reduce cost is then to use the same site for several cells. Before going further, it is important to note that the word *cell* is used here (and in GSM in general) to refer to the unit of choice as seen by the mobile station. A cell corresponds to one beacon frequency on which an SCH and an FCCH are broadcast. In the North-American literature in particular, the term cell is used to refer to the area covered by all antennas of a given site, and the term sector is used to refer to what we call "cell" here.

When a site is used for a single cell, it is usually at the centre of the cell area, and the radio antennas are omnidirectional in the horizontal plane. Now, it is possible to use directional antennas, which cover only a sector in the horizontal plane. With such antennas, several cells can be covered from the same site (3 of them typically). The impact on the spectral efficiency when measured in terms of traffic per *cell* is negative. Though there is a gain on one hand because interference is limited to a sector, on the other hand the ratio between the maximum distance between the BTS and a mobile station in the cell and the square root of the cell surface is doubled, which worsens the C statistics for a given cell area. These two effects can at best compensate for each other, but in practical situations the negative side wins. A reuse factor of 9 with 3-sector sites (hence with a reuse pattern of 3 sites) corresponds roughly to a reuse factor of 7 with omnidirectional cells. Now, if the spectral efficiency is evaluated in terms of traffic per *site*, there is a gain factor equal to the number of cells per site, and the balance is positive. In the previous example 3 sites have to be installed instead of 7, resulting in a substantial economy for the operator.

Conclusion: GSM Spectral Efficiency

All the factors introduced so far can be used in computations aiming at minimising the investment cost for a given quality threshold. Our purpose here is not to push such computations to their limit, but to show how the different factors influence the result. It should now be clear that giving a value for the reuse factor of GSM requires first to state the choices about all the identified impacting features. The value of 9 determined in the past does not take power control and mobile station assisted handover into account, nor discontinuous transmission, nor interferer diversity brought about by frequency hopping. If all these features are used, the reuse factor becomes much better, maybe below 5, making GSM with full-rate speech a system of high performance, and with half-rate speech the best system yet developed in terms of spectral efficiency.

9.4.1.4. Tasks of Cellular Planning

Let us now come back to cellular planning, and look at the different steps an operator must perform to make a cell plan. The basis of the optimisation process is the assessment of the traffic density which is aimed at. This determines, together with the available spectrum and the expected spectral efficiency, the cell sizes and capacities. At this point, the operator must choose whether the cells are omnidirectional or sectored, and in the latter case the number of sectors covered from a given site (from 2 to 6, very often 3) and the directions of the antennas.

These different data determine the density of sites. The next step, and not the least one, is to find the sites as best as possible to cover the required area. In areas where planning is constrained by coverage, the choice of the sites aims at minimising the uncovered area. In areas where planning is constrained by interference, the coverage at the street level is usually not a problem. However, an over-dimension of such an outdoor coverage is useful to improve indoor coverage. A compromise has to be found between improving the coverage of indoor premises, and limiting the interference level generated outside the cell. The goal is not so much to have a geographically extended coverage than to match the covered area with the planned cell.

Next, the following parameters must be chosen:

- the frequency allocation, including frequency hopping considerations;
- the power control parameters;
- the handover parameters, chosen to fit exactly the cell to the expected traffic.

In areas limited by the range, the goal is simply that no dysfunctioning is introduced by an incorrect choice of parameters (for instance, it must be ensured that a connection handed over to another cell will not be immediately handed back). In interference-limited areas, in addition, these parameters must be chosen so that interference is minimised and that the cumulative distribution of C/I in each cell respects the quality criterion (e.g., $n\%$ of the traffic above a given threshold). As already noted, an important factor is the cell "boundaries", which are determined by the handover parameters. Different choices will change the distribution of the traffic between cells, and change the conformation of the cells.

It should be noted that these operations are inter-dependent, and that the global optimisation problem is a very complex one. As already mentioned at the start of the system engineering section, medium- and long-term evolution also intervenes, as well as financial aspects.

A last step, of marginal impact on the spectral efficiency, is the choice of the parameters controlling how a mobile station in idle mode chooses the serving cell (for instance the parameter $C1$ described in Chapter 7). The goal is usually to have the cells as defined by the idle mode selection as close as possible as the cells determined by the handover criteria. The idle mode selection parameters also influence the quality of service, in so far as they act to prevent access when quality is so bad that a call will be dropped or handed over immediately. A correct setting of the cell selection parameters will therefore delay the first handover of a communication.

Similarly, the choice of the beacon frequencies is part of cellular planning, but does not impact spectral efficiency. The beacon frequencies are chosen so as to obtain as wide a radio separation as possible between two cells using the same beacon frequency. The "reuse" factor can be as high as the number of frequencies available to the network.

The deployment of a network is typically an on-going process. Cells are introduced progressively and the capacity is gradually increased. Cellular planning is then a continuous activity throughout the life of the system. If some of the parameters are difficult to change in time (such as the location of existing sites), many others can be modified by remote control, enabling the operator to re-configure part of a network with minimum—if any—service interruption. Such changeable characteristics include power control and handover parameters. With some BTS implementations, a remote change of frequencies is possible, enabling an operator to change the entire frequency plan overnight, for instance.

The general optimisation is very complex and requires automation and centralised computations. System observation and performance analyses need to be done regularly, leading to new configuration attempts. By such computer-assisted trial and error methods, the cell plan approaches its optimum, thus leading to cost reduction or conversely traffic increase.

9.4.2. CELL CONFIGURATION

In this section and in the next, devoted to network engineering, the focus is given to the dimensioning aspects, that is to say how to choose the number of transmission devices, the switching capacity, and so on, of the different pieces of equipment. These activities require some knowledge about the behaviour of the subscribers: how many there are; where they are when communicating; their movements; when do they attempt calls; what is the duration of a call; and so on. All this information is of a statistical nature. It is part of a traffic model, and any network engineering activity implies such a model. So, before looking to the detailed points, we will present a typical traffic model, which will be used afterwards for providing examples.

Traffic Model

A traffic model is a mixture between observations in the field, either on the target network, or on functionally close networks, and of assumptions, some of them quite arbitrary. The model presented in this section is by no means the ultimate traffic model for a GSM network. The figures most often found in the literature come originally from analog cellular networks, and may be quite obsolete already now or at least in a few years time. It should be understood that the goal here is not to replace the traffic modelling work which is the first step of GSM network engineering. It is rather to give some general ideas on the process, and we ask for the understanding of the readers who might consider some of these figures inadequate.

A general point first. All traffic data varies in time, and the statistical characterisation of such phenomena is not easy. A general evolution exists on a time scales of years. This will not be our concern here. Other variations exist at the time scale of the month (seasonal variations) and at the time scale of the hour (daily variations). A network is designed to cope with a peak traffic situation. The traffic model is given for a peak hour, chosen by consideration of quality of service. Variations also exist at a small time scale, over minutes or less. They are considered as statistical noise, and averaged.

In a cellular system there are two main categories of traffic data. The first batch relates to the communications, and the second to the users movements.

Communications

Beside the number of subscribers, two values are used to estimate the communication traffic. One is the traffic per subscriber, defined as the average probability that a given user is engaged in a conversation at a given moment during the peak hour. As for all occupancy averages, the unit is the Erlang. The value used in GSM studies around 1988 was 0.025 Erlang, a value observed with cellular subscribers at this time. This value will necessarily increase with the success of cellular telephony. We will take here, arbitrarily, 0.05 Erlang. The other basic data is the mean duration of an effective communication. The usual value is 120 seconds, i.e., 2 minutes. The traffic evaluation in terms of Erlang is well adapted to transmission, when the number of channels or of circuits must be chosen.

In switching arenas, where the stress is more on the signalling load than on transmission line occupancy, another related value is used to assess traffic: the number of call attempts per hour, measured in BHCA (Busy Hour Call Attempts). This includes all attempts, whether coming from or targeted to mobile stations, and whether successful or not. This figure in particularly useful to estimate the load in terms of signalling processors. To derive the number of BHCA per subscriber, more parameters are needed in addition to the average call duration and the traffic per subscriber. It is usually assumed that in public cellular telephony there are more effective communications sent from the mobile station than towards it. This is because people rarely give their cellular phone number as the first number to try to reach them. Typical values are

60% or 70% of effective calls being mobile originating calls. We will take here 60%.

Now comes the relationship between call attempts and effective communications. For mobile originating calls, the PSTN figures will do. For 100% call attempts, 60% succeed, 35% fail by lack of answer, and 5% fail before alerting starts. For calls toward mobile stations, a first point is the probability for the *mobile station* to answer when called during the peak hour, that is to say to be switched-on and within coverage. We will assume a 60% figure for this probability. We will also deem (this is an approximation) that the *user* always answer if his mobile station indicates a call arrival. With such hypotheses, there are per subscriber 1.5 mobile originating BHCA, 0.6 mobile terminating and successful BHCA, and 0.4 mobile terminating BHCA towards non reachable mobile stations.

User Mobility

Handover rates and location updating rates depend on the movements of the users. The estimation of this signalling load must be based on statistics concerning these movements. Little is provided in the general literature to quantify the movements of mobile telecommunications users. Moreover, the rapid increase of the relative proportion of handheld mobile terminals compared with vehicle-mounted equipment makes past experience rapidly obsolete.

To give an idea on the order of magnitude, we can make very simple assumptions. First we will take the assumption that the speed of 70% of the users is zero, and that the speed of the other 30% is 30 km/h. Then, we will assume an average cell diameter of 3 km, and translate this into a mean lifetime in a cell for the moving users of 4.5 minutes, that is to say an average of around one handover every two communications.

A related point is the location updating traffic. Different reasons may lead to location updatings, as explained in Chapter 7: movements of users between cells, switch-on and off, periodic updating. If the two last terms can be considered roughly proportional to the traffic in the cell (within a given traffic model), the first one varies from 0 to a high value depending on the proportion of the boundary of the cell which corresponds to a boundary between location areas. We will make the simple assumption that the location areas are big enough, and their boundaries carefully chosen so that we will neglect the location updating traffic due to movements. We will make the assumption that the period for periodic location updating is half an hour, and we will neglect the traffic due to switching on and off.

The aim of this presentation was to give some insight on the issues. For the next sections, we will not go too much into the fine parts of the model. The main parts of the traffic can be described with a few figures, related to a number of events during the peak hour as follows:

Number of mobile originating calls	1.5	
- successful		1.0
- unsuccessful		0.5
Number of mobile terminating calls	1.0	
- successful		0.6
- mobile station not reachable		0.4
Number of location updating	2.0	

9.4.2.1. Dimensioning of Radio Channels

With cellular planning, the main topic is radio coverage. With cell configuration, the main issue is to choose the channel configuration in each cell, given the traffic it must support. As explained in detail in Chapter 4, a cell includes a set of common channels and dedicated channels. We will look at both aspects in turn.

Dedicated Channels

Several dedicated channel types exist in GSM. In phase 1, the radio capacity must be distributed among TACH/Fs (full-rate traffic channels) and TACH/8s (channels eight times smaller, used for signalling). Some machines enable a dynamic distribution between these types of channels (one TACH/F = 8 TACH/8), but in other cases the allocation is more static, and can only be changed by some intervention on the configuration parameters.

The number of TACH/F required in a cell can be directly evaluated from the planned traffic in the cell, through the application of the "Erlang B formula". An important parameter is the planned probability that all the channels are used at a given moment during the peak hour. This blocking probability measures the probability that a call attempt fails by lack of a free radio channel. The Erlang B formula relates the average channel occupancy (in Erlang), the number of channels (an integer) and the blocking probability, under the assumption that the instants of call establishments and the duration of the calls follow Poisson processes. Table 9.2 gives the capacity in Erlang for some number of channels (corresponding to representative cell configurations) and a blocking factor of 2%, a typically acceptable value.

TRX in cell	1	2	3	4	5	6	7
Channels	7	14	22	30	37	45	53
Capacity	**2.9**	**8.2**	**15**	**22**	**28**	**35.5**	**43**
ratio	0.41	0.57	0.68	0.73	0.76	0.79	0.81

Table 9.2 - Capacity (in Erlang) of a set of full rate traffic channels
(Erlang B formula, 2% blocking)

A blocking probability of, here, 2% results in an average load
which is not proportional to the number of traffic channels.

The needs in terms of TACH/8 capacity come from two main signalling requirements (not including short message transmission needs): communication establishment attempts and location updating attempts. The dimensioning must be done for full-load conditions, that is to say when the cell is congested and all TACH/F are used. The maximum useful rate of communication establishment is then obtained when the congestion state is maintained, and is roughly proportional to the number of TACH/Fs, given the mean duration of calls and the ratio between successful and unsuccessful call attempts. It is worth noting that the channel allocated to a mobile station upon access may not always be a TACH/8, as explained in Chapter 6. In particular, the Very Early Assignment strategy does not make use of TACH/8s for call establishment (but alternatively increases the number of TACH/Fs required for a given amount of traffic).

The second signalling flow impacting on TACH/8s is the location updating flow. As an example, with the traffic model presented in the previous section, a cell of 30 traffic channels will support at congestion 22 Erlang and thus 440 subscribers, among which around 260 on average have a switched-on mobile station. The access rate is 0.27 access per second for successful communications including mobile originating and mobile terminating, plus 0.11 per second for unsuccessful mobile originating attempts, and 0.14 for location updatings. The total is close to 0.5 accesses per second. Assuming that any of these events makes use of a TACH/8 for 4 seconds, we obtain a TACH/8 average occupation of 2, which requires at least 6 TACH/8s to obtain a blocking rate less than 2%.

Consequently, many parameters influence the dimensioning of the TACH/8s, such as the choice to use periodic location updating or not (and the corresponding period), the choice to use "IMSI attach/detach" or not and the planning of location areas.

Common Channels

The dimensioning of the common channels has much in common with the choice of the number of TACH/8s. The BCCH, FCCH and SCH raise no engineering problem. On the other hand, the configuration of the PAGCH and of the RACH can be chosen among several cases. The common channels have to support two kinds of traffic, not proportional one to the other. The first is the access traffic, which is measured as a number of accesses per unit of time, which we have already assessed to evaluate the needs for TACH/8s. The second source of traffic is paging, measured as a number of pagings per unit of time. The number of paging messages in a given cell depends not on the traffic in the cell, but on the traffic in the location area.

The lowest paging load corresponds to the case of a one-cell location area, and is comparable to the access load. On the other hand, in such a configuration the whole cell boundary is a location area boundary, and location updatings due to movements lead to an important location updating load. At the other extreme allowed by the *Specifications*, a location area can include all cells connected to one MSC. This leads to an important increase of the paging load. However, if the TMSIs (temporary identities shorter than the IMSI) are used and if message scheduling is optimised on the downlink common channels, as explained in Chapter 6, the corresponding increase can be supported easily.

There are several possibilities for the configuration of the common channels, as described in Chapter 6. Depending on the capacity requirement for access, five configuration choices exist for the RACH and PAGCH. For each of these configurations, table 9.3 shows the maximum access rate, calculated as the lowest of the two following capacities:

- the access capacity on the RACH, evaluated at about one fourth of the RACH burst rate (see the section dealing with the RACH throughput, in Chapter 4);

- the assignment capacity, calculated here based on the assumptions that each message contains two assignments, and that the channels are entirely used for assignments. This is an upper bound, as it is not very realistic not to count any paging!

The table also shows the maximum paging rate for each case, based on the assumption that the channels are entirely used for paging, and that paging is performed solely on TMSI (this enables the packing of up to four requests in a single message). The maximum number of pagings and the maximum number of accesses are mutually exclusive,

CCCH capacity	maximum RACH access rate	maximum paging rate	maximum assignment rate	maximum channel access rate
(equiv. in TACH/F)	(accesses per second)	(mobile stations paged per second)	(assignments per second)	(accesses per second)
1/2 (*)	29	50	25	25
1	54	152	76.5	54
2	108	306	153	108
3	162	458	229	162
4	217	611	306	217

Table 9.3 – Access channel capacities

Depending on the capacity to be offered by a given cell,
the access channels can be configured in several different sizes.
(*) the other half can be used only for 4 TACH/8.

and the choice of common channel configuration has to be done taking the two factors into account.

As an example, with the same assumptions as for the computations for the TACH/8 requirements, a cell of 30 traffic channels experiences about 0.5 accesses per second. For pagings we have to assume some size of location area, say 20 such cells, and the paging repetition rate, for which we will take 3. This leads to around 3.5 pagings per second. Table 9.3 shows that with such values, even the smaller common channel configuration supports these requirements with ample margins.

9.4.2.2. Other Cell Configuration Parameters

Beside the dimensioning of channels, system engineering must also deal with other cell characteristics. A set of parameters, which has been described in Chapter 6, is used to control the access of mobile stations in a cell. They represent in fact one facet of a more general issue, which is the management of congestion. The respective priorities given to handovers, location updating, call establishments, etc. must also be managed by system engineering through BSC parameters.

Another cell parameter to deal with is the BSIC. As described in Chapter 7, its allocation must take into account the beacon frequency

Area	Objects
Cell planning	Frequencies Beacon frequencies Hopping sequences Power control parameters Handover parameters Cell selection parameters BSIC
Dimensioning	Common channels TACH/8s Location areas Periodic location updating IMSI attach/detach
Load control	Overload control parameters

Table 9.4 – Cell characteristics

System engineers must manage many different cell characteristics
in a co-ordinated manner, in order to ensure proper quality of service
in terms of coverage, dimensioning and congestion control.

allocation, and must be co-ordinated near the boundaries with PLMNs using common frequencies.

Table 9.4 summarises the objects which system engineering must manage to optimise the network configuration for a given quality of service to an expected population of users.

9.4.2.3. Inter-Cell Synchronisation

A last aspect of system engineering is "time engineering": the synchronisation between cells. This has some impact on the quality of service, in the area of handover performances. The notion of synchronisation is taken here at a general sense. It includes also the *de*-synchronisation of the cells, as we will see that full synchronisation can be very detrimental to some aspects of system performance. Best performance is obtained when time bases in neighbour cells are synchronised so that burst emissions are synchronous, but de-synchronised so that in particular multiframes are not synchronous.

The time division multiplexing scheme on the radio interface requires a complex time management, resulting in a series of clocks to be managed at the level of bursts, TDMA frames, multiframes, superframes and hyperframes. GSM design enables the clocks in different cells to be run independently. Even their frequencies in the long run may differ somewhat, each frequency being within a 5.10^{-8} tolerance. However, some performances are improved when burst emissions in neighbour cells are finely synchronised (and this requires equal long term clock frequencies). The main one is handover execution time. We will now look at what is exactly to be synchronised and to what accuracy.

The interruption time during a handover execution can be improved (from roughly 200 ms down to 100 ms) if the "synchronous" handover procedure is used, and this requires synchronised clocks. In this case, the only problem is timing advance. What needs to be synchronised is therefore the long term frequency and the phasing of the burst clock. Unphasing of higher rank clocking should not cause any problem, as long as the return propagation time in a cell does not exceed 577 µs (which would correspond to a cell radius of 90 km). However, the *Specifications* do not explicitly state whether mobile stations should behave correctly if asked to perform a synchronous handover and *only* the burst clocks are in phase. This imposes an all-clock phasing between these cells, the drawbacks of which will be explained later. From the accuracy point of view, synchronous handovers require simply that the timing advance evaluation is correct with an accuracy of a few microseconds. Though not explicitly stated in the *Specifications*, this value comes from the duration of the guard time. The phasing accuracy between the two cells must therefore be of the same order of magnitude.

Synchronisation between cells, if limited to bursts, can also be useful for pre-synchronisation (see Chapter 6). It improves the search time for neighbour cells, though not in an obvious way. In fact, all-clock phasing is the worst imaginable case for pre-synchronisation performance. Let us recall that the mobile station uses the long intervals in its traffic channel cycle to decode the bursts of neighbour beacon frequencies. If all SCH, FCCH and BCCH bursts are exactly synchronised in all cells, the time between reception of a burst of a given type on two neighbour cells will be *maximised*: about 1.3 seconds for an SCH or FCCH burst, and up to 3 seconds for a given BCCH burst. On the contrary, if cells are carefully de-phased, simultaneity can be avoided, and the average time described above will be statistically reduced to half of this value. The best scheme for pre-synchronisation is when cell clocks are organised to minimise the probability of simultaneity between FCCH, SCH or BCCH bursts in two adjacent cells. This kind of "offset" synchronisation is of course more complex to implement than an all-clock phasing synchronisation.

Another aspect of the relationship between neighbour cell monitoring and cell synchronisation is the initial search for FCCH bursts. This is a costly part of the monitoring, because the mobile station must scan all time slots to find such a burst, which is transmitted 5 times every $51 \times 8 = 408$ burst periods, i.e., once every 82 bursts on average. If the mobile station was to have information on the relative synchronisation of the cells, the search would be sped up considerably. However, the phase 1 *Specifications* do not include any means to indicate to mobile stations in idle mode whether surrounding cells are synchronised or not. Synchronisation of base stations has therefore no influence on the process. If this application were introduced, it would require synchronisation up to the 51×8 burst cycle, to a low accuracy (100 µs would not be a problem, since it is already the order of magnitude of the spread due to propagation times).

A last application of cell synchronisation is distance measurement. With pre-synchronisation, a mobile station is already able to know the difference in propagation times between its serving cell and the each of the neighbour cells it is pre-synchronised with. This information is not exploited in the phase 1 *Specifications*, but will be in phase 2. It can be used to several ends. Let us cite handover preparation based on distance measurements, such as in the German C-network, or geographical localisation of mobile stations. Phasing constraints for such applications are the same as for synchronous handover. But specific localisation applications may be more demanding, since the accuracy of the distance estimation derives immediately from the synchronisation accuracy.

9.4.3. NETWORK ENGINEERING

Network engineering in general consists in choosing the place and capacity of the various network nodes, the capacity of the intervening transmission links, and a variety of parameters. For GSM, the nodes under consideration are the BSCs and the MSCs. The HLR/AuC capacity will also be looked at. The links under consideration are those on the A and Abis interface, and those between the MSCs and the external networks.

9.4.3.1. The Abis Interface

The capacity required on the Abis interface in terms of traffic circuits is equal to the number of TACH/F of the cell. The capacity requirement for signalling is dominated by the measurement messages. Without pre-processing in the BTS, there are around 2 messages per

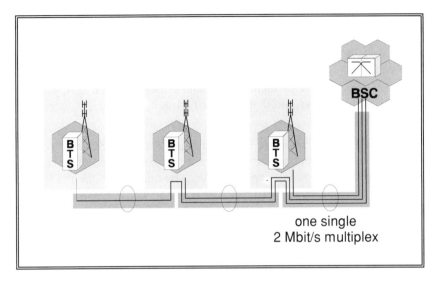

Figure 9.17 – A drop-and-insert configuration

Such configurations may be useful to reduce the number of 2 Mbit/s multiplexes
necessary on the Abis interface, especially for small capacity cells,
compared with a star configuration where each BTS requires at least one multiplex.

TACH and per second. For instance, a cell of 30 TACH/F and 12 TACH/8 will generate at congestion (i.e., when fully loaded) 88 measurement messages per second, that is to say a load close to 16 kbit/s. Thus a medium cell will have little spare capacity with one 64 kbit/s circuit for signalling transfer with the BSC.

Abis links can represent a substantial part of the running costs of a PLMN. One of the problems is that each BTS site requires a relatively small number of circuits (in comparison with the usual figures inside public telecommunications networks such as the PSTN). The transmission unit being the 2 Mbit/s multiplex, economies can be obtained if the *drop-and-insert* connection method can be used at the BTS. This technique provides the ability to share a 2 Mbit/s multiplex between several BTS sites, and to decrease the number of leased or installed transmission links. Figure 9.17 shows an example of a drop-and-insert configuration for the Abis interface.

9.4.3.2. The BSC

Two aspects of the BSC have to be looked at: the switching capacity, and the computation load. The switching capacity is measured as the total number of circuits with the BTSs and the MSC. On the A interface side, things are simple. The required number of circuits can be computed with the Erlang formula from the total planned traffic in the cells managed by the BSC. The blocking factor is chosen to be much smaller than the one which constrains the capacity of a cell, for instance it can be set at 0.2 %. With such a value and a total planned BSC capacity of 500 Erlang, the required number of circuits on the A interface is roughly 550.

From the Abis interface side, constraints on the BSC vary whether the total planned capacity is spread over a few large capacity cells, or on the contrary on many small capacity ones. The number of circuits is equal indeed to the number of traffic channels in the cells, not to the traffic. Moreover, redundancy considerations may ask for additional circuits on a cell per cell basis. We will take two extreme examples. If the traffic is handled by a single cell, the number of circuits is 513 (applying the Erlang law with 2% blocking), and the required switching capacity, adding circuits from both directions, is 1060 circuits (that is to say with multiplexed 16 kbit/s circuits and a little redundancy, around 12 multiplexes of 2 Mbit/s each, 6 on each interface). At the other extreme, if we imagine that each cell generates only 1 Erlang of traffic, there are 500 cells connected to the BSC, and each cell has 7 TACH/Fs, out of which 4 are needed for this traffic. Assuming in addition a redundancy at the 2 Mbit/s level, the number of circuits is doubled, and the total number of circuits entering the BSC from the BTSs is 4000; the total number of circuits entering the BSC is then 4550 (including the A and Abis interface circuits). This shows that the switching requirement for a given total planned traffic can vary in the ratio of 4, or reciprocally that the actual traffic handled by a BSC of some make can vary in the same proportions according to the average cell capacity.

The location and capacity range of the BSCs is a debated point. Some operators want small BSCs on the BTS sites. At the other end, some operators want big BSCs on the MSC sites, even possibly a single BSC per MSC. Others want independent BSCs with a capacity intermediate between a BTS's and an MSC's, and which can potentially be sited in any location, not necessarily with a BTS or an MSC. Various considerations will dictate the choice. A BSC has three main functions: it acts as a circuit concentrator, and as such its position impacts the running costs of the transmission lines between BTSs and MSCs. A BSC is also an operation and maintenance agent; we will see that the BTSs are not linked directly to the OSS, but through their BSC. Finally, a BSC is where handovers are controlled. Bigger BSCs lead to a smaller number of handovers which must be handled by the MSC, and the bigger the BSC the wider the knowledge concerning the traffic used to decide on handovers.

The A Interface

The capacity of the link between a BSC and an MSC has already been looked at, in so far as the circuits carrying user data are concerned. The signalling load on an A interface depends on the traffic handled by the BSC. The dominant factor comes from the messages exchanged during call establishment. The analysis of the protocols shows that a typical call establishment requires around six messages in each direction. A call release corresponds to, say, five messages counting both directions. The traffic handled by a BSC of 500 Erlang corresponds to 550 simultaneous communications at congestion, and with our traffic model to some 9 calls established and released per second, that is around 75 messages per second in each direction. Assuming an average message size of 25 octets, we obtain a load of 15 kbit/s. One 64 kbit/s link is then sufficient, including the location updating and paging traffic. However, the current practice in such a case is to install two links, for redundancy against failures.

The MSC

The trend is to have MSCs of as high a capacity as possible with the present switch technology. At the date of writing, the order of magnitude of an MSC capacity is one to several thousands of Erlang. For a network with a 5% penetration of the population and 0.05 Erlang per subscriber, a 2000 Erlang MSC is suitable for an area with 800 000 inhabitants. This is commensurate with the present density of PSTN switch locations. MSCs can then be sited in rather important towns, and will cover a part of the biggest towns, or a medium town and the surrounding area.

HLR and AuCs

An HLR machine and the corresponding AuC must support a load which is dominated first by the GMSC-HLR and MSC-HLR exchanges which occur for each mobile terminating call attempt, and possibly by the authentication flow, when this procedure is done as often as once per call.

Let us take a numerical example to show the importance of authentication. Aiming at an HLR capacity of the order of half a million subscribers seems reasonable. This means that a typical national GSM network will have no more than a few HLR machines. This corresponds with our traffic model to 400 call attempts involving the corresponding mobile stations per second. An authentication triplet has 224 bits, say 250 to include it in a signalling message. The AuC must then generate 100 000 bits per second, and the HLR has to transmit this volume of information to the MSC/VLRs. This figure climbs up if authentication is required at each location updating procedure.

With the same HLR capacity, there are 200 mobile terminating call attempts per second, each involving two messages to the HLR, and two messages from the HLR. The order of magnitude of this traffic is comparable to the one caused by authentication. Of course, the HLR must also deal with the signalling caused by location updatings when the subscriber changes from MSC/VLR area, but this load can be considered to be much smaller than the two previous ones.

The NSS Configuration

Up to now, we have focused on the Base Station Sub-system and on the dimensioning of MSCs and HLRs. There is another part of network engineering as far as the Network and Switching Sub-System is concerned, which is closer to fixed network engineering than any of the other items. It deals with the network architecture for traffic and for signalling. The operator may or may not, depending on the terms of its licence, have the right to mesh its MSCs and GMSCs and have its own transit exchanges. Similarly, the operator may have the right to set up its own signalling links between NSS machines, and have its own Signalling Transfer Points (STPs). When the operator has the freedom to set up its own traffic network and/or signalling network, it must choose the number and location of the transit exchanges and/or signalling transfer points, and it must optimise the configuration of the links between all the machines of the network.

In all cases, it must decide on the number and location of the GMSCs (e.g., in the same machine as an MSC or not), of the interworking functions with the fixed networks, of the gateway-MSCs for short messages, etc. This optimisation problem is not specific to GSM and is one of the tasks of every operator of a large network.

9.4.4. OBSERVATIONS

As seen in the engineering section, the dimensioning of a system and the setting of its multitudinous parameters are based on a number of assumptions concerning propagation or the behaviour of the subscribers. Moreover, even the effect of some choices on the internal functioning of the system is not easy to predict. With a system in operation, this data, as well as the results in term of quality of service for instance, can be observed directly.

The observation activity consists in gathering statistical data in the traffic handling machines, and in analysing them for different ends. One goal is to determine the bottlenecks in the system, so as to know where upgrading is needed. This allows an increase the network capacity in the most efficient way. More generally, one goal of the analysis is to streamline the network so as to get the maximum capacity, and revenue, out of it. We have seen another purpose when dealing with maintenance: the a priori model for performance can be compared with effective results. A discrepancy can be the symptom of a dysfunctioning otherwise undetected.

These different goals direct what should be observed. Observation can be quite load consuming, and it is important to select correctly what has to be observed, and when. Unfortunately, most of the performance observations are interesting at the peak hour, i.e., at the moment when the machines are responding to their greatest demand for traffic handling tasks and have little extra time to devote to observations. A way out is to make some observations on some machines only, changing them for instance each day.

The amount of information which may be collected is very high. It can be classed roughly as follows.

Traffic measurements aims at the evaluation of the load on the various interfaces at the peak hour. The goal is to know the mean peak traffic on a group of circuits, on a cell, or for signalling at a given node. A finer knowledge is in fact needed, with the distribution of the load between for instance mobile originating calls, mobile terminating calls, location updatings, and so on. The traffic model expounded on in a previous section is a good example of the information sought.

Quality of service measurements aim at evaluating a number of indicators of the quality as perceived by the users. This includes congestion rates, for instance the proportion of call attempts which failed because of lack of resources. It includes also delay statistics, as for instance the call setup time, or the delay between the decision to make a handover and the effective transfer, and so on. Such observations can include error rate statistics, in particular on the radio interface.

Availability measurements aim at evaluating the failure rates of the components of the system, and the time between a failure and its repair.

It is clear that observation is a function of all the machines of the system, from the BTS (which is the only point where detailed measurements can be made on the radio interface) up to OSS machines, which must observe their own availability. The raw observation data is usually kept rather simple, such as the value of an event counter. Such raw data is stored in a local file, and transferred to a place where they can be analysed. In the past (not so long ago), this was done simply by a local printout, analysed afterwards directly by human beings. Progress was made when the transfer was done by magnetic tapes, and a first level of analysis was performed by a computer. With an integrated telecommunication management network, electronic file transfer is used.

In practice, a measurement is obtained via the OSS. Such a request includes the type of data to gather, the machine which will do it, its schedule (it may be a regular observation, e.g., every Monday from 6 p.m. to 8 p.m., or a point observation, at the request of operator personnel or as a follow-up of some event) and where to forward the result. This activity requires signalling functions between the OSS and all the machines in the network.

After the transfer of the results, the raw data must go through some statistical analysis, possibly compiling data from many sources, so as to provide meaningful and accurate indicators. This function is located in the OSS. Finally, the results have to be used. An example is the access to the statistics by operator personnel, through workstations.

9.4.5. NETWORK CHANGE CONTROL

A network is not designed on paper and implemented once and for all. It evolves, to follow the progress of technology, to remedy errors or inefficiencies detected by maintenance or observations, or to follow the planned evolution of the handled traffic. These modifications could be done easily by stopping the network, installing the new configuration, testing it and restarting the normal functioning. This approach, if comfortable for the operator, is less so for the users, and technical means are introduced in the system so as to allow these changes to be implemented but with minimum disturbance for the traffic handling. Let us see what can be changed, and the difficulties for performing the modification without traffic interruption. Three general categories of modifications can be identified.

Hardware Modifications

Some changes affect the hardware. This includes the installation of a new piece of equipment, to increase the traffic capacity (e.g., a new radio site), the replacement of some machine by a more advanced one, or also to change some functionality such as for instance the introduction of half rate transcoders.

Such changes cannot be done remotely. To avoid a disturbance of the traffic, the machine to be replaced, if any, is put out of operation when the new one is put in. A central control can be helpful in the testing phase which precedes the effective operation of the newly installed hardware. It can also be of interest to speed up the transition. The introduction of a new node in a network, and even more of a new cell in a cellular network, requires the changing of a number of parameters in neighbouring nodes. This can be commanded at the right moment through the management network, to minimise the period during which the network parameters are inconsistent.

An example of hardware modifications is the installation of new switching nodes (MSCs or BSCs). When this is done for coverage extension into an area not yet covered, there is no problem of service interruption. The impact of the machines already in operation is the parameter modifications needed for them to take into account the network extension. The other case is when for traffic load reasons a new switching node is introduced to take over a part of the coverage of an existing node of the same nature. The modification has many more impacts, since transmission routes are modified.

Another example of system extension is the addition of a new cell. One of the difficult points is the testing of the new cell prior to its opening to commercial operation. To forbid such a cell during the test period, it must be barred (this state is indicated in messages sent on the BCCH). Special mobile stations, disregarding this indication, are useful to test the cell. However, the complete new set of the parameters which control the choice of cell (in idle mode as well as for handover preparation) cannot be introduced and tested, since they must be consistent with those in the neighbour cells already in operation.

Software Modifications

Another category of change is the replacement of software modules. Software is much easier to replace than hardware, and some software modules are updated regularly, for error correction or because of the progressive introduction of new functionalities. With digital technology, more and more functions are easier to implement in software than in hardware, including possibly transmission functions such as coding/decoding or modulation/demodulation. Easy means to change software modules can then provide a very powerful evolution mechanism, for a better optimisation of the system, and hence for higher revenues for the operator. The OSS will need to monitor the subsequent performance of the network to ensure that the new software does not subtly degrade the network performance.

The replacement of a software module with minimal traffic interruption is possible with redundant machines. The new module is loaded and tested in the redundant part. If the test fails, the old software can be re-loaded, or, best, kept on the machine to be restored to operation if need be. If tests indicate all is right, a change of configuration puts the modified module in operation, and so on.

The change of a software module does not intrinsically require an on-site action, as for hardware. Such changes can be done remotely (software downloading), through the management network. A file transfer protocol is needed, as well as procedures to control the activation and deactivation of the modules.

Parameter Modifications

We have seen in the engineering section that a network can be tuned by a variety of parameters, affecting for instance the cellular planning. Some parameters in a given machine are simply the description of the surrounding parts of the system, and are modified when those parts are changed. Other parameters can be chosen between several values for a given system configuration. Some choices are arbitrary (e.g., PCM link numbering), and the only issue is to ensure the consistency between several machines; such parameters are rarely modified once set. In other cases, the choice of value impacts in some way the behaviour of the network, and is done according to some optimisation criterion. It is important that they can be modified easily during operation, with minimum impact on the service. This allows the operator to dynamically refine the optimisation of the system, when observations enable a better choice of parameters. In wireline systems, and in the switching field, there are few such parameters (an example is the routing tables). On the other hand, cellular planning is a field full of them. Let us see some of them, and how they can be modified.

The frequency plan has an important impact on the system capacity, and is difficult to optimise before full scale operation. Moreover, any addition or removal of a cell, or any change of the capacity of one cell may change the optimal frequency plan for many cells. The possibility to change the frequency plan during operation is then an important asset for an operator, and even more if no on-site action is needed.

There are several difficulties. The major one comes from the radio frequency part of the BTS, and more exactly from the coupling device, that is to say the piece of hardware which combines the output of several amplifiers to feed a single antenna. 900 or 1800 MHz signals cannot be simply added by connecting wires. The power emitted by one amplifier would be dissipated in the other ones: the antenna would radiate nothing, and (worse!) the amplifiers would simply burn. Special devices are needed to force the energy to go into the antenna (and overall not into the other amplifiers). Two major techniques exist. A *hybrid* coupling device has the advantage that it does not require special tuning depending on the emitted frequency. Changing the operation frequency remotely does not raise any problem. Unfortunately, such devices introduce an important power loss, proportional to the number of combined outputs. Hybrid coupling is typically unacceptable above 4 combined outputs. With TDMA, even a one-output cell is meaningful (with half-rate coders, this corresponds to around 15 voice channels), and hybrid coupling appears as a good solution for low to average density areas, but not to the cells of highest capacity. The other coupling technique, called *cavity* coupling, makes use of tuned filters put between the amplifier output and the common point, to prevent the energy from the other amplifiers (which is at different frequencies) to go to the amplifier. The insertion loss is smaller than with the hybrid technique, and moreover increases slowly with the number of combined outputs. The main drawback of tuned filters (sometimes known as cavities) is that the frequency tuning involves moving mechanical parts, and is not an easy task. Remotely tuneable cavities, with small motors, now exist, but still seem to have reliability problems (as with most mechanical devices). Otherwise the frequency tuning requires the intervention of a screwdriver!

Providing the combining method does not prevent a remotely and rapid change of the transmission frequencies, a second difficulty comes from frequency hopping. With frequency hopping, the change of one frequency in the set allocated to a cell impacts many (if not all) traffic channels of the cell. This makes such a change difficult if the goal is to have no impact on the existing traffic. We have seen in Chapter 6 that special procedures have been introduced for this purpose. Another way is simply to stop the cell operation in order to change the frequency configuration, for instance at three o'clock in the morning.

A second example of cellular planning parameters for which traffic handling procedures enable a change with limited traffic interruption are those concerning the control channel configuration. The number of PAGCHs in a cell (1, 2, 3 or 4), or the organisation of the paging sub-channels, can be modified during operation to match the access and paging load. We have seen in Chapter 6 how the paging mode can be handled to ensure that the mobiles stations in idle mode in the cell do not miss paging messages.

The last example we choose to give here concerns the parameters controlling the handover preparation. We have seen how some of these parameters control the true borders of the cells, and hence the distribution of the traffic between adjacent cells. This is useful to refine the cellular planning, or even on a shorter time scale, to limit the effect of a local traffic overload. Handover preparation parameters can be modified at any moment, without more precaution than ensuring the consistency between the set of parameters used in each cell.

These examples are only a few among many others. Priority mechanisms, access classes, emission power, barring status, location areas, etc., are other parameters on which operators can play to drive their network and get the maximum out of it.

9.5. ARCHITECTURE AND PROTOCOLS

Maintenance, monitoring and system modification can be done from remote centralised machines to a large extent. The main requirement is to connect all the traffic handling machines to one or several operation and maintenance centres, from which commands can be sent, to change parameters, to download software, to start tests or observations, to gather the results of the tests or observations, or to receive alarms. This calls for machines devoted to these tasks, and for a network to link them between themselves as well as with the traffic handling machines and work stations from which the operating staff can act. These connections between the machines call for signalling protocols to structure their exchanges.

9.5.1. MANAGEMENT NETWORK ARCHITECTURE

The operation, maintenance and administration functions we have described are not really GSM specific. They exist in any telecommunications system. Historically, these tasks were first performed manually, with only local access to each machine. The next step was characterised by a proliferation of ad hoc systems, for instance one for observations, another one for failure reporting from transmission equipment, a third one for the configuration of another type of equipment, and so on, all of these systems using different proprietary protocols.

Such an approach can still be found in many networks. However, the *Specifications* include some specifications to guide the setup of an integrated management network, according to the Telecommunication Management Network concept (or TMN) developed by CCITT, which aims at providing bases for integrated management networks. These GSM specifications are mainly a template, a conceptual framework underlying the development of equipment by different manufacturers. When applied, this framework will make the integration of machines of different make much easier for the operator.

An integrated management network is linked, on one side, to the traffic handling machines, and on the other side, to the operating staff

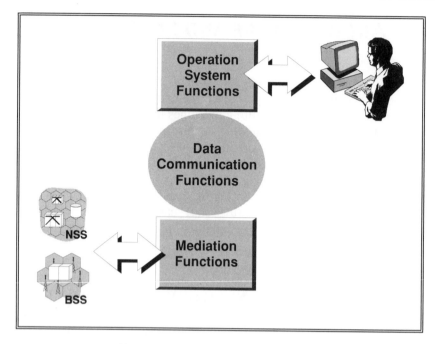

Figure 9.18 – TMN categories of functions

The TMN concept distinguishes three main functions:
the Operation System Functions (the centre of the management network),
the Mediation Functions (adaptation functions for specific equipment) and
the Data Communication Functions linking them.

through workstations. Between these interfaces, TMN distinguishes three main categories of functions, as shown in figure 9.18:

- the *Operation System Functions*, which are the management applications proper, connected to workstations from which wide-scope applications can be conducted by operating personnel;

- the *Mediation Functions*, which are the intermediaries between the operation system functions and the traffic handling machines. Mediation functions are introduced for concentration purposes but also for adaptation of the generic operation system functions to specific machines. Typically, operators introduce different mediation functions for every subsystem (the BSS or the NSS) and for every manufacturer of the machines. In the TMN architecture, the mediation functions may or may not appear. Traffic handling machines can be connected directly to a machine implementing operation system functions;

- the *Data Communication Functions*, which are the transmission means used to link the operation system functions, the mediation functions, the traffic handling equipment and the workstations. These data communication functions can be point-to-point leased lines, or switched links provided by an X.25 network or an SS7 network. The choice of the data communication structure is entirely the operator's, and does not impact on the possibility to harmonise the applicative functions.

The *Specifications* do not give much direction of the structure of the management network, except on one point: the BTSs are connected to the management network only through their BSC, which then acts as a mediation function for all the BTSs under its control. The data connection between BSC and BTS for network management needs is specified in the *Specifications*, in particular in TS GSM 08.59 (for the transport mechanism) and in TS GSM 12.21 (for the semantics).

The *Specifications* introduce other architectural terms than the TMN terminology cited so far. In particular, the OMC, or Operation and Maintenance Centre, is a term used to describe the management network as seen by a given traffic handling machine. It may stand for a single piece of equipment responsible for operation and maintenance of a given part of the network, or alternatively for a mediation device party of a centralised management network. An OMC is typically considered to be in charge of a subsystem, such as the BSS (the term OMC-R, with "R" standing for "radio", is used by some manufacturers) or the NSS (OMC-S, with "S" for "switching"). Besides the OMCs, the *Specifications* also refer to the NMC, or Network Management Centre, which is concerned with the management of a whole PLMN, and hence has a wider scope than a given OMC. In the rest of the chapter, we will stick to the TMN terminology, to be more general.

9.5.1.1. TMN Interfaces

The goal of TMN is to harmonise the applicative management protocols. For this objective, a common language must be used regardless of the type and make of the equipment to manage. But the internal organisation of each machine is different. It would be costly and non-efficient to have remote operation system functions taking care of irrelevant implementation characteristics. Therefore, the TMN model defines one or more *Network Element Functions* in each machine; these functions are the sole correspondents of the rest of the management network, and they are in charge of the dispatching of commands toward

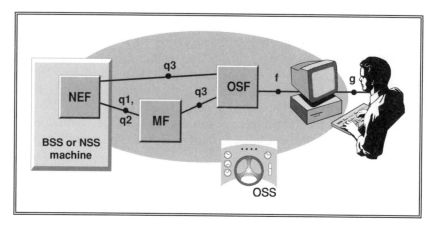

Figure 9.19 – TMN reference points

Applicative protocols are defined at specific reference points (q, f, g) defined between
Network Element Functions (which are part of the machine to be managed),
Mediation Functions, Operation System Functions,
Workstations and the operating staff.

the different specific modules of the machine, as well as gathering information from these modules.

The structure of a Telecommunication Management Network, based on the functions we have identified so far, determines several classes of applicative protocols, which are defined at different reference points. These "reference points" between functions are shown in figure 9.19. The major points at which exchanges are modelled by TMN are the q points. The more general protocols are found at the $q3$ reference point for connection of the Operation System Functions, whereas protocols of less generality, e.g., specific to some manufacturer, are found at the $q1$ and $q2$ reference points involving Network Element Functions and Mediation Functions. Other reference points than the q points are also defined in TMN, such as the f reference points between workstations and other network management entities; the g points between workstations and the operating staff; or the x points between different management networks.

The management exchanges between different PLMNs must not be neglected. They concern the accounting procedures as well as the exchange of information regarding mobile station equipment (e.g., black-listed IMEIs).

Most of the protocols on the reference points identified here are left for the operators to specify, and we will not deal with them further. But some protocol aspects are covered in the *Specifications*: the usage of traffic handling protocols for operation and maintenance activities, the BTS management protocol on the Abis interface, as well as a template for a Q3 protocol specific to GSM. We will now describe these three points in more detail.

9.5.2. OPERATION AND MAINTENANCE IN THE TRAFFIC HANDLING PROTOCOLS

The radio interface protocols are hardly impacted by operation and maintenance functions (the interrogation of the IMEI may be considered one of the few functions in the management domain on this interface). On the contrary, the internal protocols such as the BSSMAP (between BSC and MSC) and the MAP (between NSS entities) include a number of management functions. Such functions are grouped with the traffic handling protocols for two main reasons: either they correspond to actions of limited scope impacting only the two ends of a liaison, or they are so entangled with traffic handling operations that they warrant separate handling by network management entities.

The first category of functions includes the management of circuits between BSC and MSC, and between switching machines. Both the BSSMAP and the ISUP include procedures to **block** and **unblock** circuits, i.e., to inform the other end not to allocate a given circuit (e.g., because of a failure detected on this circuit), and conversely to set it back to normal use. This is indeed the role of the BSSMAP BLOCK, BLOCKING ACKNOWLEDGE, UNBLOCK and UNBLOCKING ACKNOWLEDGE messages which enable a BSS to advise the MSC about the status of its circuits.

Another procedure for the local management of the A interface is the reset procedure, which enables one of the machines (either the BSC or the MSC) to restart and indicate it to the other end. In such cases, all communications in progress are lost, making it a last resource mechanism. The procedure makes use of the BSSMAP RESET and RESET ACKNOWLEDGE messages.

A related procedure exists at the level of individual circuits: it uses the BSSMAP RESET CIRCUIT and RESET CIRCUIT ACKNOWLEDGE messages. It is used to release a possible connection using a given circuit, when normal means cannot be applied because some of the knowledge of the relationship between SCCP connections and terrestrial circuits has been lost.

An example of a procedure belonging to the network management scope, but so entangled with the traffic handling protocols that it cannot be handled by other means, is the trace procedure. The aim of this procedure is to start a log of all or some events happening during the lifetime of a communication. This may be particularly useful when the mobile station is known to be dubious, or to help the diagnosis of some dysfunctioning. When it is felt useful to record both radio interface aspects (channel allocation, handovers, etc.) and communication management aspects, the BSC and the MSC must co-operate. Similarly, when a relay MSC is part of the chain, it is party to the trace generation. GSM also provides means to trace all activities of a given subscriber. This may involve several MSC/VLRs, depending on the movements of the given subscriber. The HLR acts in this case as a co-ordinator, and signalling means are provided to achieve this.

All the entities involved in the tracing of events must use a common reference, so that all logs can later on be analysed together by the management network. On the A interface, the trace number is exchanged by the BSSMAP TRACE INVOCATION message, either from BSC to MSC or in the reverse direction. The corresponding MAP/E message (between relay MSC and anchor MSC) is the MAP/E TRACE SUBSCRIBER ACTIVITY message, while the triggering by the HLR of the tracing activities in the MSC/VLRs is supported by the MAP/D ACTIVATE TRACE MODE and DEACTIVATE TRACE MODE messages together with their respective results.

The MAP protocols provide the support for many other management related functions. The MAP/F protocol enables the MSC/VLRs to get the status of a given IMEI from the EIR, as explained in the mobile stations management section. We have also seen in Chapter 7 that procedures exist in MAP/D to deal with the recovery of information after a database failure and restart. These maintenance functions are mixed with traffic handling functions, and would be less efficient if managed independently by the management network alone.

9.5.3. THE BTS MANAGEMENT PROTOCOL

The Abis interface appears as a singularity for the operation and maintenance protocol structure. It is the only interface between GSM traffic handling machines used to transfer all OAM information pertaining to a machine, namely the BTS. BTS management procedures can be found in a protocol specified in the *Specifications* (in TS GSM 12.21), but of optional application, and which has no specific name: we will call it the BTSM (for BTS management).

The point here is not to detail all the procedures, but just to use these protocols to exemplify the functions described in the previous sections.

Maintenance is supported by a number of procedures in the BTSM dealing with event reports (CHANGED STATE EVENT REPORT, FAILURE EVENT REPORT, STOP SENDING EVENT REPORT, RESTART SENDING EVENT REPORT[1]), as well as procedures to manage tests, for fault localisation or preventive maintenance (SET ERROR PARAMETERS, GET ERROR PARAMETERS, PERFORM TEST, TEST REPORT, SEND TEST REPORT, STOP TEST).

The majority of the BTSM procedures are related to operation. No procedure deals explicitly with observations. These tasks are in fact included in the tests, together with those needed for maintenance purpose. The operation procedures mostly deal with the BTS configuration. This includes the configuration of the terrestrial links between the BSC and the BTS site. An example are the procedures to connect and disconnect links. Another procedure in the same area is more specific: the ESTABLISH TEI message is used to set the TEI (see Chapter 5) used by a given TRX for instance.

The configuration of the radio interface calls for a number of procedures. A number of them, defined as part of the BTSM protocol, are related to the radio transmission parameters (SET RF PARAMETERS, SET HOPPING PARAMETERS, SET CARRIER 0 FILLING). The SET HOPPING PARAMETERS is a first example of a procedure at the border of traffic handling and operation. This procedure can be used to change the time-domain parameters of a channel at a specific time: it is designed to be used during a frequency redefinition, which also involves traffic handling procedures. Another BTSM procedure which is used for frequency redefinition (and maybe only for that) is the GET GSM TIME procedure, which enables the BSC to know the current position of the BTS clock within the 3.5 hours transmission cycle (the hyperframe cycle).

A difficult area is the common channel configuration, and the contents of the signalling messages sent on the SACCH and the BCCH (the RIL3-RR SYSTEM INFORMATION TYPE 1 to 6 messages). There is a need for consistency between some of the contents of these messages and the configuration of the common channels. On the other hand, a part of the messages broadcast on the BCCH, for instance the RACH control parameters, is under control of traffic handling functions. The split between the RSM and the BTSM protocol is as follows: the message contents are set by the BSC through RSM procedures (using the RSM BCCH INFORMATION and RSM SACCH FILLING messages). The common channel configuration (number of PAGCH/RACH, paging sub-channels

---

1   The procedures are referred to in the BTSM Technical Specification by the name of the first message; we will do the same.

structure, ...) is set through BTSM messages. The BSC is responsible for the overall consistency.

Other sets of BTSM procedures support the downloading of BTS software from the BSC, and different other aspects of operation.

9.5.4. THE GSM Q3 PROTOCOL

It is now time to move closer to the management network itself. The *Specifications* include a basis for a GSM Q3 protocol, i.e., a protocol for network management between Operation System functions and Mediation Functions (or some traffic handling machines). This basis is of general application to any GSM network, and must be refined and tailored to each operator's needs. The conception of such a generic GSM Q3 protocol was aimed at minimising the needs for adaptation, while providing the flexibility for them. Two aspects of the GSM Q3 protocol are important to deal with. The first one is that protocols derived from this basis use a number of standardised data communication protocols on the application level, for instance for file transfer. The stack of protocols underlying the GSM Q3 application will therefore be shortly described. The second major aspect concerns the network modelling embodied in the GSM Q3 protocol, and we will then deal with this information model.

9.5.4.1. The GSM Q3 Protocol Stack

As already mentioned, connections between the entities involved in network management may be supported by various network protocols, specific to the networking choices (X.25, SS7, ...). In the TMN environment, a whole OSI stack is used, and this includes a transport protocol, a session protocol, a presentation protocol and various application protocols (layer 7 in the OSI stack). The latter are to be distinguished from the applicative protocol itself (the end user), which is GSM Q3.

In the GSM TMN application, these protocols are among CCITT or ISO standardised protocols. FTAM is a file transfer protocol used when large amounts of information must be exchanged, such as observation logs or software versions for downloading. In parallel with FTAM, usual signalling exchanges make use of a protocol defined between so-called ACSE (Association Control Service Elements). This protocol structures exchanges into full transactions having a beginning, an end, and things in between. It also manages the existence of several such "associations" in parallel. Beside ACSE, the concept of ROSE

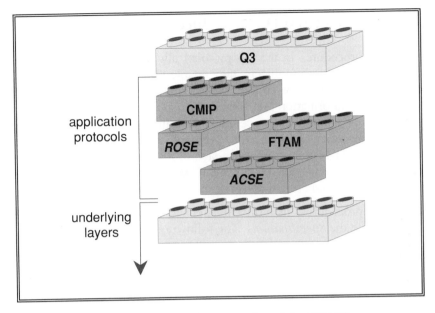

Figure 9.20 – Application protocols stack for GSM Q3

CMIP provides the common procedural basis for the management of objects,
and makes use of FTAM for large file transfers,
as well as of other CCITT defined protocols
for structuring the dialogue between management entities.

(Remote Operations Service Elements) structures the basic exchanges into operations and their results. One may well compare inter-ACSE and inter-ROSE protocols[1] with the transaction and the component entities used in TCAP, and described in Chapter 5.

The whole protocol stack underlying the GSM Q3 protocol is shown in figure 9.20. On top of the inter-ACSE and inter-ROSE protocols, the CMIP (the Common Management Information Protocol) between CMISE (Common Management Information Service Elements) provides generic procedures for the management of "objects" between distributed data bases. It includes procedures for object creation or deletion, for the modification of their attributes, for actions directed to these objects and for the transfer of reports concerning these objects. The notion of object as managed by CMIP will be described in a later section; objects range from a full network to a software module, a transmission link or an alarm.

---

1   It seems that the terms ROP (Remote Operations Protocol) and ACP (Association Control Protocol), coined on the pair CMISE/CMIP, are not used.

9.5.4.2. The GSM Q3 Procedures

The procedures in the Q3 protocol always refer to objects. They create, delete or modify objects, they enable the management network to ask for "actions" from or on the objects, and they enable objects to send "events", i.e., information that something happened, or data. Actions and events are in some way procedures, and they are also properties of objects. The main part of the protocol specification is organised according to classes, describing for each of their attributes (which can be the subject of attribute management procedures), and the specific procedures pertaining to the object of the class, the actions and events.

The procedures, in terms of exchanged messages, are described generically in the *Specifications*. The following aspects are covered:

- the generic object procedures, coming from CMIP, include means to create and delete objects in the databases of the network element functions, and to set and read the attributes of these objects;

- other generic procedures include means for state handling and failure reporting; others cover the basic file transfer mechanisms using FTAM;

- procedures specific to some of the common objects are defined for given classes of objects. These procedures include:

  * the measurement object procedures, for reporting measurements, forcing their execution or inhibiting them;

  * the test object procedures, for starting tests and reporting their result;

  * the software object procedures, to download software.

9.5.4.3. The GSM Q3 Information Model

The general philosophy of TMN is that dialogues between management entities pertain to modelled abstract representations of the network to manage. This representation is the basic information medium on which all management operations will take place. What the operating staff see on their workstation displays are not actual BSCs or HLRs, but text or drawings representing them, with an important level of abstraction: only those aspects relevant to network management will be considered (this is already a lot!). This representation implies that an information model, representing and abstracting the network, be defined and stored in a management database. This model must list the different components of the network, their relationships and their attributes. The concept of attribute has to be taken in the widest sense, covering static aspects (localisation of an equipment, manufacturer, ...), less static ones (software versions, parameters influence traffic handling functions, ...) as well as real-time information such as the operational state (correct, or in failure) or current observations. Network components are objects, a more general concept encompassing anything which is managed. We will see that the information model covers also objects such as tests or observations.

This information model serves as a template for the GSM Q3 protocol. Actions done on the representation of a BSC in this model will be conveyed by messages up to the network element function in the BSC. This agent holds the part of the model representation which is directly relevant to it, and modifies it consistently. It must also (otherwise all this would be quite in vain!) translate the modifications of the abstract representation into concrete actions on the machine itself.

The specification of a GSM Q3 protocol appears then as the design of an information model, more precisely as the conception of object *classes*. The concepts of object, class, and inheritance, which we will see soon, are part of a general approach to the design of the information model. These concepts are, both in name and in reality, very close to "object oriented programming" in the field of software languages. A class is a description of the attributes, properties, and actions which are common to all objects belonging to the class. The objects themselves are instances of classes, and are dynamically created to build up the representation of the network. The classes are chosen during the model design and may cover many things. Typical managed objects are sites, machines such as MSC/VLR, BSC, HLR, hardware modules inside these machines, transmission links, software packages, but also activities such as observations, tests and so on. The management of objects, as mentioned earlier, is supported by CMIP for its basic part. Additional operations can be added specifically to each object class. This concept of object class (e.g., BSC is an object class, each instance of the class representing one concrete BSC) is very important for obtaining both the universality and the flexibility sought in management protocols.

Universality is achieved by a suitable choice of objects, of their attributes and of their relationships. The detailed specifications of the GSM canonical architecture give the ability to identify object classes which will apply to all GSM PLMNs, such as BTSs, radio channels, MSC/VLRs, HLRs, links to PSTN or ISDN, BSCs, remote TRAUs, AuCs, network management machines, and you name it. A number of attributes and properties of these classes can be introduced in the model, independently of the actual implementation.

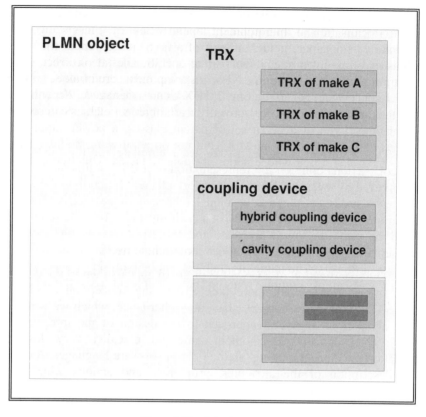

Figure 9.21 – Inheritance trees

The characteristics of an object class may be derived from another one,
building up trees of inheritance relations between object classes.

Flexibility is obtained by the notion of **inheritance**, which covers the possibility to derive a class from a parent, using all of the attributes of the parent while adding specific operations or attributes, enhancing or adapting the inherited operations and attributes. By this method, it is for instance possible to derive from a generic BTS object class (describing the universal aspects of a BTS) a more specific class corresponding to the BTS of a given manufacturer (by adding specific attributes or operations). The inheritance method is also used for a stepwise approach to the design of the classes. For instance BTS, BSC and MSC are machines, and as such share some properties. This commonality can be expressed by a class from which the specific BTS, BSC or MSC classes will be derived. This avoids the need to redefine several times the same properties or actions. The relations of inheritance give the ability to represent the set of the classes by trees as shown in figure 9.21. In such trees, child classes are derived from the parent class.

Another type of relationship which leads to a different tree representation is the hierarchical dependencies of objects inside a network. For instance, a TRX is part of a BTS, itself part of the set of machines on a given site, itself part of a BSS, and so on. This object hierarchy must obey some rules. For instance, only certain object classes can depend on another given one: a TRX or an antenna can depend on a BTS, but an MSC cannot. Reciprocally, certain object classes can only be found as children of a given set of parent classes: a power supply can depend on e.g., a BSC or an MSC, but not on a software module. These relations define a tree whose nodes are object classes. This tree is called the **containment tree**, and is not to be confused with the inheritance trees we have just described. The containment tree between classes represents a generic template for the containment tree of individual objects in the management database.

An important use of the object containment tree is the unequivocal naming of objects. The name of an object can be built up by a succession of identifiers along a unique route in the containment tree, starting at the

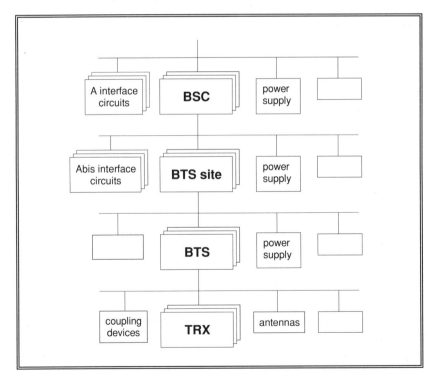

Figure 9.22 – Containment tree

The hierarchical dependencies of objects in a network can be represented
by a containment tree, used as a model for naming objects by their position in the tree
(e.g., TRX 3 within BTS 2 within BTS site 9 within BSC 56...).

root and adding at each step an identifier relative to the previous object encountered. With the example of a containment tree shown in figure 9.22 (covering only a part of the classes needed for the representation of a GSM network), a BTS number would identify a given BTS among those connected to a given BSC, or even those of a given site, the site being identified among all sites connected to a given BSC, and the naming of a BTS inside a given PLMN would include its BSC number, site number and BTS number, instead of just a global and not so meaningful number.

Having thus set the scene for the model, the *Specifications* list an impressive number of generic GSM object classes, with their generic attributes. There is no point here going into the very detail of this information model. It can be found in TS GSM 12.20, and the phase 1 version of this specification represents a first attempt at defining a GSM generic inheritance tree and containment tree. Substantial work is still going on at the date of writing, and the phase 2 *Specifications* will undoubtedly present many improvements in this area, with a detailed information model rendered possible by the extensive standardisation of the GSM system.

SPECIFICATIONS REFERENCE

The bulk of the GSM Specifications related to network management is the 12 series. However, a number of other specifications also contain material related to the substance of this chapter. Let us list them according to the outline of the chapter.

The subscriber administration aspects are introduced in **TS GSM 12.02**, but the more technical description of the subscriber-related data to be handled by networks is given in **TS GSM 12.05**. As far as charging is concerned, the tariff structure and the requirements on interactions between networks are listed in **TS GSM 02.20**, whereas the transferred account procedure is described in **TS GSM 02.21**.

Maintenance is the subject of the specifications in the 12.1x series: **TS GSM 12.10** is concerned with mobile stations, **TS GSM 12.11** with the BSS, **TS GSM 12.13** with the MSC and **TS GSM 12.14** with the HLR and VLR.

The control of mobile equipment identities is described in **TS GSM 02.16**, and the structure of the IMEI is given in **TS GSM 03.03**.

System operation is dealt with from the procedural point of view, but obviously cellular engineering is a matter left to the operator. A report, TS GSM 03.30, is also available, though not part of the official list. It addresses the radio network planning aspects. **TS GSM 12.07** gives general requirements on network performance control and configuration. The management, collection and transfer of statistics generated by the traffic handling machines is described in **TS GSM 12.04**, and the data required for network change control is identified in **TS GSM 12.06**.

The protocol dedicated to operation, administration and maintenance on the Abis interface, which we have called BTSM, is specified in **TS GSM 12.21**. A number of traffic handling protocols also include relevant procedures, such as the A interface BSSMAP (specified in **TS GSM 08.08**) or the MAP (specified in **TS GSM 09.02**).

A general overview of the TMN principles can be found in **TS GSM 12.00**. This specification, as well as **TS GSM 12.01**, is an introduction to the generic GSM Q3 protocol. The containment tree and inheritance rules for the GSM application, as well as the procedures of the Q3 protocol, are defined in the bulkiest of the 12 series specifications, **TS GSM 12.20**.

THE LIST OF THE GSM TECHNICAL SPECIFICATIONS

The whole set of official phase 1 *Specifications* is structured in 12 series, one of which is empty (the 10 series, originally planned for service interworking specifications). At the end of each chapter in this book, we have introduced a section summarising the Technical Specifications applicable, domain by domain. The reader will now find here the list of the *Specifications* sorted by serial number, presented in such a way as to provide a quick reference.

For each GSM Technical Specification, the entry lists its number, title, number of pages and when relevant the existence of a separate addition for the DCS1800 standard (Δ DCS). Below each *Specification* number, we have listed the parts of a network concerned by the *Specification*. Moreover, for each *Specification*, a short summary of its scope is given, with a reference to the chapters of the book where relevant information can be found.

SS series: Title of the series	number of pages

SS.NN Title of the *Specification*	number of pages
impact Summary and reference to relevant chapters of this book.	(+ Δ DCS)

01 series: General — 54 p.

01.04 Vocabulary in a GSM PLMN — 51 p.

Glossary of terms and acronyms used in the GSM Technical Specifications with their definitions and references when need be to the CCITT recommendations.

01.06 Service implementation phases and possible further phases in the GSM PLMN — 3 p.

Defines a classification of services, which is used in the 02 series, in terms of an implementation calendar.

02 series: Service Aspects — 276 p.

02.01 Principles of telecommunication services supported by a GSM PLMN — 18 p.

MS, NSS Presents the description method used in TS GSM 02.02 and 02.03 to define services. This method uses "attributes" (i.e., characteristics in terms of rate, connecting network, etc.) in the same way as ISDN specifications. See Chapter 1.

02.02 Bearer services supported by a GSM PLMN — 38 p.

MS, NSS Lists the bearer services, that is to say the services limited to information transport, and describes their main characteristics. See Chapter 1.

02.03 Teleservices supported by a GSM PLMN — 23 p.

MS, NSS Lists the teleservices, that is to say the services including terminal functions, and describes their main characteristics. See Chapter 1.

02.04 General on supplementary services — 18 p.

MS, NSS Lists the supplementary services to be described in the 02.8x series, that is to say the services modifying the provision of bearer or teleservices, and provides a set of definitions applicable to them. See Chapter 1.

02.05 Simultaneous and alternate use of services — 3 p.

Specifies that, except for short messages, the MS may use only one (bearer or tele-) service at a time, but may use two different services in an alternate manner inside the same communication, if indicated as such at call establishment.

02.06 Types of mobile stations — 4 p.

MS	Distinguishes vehicle-mounted, portable and handheld mobile stations, and defines their transmission power capabilities.	+ Δ DCS

02.07 Mobile station features

15 p.

MS   Gives a non-exhaustive list of features which may be offered locally by the mobile station. For each of them, gives their status as mandatory or optional, and specifies restrictions on automatic calling facilities.

02.09 Security aspects

8 p.

Presents the method used in GSM to ensure the privacy of user data and of user location and to guarantee the veracity of subscriber identities. These methods must mandatorily be supported both by mobile station and network. See Chapter 7.

02.10 Provision of telecommunication services

3 p.

Recommends that there be operators in all CEPT countries, offering a Europe-wide service to subscribers.

02.11 Service accessibility

7 p.

MS   Sets up the requirements for the selection of a network by the mobile station and defines how subscribers are spread among classes for selective access in cases of congestion. See Chapter 7.

+ Δ DCS

02.12 Licensing

2 p.

Recommends that no individual license be required for the transportation or use of type approved mobile stations.

02.13 Subscription to the services of a GSM PLMN

4 p.

OSS   Describes different geographical entitlements for subscriptions.

02.14 Service directory

2 p.

OSS   Recognises the existence of directories.

02.15 Circulation of mobile stations

2 p.

Recommends that Administrations make provisions for the free circulation and use of type approved mobile stations.

02.16 International MS equipment identities

5 p.

MS, OSS   Introduces the concept of IMEI, as an identity for the mobile equipment, and its use to control the behaviour of mobile stations through white, grey and black lists of IMEIs.

02.17 Subscriber identity modules, functional characteristics

11 p.

SIM   Defines the concept of subscriber module, or SIM; as an IC-card or a cut-out thereof called the plug-in SIM. Describes the life cycle of a SIM, and lists the items for which a non-volatile storage must be provided in the SIM. See Chapter 7.

Note: Other phase 2 *Specifications* dealing with supplementary services should also be cited here:

02.87 Additional information transfer supplementary services

Defines a supplementary service enabling user-to-user signalling transfer.

03 series: Network Aspects 636 p.

03.01 Network functions 10 p.

Lists some of the functions to be provided by the network to handle calls in a cellular environment.

03.02 Network architecture 17 p.

Introduces the "canonical architecture" of a GSM network, with an emphasis on the switches (MSCs) and data bases, and lists the interfaces between them. See Chapter 2.

03.03 Numbering, addressing and identification 14 p.

Specifies the structure of identities for mobile stations (subscription, equipment), location areas and infrastructure entities. Also specifies the numbering plan for GSM users.

03.04 Signalling requirements relating to routing of calls to 11 p.
mobile subscribers

Describes the different routing configurations for mobile terminating calls, based on the relative positions of the interrogating exchange. See Chapter 8.

03.05 Technical performance objectives 23 p.

BSS, NSS Specifies performances to be aimed at when implementing infrastructure equipments, in terms of probability of failure, processing delays and transmission delays. Resulting delay diagrams for speech and data transmission are included. See Chapter 3.

03.07 Restoration procedures 11 p.

NSS Defines procedures for restoring the consistency between network data bases (HLRs, VLRs) after a failure. See Chapter 7.

03.08 Organisation of subscriber data 18 p.

NSS Summarises the data to be stored in the network data bases (HLRs, VLRs) with the status of each parameter (mandatory/optional, permanent/temporary). see Chapters 7 and 8.

04.11 Point-to-point short message service support on mobile radio interface 72 p.

MS, MSC Specifies the type of radio channel on which short messages can be transported between the MSC and the mobile station, as well as the transfer protocols between these two entities. See Chapter 5.

04.12 Cell broadcast short message service support on mobile radio interface 4 p.

MS, BSS Specifies the segmentation of cell broadcast short messages into blocks on the radio path.

04.21 Rate adaptation on MS-BSS interface 12 p.

MS, TRAU Specifies the different steps of rate adaptation (inspired from the ISDN rate adaptation schemes) which are used to support user data between the mobile station and rate adaptation equipment on the network side. See Chapter 3.

04.22 Radio link protocol for data and telematic services on the MS-BSS interface 59 p.

MS, IWF Specifies completely the link protocol (RLP) used to support data services on "NT connections" inside GSM. See Chapters 3 and 5.

04.80 Mobile radio interface layer 3 – supplementary services specification – formats and coding 38 p.

MS, MSC, HLR Specifies the coding of messages concerning the management of supplementary services and their impact on calls. See Chapter 8.

04.82 Mobile radio interface layer 3 – call offering supplementary services specification 34 p.

MS, MSC, HLR Specifies the procedures to manage the call forwarding services, using messages defined in TS GSM 04.08 and TS GSM 04.80. See Chapter 8.

04.88 Mobile radio interface layer 3 – call restriction supplementary services specification 12 p.

MS, MSC, HLR Specifies the procedures to manage the call barring services, using messages defined in TS GSM 04.08 and TS GSM 04.80. See Chapter 8.

05 series: Physical Layer on the Radio Path 135 p.

05.01 Physical layer on the radio path (general description) 11 p.

MS, BSS Introduces the contents of the 05 series. Summarises the channel types and the hierarchy of time cycles. See Chapter 4. + Δ DCS

06.12

Comfort noise aspects for full rate speech traffic channels   6 p.

MS,
TRAU

Specifies how the background noise should be evaluated and information transmitted for its regeneration in the absence of speech. See Chapter 3.

06.31

Discontinuous transmission (DTX) for full rate speech traffic channels   13 p.

MS, BTS,
TRAU

Specifies how to reduce the amount of transmission when no speech signal needs to be sent by the mobile station or by the network. See Chapters 3 and 6.

06.32

Voice activity detection   37 p.

MS
TRAU

Specifies the bit-exact algorithm for detecting the presence of speech at the output of the speech coder (with the aim to benefit from DTX, as specified in TS GSM 06.31). See Chapter 3.

07 series: Terminal Adaptors for Mobile Stations   102 p.

07.01

General on terminal adaptation functions for MSs   49 p.

MS

Presents the functions required in the mobile station to accommodate off-the-shelf terminal equipments. Also lists the values of the compatibility characteristics for all services defined in TS GSM 02.02 and 02.03. See Chapter 3.

07.02

Terminal adaptation functions for services using asynchronous bearer capabilities   17 p.

MS

Specifies the adaptation functions required in the mobile station to support terminals with an asynchronous interface. Includes the specification of the data link protocol used in connection with RLP (TS GSM 04.22) to replace the terminal link protocol. See Chapter 3.

07.03

Terminal adaptation functions for services using synchronous bearer capabilities   36 p.

MS

Specifies the adaptation functions required in the mobile station to support terminals with a synchronous interface. Includes the specification of the data link protocol used in connection with RLP (TS GSM 04.22) to replace the terminal link protocol. See Chapter 3.

08 series: BS to MSC Interfaces | 297 p.

08.01 General aspects on the BSS-MSC interface | 5 p.

Introduces the *Specifications* in the 08.0x series, as specifications of the A interface.

08.02 BSS-MSC interface – interface principles | 17 p.

Defines the split of functions between the base station sub-system and the MSC, and describes the protocol stack applicable at the A interface. See Chapters 6, 7 and 8.

08.04 BSS-MSC layer 1 specification | 2 p.

BSC,
MSC

Specifies the structure of 64 kbit/s digital circuits on the A interface, based on the CCITT Recommendations of the G series. See Chapter 5.

08.06 Signalling transport mechanism specification for the BSS-MSC interface | 29 p.

BSC,
MSC

Specifies the applicability of the CCITT signalling system n°7 lower layer protocols for the transfer of signalling messages on the A interface. Also specifies a distribution function on top for identifying different flows of messages. See Chapter 5.

08.08 BSS-MSC layer 3 specification | 92 p.

BSC,
MSC

Specifies the protocol used for co-ordinating the MSC and the base station sub-system in the area of radio resource management. See Chapter 6.

08.20 Rate adaptation on the BSS-MSC interface | 5 p.

TRAU,
IWF

Specifies the rate adaptation characteristics for the transport of data services on 64 kbit/s channels in GSM. See Chapter 3.

08.51 BSC-BTS interface, general aspects | 4 p.

BTS,
BSC

Introduces the *Specifications* of the 08.5x series, and lists the reasons why the Abis interface was specified.

08.52 BSC-BTS interface principles | 15 p.

BTS,
BSC

Describes the split of functions between BTS and BSC, as well as the protocol model on the Abis interface. See Chapter 6.

08.54 BSC-TRX layer 1: structure of physical circuits | 3 p.

BTS,
BSC

Specifies the structure of 16 and 64 kbit/s digital circuits on the Abis interface, referring extensively to the CCITT Recommendations in the G series and to CCITT I.460. See Chapter 5.

| 08.56 | BSC-BTS layer 2 specification | 11 p. |

BTS, BSC
Specifies the applicability of the link layer protocol used for ISDN access (LAPD) for the transfer of signalling messages on the Abis interface. See Chapter 5.

| 08.58 | BSC-BTS layer 3 specification | 80 p. |

BTS, BSC
Specifies the protocol used for relaying radio interface messages on the Abis interface and the protocol used for co-ordinating the BSC and the BTS in the area of radio resource management. See Chapter 6.

+ Δ DCS

| 08.59 | BSC-BTS O&M signalling transport | 6 p. |

BTS, BSC
Specifies how network management messages are transported over the Abis interface (but not their semantics). See Chapter 5.

| 08.60 | In-band control of remote transcoders and rate adaptors | 28 p. |

BTS, TRAU
Specifies the format of speech and data frames between a BTS and the transcoder/rate adaptor when it is remote from the BTS, and describes how these frames are used by the BTS to control the configuration and behaviour of the remote transcoder/rate adaptor. See Chapter 3.

09 series: Network Interworking   **666 p.**

| 09.01 | General network interworking scenarios | 8 p. |

NSS
Introduces the *Specifications* of the 09 series.

| 09.02 | Mobile application part (MAP) specification | 507 p. |

NSS, MS
Specifies the application protocols between exchanges and data bases (MSCs, GMSCs, VLRs, HLRs, EIRs) for supporting call management, supplementary services management, short message transfer, location management, security management, radio resource management and mobile equipment management. Specifies the applicability of the CCITT signalling system N°7 protocols (SCCP, TCAP) to support these exchanges. See Chapters 5, 6, 7, 8 and 9!

+ Δ DCS

| 09.03 | Requirements on interworking between the ISDN or PSTN and the PLMN | 6 p. |

NSS
Presents a few considerations on the interconnection of GSM with a public switched telephone network (PSTN) or ISDN. See Chapter 3.

11.11 Specification of the SIM-ME interface
SIM, ME

131 p.
+ Δ
DCS

Specifies the structure of the subscriber module as far as its operation in a mobile equipment is concerned. Specifies the application protocol between the SIM and a mobile equipment, and the applicability of ISO standards on this interface. See Chapter 7.

11.20 The GSM base station system: equipment specification
BSS

423 p.
+ Δ
DCS

Specifies a set of tests to check the conformity of the base station system (BTS + BSC)

11.30 Mobile services switching centre (report)
MSC

55 p.

Summarises the functions of an MSC, giving their status (mandatory/optional), and lists the relevant *Specifications*.

11.31 Home location register specification (report)
HLR

9 p.

Summarises the functions of the HLR and lists the relevant *Specifications*.

11.32 Visitor location register specification (report)
VLR

13 p.

Summarises the functions of the VLR and lists the relevant *Specifications*.

11.40 System simulator specification

46 p.
+ Δ
DCS

Specifies the equipment which is used to perform type approval tests on mobile stations, according to TS GSM 11.10. See Chapter 9.

12 series: Operation and Maintenance
794 p.

12.00 Objectives and structure of network management
OSS

51 p.

Defines the objectives of network management. Introduces the TMN (Telecommunication Management Network) concepts and the relevant definitions. See Chapter 9.
Note: the phase 2 version of this specification is an excellent primer in the domain of network management.

12.01 Common aspects of GSM network management
OSS

69 p.

Describes the GSM network management functional areas, interfaces and protocols.

12.02 Subscriber, mobile equipment and services data administration
NSS, OSS

17 p.

Lists the network management functions in the domain of subscriber management and mobile equipment management.

12.21 Network management procedures and messages on the   83 p.
Abis interface

BTS, Specifies a protocol for network management of the BTS as a
BSC superset of the procedures needed for operation of any one BTS by
the BSC. See Chapter 9.

Total: 5230 pages of official phase 1 *Specifications*.

BIBLIOGRAPHY

Mobile Communications in General

E.A. LEE, D.G. MESSERSCHMITT, *Digital Communication*, Kluwer Academic Publishers (Boston), 1988

J.D. PARSONS, D. JARDINE, J.G. GARDINER, *Mobile Communication Systems*, Blackie (Glasgow), Halsted (New York), 1989

G. CALHOUN, *Digital Cellular Radio*, Artech House, 1988

GSM in General

Most of the existing literature about GSM consists of the proceedings of seminars or conferences. Two seminars, organised by the MoU and devoted to a general presentation of GSM, were held in Europe, at Hagen (FRG), October 1988, and in Budapest, October 1990. Besides, there are regularly, about once a year since 1985, a technical conference in Europe about digital mobile communications. These are the forums where most technical papers directly related to GSM are published. A number of such papers, among the more recent, are listed in the bibliography hereinafter. The full list of these conferences is given here, for those interested in what was said on the subject in the first years:

- DMR I (Nordic Conference on Digital Land Mobile Radio Communication), Espoo (Finland), February 1985 (-20° C!)
- DMR II, Stockholm, October 1986;
- DMRC (International Conference on Digital Land Mobile Radio Communications), Venice (Italy), July 1987;
- DMR III, Copenhagen, September 1988;
- DMR IV, Oslo, June 1990;
- MRC (Mobile Radio Conference), Nice (France), November 1991;
- DMR V, Helsinki, December 1992 (not held when printed).

Chapters 1 and 2

T. HAUG, "The GSM Programme, a pan-European effort", *Proceedings of the Mobile Radio Conference*, Nice, November 1991

A. FOXMAN, "Activities of the group of Signatories of the GSM MoU", *Proceedings of the Mobile Radio Conference*, Nice, November 1991

A. HADDEN, "Development of the DCS 1800 standard", *Proceedings of the Mobile Radio Conference*, Nice, November 1991

B. MALLINDER, "An overview of the GSM system", *Proceedings of the third Nordic Seminar on Digital Land Mobile Radio Communication*, Copenhagen, September 1988

Chapter 3

R. MONTAGNA, "The half rate speech voice codec for the GSM system", *Proceedings of the Mobile Radio Conference*, Nice, November 1991

I. DITTRICH et al., "Implementation of the GSM data services into the radio mobile radio system", *Proceedings of the Mobile Radio Conference*, Nice, November 1991

G. CRISP, A. EIZENHÖFER, "Architectural aspects of Data and Telematic Services in a GSM PLMN", *Proceedings of the third Nordic Seminar on Digital Land Mobile Radio Communication*, Copenhagen, September 1988

A. CLAPTON et al., "Supporting Facsimile in the GSM PLMN", *Proceedings of the fourth Nordic Seminar on Digital Land Mobile Radio Communication*, Oslo, June 1990

P. VARY, R. HOFMANN, "Sprachcodec für das Europaische Funkfernsprechnetz", *Frequenz*, vol. 42, no. 2-3, February 1988

Chapter 4

A. MALOBERTI, "Some aspects of the GSM radio interface", *Proceedings of the third Nordic Seminar on Digital Land Mobile Radio Communication*, Copenhagen, September 1988

M. BERTELSMEIER et al., "Design and Performance of Punctured Convolutional Codes for the Pan-European Mobile Radio System", *Proceedings of the third Nordic Seminar on Digital Land Mobile Radio Communication*, Copenhagen, September 1988

G.D. FORNEY, "The Viterbi Algorithm", *Proceedings of the IEEE*, 61, March 1973

A. KRANTZIK, D. WOLF, "Statistische Eigenschaften von Fadingprozessen zur Beschreibung eines Landmobilfunkkanals", *Frequenz*, vol.44, no.6, June 1990

G. CASTAGNA, A. COLAMONICO, R. FAILLI, R. MONTAGNA et al., "Italian experimental activity on the European digital land mobile system", *CSELT Tech. Rep.*, vol. 17, no. 2, April 1989

T. MASENG, "On selection of system bit-rate in the mobile multipath channel", *IEEE Trans. VT*, Vol. 36, no. 2, May 1987

W. W. PETERSON, *Error-Correcting Codes*, M.I.T. Press, Cambridge, Mass. (1961, 1970)

CHAPTERS 6, 7 AND 8

U. JANSSEN, P. BRUNE, "The Mobile Application Part protocol", *Proceedings of GSM Seminar*, Budapest, October 1990

M. B. PAUTET, M. MOULY, "GSM protocol Architecture: Radio Subsystem Signalling", *Proceedings of VTC'91*, Saint-Louis, Missouri, May 1991

Chapter 6

I. BRINI et al., "European roaming related technical problems", *Proceedings of the Mobile Radio Conference*, Nice, November 1991

Chapter 7

B. CHATRAS, C. VERNHES, "The European mobile service: an application of the intelligent network concept", *Proceedings of the 1st international seminar on intelligent networks*, Bordeaux, March 1989

M. BALLARD, D. VERHULST, "ECR900 : la radiotéléphonie numérique européenne d'Alcatel", *Commutation & Transmission*, no. 4, 1988

G. MAZZIOTTO, "The Subscriber Identity Module for the European Digital Cellular System GSM and other Mobile Communication Systems", *Proceedings of XIV International Switching Symposium*, Yokohama, October 1992

CHAPTER 9

R. HAGEDOORN, "Conformity Testing of GSM mobile stations", *Proceedings of GSM Seminar*, Budapest, October 1990

A. BERGMANN et al., "Protocol conformance testing of a GSM mobile station", *Proceedings of the Mobile Radio Conference*, Nice, November 1991

R. KÖSTER et al., "ISO test methods and their applicability to the GSM mobile network system", *Proceedings of the Mobile Radio Conference*, Nice, November 1991

P. GUILLIER et al., "Implementation and operation of the GSM network", *Proceedings of the Mobile Radio Conference*, Nice, November 1991

H. PERSSON, "Performance of GSM options in a cellular environment", *Proceedings of the Mobile Radio Conference*, Nice, November 1991

R. THOMAS, M. MOULY, et al., "Performance evaluation of common control channels in the European digital cellular system", *Proceedings of the Digital Mobile Radio Communications Conference*, Venice, July 1987

R. THOMAS et al., "Influence of the moving of the mobile stations of a radio mobile cellular network", *Proceedings of the third Nordic Seminar on Digital Land Mobile Radio Communication*, Copenhagen, September 1988

D. VERHULST, "High performance cellular planning with Frequency Hopping", *Proceedings of the fourth Nordic Seminar on Digital Land Mobile Radio Communication*, Oslo, June 1990

R. WYRWAS, J. C. CAMBELL, "Radio topology design with slow frequency hopping for interference limited digital cellular systems", *Proceedings of VTC'91*, Saint-Louis, Miss., May 1991

INDEX

Signalling messages are referred to in this index by alphabetical order of the message name. A separate message index, sorted by protocols, follows the general index. Message names are in small capitals, and message field names are in italic small capitals.

C

I

M

MESSAGE INDEX

The following index includes the messages referred to in the book within the description of the applicative protocols. This does not cover exhaustively the messages specified for these protocols. The main exceptions are the messages used to indicate protocol errors and the messages not used in phase 1.

INDEX OF FIGURES

EUROPE MEDIA DUPLICATION S.A.
53110 Lassay-les-Châteaux
N° 2871 – Dépôt légal : août 1993
Imprimé en France